Lecture Notes in Mathematics

Edited by A. Dold and B. Eckmann

1274

N. Christopher Phillips

Equivariant K-Theory and Freeness of Group Actions on C*-Algebras

Springer-Verlag

Berlin Heidelberg New York London Paris Tokyo

Author

N. Christopher Phillips
Department of Mathematics, University of California
Los Angeles, CA 90024-1555, USA

Mathematics Subject Classification (1985): Primary: 46L55, 46L80, 46M20
Secondary: 19K33, 19K99, 19L47

ISBN 3-540-18277-2 Springer-Verlag Berlin Heidelberg New York
ISBN 0-387-18277-2 Springer-Verlag New York Berlin Heidelberg

Library of Congress Cataloging-in-Publication Data. Phillips, N. Christopher (Norman Christopher), 1956-. Equivariant K-theory and freeness of group actions on C*-algebras. (Lecture notes in mathematics ; 1274) Bibliography: p. Includes index. 1. K-theory. 2. C*-algebras. 3. Lie groups. I. Title. II. Series: Lecture notes in mathematics (Springer-Verlag) ; 1274. QA3.L28 no 1274 510 s 87-23345 [QA612.33] [512'.55]
ISBN 0-387-18277-2 (U.S.)

Printing and binding: Druckhaus Beltz, Hemsbach/Bergstr.
2146/3140-543210

To

汪開祥
Wang Kai-Shyang

Acknowledgments

I would like to thank a number of people for their help and encouragement during the preparation of this book. Graeme Segal first called my attention to the paper from which I learned about the relation between equivariant K-theory and freeness of actions. I have had valuable discussions with, among others, Claude Schochet concerning Künneth theorems, I. M. Singer concerning pseudodifferential operators, Nigel Higson concerning extendible prequasihomomorphisms, and Jonathan Rosenberg concerning projective unitary representations. David Handelman suggested the main result of section 9.2 and its proof.

Most of all, however, I would like to thank Marc Rieffel. Many of the results in this book first appeared in my Ph.D. thesis written under his direction, and many others were first proved while I was working on my thesis. His patience, encouragement, and suggestions, both before and after I received my Ph.D., have been extremely valuable, and he has helped eliminate many obscurities in the exposition. He was also kind enough to show me some of his unpublished research notes, and to allow me to use some of the material from them in chapter 7.

This book is based on the author's Ph.D. thesis written at the University of California at Berkeley. However, it incorporates substantial revisions and many additional results. The research reported here has been partially supported by a National Science Foundation Graduate Fellowship, by the Mathematical Sciences Research Institute (NSF Grant 8120790), and by a National Science Foundation Postdoctoral Fellowship. All of this support is gratefully acknowledged. Thanks are also due to Marsha Colby, who converted an enormous number of sometimes rather illegible handwritten pages into the computer files from which this manuscript was produced, and who, with rare exceptions, was able to persuade a University of California computer to print many symbols that it had never been intended to make.

Preface

This book is about equivariant K-theory and K-theoretic conditions for freeness of actions of compact Lie groups on C^*-algebras. The introduction, explaining in detail the motivation for this work, is followed by two primarily expository chapters, one each on equivariant K-theory for C^*-algebras and equivariant KK-theory for C^*-algebras. The remaining six chapters contain the results of the author's research on K-theoretic conditions for freeness of actions on C^*-algebras. We assume throughout familiarity with the theory of C^*-algebras, including crossed product C^*-algebras and ordinary (that is, not equivariant) K-theory of C^*-algebras.

Our work is motivated by the observation that, given an action of a compact Lie group on a compact Hausdorff space, one can determine whether the action is free solely by examining the equivariant K-theory of the space for the given action. (Details, with proofs, are given in chapter 1.) Now the category of compact Hausdorff spaces is contravariantly equivalent to the category of commutative unital C^*-algebras via the functor assigning to each space the algebra of continuous complex-valued functions on the space. Therefore the relation between equivariant K-theory and freeness can be interpreted as an assertion about actions of compact Lie groups on commutative unital C^*-algebras. In view of the recent successes of noncommutative algebraic topology, in which general C^*-algebras are regarded as "noncommutative locally compact spaces," we are naturally led to try to generalize the results mentioned above to general C^*-algebras. However, no completely satisfactory notion of freeness of an action on a C^*-algebra is known. We therefore define and study conditions on the equivariant K-theory of a C^*-algebra which, in case the algebra is commutative, imply that the action on the underlying space is free.

Chapters 2 and 3 develop the necessary background material on C^*-algebraic equivariant K-theory and KK-theory respectively. Neither chapter depends significantly on the rest of the book. Chapter 2 consists mostly of material which has previously appeared only in somewhat condensed form, or which has never been published but straightforwardly generalizes ordinary K-theory. It is fairly comprehensive; however, there are some results purely about equivariant K-theory elsewhere, especially in sections 5.1 and 6.1-6.4. Previous knowledge of ordinary K-theory for C^*-algebras is assumed, but is used only at a few points, most notably for the six term exact sequence and Bott periodicity. Chapter 3 develops equivariant KK-theory from Cuntz' quasihomomorphism point of view. Most of the material has appeared previously but again only in very condensed form. We prove only the basic facts, up to the product and Bott periodicity, omitting however the construction of the six term exact sequences, for which we refer to a paper of Cuntz and Skandalis. Again, some additional material can be found elsewhere, particularly in sections 5.1 and 9.7.

In chapter 4, we define our K-theoretic notions of freeness. There and in the next two chapters we consider consistency theorems, the analogs of such facts as the freeness of the restriction of a free action to an invariant subspace or a closed subgroup. Most of the appropriate statements are either easily proved or, in bad cases, easily disproved. Two topics, namely actions of subgroups and actions on tensor products, present greater difficulties. Each of these topics gets a chapter to itself, and the gaps between our theorems and our counterexamples are larger than was generally the case in chapter 4.

Chapter 7 is devoted to the relation between our K-theoretic notions of freeness and previously known measures of freeness, especially Kishimoto's strong Connes spectrum. Our conditions are, unfortunately, trivially satisfied by a trivial action on a simple C^*-algebra whose K-groups are all zero. (This kind of difficulty cannot arise in the context of spaces.) However, if the K-theory of the C^*-algebra is sufficiently nontrivial, and if the group is sufficiently small, then K-theoretic freeness does imply other forms of freeness.

The remaining two chapters consider the implications of our K-theoretic conditions for freeness for actions on two special classes of C^*-algebras, namely type I and AF algebras. In both cases, we obtain analytic characterizations of several of our K-theoretic freeness conditions. In the type I case, one of our conditions is shown to be equivalent to freeness of the corresponding action on the primitive ideal space. For AF algebras, we obtain results of a somewhat different nature but which should not be surprising in view of known results about the ordinary K-theory of AF algebras. Our results here have led us to hope that there might be a good analytic notion of freeness which implies K-theoretic freeness in general and coincides with it on these special classes of C^*-algebras. However, we have made no attempt to investigate this question.

Table of Contents

Chapter 1

Introduction: The Commutative Case

1.1. Introduction

Recall that an action of a group G on a space X is said to be free if for $g \in G$, $g \neq e$, and $x \in X$, one has $gx \neq x$. Since the commutative C^*-algebras are exactly the algebras $C_0(X)$ of continuous complex valued functions vanishing at infinity on locally compact Hausdorff spaces X, we can transfer the definition of freeness to commutative C^*-algebras as follows: we say that a continuous action of a group G on $C_0(X)$ is free if the corresponding action on X is free. Viewing general C^*-algebras as "noncommutative topological spaces," we now seek generalizations of the concept of freeness to the case of actions of groups on noncommutative C^*-algebras.

The starting point of this work is a relation between freeness of a continuous action of a compact Lie group G on a compact space X and the equivariant K-theory of X. Before stating it, we introduce some notation. Recall ([98]) that equivariant K-theory is a generalized cohomology theory on locally compact G-spaces (that is, spaces carrying continuous actions of G), where G is a compact group. This theory is analogous to ordinary K-theory of locally compact spaces. The equivariant K-groups of a locally compact G-space X are written $K_G^0(X)$ and $K_G^1(X)$, and as usual we let $K_G^*(X)$ stand for the direct sum $K_G^0(X) \oplus K_G^1(X)$. These groups are in fact modules over the representation ring $R(G)$ of G, as defined in [97]. The ring $R(G)$ contains a distinguished ideal, namely the augmentation ideal $I(G)$. (It is defined in the example preceding proposition 3.8 of [97]. The definitions of $K_G^*(X)$, $R(G)$, and $I(G)$ will all be given in detail in chapter 2.)

We are now in a position to state the relations between freeness and equivariant K-theory. This result is essentially due to Atiyah and Segal ([6], proposition 4.3, and [98], proposition 4.1).

1.1.1 Theorem. Let G be a compact Lie group, and let X be a compact Hausdorff G-space. Then the following are equivalent:

(1) G acts freely on X.

(2) The natural map $K^*(X/G) \longrightarrow K_G^*(X)$ is an isomorphism.

(3) $K_G^*(X)$ is complete in the $I(G)$-adic topology.

(4) $K_G^*(X)$ is complete and Hausdorff in the $I(G)$-adic topology.

(5) $K_G^*(X)$ is discrete in the $I(G)$-adic topology.

(6) For all prime ideals $P \subset R(G)$ such that $P \not\supset I(G)$, the localization $K_G^*(X)_P$ is zero.

In condition (2), the space X/G is the space of orbits of the action of G on X. Since G is compact, X/G is a compact Hausdorff space with the quotient topology. The map sends the class $[E]$ of a vector bundle E over X/G to the class of the pull-back of E via the quotient map. More details will be given in the next section.

The methods used to generalize ordinary K-theory to C^*-algebras can also be used to generalize equivariant K-theory to C^*-algebras carrying continuous actions of compact groups. (The resulting theory is defined in, for example, [51]. See also [55]. We will develop its properties in detail in chapter 2.) Since equivariant K-theory is a covariant functor on C^*-algebras, we denote the equivariant K-theory of a C^*-algebra A carrying a continuous action of a compact group G by $K_*^G(A)$. We then have $K_*^G(C_0(X)) \simeq K_G^*(X)$ for locally compact G-spaces X.

The original motivation for the work reported here was an attempt to generalize theorem 1.1.1 to noncommutative C^*-algebras. Lacking an adequate notion of freeness of a group action on a C^*-algebra, we were led to define a concept of K-theoretic freeness in terms of one of the other conditions of this theorem. Our choice is condition (6). One reason for choosing this condition is that localization is an exact functor, so that exact sequences in equivariant K-theory yield exact sequences for the localized modules $K_*^G(A)_P$. In particular, if P is a prime ideal in $R(G)$, then $K_*^G(\cdot)_P$ is a generalized homology theory on C^*-algebras carrying continuous actions of the compact group G.

Theorem 1.1.1 fails for general locally compact G-spaces. Indeed, if X is a contractible space, such as $[0,1)$, and the action of G on X is trivial, then conditions (2) through (6) are satisfied, since the K-groups which appear are all zero, but condition (1) fails. The situation can be saved by observing that a continuous action of a compact group G on a topological space X is free if and only if the restriction of the action to every G-invariant compact subset of X is free. Since compact subsets of Hausdorff spaces are closed, and since the C^* equivalent of a closed subset is the quotient by an ideal, we are led to the following definition: a continuous action of a compact Lie group G on a C^*-algebra A is called K-free if for every G-invariant ideal I in A and every prime ideal P in $R(G)$ not containing $I(G)$, we have $K_*^G(A/I)_P = 0$. This work, then, is devoted to the study of actions which are K-free or satisfy certain closely related conditions.

We have defined K-freeness only for compact Lie groups. We restrict ourselves to compact groups because equivariant K-theory is only defined for actions of compact groups. (Some work has been done toward a more general definition — see [56] and [8].) Furthermore, some of our work involves condition (2) of theorem 1.1.1 and its C^* analog. If G is not compact, then X/G need not be Hausdorff, and the C^* analog of X/G, namely the algebra A^G of fixed points under the action of G on the C^*-algebra A, is often so small as to be useless. The restriction that G be a compact Lie group is necessary to ensure that $R(G)$ is a Noetherian ring. (We do not require that Lie groups be connected. In particular, finite groups are not excluded.) The fact that $R(G)$ is Noetherian is used in the proof of theorem 1.1.1, and also in the proof of one of the basic lemmas on K-freeness, proposition 4.1.3. Finally, many of the most interesting results have actually been proved only for various classes of finite groups. Such restrictions are presumably not always necessary, but the generalizations of many of our results to infinite compact groups appear to be substantially more difficult.

We now outline the organization of this work. The remainder of this chapter is devoted to the proof of theorem 1.1.1 and several related results, assuming the basic properties of equivariant K-theory. This theorem is not merely motivation — the proofs of many of our results ultimately depend on it.

In chapter 2 we develop the properties of equivariant K-theory of C^*-algebras. The theorem of Julg ([51]), according to which there is a natural isomorphism $K_*^G(A) \simeq K_*(C^*(G,A))$, where $C^*(G,A)$ is the crossed product C^*-algebra, plays a central role in our development. Thus, we assume some knowledge of ordinary K-theory of Banach algebras. (See for example [26], [32], [46], and [105]; note that the hypothesis in [105] that all algebras are commutative is unnecessary.) We do not, however, assume any knowledge of equivariant K-theory of spaces. In particular, the results on equivariant K-theory used in the proof of theorem 1.1.1 are all proved in chapter 2. Little of this material, if any, is really new.

Chapter 3 is an introduction to the equivariant version of Kasparov's KK-theory [55], using the approach of Cuntz [25]. We use KK-theory in several ways: both as a useful technical device for proving things about K-theory, and because the analog of K-freeness using KK-theory is interesting in its own right. Again, the material is not new, although Cuntz does not deal with the equivariant theory in his account. This chapter is rather technical; we fill in many of the details omitted from [25].

In chapter 4, we formally define K-freeness and two related concepts, namely total K-freeness and KK-freeness. An action α of a compact Lie group G on a C^*-algebra A is called totally K-free if the restriction $\alpha|_H$ of α to every closed subgroup H is K-free. We find it necessary to introduce this concept because a simple example shows that K-freeness, unlike freeness, is not inherited by restrictions of actions. KK-freeness is the analog of K-freeness using KK-theory. We prove the basic properties of these concepts, showing, for example, that they behave well with respect to passage to G-invariant ideals and quotients by them. We also show that KK-freeness implies K-freeness. Perhaps the most significant results of chapter 4 are the following two results, showing that certain actions, which one would expect to be totally K-free or KK-free, in fact are. We prove that if α is a continuous action of a finite group G on a separable C^*-algebra A such that the induced action on the primitive ideal space of A is free, then α is totally K-free. We also prove that if α is an action of a compact Lie group on a separable unital C^*-algebra A such that the induced action on the maximal ideal space of the center of A is free, then α is KK-free. Unfortunately, this last result fails for nonunital algebras: there is a free action on a locally compact topological space X such that the corresponding action on $C_0(X)$ is not KK-free.

The next two chapters are devoted to two particularly difficult topics of the same general nature as those of chapter 4, namely restrictions of actions to subgroups and actions on tensor products. As mentioned before, the restriction of a K-free action to a closed subgroup need not be K-free. Nevertheless, in chapter 5 we prove that the restriction of a KK-free action of a finite group on a nuclear C^*-algebra to a subgroup *is* KK-free. The proof requires a technical lemma, the Ideal Decomposition Lemma, which is used to express an ideal invariant under the action of a subgroup in terms of ideals invariant under the action of the group. Going in the other direction, we are able to prove that if α is an action of a finite p-group G (a group whose order is a power of the prime number p) on a C^*-algebra A, and if $\alpha|_S$ is K-free for every cyclic subgroup S of G, then α is totally K-free. This is the best generalization we have of the fact that a continuous action of a compact Lie group on a space is free if the restriction of the action to every finite subgroup of prime order is free. An example given in chapter 9 shows that "cyclic" cannot be replaced by "prime order" in our theorem, but I do not know if the other hypotheses can be weakened.

In chapter 6, we turn to tensor products. Here the motivation is the fact that if X is a free G-space and Y is any G-space, then the diagonal action of G on $X \times Y$ is free. We prove that if G is a finite p-group which acts totally K-freely on a C^*-algebra A and arbitrarily on a C^*-algebra B, and if one of the actions satisfies certain technical conditions (including nuclearity of the algebra), then the diagonal action of G on $A \otimes B$ is totally K-free. The proof uses a Künneth theorem, similar to the one in [95], for the localized homology theories $K_*^H(\cdot)_P$ for certain groups H and certain prime ideals P in $R(H)$. The proof also uses the Ideal Decomposition Lemma from the previous chapter. I do not know if the hypothesis on G can be weakened, but examples show that total K-freeness cannot be replaced by K-freeness. Indeed, we produce an example in which the actions on *both* A and B are K-free, but the diagonal action is not K-free.

Chapter 7 examines the relations between K-freeness and other conditions for freeness of a group action on a C^*-algebra. One of the conditions considered is the C^* analog of condition (2) of theorem 1.1.1. Another type of condition, involving the Connes spectrum and its variants, has been investigated by Olesen and Pedersen in [71], [72], and [73], and by Kishimoto in [58], in connection with the problem of determining when crossed products of C^*-algebras by abelian groups are simple or prime. We obtain theorems along the following lines: if a finite abelian group G acts K-freely on a C^*-algebra A whose K-theory is sufficiently nontrivial in an appropriate sense, then the strong Connes spectrum ([58]) is the full dual group $\hat{G}$. We are then able to conclude that the fixed point algebra A^G is strongly Morita equivalent ([87]) to the crossed product $C^*(G,A)$. If A is separable, it follows that there is a canonical isomorphism from $K_*(A^G)$ to $K_*^G(A)$. We then devote some space to showing that many C^*-algebras do indeed have sufficiently nontrivial K-theory. If $G = \mathbf{Z}/2\mathbf{Z}$, we can furthermore show that for a K-free action of G on an arbitrary C^*-algebra A, one has $K_*(A^G) \simeq K_*^G(A)$ up to 2-torsion.

The last two chapters examine the implications of K-freeness for two special classes of C^*-algebras. In chapter 8, we study type I algebras. We obtain a converse to one of the results of chapter 7: if α is an action of a compact abelian Lie group G on a separable type I algebra A, and if the strong Connes spectrum of α is the entire dual group $\widehat{G}$, then the action is K-free. We also obtain a converse to a theorem in chapter 4: if a compact Lie group G acts totally K-freely on a type I algebra A, then the induced action on the primitive ideal space of A is actually free. If the group G is finite cyclic, then K-freeness of an action on a separable type I algebra actually implies total K-freeness, and we thus obtain a number of equivalent conditions for such an action to be K-free.

Finally, in chapter 9 we specialize to AF algebras. It is not known whether an action of a finite group on an AF algebra leaves invariant an increasing sequence of finite dimensional subalgebras whose union is dense in A; in order to obtain results about AF algebras we must assume that all actions considered do indeed satisfy this property. (The results in chapter 7 apply to AF algebras without any such assumption on the actions.) An action α of a finite abelian group G on an AF algebra A is shown to be K-free if $K_0^G(A)_P = 0$ for all primes $P \subset R(G)$ such that $P \not\supset I(G)$; thus one does not need to look at the equivariant K-theory of the quotients A/I. We furthermore show that if α is K-free, then in fact $I(G)K_*^G(A) = 0$, which is a strong version of condition (5) of theorem 1.1.1. These results simplify considerably the verification that an action is or is not K-free, and we use them to compute some of our most interesting examples. In particular, we construct a K-free action of $\mathbf{Z}/2\mathbf{Z}$ on a C^*-algebra A which induces the trivial action on the primitive ideal space of A, an action of $\mathbf{Z}/2\mathbf{Z}$ such that the strong Connes spectrum is the full dual group but which is not K-free, and an action of $\mathbf{Z}/4\mathbf{Z}$ which is not K-free but whose restriction to the two element subgroup is K-free. We then prove the main theorem on actions on AF algebras: we show that, for actions of finite abelian groups, KK-freeness, total K-freeness, and certain homotopy conditions are all equivalent. We are able to generalize part of this theorem to actions of arbitrary finite groups. The proof of the generalization uses a general method for computing $KK_G^*(A, B)$, where G is a finite group which acts on the AF algebras A and B in such a way that the assumption made above on invariant finite dimensional subalgebras is satisfied. We also examine the special case of locally representable actions.

It should be mentioned that the later chapters are not dependent on all of the preceding ones. They all, of course, depend on chapter 4, while chapter 6 also requires results from chapter 5. Chapter 7 depends only on chapter 4, except that it uses one lemma, not involving K-theory, from chapter 6. Chapter 8 uses material from chapter 7 but not from chapters 5 and 6. Chapter 9 depends only on chapters 4 and 5, except for one example which is related to chapter 7.

Some of the results of this work were announced in [81].

1.2. Proof of Theorem 1.1.1

In this section, we prove theorem 1.1.1 and an algebraic lemma, used in its proof, which will be needed later. We also prove, again for later use, the Localization Theorem ([98], proposition 4.1), which is a generalization of one of the implications in theorem 1.1.1.

Before proving theorem 1.1.1, we give a precise definition of the map appearing in condition (2) and recall a few facts about localization. If G is a compact group, and X is a compact G-space, then the map from $K^0(X/G)$ to $K_G^0(X)$ is defined as follows. Let π be the canonical identification map from X to X/G, and let E be a vector bundle over X/G with projection $p : E \longrightarrow X/G$. Then the image of the class $[E]$ of E in $K_G^0(X)$ is the class of the G-vector bundle

$$\pi^*E = \{(x,v) \in X \times E : \pi(x) = p(v)\},$$

where the G-action is $g \cdot (x,v) = (gx,v)$. It is clear that the resulting map from $K^0(X/G)$ to $K_G^0(X)$ is well defined. It is also easily seen to be a ring homomorphism. If X is not compact, then there is a canonical homeomorphism $X^+/G \simeq (X/G)^+$, where X^+ and $(X/G)^+$ are the one point compactifications of X and X/G. (The action of G on X^+ fixes the point at infinity.) It is clear that the image of $K^0(X/G)$ under the map from $K^0((X/G)^+)$ to $K_G^0(X^+)$ actually lies in $K_G^0(X)$, so that we have a homomorphism from $K^0(X/G)$ to $K_G^0(X)$ for arbitrary locally compact G-spaces X. By taking suspensions, we also obtain a homomorphism from $K^1(X/G)$ to $K_G^1(X)$.

We now recall the definition of localization. (See chapter 3 of [5] for details.) Let R be a commutative ring with identity, and let S be a multiplicative system in R, that is, a subset of R which contains 1, does not contain 0, and is closed under multiplication. Then there is a commutative ring $S^{-1}R$ consisting of all fractions a/s with $a \in R$ and $s \in S$. Two fractions a/s and b/t are equal if and only if there is $u \in S$ such that $u(ta - sb) = 0$. Similarly, if M is an R-module, then there is an $S^{-1}R$-module $S^{-1}M$ consisting of all fractions m/s with $m \in M$ and $s \in S$; the condition for equality of two such fractions is similar. The assignment $M \longrightarrow S^{-1}M$ is in fact a functor from R-modules to $S^{-1}R$-modules, and it can be easily shown ([5], proposition 3.3) that this functor preserves exactness. Now let $P \subset R$ be a prime ideal, that is, an ideal such that if $ab \in P$ then $a \in P$ or $b \in P$. Then $S = R - P$ is a multiplicative system. The ring $S^{-1}R$ is called the localization of R at P, and is written R_P. Similarly, $S^{-1}M$ is called the localization of M at P and written M_P.

We need one lemma before starting the proof of theorem 1.1.1.

1.2.1 Lemma. Let X be a locally compact G-space. Then there is a natural isomorphism of $R(G)$-modules $K_G^*(X) \simeq K_G^0(S^1 \times X)$, where S^1 is the unit circle, and the G-action on $S^1 \times X$ is given by $g(z,x) = (z,gx)$.

Proof. We identify S^1 with the one point compactification of $(0,1)$. We obtain a natural identification of the suspension $(0,1)\times X$ of X with an open subset U of $S^1\times X$. The complement of U is just a copy of X, and furthermore the inclusion of the complement has a left inverse, namely the projection on the second factor. We obtain an exact sequence in K-theory

$$K_G^1(S^1\times X)\longrightarrow K_G^1(X)\longrightarrow K_G^0(U)\longrightarrow K_G^0(S^1\times X)\longrightarrow K_G^0(X)\longrightarrow K_G^1(U).$$

The maps from $K_G^i(S^1\times X)$ to $K_G^i(X)$ have right inverses, so there is actually a split exact sequence

$$0\longrightarrow K_G^0(U)\longrightarrow K_G^0(S^1\times X)\longrightarrow K_G^0(X)\longrightarrow 0.$$

Since $K_G^0(U)$ is naturally isomorphic to $K_G^1(X)$, this completes the proof. Q.E.D.

Proof of theorem 1.1.1. (1)$\Longrightarrow$(2). (This proof is taken from [98], proposition 2.1.) If G acts freely on X then G acts freely on $S^1\times X$. The previous lemma now implies that we can replace X by $S^1\times X$ and then only prove that $K^0(X/G)\longrightarrow K_G^0(X)$ is an isomorphism. Let E be a G-vector bundle over X. We claim E/G is a vector bundle over X/G, with the obvious projection map. The only part which is not obvious is showing that E/G is locally trivial. For this, let $x\in X$. Then by [14], theorem II.5.4, there is a subset Z of X with $x\in Z$ such that the map $(g,z)\longrightarrow gz : G\times Z\longrightarrow X$ defines a homeomorphism from $G\times_{G_x}Z$ onto a neighborhood $U=G\cdot Z$ of the orbit Gx. (Here G_x is the stabilizer group of the point x. The notation $G\times_{G_x}Z$ denotes the twisted product. See the discussion preceding definition 2.9.2 for details.) Since the action is free, $G\times_{G_x}Z=G\times Z$. Since E is locally trivial, we may assume, by choosing Z small, that the restricted bundle $E|_Z$ is trivial. Let $\pi : X\longrightarrow X/G$ be the identification map. Since $U\simeq G\times Z$, we obtain isomorphisms $(E/G)|_{\pi[U]}\simeq(E|_U)/G\simeq E|_Z$. Since $\pi[U]$ is a neighborhood of $\pi(x)$ in X/G, it follows that E/G is locally trivial, and hence a vector bundle.

We now claim that $[E]\longrightarrow[E/G]$ defines an inverse for $[F]\longrightarrow[\pi^*(F)]$. Let $p : E\longrightarrow X$ and $p_0: E/G\longrightarrow X/G$ be the projections; then recall that

$$\pi^*(E/G)=\{(x,v)\in X\times E/G : \pi(x)=p_0(v)\}.$$

The function from E to $\pi^*(E/G)$ defined by $e\longrightarrow(p(e),\pi(e))$ is then obviously an isomorphism of G-vector bundles. In the other direction, if F is a vector bundle over X/G, then the map sending the class $[x,v]\in\pi^*(F)/G$ of $(x,v)\in\pi^*(F)$ to v is clearly an isomorphism. It follows that $[F]\longrightarrow[\pi^*(F)]$ is an isomorphism from $K^0(X/G)$ to $K_G^0(X)$ as desired.

(2)$\Longrightarrow$(5). (This is from [6], proof of proposition 4.3.) Again, by lemma 1.2.1, we can replace X by $S^1\times X$, and consider only K^0 and K_G^0. The ring $R(G)$ is noetherian ([97], corollary 3.3), hence $I(G)$ is finitely generated. Since $K^0(X/G)\simeq K_G^0(X)$, it follows that $K^0(X/G)$ is also an $R(G)$-module. (This module structure is not particularly easy to describe.) Since it is also a commutative ring with identity, the ideal $I(G)K^0(X/G)$ is also finitely generated. Each element in it has the form $[E]-[F]$, where E and F are vector bundles such that for $x \in X/G$, the fiber dimensions $dim(E_x)$ and $dim(F_x)$ are equal. (This follows from the fact that elements of $I(G)$ have the form $[V]-[W]$, where V and W are representation spaces of G with $dim V = dim W$.) By [2], proposition 3.1.6, all such elements are nilpotent. Since $I(G)K^0(X/G)$ is finitely generated, there is an integer n such that $(I(G)K^0(X/G))^n = 0$, and since $K^0(X/G)$ is unital it follows that $I(G)^n K^0(X/G) = 0$. Thus $I(G)^n K_G^0(X) = 0$, as desired.

(5)$\Longrightarrow$(4) $\Longrightarrow$(3): trivial.

(3)$\Longrightarrow$(1): (This is also from [6], although our argument below that $R(H)$ is not complete is different.) Suppose G does not act freely on X. Then there is $x \in X$ and a subgroup H of G of prime order p such that $H \subset G_x$. The composite ring homomorphism

$$K_G^*(X) \longrightarrow K_H^*(X) \longrightarrow K_H^*(\{x\}) \simeq R(H)$$

makes $R(H)$ into a topological $K_G^*(X)$-module, where everything has the $I(G)$-adic topology. Now $R(H)$ is finitely generated as a module over $K_G^*(X)$, since by [97], proposition 3.2, it is already finitely generated over $R(G)$ and $K_G^*(X)$ is a unital $R(G)$-algebra. By [5], theorem 10.13, $R(H)$ is complete in the $I(G)$-adic topology if $K_G^*(X)$ is. Then $R(H)$ would also be complete in the $I(H)$-adic topology, since by [97], corollary 3.9, it is the same as the $I(G)$-adic topology. We will show that this is not the case.

We have $R(G) \simeq \mathbf{Z}[x]/<1-x^p>$ (the quotient of the polynomial ring in one variable over $\mathbf{Z}$ by the ideal generated by $1-x^p$), and $I(H) = <1-x>$. Clearly $\bigcap_n I(H)^n = \{0\}$, whence $R(H)$ is Hausdorff. It is also metrizable. (Set $d(\eta,\lambda) = 2^{-n}$ where n is the largest integer such that $\eta - \lambda \in I(H)^n$.) One has $(1-x)^n \neq 0$ for any n, but $(1-x)^n \longrightarrow 0$ as $n \longrightarrow \infty$, so that the open set $R(H)-\{0\}$ is dense; therefore all sets $R(H)-\{\eta\}$ for $\eta \in R(H)$ are dense and open. We have

$$\emptyset = \bigcap_{\eta \in R(H)} R(H)-\{\eta\}.$$

Since $R(H)$ is countable, we have shown that the empty set is a countable intersection of dense open subsets. By the Baire Category Theorem, $R(H)$ is therefore not complete. So $K_G^*(X)$ cannot be complete.

(5)$\Longleftrightarrow$(6). $R(G)$ is a noetherian ring, $I(G)$ is a prime ideal in it, and $K_G^*(X)$ is a unital algebra over $R(G)$, since X is compact. The equivalence of condition (5) and (6) then follows immediately from the the following lemma, which for later use is stated in greater generality than needed here.

1.2.2 Lemma. Let R be a Noetherian ring, let I be a prime ideal in R, and let M be an R-module. Then the following are equivalent:

(1) $M_P = 0$ for all prime ideals $P \subset R$ such that $P \not\supset I$.

(2) For every $\eta \in M$ there is $n \in \mathbf{Z}$ such that $I^n \cdot \eta = 0$.

If M is either finitely generated or a unital algebra (not necessarily commutative) over R, then these are equivalent to:

(3) There is $n \in \mathbf{Z}$ such that $I^n \cdot M = 0$.

Proof. (1)$\Longrightarrow$(2): Let $\eta \in M$. Let $J = \{r \in I : r\eta = 0\}$ and let

$$rad(J) = \{a \in R : a^n \in J \text{ for some } n\}$$

be the radical of J. By [5], proposition 1.14, $rad(J)$ is the intersection of all prime ideals of R which contain J.

Let P be a prime ideal of R which does not contain I. Then by hypothesis, the image of η in M_P is zero, so there is $r \notin P$ such that $r\eta = 0$. Let s be any element of I not in P. Then $rs\eta = 0$ and $rs \in I$, that is, $rs \in J$. Also $rs \notin P$ because P is a prime ideal. So $J \not\subset P$. Thus, any prime ideal containing J also contains I, and hence $rad(J) \supset I$. Since $J \subset I$ and I is prime, we actually have $rad(J) = I$. Since R is noetherian, proposition 7.14 of [5] implies that there is n such that $I^n \subset J$, whence $I^n \cdot \eta = 0$.

(2)$\Longrightarrow$(1). Let $P \subset R$ be a prime ideal with $P \not\supset I$, and let $r \in I, r \notin P$. For every $\eta \in M$, there is n such that $I^n \cdot \eta = 0$, and in particular $r^n \eta = 0$. Since P is prime, $r^n \notin P$ for any n, and therefore the image of η in M_P is zero. Since this is true for all η, we obtain $M_P = 0$.

(3)$\Longrightarrow$(2) is obvious for any M.

(2)$\Longrightarrow$(3) for finitely generated modules: Let $\eta_1, \dots, \eta_k$ generate M, and find n_i such that $I^{n_i} \cdot \eta_i = 0$. Then clearly $I^n M = 0$ with $n = \max_i (n_i)$.

(2)$\Longrightarrow$(3) for unital algebras: Choose n such that $I^n \cdot 1 = 0$, where 1 is the unit of M. Then for every $\eta \in M$, $I^n \cdot \eta = I^n \cdot 1 \cdot \eta = 0$, so that $I^n M = 0$.

This completes the proof of the lemma, and thus also of the theorem. Q.E.D.

1.2.3 Corollary (of theorem 1.1.1). Let G be a compact Lie group, and let X be a locally compact free G-space. Then the map $K^*(X/G) \longrightarrow K^*_G(X)$ is an isomorphism.

Proof. Let $(X_\alpha)_{\alpha \in I}$ be an increasing family, indexed by a directed set I, of open G-invariant subsets of X with compact closures $\bar{X}_\alpha$, such that $X = \bigcup_{\alpha \in I} X_\alpha$. Then the boundaries ∂X_α are also compact G-invariant subsets, and there is a commutative diagram with exact rows:

$$\begin{array}{ccccccccc} K^{1-i}(\bar{X}_\alpha/G) & \longrightarrow & K^{1-i}(\partial X_\alpha/G) & \longrightarrow & K^i(X_\alpha/G) & \longrightarrow & K^i(\bar{X}_\alpha/G) & \longrightarrow & K^i(\partial X_\alpha/G) \\ \downarrow & & \downarrow & & \downarrow & & \downarrow & & \downarrow \\ K_G^{1-i}(\bar{X}_\alpha) & \longrightarrow & K_G^{1-i}(\partial X_\alpha) & \longrightarrow & K_G^i(X_\alpha) & \longrightarrow & K_G^i(\bar{X}_\alpha) & \longrightarrow & K^i(\partial X_\alpha). \end{array}$$

Since $\bar{X}_\alpha$ and ∂X_α are compact free G-spaces, the implication (1)$\Longrightarrow$(2) of theorem 1.1.1 implies that all the vertical arrows except the middle one are isomorphisms. By the Five Lemma, the middle vertical arrow is also an isomorphism. Now

$X = \varinjlim X_\alpha$ (in the sense that $C_0(X) = \varinjlim C_0(X_\alpha)$) and correspondingly $X/G = \varinjlim X_\alpha/G$. Since K-theory commutes with direct limits, it follows that $K^i(X/G) \longrightarrow K^i_G(X)$ is an isomorphism for $i = 0,1$. Q.E.D.

The implication (1)⟹(6) of theorem 1.1.1 admits a generalization which will be needed later. To state it, we need to introduce the concept of the support of a prime ideal of $R(G)$, as defined in [97]. Let P be a prime ideal of $R(G)$. Then its support is a topologically cyclic subgroup of G (that is, one of the form $\overline{\{g^n : n \in \mathbf{Z}\}}$ for some $g \in G$, where the bar indicates closure), which is minimal among all closed subgroups H of G such that P is the inverse image of a prime ideal in $R(H)$ under the restriction map from $R(G)$ to $R(H)$ (obtained by restricting representations of G to the subgroup H.) It is unique up to conjugation by elements of G. (For a proof that such a thing exists, see [97], proposition 3.7.) It is easily seen that a prime ideal P in $R(G)$ has support equal to the trivial subgroup $\{e\}$ if and only if P contains the augmentation ideal $I(G)$.

For convenience, if X is a locally compact G-space and S is a closed subgroup of G, we let

$$X^{(S)} = \{x \in X : G_x \text{ contains some conjugate of } S\}.$$

Then we can also write

$$X^{(S)} = G \cdot \{x \in X : sx = x \quad \text{for } s \in S\}.$$

If G is compact, it follows that $X^{(S)}$ is closed in X. We can now state the theorem:

1.2.4 Localization Theorem. ([98], proposition 4.1.) Let $P \subset R(G)$ be a prime ideal with support S. Then the restriction map $K^*_G(X)_P \longrightarrow K^*_G(X^{(S)})_P$ is an isomorphism.

Proof ([98]). Since localization is exact ([5], proposition 3.3), there is an exact sequence ($i = 0,1$):

$$K^i_G(X - X^{(S)})_P \longrightarrow K^i_G(X)_P \longrightarrow K^i_G(X^{(S)})_P \longrightarrow K^{1-i}_G(X - X^{(S)}).$$

So it is sufficient to show that $K^i_G(X - X^{(S)})_P = 0$, that is, we may assume $X^{(S)} = \emptyset$ and prove $K^*_G(X)_P = 0$. The argument used in the proof of corollary 1.2.3 now shows that we may assume X is compact.

Let $x \in X$. By [14], theorem II.5.4, there is a G_x-invariant subset Z of X which contains x such that the map $(g, z) \longrightarrow gz : G \times Z \longrightarrow X$ defines a homeomorphism from $G \times_{G_x} Z$ onto an open neighborhood $U = GZ$ of the orbit Gx. Then $K^*_G(U) \simeq K^*_{G_x}(Z)$ as $R(G)$-modules (this is proved in chapter 2, proposition 2.9.4). Since G_x does not contain any conjugate of S, it follows from [97], proposition 3.7, that $R(G_x)_P = 0$ as an $R(G)$-module. Therefore $K^*_G(U)_P = K^*_{G_x}(Z)_P = 0$. Furthermore, if V is any open G-invariant subset of U, then $V = G(V \cap Z) \simeq G \times_{G_x}(V \cap Z)$, so that the same reasoning implies $K^*_G(V)_P = 0$.

Since X is compact, we may choose $x_1, \ldots, x_n \in X$ such that the corresponding neighborhoods $U_1, \ldots, U_n$ constructed as above cover X. We claim that $K_G^*\left(\bigcup_1^k U_j\right)_P = 0$, and we will prove this by induction on k. By the previous paragraph, this is true for $k = 1$, so assume it is true for some $k \geq 1$. Let $V = \left(\bigcup_1^k U_j\right) \cap U_{k+1}$. From the exact sequence

$$\cdots \longrightarrow K_G^i\left(\bigcup_1^k U_u\right)_P \longrightarrow K_G^i\left(\left(\bigcup_1^k U_j\right) - V\right)_P \longrightarrow K_G^{1-i}(V)_P \longrightarrow \cdots$$

and the fact that $K_G^*(V)_P = 0$ (since $V \subset U_{k+1}$), we obtain $K_G^i\left(\left(\bigcup_1^k U_j\right) - V\right)_P = 0$. Now the exact sequence

$$\cdots \longrightarrow K_G^i\left(\bigcup_1^k U_j\right)_P \longrightarrow K_G^i\left(\bigcup_1^{k+1} U_j\right)_P \longrightarrow K_G^i\left(\left(\bigcup_1^{k+1} U_j\right) - U_{k+1}\right)_P \longrightarrow \cdots$$

together with the equality

$$\left(\bigcup_1^{k+1} U_j\right) - U_{k+1} = \left(\bigcup_1^k U_j\right) - V,$$

implies that $K_G^*\left(\bigcup_1^{k+1} U_j\right)_P = 0$. It therefore follows by induction that $K_G^*(X)_P = 0$, as desired. Q.E.D.

Chapter 2

Equivariant K-Theory of C^*-Algebras

This chapter is an introduction to the equivariant K-theory of C^*-algebras. We obtain the equivariant K-theory of locally compact spaces (see [98]) as a special case. We derive the long exact sequence of K-groups associated to a short equivariant exact sequence of C^*-algebras using the theorem of Julg [51], according to which the equivariant K-theory of a C^*-algebra A with a continuous action α of G is the same as the ordinary K-theory of the crossed product $C^*(G,A,\alpha)$. We therefore assume some familiarity with the ordinary K-theory of Banach algebras (see [26], [32] and [46] for C^*-algebras, and [105] for Banach algebras; note that the hypothesis of commutativity in [105] is unnecessary).

In the first section of this chapter, we define $K_G^0(X)$ for locally compact G-spaces X, using vector bundles. In the next section, we define $K_0^G(A)$ for Banach algebras A carrying continuous actions of the compact group G, in terms of finitely generated projective modules. The ordinary K-theory of Banach algebras is a special case, obtained by taking G to be the one element group. The third section is devoted to the appropriate version of Swan's theorem, which shows that we do in fact have $K_0^G(C_0(X)) \simeq K_G^0(X)$ (where $C_0(X)$ is the algebra of all continuous complex valued functions vanishing at infinity on the locally compact Hausdorff space X).

Section four gives a different construction of $K_0^G(A)$, using idempotents in matrix algebras over A rather than finitely generated projective modules. If A is a C^*-algebra and the action of G is by $*$-automorphisms, we need consider only projections, that is, selfadjoint idempotents. The fifth section then applies this construction to the equivariant K-theory of direct limits. In order to ensure the existence of direct limits, we restrict ourselves here to the category of C^*-algebras, although we are able to say one or two useful things about more general Banach algebras. The crucial lemma in this section, lemma 2.5.3, asserts that sufficiently close G-invariant projections are equivalent. This lemma will be used frequently in later chapters. Here, we also use it to give a quick proof of the homotopy invariance of equivariant K-theory.

In section six, we prove Julg's theorem. The proof has two steps: $K_0^G(A) \simeq K_0(L^1(G,A))$, where $L^1(G,A)$ is the set of all L^1 functions from G to A with twisted convolution as multiplication; and $K_0(L^1(G,A)) \simeq K_0(C^*(G,A))$ (recall that $C^*(G,A)$ is the universal enveloping C^*-algebra of $L^1(G,A)$.) The first step is actually valid for general Banach algebras, and many later results can be generalized to the equivariant K-theory of Banach algebras simply by using $L^1(G,A)$ in place of $C^*(G,A)$. However, if we start with a C^*-algebra then we want to end up with a C^*-algebra. Therefore we abandon the equivariant K-theory of Banach algebras at this point, and work in terms of $C^*(G,A)$.

The group $K_0^G(A)$ is actually a module over the representation ring $R(G)$ of G, and we need to determine how to compute the corresponding module structure on $K_0(C^*(G,A))$. This is done in section seven. The next section is devoted to the proof that K_0^G is part of a periodic (of period 2) equivariant homology theory on C^*-algebras. In particular $K_1^G(A)$ is defined and the long exact sequence is constructed. Finally, in section nine, several miscellaneous results are collected for later use.

Throughout this chapter, G is a compact group unless otherwise specified.

2.1. The Equivariant K-Theory of Spaces

In this section we define the equivariant K-theory of locally compact G-spaces for a compact group G. Recall that if G is a topological group, then a G-space X is a topological space X with a continuous action of G. That is, there is a jointly continuous function $(g,x) \longrightarrow gx$ from $G \times X$ to X such that $(gh)x = g(hx)$ and $ex = x$ for $g, h \in G$ and $x \in X$, where e is the identity element of G. Morphisms of G-spaces are of course the continuous equivariant maps, where we say a map $f : X \longrightarrow Y$ of G-spaces is equivariant if $g \cdot f(x) = f(gx)$ for $g \in G$ and $x \in X$. When dealing with locally compact G-spaces we will usually in addition assume that maps are proper, that is, inverse images of compact sets are compact. We assume throughout this chapter, and in general unless otherwise specified, that all topological spaces are Hausdorff.

2.1.1 Definition. ([98], page 129.) Let X be a compact G-space. A G-vector bundle over X is a complex vector bundle E over X which is also a G-space, in such a way that the projection from E to X is equivariant, and for $g \in G$ and $x \in X$, the action of g from E_x to E_{gx} is linear. (Here E_x is the fiber of E over x.)

2.1.2 Definition. ([98], page 132.) Let $[E]$ and $[F]$ be isomorphism classes of G-vector bundles E and F over X. We define $[E]+[F] = [E \oplus F]$, where $E \oplus F$ is the Whitney sum (satisfying $(E \otimes F)_x = E_x \otimes F_x$) with the obvious G-action. With this operation, the collection of isomorphism classes of G-vector bundles over X becomes a commutative semigroup $Vect_G(X)$ with identity. Then $K_G^0(X)$ is the Grothendieck group of this semigroup. We also write $[E]$ for the class of E in $K_G^0(X)$. Define $[E] \cdot [F] = [E \otimes F]$, where $E \otimes F$ is the tensor product bundle (satisfying $(E \otimes F)_x = E_x \otimes F_x$) with the diagonal G-action. This operation extends to all of $K_G^0(X)$ by bilinearity, and evidently makes $K_G^0(X)$ a commutative ring, with identity given by the class of the trivial one-dimensional bundle $\mathbf{C} \times X$ with the G-action $g \cdot (x, \lambda) = (gx, \lambda)$.

2.1.3 Definition. ([97], page 113.) Let G act trivially on a one point space $\{x_0\}$. Then $K_G^0(\{x_0\})$ is a commutative unital ring, called the representation ring of G and written $R(G)$. Rephrasing the definition, it is the Grothendieck group of the semi-group of isomorphism classes of finite dimensional complex representations of G, with addition $[V]+[W]=[V\oplus W]$ and multiplication $[V][W]=[V\otimes W]$ for complex representation spaces V and W of G. From the representation theory of compact groups, we actually have $[V]=[W]$ in $R(G)$ if and only if the representations of G on V and W are equivalent. We furthermore may and will assume that all representations are actually unitary representations on (finite dimensional) Hilbert spaces; we then obtain $[V]=[W]$ if and only if the representations are unitarily equivalent.

There is a canonical ring homomorphism from $R(G)$ to $\mathbf{Z}$ which sends $[V]$ to $dim\,(V)$. Its kernel, which is the set of elements of $R(G)$ of "virtual dimension" 0, is called the augmentation ideal and written $I(G)$. The structure of $R(G)$ will play an important role, but we do not investigate it here. The general structure theorems we need can be found in [97] and the results we need for special classes of groups will be proved where we use them.

Let X be a compact G-space, E a G-vector bundle over X, and V a representation space of G. Then the bundle $V\otimes E$, having fibers $(V\otimes E)_x = V\otimes E_x$, and equipped with the diagonal G-action, is again a G-vector bundle over X. The formula $[V]\cdot[E]=[V\otimes E]$ then makes $K_G^0(X)$ into an $R(G)$-module, in fact an $R(G)$-algebra.

2.1.4 Proposition. ([98], page 132.) K_G^0 is a contravariant functor from compact G-spaces and equivariant maps to commutative unital $R(G)$-algebras and homomorphisms.

Proof (outline). Let $f : X \longrightarrow Y$ be an equivariant map of compact G-spaces, and let E be a G-vector bundle over Y with projection $p : E \longrightarrow Y$. Then the pullback

$$f^*(E)=\{(x,v)\in X\times E : f(x)=p(v)\},$$

with projection $q : f^*(E) \longrightarrow X$ given by $q(x,v)=x$ and with G-action $g(x,v)=(gx,gv)$, is a G-vector bundle, and the map $f^* : K_G^0(Y) \longrightarrow K_G^0(X)$ is given by $f^*([E])=[f^*(E)]$. One checks that it is a ring homomorphism and functorial. To show it is an $R(G)$-module homomorphism, notice that the $R(G)$-algebra structure on $K_G^0(X)$ is induced by the ring homomorphism $t^* : R(G) \longrightarrow K_G^0(X)$ coming from the map t from X to the one point space. Q.E.D.

To deal with G-spaces which are locally compact but not compact, we introduce the one point compactification $X^+ = X\cup\{\infty\}$ of the locally compact space X. If X is a G-space, then X^+ becomes a G-space if we define $g\cdot\infty=\infty$ for $g\in G$. Notice that if X is already compact, then X^+ still makes sense: it is the topological disjoint union of X and a one point space.

2.1.5 Definition. Let X be a locally compact G-space. Then $K_G^0(X)$ is the kernel of $r^* : K_G^0(X^+) \longrightarrow K_G^0(\{\infty\})$, where r is the map from the one point space $\{\infty\}$ to X^+ sending ∞ to the point at infinity in X^+. Since $K_G^0(X)$ is the kernel of an $R(G)$-algebra homomorphism, it is a (usually nonunital) $R(G)$-algebra. If X is in fact already compact, then there is a canonical identification of the kernel of r^* with $K_G^0(X)$ as already defined. (If E is a G-vector bundle on X, then E can be extended to a G-vector bundle over X^+ by setting the fiber over the point at infinity equal to 0.) This definition is therefore a consistent extension of the definition of $K_G^0(X)$ for compact X.

2.1.6 Proposition. K_G^0 is a contravariant functor from locally compact G-spaces and proper equivariant maps to commutative $R(G)$-algebras and homomorphisms.

Proof (outline). If $f : X \longrightarrow Y$ is a proper equivariant map, then it has a continuous equivariant extension to a map $f^+ : X^+ \longrightarrow Y^+$ which sends to point at infinity in X^+ to the point at infinity in Y^+. We therefore obtain a homomorphism $(f^+)^* : K_G^0(Y^+) \longrightarrow K_G^0(X^+)$. Set f^* equal to the restriction $(f^+)^* |_{K_G^0(Y)}$. Then it is easy to check that the image of f^* is in $K_G^0(X)$, that $f \longrightarrow f^*$ is functorial, etc.

Q.E.D.

One can now define $K_G^1(X) = K_G^0(SX)$, where SX is the suspension $(0,1) \times X$ of X, with the G-action $g(t,x) = (t,gx)$ for $t \in (0,1)$, $x \in X$, and $g \in G$. Then K_G^1 is obviously a functor from locally compact G-spaces and proper G-maps to $R(G)$-modules and homomorphisms. However, rather than continuing the development of the equivariant K-theory of spaces in this way, we turn to the equivariant K-theory of C^*-algebras.

2.2. The Definition of Equivariant K-Theory for Banach Algebras

An action of a group G on a Banach algebra A is a group homomorphism $g \longrightarrow \alpha_g$ from G to the group $Aut(A)$ of automorphisms of A. The action is said to be continuous if $g \longrightarrow \alpha_g(a)$ is a continuous function from G to A for each $a \in A$. We refer to A, or more properly the triple (G,A,α), as a G-Banach algebra. If A is a C^*-algebra, then we write $Aut(A)$ for the set of all $*$-automorphisms of A, and we call (G,A,α) simply a G-algebra. Our main reference for general facts about G-algebras is chapters 7 and 8 of Pedersen's book [79]; note that he writes a G-algebra in the order (A,G,α), and calls it a C^*-dynamical system. The right sort of morphism is of course an equivariant homomorphism, where if (G,A,α) and (G,B,β) are G-Banach algebras then a function $\varphi : A \longrightarrow B$ is equivariant if $\varphi(\alpha_g(a)) = \beta_g(\varphi(a))$ for $a \in A$ and $g \in G$. We will assume that all homomorphisms of C^*-algebras are $*$-homomorphisms.

The correct noncommutative analog of a vector bundle is a finitely generated projective module. (The justification for this is Swan's theorem, which is the subject of the next section.) We therefore begin by recalling a few facts about such modules. In order to avoid certain problems with notation, such as having matrices act on the right, we consider only *right* modules.

Let A be a unital ring. Then a module E over A is called projective if, whenever we are given a surjective module homomorphism $\varphi : M \longrightarrow N$ and a module homomorphism $\psi : E \longrightarrow N$, there is a module homomorphism $\eta : E \longrightarrow M$ such that $\varphi \cdot \eta = \psi$, as in the following diagram with exact row:

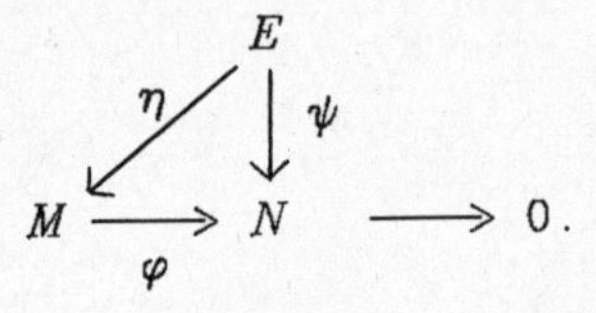

It is well known that a module is projective if and only if it is a direct summand of a free module.

Now let A be a unital Banach algebra. Then a finitely generated projective A-module E is a direct summand in A^n for some (finite) n, and hence inherits a norm. Any A-module endomorphism of A^n is continuous, as can be seen by representing it as multiplication on the left by an $n \times n$ matrix with entries in A. The continuity of the projection from A^n onto E then implies that E is closed in A^n and hence complete; furthermore, multiplication is continuous: there is a constant C such that $\| ea \| \leq C \| e \| \| a \|$ for $a \in A$ and $e \in E$. (We have not specified which norm we use on A^n. There are many equivalent choices, and we have no need to single out any particular one. For convenience, however, we will always assume that a norm has been chosen for which we can take $C = 1$ above.) It follows easily that if multiplication on the modules M and N of the diagram above is continuous, and if ψ and φ are continuous, then η can be chosen to be continuous. Furthermore, the topology defined on E by embedding it as a direct summand in some A^n doesn't depend on the embedding. When continuity of functions to or from finitely generated projective modules is assumed, it is always relative to this topology.

2.2.1 Definition. ([51], definition 2.1.) Let (G,A,α) be a unital G-Banach algebra. Then a finitely generated projective (G,A,α)-module is a triple (G,E,λ), where E is a finitely generated projective A-module, and $\lambda : G \longrightarrow L(E)$ is a continuous representation (not unitary, since E is not a Hilbert space) of G on E, such that $\lambda_g(ea) = \lambda_g(e)\alpha_g(a)$ for $g \in G$, $a \in A$, and $e \in E$. Here $L(E)$ is the space of all bounded operators on the Banach space E, and "continuous" means "strong operator continuous", that is, for $e \in E$ the function $g \longrightarrow \lambda_g(e)$ is continuous from G to E. This formula is the analog for right modules of the condition for a representation (u,π) of (G,A,α) on a Hilbert space to be covariant, namely $u_g \pi(a) u_g^* = \pi(\alpha_g(a))$ for $a \in A$ and $g \in G$. Notice that λ_g is *not* an A-module homomorphism unless the action of G on A is trivial. We will sometimes tacitly assume that (G,A,α)-modules are finitely generated and projective. We will also sometimes refer to E as a (G,A,α)-module when λ is understood.

2.2.2 Definition. ([51], definition 2.2.) Let $[E]$ and $[F]$ be isomorphism classes of (G,A,α)-modules. We define $[E]+[F]=[E\oplus F]$, where $E\oplus F$ has the obvious G-action. With this operation, the collection of isomorphism classes of finitely generated projective (G,A,α)-modules becomes a commutative semigroup $Proj_G(A)$ with identity. Then $K_0^G(A)$ is the Grothendieck group of this semigroup. If V is a finite dimensional representation space of G, and E is a (G,A,α)-module, we define $[V]\cdot[E]=[V\otimes E]$, where the G-action on $V\otimes E$ is the diagonal action and the A-module structure is given by $(v\otimes e)a=v\otimes ea$ for $a\in A$, $v\in V$, and $e\in E$. It is readily verified that $V\otimes E$ is a (G,A,α)-module, and that this multiplication makes $K_0^G(A)$ into an $R(G)$-module. Unlike $K_G^0(X)$, however, $K_0^G(A)$ cannot be made a ring in a natural way.

In the ordinary K-theory of Banach algebras, functoriality can be proved as follows. Let $\varphi: A \longrightarrow B$ be a unital homomorphism of C^*-algebras, and let E be a finitely generated projective A-module. Then $\varphi_*([E])=[E\otimes_A B]$, where the left A-module structure on B is given by the homomorphism φ. The module $E\otimes_A B$ is projective because, if E is a direct summand of A^n, then $E\otimes_A B$ is a direct summand in $A^n\otimes_A B$, which is just B^n. We want to use a similar construction to define $\varphi_*: K_0^G(A) \longrightarrow K_0^G(B)$. If E is actually a (G,A,α)-module, then the construction above certainly produces a finitely generated B-module with a compatible action of G. Unfortunately, the topology on $E\otimes_A B$ is not manufactured directly from the topologies on E and B; rather, $E\otimes_A B$ is topologized as a subspace of a free B-module B^n. Thus, it is not clear that the action of G on $E\otimes_A B$ is continuous. We must therefore show that E is an equivariant direct summand in some $V\otimes A$, which is the analog of proposition 2.4 of [98]. We can then use the isomorphism $(V\otimes A)\otimes_A B \simeq V\otimes B$ to show that $E\otimes_A B$ is a direct summand in a (G,B,β)-module.

The proof that E is a direct summand in some $V\otimes A$ uses several lemmas. The first is the analog of the integrated form of a covariant representation of (G,A,α), as in, for example, [79], 7.6.4. The integrated form is a representation, or in our case a module action, of the algebra $L^1(G,A,\alpha)$, whose definition we recall. (See [79], 7.6.1, where it is called $L^1(G,A)$.) Since we will need it, or rather its universal enveloping C^*-algebra $C^*(G,A,\alpha)$, for some noncompact groups, we assume that G is a locally compact unimodular group. (We never consider groups which are not unimodular, and this assumption simplifies the formulas.) We first introduce some notation. If X is any locally compact space and A is a C^*-algebra, we write $C_c(X,A)$ for the set of continuous functions from X to A with compact support, and $C_0(X,A)$ for the set of continuous functions from X to A which vanish at infinity. If X is compact, then both of these reduce to the set of all continuous functions from X to A, which we denote by $C(X,A)$.

Define multiplication in $C_c(G,A)$ by

$$(ab)(g)=\int_G a(h)\,\alpha_h(b(h^{-1}g))\,dh,$$

for $a, b \in C_c(G,A)$ and $g \in G$. Integrals are always taken with respect to Haar measure on G, which is assumed to be normalized so that the measure of G is 1 if G is compact, and so that the measure of $\{e\}$ is 1 if G is discrete. (Which convention is used for finite groups G will usually be clear from the context.) If A is a C^*-algebra, we further define an involution on $C_c(G,A)$ by

$$a^*(g) = \alpha_g(a(g^{-1})^*).$$

Then $L^1(G,A,\alpha)$ is the completion of $C_c(G,A)$ with respect to the norm

$$\| a \|_1 = \int_G \| a(g) \| \, dg .$$

$L^1(G,A,\alpha)$ is a Banach algebra, using the extensions of the operations defined above. If A is a C^*-algebra, then $L^1(G,A,\alpha)$ has a continuous involution and a bounded approximate identity, so it has a universal enveloping C^*-algebra, which is the crossed product $C^*(G,A,\alpha)$.

We now return to our standing assumption that G is compact.

2.2.3 Lemma. ([51], section 4.) Let (G,E,λ) be a (G,A,α)-module, where A is unital. Then the formula

$$ea = \int_G \lambda_g^{-1}(e)\alpha_g^{-1}(a(g))\, dg ,$$

for $a \in L^1(G,A,\alpha)$ and $e \in E$, makes E an $L^1(G,A,\alpha)$-module.

Proof. Let $\| \lambda \|_\infty = sup\{\| \lambda_g \| : g \in G\}$. Since G is compact, the sets $\{\lambda_g(e) : g \in G\}$ are bounded for $e \in E$, and the Uniform Boundedness Principle implies that $\| \lambda \|_\infty$ is finite. Now let $a \in C_c(G,A)$ and $e \in E$. Then ea as defined in the statement of the lemma clearly exists and satisfies

$$\| ea \| \le \int_G \| \lambda_{g^{-1}}(e) \| \, \| a(g) \| \, dg \le \| \lambda \|_\infty \| e \| \, \| a \|_1 .$$

Therefore $(e,a) \longrightarrow ea$ extends by continuity to a map from $E \times L^1(G,A,\alpha)$ to E. If $a,b \in L^1(G,A)$ and $e \in E$, then a straightforward calculation shows that $e(ab) = (ea)b$. Thus, E is an $L^1(G,A)$-module. Q.E.D.

We now introduce the following convention: an idempotent is any element e of a ring such that $e^2 = e$, and a projection in a ring with involution $a \longrightarrow a^*$ is an idempotent e which also satisfies $e^* = e$. We also say that a net $\{e_n\}_{n \in D}$ of idempotents is increasing if $e_n e_m = e_m e_n = e_m$ for $m \le n$. (If the e_n are projections in a C^*-algebra, this is equivalent to the condition $e_m \le e_n$ for $m \le n$.)

2.2.4 Lemma. There is an increasing net $(p_n)_{n \in D}$ of central projections in $L^1(G)$ such that, for any unital G-Banach algebra (G,A,α), one has

$$\overline{\bigcup_{n \in D} p_n L^1(G,A,\alpha) p_n} = L^1(G,A,\alpha).$$

Furthermore, for any unital G-algebra (G,A,α), the net $(p_n)_{n\in D}$ constitutes an increasing approximate identity of projections in $C^*(G,A,\alpha)$. (Here of course we are identifying $L^1(G)$ with a subalgebra of $L^1(G,A,\alpha)$ by identifying $\lambda \in \mathbf{C}$ with $\lambda \cdot 1 \in A$.)

Proof. Since G is compact, $L^1(G)$ is the closure of the direct sum of its minimal ideals, which are exactly the subalgebras $\chi L^1(G)\chi$ for χ the appropriately normalized character of an irreducible representation of G. (This is well known for $L^2(G)$ – see for example chapter 15 of [29]. The result for $L^1(G)$ now follows because the compactness of G implies that $L^2(G)$ is a dense subalgebra of $L^1(G)$.) Let D be the set of finite subsets of the set $\widehat{G}$ of equivalence classes of irreducible representations of G. For $n \in D$, let p_n be the sum of the normalized characters of the representations in n. Then p_n is a central projection in $L^1(G)$. (It is selfadjoint for the usual involution in $L^1(G)$ because characters are.) We clearly have

$$\overline{\bigcup_{n\in D} p_n L^1(G) p_n} = L^1(G), \tag{*}$$

because the left hand side contains all minimal ideals of $L^1(G)$.

Now let (G,A,α) be a unital G-Banach algebra. Then $L^1(G,A,\alpha)$ is equal to the closed linear span of the set

$$B_0 = \{ fa : f \in L^1(G), a \in A \}.$$

Here $fa \in L^1(G,A,\alpha)$ is defined by $(fa)(g) = f(g)a$ for $g \in G$. The equation (*) implies that

$$B_0 \subset \overline{\bigcup_{n\in D} p_n L^1(G,A,\alpha) p_n},$$

from which it follows that

$$\overline{\bigcup_{n\in D} p_n L^1(G,A,\alpha) p_n} = L^1(G,A,\alpha),$$

as desired.

Finally, assume that (G,A,α) is a G-algebra. Since $L^1(G,A,\alpha)$ is dense in $C^*(G,A,\alpha)$, we clearly have

$$\overline{\bigcup_{n\in D} p_n C^*(G,A,\alpha) p_n} = C^*(G,A,\alpha).$$

As elements of $C^*(G,A,\alpha)$, the projections p_n satisfy $\|p_n\| = 1$. It follows that they constitute an approximate identity in $C^*(G,A,\alpha)$. Q.E.D.

Of course, if G is a compact Lie group, then it is second countable, and we can take a sequence rather than a net. We also point out that the net $(p_n)_{n\in D}$ is not in general an approximate identity for $L^1(G)$. Indeed, if G is the circle group S^1, then the assertion that $(p_n)_{n\in D}$ is an approximate identity for $L^1(G)$ can be shown, without much difficulty, to be equivalent to the assertion that every function in $L^1(S^1)$ is the limit in the norm of $L^1(S^1)$ of the partial sums of its Fourier series. This assertion, however, is known to be false.

2.2.5 Proposition. ([51], proposition 4.2.) Let (G,A,α) be a unital G-Banach algebra, and let (G,E,λ) be a finitely generated projective (G,A,α)-module. Then there is a finite dimensional representation space V of G such that E is equivariantly isomorphic to a direct summand of $V\otimes A$.

Proof. Let $e_1,\dots,e_k$ generate E as an A-module. We claim that there is $\varepsilon>0$ such that if $\| d_i - e_i \| < \varepsilon$ for $i=1,\dots,k$, then $d_1,\dots,d_k$ also generate E as an A-module. To see this, define $\varphi: A^k \longrightarrow E$ by $\varphi(a_1,\dots,a_k)=\sum_1^k e_i\, a_i$. Then φ is a continuous surjective homomorphism of A-modules. Since E is projective, it has a continuous right inverse ψ, and $\psi\cdot\varphi$ is an idempotent in the Banach algebra of all bounded A-module endomorphisms of A^k. (The formula $\varphi\cdot\psi = id_E$ readily implies that $(\psi\cdot\varphi)^2=\psi\cdot\varphi$.) For $d_1,\dots,d_k\in E$ define $\varphi_0(a_1,\dots,a_k)=\sum_1^k d_i\, a_i$. If $\| d_i - e_i \| < \varepsilon$ for all i, then $\| \varphi - \varphi_0 \| < k\varepsilon$, and

$$\| ((id_{A^k} - \psi\cdot\varphi) + \psi\cdot\varphi_0) - id_{A^k} \| = \| \psi\cdot\varphi_0 - \psi\cdot\varphi \| \le \| \psi \| \, k\varepsilon .$$

Therefore, if $\varepsilon < (k\ \| \psi \|)^{-1}$, then $(id_{A^k} - \psi\cdot\varphi) + \psi\cdot\varphi_0$ is invertible. Since A^k is the direct sum of the range of $id_{A^k} - \psi\cdot\varphi$ and the range of ψ, it follows that the range of $\psi\cdot\varphi_0$ equals the range of ψ. Because ψ is injective we can now conclude that φ_0 is surjective. The surjectivity of φ_0 is exactly the statement that $d_1,\dots,d_k$ generate E, so the claim is proved.

Now let $(p_n)_{n\in D}$ be the increasing net of projections in $L^1(G)$ constructed in the previous lemma. The next step is to show that for $e\in E$ and $\varepsilon>0$ there are $d\in E$ and $n\in D$ such that $dp_n = d$ and $\| d - e \| < \varepsilon$. We start by choosing an open subset V of G such that $\| \lambda_g^{-1}(e) - e \| < \frac{\varepsilon}{2}$ for $g\in V$. Letting $f = r^{-1}\chi_V$, where χ_V is the characteristic function of V and r is the measure of V, we then have $f\in L^1(G)$ and

$$\| ef - e \| = \| \int_G \lambda_g^{-1}(e) f(g)\, dg - e \| \le \frac{\varepsilon}{2}.$$

Now choose $f_0 \in \bigcup_{n\in D} p_n L^1(G) p_n$ such that $\| f_0 - f \|_1 < \frac{\varepsilon}{2} \| \lambda \|_\infty^{-1} \| e \|^{-1}$, and set $d = ef_0$. Then $f_0 \in p_n L^1(G) p_n$ for some n, whence $dp_n = d$. Furthermore,

$$\| d - e \| \le \| e(f_0 - f) \| + \| ef - f \| \le \sup_{g\in G} \| \lambda_g^{-1}(e) \| \ \| f_0 - f \|_1 + \frac{\varepsilon}{2} < \varepsilon .$$

This proves the claim. The result of the first paragraph now implies that there exist $d_1,\dots,d_k\in E$ and $n_1,\dots,n_k\in D$ such that $d_1,\dots,d_k$ generate E as an A-module and $d_i p_{n_i} = d_i$ for each i. Choosing $n\in D$ such that $n \ge n_i$ for all i, we may assume that $d_i p_n = d_i$ for all i.

We now claim that for any $d\in E$ and $n\in D$, the subspace $dp_n L^1(G) p_n$ of E is finite dimensional and G-invariant. Finite dimensionality follows from the fact that

$p_n L^1(G)p_n$ is finite dimensional, so we need only prove G-invariance. If $h \in G$, then for $e \in E$, we have

$$\lambda_h(ep_n) = \int_G \lambda_g^{-1}(e)p_n(gh)\,dg\,, \tag{*}$$

using the fact that the values of p_n are scalar multiples of the identity. Let $\delta_{h^{-1}}$ be the point mass measure at h^{-1}, which is an element of the measure algebra $M(G)$. Since $(p_n\delta_{h^{-1}})(g) = p_n(gh)$, the formula (*) implies that $\lambda_h(ep_n) = ep_n\delta_{h^{-1}}$. Since p_n is a central projection, $\delta_{h^{-1}}$ commutes with p_n, and therefore $p_n\delta_{h^{-1}} \in p_n L^1(G)p_n$. (Recall that $L^1(G)$ is an ideal in $M(G)$.) Taking $e = dp_n fp_n$, where $f \in L^1(G)$ is arbitrary, we now obtain

$$\lambda_h(dp_n fp_n) = dp_n fp_n\delta_{h^{-1}} \in dp_n L^1(G)p_n\,.$$

This shows that $dp_n L^1(G)p_n$ is a G-invariant subspace, and proves the claim.

Let V be the vector subspace of E spanned by the spaces $d_i p_n L^1(G)p_n$ for $1 \le i \le k$. The result of the previous paragraph implies that V is finite dimensional and G-invariant, and the choice of n and the d_i implies that $d_i \in V$ for all i. Therefore V generates E as an A-module.

Let $\mu_g = \lambda_g \otimes \alpha_g$ be the G-action on $V\otimes A$. Define $\varphi: V\otimes A \longrightarrow E$ by $\varphi(e\otimes a) = ea$. Then φ is a surjective equivariant A-module homomorphism. Since E is projective, φ has a right inverse ψ_0, and therefore an equivariant right inverse defined by

$$\psi(e) = \int_G \mu_g \circ \psi_0 \circ \lambda_g^{-1}(e)\,dg\,.$$

It follows that E is equivariantly isomorphic to a direct summand of $V\otimes A$. Q.E.D.

2.2.6 Proposition. K_0^G is a covariant functor from unital G-Banach algebras and equivariant homomorphisms to $R(G)$-modules and homomorphisms.

Proof. Let $\varphi: A \longrightarrow B$ be an equivariant homomorphism of unital G-Banach algebras (G,A,α) and (G,B,β). Let (G,E,λ) be a finitely generated projective (G,A,α)-module. Make B a left A-module by defining $ab = \varphi(a)b$ for $a \in A$ and $b \in B$. Then we can form the algebraic tensor product $E\otimes_A B$, which is a (right) B-module with $(e\otimes b)c = e\otimes bc$ for $b, c \in B$ and $e \in E$. Let μ be the diagonal G-action $\mu_g(e\otimes b) = \lambda_g(e)\otimes\beta_g(b)$, which is easily seen to be well defined and to satisfy $\mu_g((e\otimes b)c) = \mu_g(e\otimes b)\beta_g(c)$.

By the previous proposition, E is a direct summand of some $V\otimes A$. Therefore $E\otimes_A B$ is a direct summand of $(V\otimes A)\otimes_A B \simeq V\otimes B$. It follows that $E\otimes_A B$ is finitely generated and projective, and that the G-action is continuous (since the action on $V\otimes B$ is continuous). So $(G, E\otimes_A B, \mu)$ is a finitely generated projective (G,B,β)-module. We define $\varphi_*: K_0^G(A) \longrightarrow K_0^G(B)$ by $\varphi_*([E]) = [E\otimes_A B]$. It is then clear that φ_* is a well defined abelian group homomorphism. If V is a finite dimensional representation space of G, we have $(V\otimes E)\otimes_A B \simeq V\otimes(E\otimes_A B)$, so that φ_* is in fact an $R(G)$-module homorphism.

To prove functoriality let (G,C,γ) be another G-Banach algebra and let $\psi: B \longrightarrow C$ an equivariant homomorphism. Then the isomorphism $(E \otimes_A B) \otimes_B C \simeq E \otimes_A C$ implies that $(\psi \circ \varphi)_* = \psi_* \circ \varphi_*$. Q.E.D.

If A is a Banach algebra, we denote its unitization by A^+. (This is the analog of the one point compactification of a locally compact space.) As an algebra, A^+ is $\{a + \lambda 1 : a \in A, \lambda \in \mathbf{C}\}$, where 1 is the adjoined unit. If A already has a unit then of course A^+ is the direct sum of A and $\mathbf{C}$ as Banach algebras. If (G,A,α) is a G-Banach algebra, then (G,A^+,α^+) is also a G-Banach algebra, with $\alpha_g^+(a + \lambda 1) = \alpha_g(a) + \lambda 1$. If A is a C^*-algebra then so is A^+, and if (G,A,α) is a G-algebra then so is (G,A^+,α^+). With these conventions, we are ready to define $K_0^G(A)$ for nonunital A.

2.2.7 Definition. Let (G,A,α) be a (possibly nonunital) G-Banach algebra. Then $K_0^G(A)$ is the kernel of $r_*: K_0^G(A^+) \longrightarrow K_0^G(\mathbf{C})$, where r is the homomorphism from A^+ to $\mathbf{C}$ sending $a + \lambda 1$ to λ. Since $K_0^G(A)$ is the kernel of an $R(G)$-module homomorphism, it is again an $R(G)$-module. If in fact A is already unital, then there is a canonical identification of the kernel of r_* with $K_0^G(A)$ as already defined. (If E is a (G,A,α)-module, then E is also a (G,A,α)-module with the module structure given by $e(a + \lambda 1) = ea + \lambda e$. If E is a direct summand of A^n then it is also a direct summand of $(A^+)^n \simeq A^n \oplus \mathbf{C}^n$, where $\mathbf{C}$ is an A^+-module with $\eta(a + \lambda 1) = \lambda\eta$. Thus, if E is a finitely generated projective A-module, it is finitely generated and projective as an A^+-module.) Our definition is therefore a consistent extension of the definition of $K_0^G(A)$ for unital A.

2.2.8 Proposition. K_0^G is a covariant functor from G-Banach algebras and equivariant homomorphisms to $R(G)$-modules and homomorphisms.

Proof. Analogous to the proof of proposition 2.1.6. Q.E.D.

2.3. Swan's Theorem

The purpose of this section is to show that there is a natural isomorphism $K_0^G(C_0(X)) \simeq K_G^0(X)$ for locally compact G-spaces X, where $C_0(X)$ is the C^*-algebra of continuous complex valued functions on X which vanish at infinity. Recall that the functor $X \longrightarrow C_0(X)$ is a contravariant equivalence between the category of locally compact spaces and continuous maps of their one point compactifications which send ∞ to ∞, and the category of commutative C^*-algebras and homomorphisms. (If $f: X^+ \longrightarrow Y^+$ is continuous, the corresponding homomorphism $\varphi: C_0(Y) \longrightarrow C_0(X)$ is given by $\varphi(h)(x) = h(f(x))$ for $h \in C_0(Y)$. It is understood that h is considered to take the value zero at the point at infinity in Y^+.)

Let X be a locally compact G-space. (For this paragraph G can be any topological group.) Then it is well known that the definition $\alpha_g(h)(x) = h(g^{-1}x)$, for $g \in G$, $h \in C_0(X)$, and $x \in X$, yields a continuous action of G on $C_0(X)$. Similarly, the category equivalence in the previous paragraph together with an argument to establish continuity shows that every continuous action of G on $C_0(X)$ comes in this way from a unique continuous action of G on X. (Note that α_g is taken to be the inverse of the homomorphism associated to $x \longrightarrow gx$ in order to ensure that $\alpha_{g_1}\alpha_{g_2} = \alpha_{g_1 g_2}$ rather than $\alpha_{g_2 g_1}$.) Whenever X is a locally compact G-space, the algebra $C_0(X)$ will be assumed to carry this action of G. It immediately follows that $X \longrightarrow C_0(X)$ in a contravariant equivalence between the category of locally compact G-spaces and equivariant continuous maps of their one point compactification and the category of commutative G-algebras and equivariant homomorphisms.

If X is actually compact, then $C_0(X)$ consists of all continuous complex valued functions on X, and we denote it by $C(X)$. The same convention as above applies to G-actions on $C(X)$. We also point out that, if X is a locally compact G-space, then there is a natural isomorphism (as G-algebras) $C(X^+) \simeq C_0(X)^+$.

The main step in proving the isomorphism $K_0^G(C_0(X)) \simeq K_G^0(X)$ is the following equivariant version of Swan's theorem ([101]), which relates vector bundles over X to finitely generated projective $C(X)$-modules for compact spaces X. We return to our standing assumption that G is compact.

2.3.1 Theorem. Let X be a compact G-space. Then there is a natural equivalence between the category $\mathit{Vect}_G(X)$ of G-vector bundles over X and the category of finitely generated projective $(G, C(X))$-modules, given by assigning to a bundle E the module $\Gamma(E)$ of continuous sections of E, with its obvious right $C(X)$-module structure and G-action.

Our proof is basically the same as the one in [101], except for the group G. We should mention that a more C^*-algebraic proof of one part of this theorem appears in [87], but the methods there don't seem easily applicable to the other part. We start with some definitions and then several lemmas; we omit many of the details.

2.3.2 Definitions. Let $f : X \longrightarrow Y$ be a continuous map of compact spaces, let E be a vector bundle over X, and let F be a vector bundle over Y. Then a homomorphism from E to F over f is a continuous function $\varphi : E \longrightarrow F$ such that $\varphi|_{E_x}$ is a linear map from E_x to $F_{f(x)}$ for $x \in X$. We write $\mathit{Hom}_f(E,F)$ for the vector space of all such homomorphisms. If X is a G-space, and $f : X \longrightarrow X$ is given by $f(x) = gx$, for some fixed $g \in G$, then we write $\mathit{Hom}_g(E,F)$ for $\mathit{Hom}_f(E,F)$. (Notice that the action of $g \in G$ from a G-vector bundle E to itself is an element of $\mathit{Hom}_g(E,E)$.) If f is the identity on X, we write $\mathit{Hom}(E,F)$ for $\mathit{Hom}_f(E,F)$, and call its elements simply homomorphisms of bundles.

2.3.3 Definition. This is the analogous definition for modules over C^*-algebras. Let $\varphi: A \longrightarrow B$ be a unital homomorphism of C^*-algebras, let E be an A-module, and let F be a B-module. A homomorphism from E to F over φ is a continuous vector space homomorphism $\lambda: E \longrightarrow F$ such that $\lambda(ea) = \lambda(e)\varphi(a)$ for $a \in A$ and $e \in E$. We write $Hom_\varphi(E,F)$ for the vector space of all such homomorphisms. If (G,A,α) is a unital G-algebra, we write $Hom_g(E,F)$ for $Hom_{\alpha_g}(E,F)$. (Again, note that if (G,E,λ) is a (G,A,α)-module, then $\lambda_g \in Hom_g(E,E)$.) If φ is the identity, we write $Hom(E,F)$ for $Hom_\varphi(E,F)$.

2.3.4 Lemma. Let X be compact, and for $x \in X$ let $I_x = \{f \in C(X) : f(x) = 0\}$, which is a maximal ideal in $C(X)$. Then for any vector bundle E over X, there is a canonical isomorphism $\Gamma(E) / I_x \Gamma(E) \simeq E_x$.

Proof. This follows immediately from the local triviality of E near x. Q.E.D.

2.3.5 Lemma. Let $f: X \longrightarrow Y$ be a homeomorphism of compact spaces, and define $\varphi: C(X) \longrightarrow C(Y)$ by $\varphi(a)(y) = a(f^{-1}(y))$ for $a \in C(X)$ and $y \in Y$. Let E and F be vector bundles over X and Y respectively. Then there is a natural isomorphism $\Phi: Hom_f(E,F) \longrightarrow Hom_\varphi(\Gamma(E), \Gamma(F))$ given by $\Phi(\lambda)(s) = \lambda \circ s \circ f^{-1}$, for $\lambda \in Hom_f(E,F)$ and $s \in \Gamma(E)$.

Proof (sketch). From the previous lemma, it follows that there is a global section of E through every point, whence Φ is injective. To show Φ is surjective, let $\psi \in Hom_\varphi(\Gamma(E), \Gamma(F))$. Define $\lambda_x : E_x \longrightarrow F_{f(x)}$ by using the previous lemma, and set $\lambda(e) = \lambda_x(e)$ for $e \in E_x$. It is not hard to show that λ is continuous using the local triviality of E and F. Q.E.D.

We want to show that the isomorphism Φ in the previous lemma is a homeomorphism. For this, we need topologies on $\Gamma(E)$ and $\Gamma(F)$. Using a partition of unity subordinate to a cover of X by open sets over which E is trivial, it is easy to construct a continuous positive definite inner product on E. (See [101], lemma 2, or [2], lemma 1.4.10.) We can them define a norm on $\Gamma(E)$ by setting

$$\| s \|_\infty = sup\{ \langle s(x), s(x) \rangle^{1/2} : x \in X\}.$$

Since X is compact, using a different choice of inner product yields an equivalent norm. We can now prove continuity of Φ in the following sense:

2.3.6 Lemma. Let X and Y be compact spaces, and let E and F be vector bundles over X and Y respectively. Let T be a compact space, and for $t \in T$ let $f_t : X \longrightarrow Y$ be a homeomorphism such that the map $(t,x) \longrightarrow f_t(x)$ is continuous from $T \times X$ to Y. Further let $\lambda_t \in Hom_{f_t}(E,F)$ for $t \in T$. Then $(t,v) \longrightarrow \lambda_t(v)$ is continuous from $T \times E$ to F if and only if $t \longrightarrow \Phi(\lambda_t)$ is continuous from T to the space $L(\Gamma(E), \Gamma(F))$ of bounded operators from $\Gamma(E)$ to $\Gamma(F)$, where this space is given the strong operator topology.

Proof. We have to show that $(t,v) \longrightarrow \lambda_t(v)$ is continuous if and only if $t \longrightarrow \Phi(\lambda_t)(s)$ is continuous for all $s \in \Gamma(E)$. The proof is essentially the same as the proof that an action of a group G on a locally compact space X is continuous if and only if the corresponding action on $C_0(X)$ is continuous. However, we give the details because this is the only major part of the proof of theorem 2.3.1 which is not used in the proof of the usual Swan's theorem. For this proof we let $p: E \longrightarrow X$ and $q: F \longrightarrow Y$ be the projections associated with the bundles E and F.

First, suppose $(t,v) \longrightarrow \lambda_t(v)$ is continuous, and let $s \in \Gamma(E)$ and $t_0 \in T$. Let $\varepsilon > 0$, and let $W \subset F$ be an ε-neighborhood of the section $\Phi(\lambda_{t_0})(s)$, that is,

$$W = \{w \in F : \| w - [\Phi(\lambda_{t_0})(s)](q(w)) \| < \varepsilon\}.$$

(The norm is to be computed using some positive definite continuous inner product on F.) For $x \in X$, choose a rectangular open set $U_x \times V_x \subset T \times E$ such that $t_0 \in U_x$, $s(x) \in V_x$, and $\lambda_t(v) \in W$ for $(t,v) \in U_x \times V_x$. Since s is continuous, the subset $M_x = s^{-1}[V_x]$ of X is open. Since X is compact, finitely many of the sets M_x cover X, say $M_{x_1}, \ldots, M_{x_n}$. Let $U = \bigcap_k U_{x_k}$, which is a neighborhood of t_0 in T. Then it is easy to see that if $t \in U$ then $\lambda_t \cdot s \cdot f_t^{-1}(y) \in W$ for $y \in Y$. It follows that $\| \Phi(\lambda_t)(s) - \Phi(\lambda_{t_0})(s) \| < \varepsilon$. Thus, we have proved that $t \longrightarrow \Phi(\lambda_t)(s)$ is continuous.

For the other direction, suppose that for all s, the function $t \longrightarrow \Phi(\lambda_t)(s)$ is continuous. Since T is compact, these functions are bounded, and by the Uniform Boundedness Principle, there is a constant C such that $\| \Phi(\lambda_t) \| \le C$ for all t. (This is the only point where the compactness of T is used. If we were considering group actions on $C(X)$, for example, we could automatically choose $C = 1$ without assuming anything about the group at all.) It follows that if $\{s_n\}_{n \in I}$ is a net of sections of E converging to s, and if $\{t_n\}_{n \in I}$ is a net in T converging to t (with the same index set), then $\Phi(\lambda_{t_n})(s_n)$ converges to $\Phi(\lambda_t)(s)$. Indeed,

$$\| \Phi(\lambda_{t_n})(s_n) - \Phi(\lambda_t)(s) \| \le \| \Phi(\lambda_{t_n}) \| \; \| s_n - s \| + \| \Phi(\lambda_{t_n})(s) - \Phi(\lambda_t)(s) \|,$$

which converges to zero since $\| \Phi(\lambda_{t_n}) \|$ remains bounded, $s_n \longrightarrow s$, and $\Phi(\lambda_{t_n})(s) \longrightarrow \Phi(\lambda_t)(s)$.

We claim that if $\{v_n\}$ is a net in a vector bundle E, then $v_n \longrightarrow v \in E$ if and only if there are sections s_n through v_n and a section s through v such that $s_n \longrightarrow s$. Granting this, we can prove that $(t,v) \longrightarrow \lambda_t(v)$ is continuous as follows: let $(t_n, v_n) \longrightarrow (t,v)$, and let s_n be a section of E through v_n such that $s_n \longrightarrow s$, where s is a section through v. Then

$$\lambda_{t_n}(v_n) = [\Phi(\lambda_{t_n})(s_n)](f_{t_n}(p(v_n))),$$

which converges to

$$[\Phi(\lambda_t)(s)](f_t(p(v))) = \lambda_t(v),$$

as desired. To prove the claim, first observe that we may assume there are open sets U and V of X with $p(v) \in U \subset \overline{U} \subset V$, and such that E is locally trivial over V. We may assume that $p(v_n) \in U$ for all n. Multiplying sections as necessary by a continuous function which is 1 on U and 0 outside V, we reduce to the case of a bundle $U \times \mathbf{C}^n$, where the claim is easily proved. Q.E.D.

With this lemma in hand, we can now prove theorem 2.3.1 essentially as in [101] (see also the end of section 1.4 in [2]). We state the main lemmas, but give little indication of their proofs.

2.3.7 Definition. If E is a vector bundle over a space X, then a subbundle of E is a subset F of E which is a vector bundle with the restricted operations.

Observe that the range of a homomorphism of bundles may fail to be a subbundle because it may not be locally trivial.

2.3.8 Lemma. ([101], proposition 1.) Let E and F be vector bundles over a topological space X, and let $\varphi \in Hom(E,F)$. Then the following are equivalent:

(1) $Ker\,\varphi$ is a subbundle of E.
(2) $dim\,(Ker\,\varphi_x)$ is a locally constant function of $x \in X$.
(3) $Im\,\varphi$ is a subbundle of F.
(4) $dim\,(Im\,\varphi_x)$ is locally constant.

(Here $\varphi_x : E_x \longrightarrow F_x$ is the restriction of φ to the fiber over $x \in X$.)

Proof. See [101]. Most of this is also in section 1.4 of [2].

2.3.9 Lemma. ([101], proposition 2; [2], corollary 1.4.11.) Let X be a compact space, let E be a vector bundle over X, and let F be a subbundle of E. Then F is a direct summand of E.

Proof (sketch). Use a continuous positive definite inner product on E to construct an orthogonal complement $F^{\perp}$ of F. Then $F^{\perp}$ can be seen to be a subbundle by the previous lemma. Q.E.D.

2.3.10 Lemma. ([101], corollary 5; [2], corollary 1.4.14.) Any vector bundle E on a compact space X is a direct summand in a trivial bundle.

Proof (sketch). Using the compactness of X, find a finite set $s_1, \dots, s_N$ of global sections of E which span each fiber. (If a given collection of sections spans a fiber, it also spans nearby fibers.) Using these sections, define a surjective map ψ from $\mathbf{C}^N \times X$ to E. Then E is isomorphic to a complementary subbundle of $Ker\,\psi$. Q.E.D.

Proof of theorem 2.3.1 (sketch). (See [101] or [2].) Let E be a G-vector bundle over X. We can assume E is a direct summand (as a vector bundle) in some $\mathbf{C}^N$. Then $\Gamma(E)$ is a $C(X)$-module with $(sf)(x) = s(x)f(x)$ for $f \in C(X)$, $s \in \Gamma(E)$, and $x \in X$. It is finitely generated and projective because it is a direct summand in

$\Gamma(\mathbf{C}^N \times X) \simeq C(X)^N$. Define an action of G on $\Gamma(E)$ by $\lambda_g(s)(x) = g(s(g^{-1}x))$ for $s \in \Gamma(E)$, $g \in G$, and $x \in X$. This action exists and is compatible with the action of G on $C(X)$ by lemma 2.3.5. By lemma 2.3.6 it is continuous for the topology on $\Gamma(E)$ defined by a continuous positive definite inner product on E. We choose the restriction of the obvious inner product on $\mathbf{C}^N \times X$. The corresponding norm on $C(X)^N$ is

$$\| (f_1, \ldots, f_N) \| = \sup_{x \in X} \left[\sum_{k=1}^{N} | f_k(x) |^2 \right]^{1/2},$$

which does indeed define the right topology. (Observe that it is bigger than $\max\{\| f_k \| : k = 1, \ldots, N\}$ but smaller than $(\sum_{k=1}^{N} \| f_k \|^2)^{1/2}$). It follows that this norm also defines the right topology on $\Gamma(E)$, and we conclude that $(G, \Gamma(E), \lambda)$ is a finitely generated projective $(G, C(X))$-module.

Now let (G, P, λ) be a finitely generated projective $(G, C(X))$-module. We may assume that P is a summand in some $C(X)^N$ (ignoring G). Let $p \in Hom(C(X)^N, C(X)^N)$ be the projection onto P, and let ψ be the corresponding element of $Hom(\mathbf{C}^N \times X, \mathbf{C}^N \times X)$ obtained from lemma 2.3.5. Using lemma 2.3.8 we can prove that ψ is the projection onto a subbundle E of $\mathbf{C}^N \times X$, and it is clear that $\Gamma(E) \simeq P$. Lemma 2.3.5 yields a G-action on E which is continuous by lemma 2.3.6. It is also easy to check that applying this construction to a bundle of the form $\Gamma(F)$ yields F again. We thus have a bijection (up to isomorphism) of the objects of $Vect_G(X)$ and $Proj_G(C(X))$. Functoriality follows from lemma 2.3.5, and the category equivalence is proved. Q.E.D.

2.3.11 Corollary. Let X be a locally compact G-space. Then there is a natural isomorphism of $R(G)$-modules $K_G^0(X) \simeq K_0^G(C_0(X))$.

Proof. If X is compact, then $K_G^0(X)$ is the Grothendieck group of isomorphism classes in $Vect_G(X)$ and $K_0^G(C(X))$ is the Grothendieck group of isomorphism classes of finitely generated projective $(G, C(X))$-modules. Since the categories are equivalent, and the equivalence obviously preserves direct sums, it follows that the groups are isomorphic. It is furthermore clear that if E is a G-vector bundle and V a finite dimensional representation space, then $\Gamma(V \otimes E) \simeq V \otimes \Gamma(E)$. Therefore the isomorphism is an $R(G)$-module isomorphism.

If $f : X \longrightarrow Y$ is an equivariant map of compact spaces, and if E is a G-vector bundle over Y, then $\Gamma(f^*(E))$ and $\Gamma(E) \otimes_{C(Y)} C(X)$ are both isomorphic to

$$\{s : X \longrightarrow E : s \text{ is continuous and } s(x) \in E_{f(x)} \text{ for } x \in X\}.$$

(To check this, write E as a summand of a trivial bundle, ignoring G, and then check that the G-actions correspond.) Therefore the isomorphism is natural for compact spaces. The natural isomorphism $C(X^+) \simeq C_0(X)^+$ can now be used to extend the isomorphism naturally to locally compact spaces. Q.E.D.

2.4. Equivariant K-Theory in Terms of Idempotents

In this section we give an alternate construction of $K_0^G(A)$ in terms of idempotents in matrix algebras over A. This definition is much more concrete, and it is therefore often easier to work with. We start off with some notation.

2.4.1 Definition. Let (G,V,μ) and (G,W,λ) be finite dimensional representation spaces of G. Let $L(V,W)$ be the set of all linear maps from V to W, with the G-action $g \cdot t = \lambda_g \cdot t \cdot \mu_g^{-1}$ for $g \in G$ and $t \in L(V,W)$. If (G,A,α) is a G-Banach algebra, we identify $L(V,W) \otimes A$ with a set of A-module homomorphisms from $V \otimes A$ to $W \otimes A$, by the formula $(t \otimes a)(\xi \otimes b) = t\xi \otimes ab$ for $t \otimes a \in L(V,W) \otimes A$ and $\xi \otimes b \in V \otimes A$. Give $L(V,W) \otimes A$ the diagonal action, so that if $t \in L(V,W) \otimes A$ is regarded as a homomorphism from $V \otimes A$ to $W \otimes A$, we then have $g \cdot t = (\lambda_g \otimes \alpha_g) \cdot t \cdot (\mu_g \otimes \alpha_g)^{-1}$. Define the product of $s \in L(V_2,V_1) \otimes A$ and $t \in L(V_3,V_2) \otimes A$ to be their product as A-module homomorphisms. Then one has $(s \otimes a)(t \otimes b) = st \otimes ab$ for $s \in L(V_2,V_1)$, $t \in L(V_3,V_2)$, and $a, b \in A$. Furthermore, the actions of G preserve multiplication. If A is actually a C^*-algebra, we will tacitly assume that all representations of G are unitary. We also define an adjoint operation from $L(V,W) \otimes A$ to $L(W,V) \otimes A$ by $(t \otimes a)^* = t^* \otimes a^*$.

We write $L(V)$ for $L(V,V)$. Then $L(V) \otimes A$ is just the tensor product of Banach algebras (there is no problem with norms, since $L(V)$ is finite dimensional). If A is a C^*-algebra, then $L(V) \otimes A$ is just the tensor product C^*-algebra.

We point out that if A is unital, then all A-module homomorphisms from $V \otimes A$ to $W \otimes A$ come from elements of $L(V,W) \otimes A$.

2.4.2 Definition. Let (G,A,α) be a G-Banach algebra. Let $P_G(A)$ be the set of all G-invariant idempotents in all of the algebras $L(V) \otimes A$, where V is some $\mathbf{C}^n$ with a unitary representation of G. If $p, q \in P_G(A)$, say $p \in L(V) \otimes A$ and $q \in L(W) \otimes A$, we define their direct sum $p \oplus q$ to be the matrix $\begin{pmatrix} p & 0 \\ 0 & q \end{pmatrix}$, which is an element of $L(V \oplus W) \otimes A$. Clearly $p \oplus q \in P_G(A)$. (Actually, we use the notation $p \oplus q$ for this matrix even if p and q are not idempotents.)

Two G-invariant idempotents $p \in L(V) \otimes A$ and $q \in L(W) \otimes A$ are called Murray-von Neumann equivalent if there are G-invariant elements $u \in L(V,W) \otimes A$ and $v \in L(W,V) \otimes A$ such that $vu = p$ and $uv = q$. This is the analog for G-Banach algebras of one of the equivalence relations used in ordinary K-theory. Also notice that u is an isomorphism of (G,A,α)-modules from $p[V \otimes A]$ to $q[W \otimes A]$.

We let $S_G(A)$ be the set of equivalence classes in $P_G(A)$. Define an operation in $S_G(A)$ by $[p] + [q] = [p \oplus q]$, where brackets denote equivalence classes as usual. If $\varphi : A \longrightarrow B$ is an equivariant homomorphism of G-Banach algebras, then we define $\varphi_* : S_G(A) \longrightarrow S_G(B)$ by $\varphi_*([p]) = [(id_{L(V)} \otimes \varphi)(p)]$ for a G-invariant idempotent $p \in L(V) \otimes A$.

2.4.3 Lemma. $S_G(A)$ is a covariant functor from G-Banach algebras to abelian semigroups with identity.

Proof. We first show that Murray-von Neumann equivalence really is an equivalence relation. The only nontrivial point is transitivity. So let $uv = p$, $vu = q$ and $xy = q$, $yx = r$. Then

$$(ux)(yv) = uqv = u(vu)v = (uv)^2 = p^2 = p.$$

Similarly $(yv)(ux) = r$, so transitivity holds.

Addition is well defined, since if $u_i v_i = q_i$ and $v_i u_i = p_i$ for $i = 1, 2$, then $u_1 \oplus u_2$ and $v_1 \oplus v_2$ define a Murray-von Neumann equivalence between $p_1 \oplus p_2$ and $q_1 \oplus q_2$. Associativity follows from associativity of direct sums of representations, as follows: Let $u \in L(V, W)$ be an equivalence between two representations of G (in this case, $V = (V_1 \oplus V_2) \oplus V_3$ and $W = V_1 \oplus (V_2 \oplus V_3)$), such that $(u \otimes 1) p (u \otimes 1)^{-1} = q$ in $L(W) \otimes A^+$ for idempotents $p \in L(V) \otimes A$ and $q \in L(W) \otimes A$ (here, say, $p = (p_1 \oplus p_2) \oplus p_3$ and $q = p_1 \oplus (p_2 \oplus p_3)$). Then we can set $x = q(u \otimes 1)p$ and $y = p(u \otimes 1)^{-1} q$, which are G-invariant elements of $L(V, W) \otimes A$ and $L(W, V) \otimes A$ respectively. We obtain $xy = q$ and $yx = p$; thus p and q are Murray-von Neumann equivalent. Commutativity of addition follows similarly.

The identity element is $[0]$, where 0 is the zero element of any $L(V) \otimes A$. Indeed, if $p \in L(W) \otimes A$ is a G-invariant idempotent, then we let u be the 1×2 matrix $(p \;\; 0)$, which is an element of $L(W \oplus V, W) \otimes A$, and we let V be its transpose $\begin{pmatrix} p \\ 0 \end{pmatrix} \in L(W, W \oplus V) \otimes A$. It is easily checked that $uv = p$ and $vu = \begin{pmatrix} p & 0 \\ 0 & 0 \end{pmatrix}$. Thus $[p] = [p \oplus 0]$, as required.

The only nonobvious statement about the maps is that they are well defined. So let $\varphi : A \longrightarrow B$ be an equivariant homomorphism. Then the maps

$$id_{L(V,W)} \otimes \varphi : L(V, W) \otimes A \longrightarrow L(V, W) \otimes B$$

are equivariant and preserve multiplication. It is now clear that φ_* is well defined.

Q.E.D.

2.4.4 Theorem. Let (G, A, α) be a unital G-Banach algebra. Then $S_G(A)$ is canonically isomorphic to $Proj_G(A)$.

Proof. Let $p \in L(V) \otimes A$ be a G-invariant idempotent. We regard p as an A-module endomorphism of $V \otimes A$, as in definition 2.4.1. Then $p[V \otimes A]$ is a (G, A, α)-module because p is G-invariant, and $p[V \otimes A]$ is finitely generated and projective because it is clearly a summand of $V \otimes A$. If $q \in L(W) \otimes A$ is Murray-von Neumann equivalent to p, then there are G-invariant elements $x \in L(V, W) \otimes A$ and $y \in L(W, V) \otimes A$ such that $yx = p$ and $xy = q$. The condition that x and y be G-invariant is exactly the condition that they define homomorphisms of (G, A, α)-modules and thus $x : p[V \otimes A] \longrightarrow q[W \otimes A]$ is an isomorphism of (G, A, α)-modules with inverse y. So if two idempotents are equivalent, the corresponding modules are isomorphic.

Obviously the module corresponding to $p \oplus q$ is the direct sum of the modules corresponding to p and q.

Now let E be a finitely generated projective (G,A,α)-module. By lemma 2.2.5, E is a direct summand in $V \otimes A$, for some finite dimensional representation space of G. There is therefore an equivariant projection from $V \otimes A$ to E, which defines a G-invariant idempotent $p \in L(V) \otimes A$ such that $E \simeq p\,[V \otimes A]$. If $p\,[V \otimes A] \simeq q\,[W \otimes A]$, then it is easy to see that the isomorphism gives a weak Murray-von Neumann equivalence between $p \in L(V) \otimes A$ and $q \in L(W) \otimes A$.

We now have well defined maps from equivalence classes of idempotents to isomorphism classes of modules and vice versa. If $p \in L(V) \otimes A$ is an idempotent, then p is certainly in the equivalence class of idempotents assigned to the module $p\,[V \otimes A]$. Conversely, if E is a finitely generated projective module, and $p \in L(V) \otimes A$ is a corresponding idempotent, then by definition $p\,[V \otimes A] \simeq E$. Our two maps are therefore each other's inverses, and we thus have an isomorphism of semigroups as desired.

If $\varphi : A \longrightarrow B$ is an equivariant unital homomorphism, then for naturality we need to show that

$$p\,[V \otimes A] \otimes_A B \simeq (\varphi \otimes id_{L(V)})(p)[V \otimes B].$$

We omit the easy calculation. Q.E.D.

2.4.5 Corollary. If A is unital, then the Grothendieck group of $S_G(A)$ is naturally isomorphic to $K_0^G(A)$ as an abelian group.

2.4.6 Remark. It is now easy to see how to define the $R(G)$-module structure on the Grothendieck group of $S(A)$: if $p \in L(V) \otimes A$ is G-invariant, and W is a finite dimensional representation space of G, then $[W] \cdot [p]$ is represented by $1 \otimes p \in L(W) \otimes L(V) \otimes A$.

2.4.7 Warning. When computing $K_0^G(A)$ using $S_G(A)$, it is important to identify $L(V) \otimes A$ and $L(W) \otimes A$ only when the representations of G on V and W are equivalent. There are many situations in which there is an equivariant isomorphism $\varphi : L(V) \otimes A \longrightarrow L(W) \otimes A$ and an idempotent $p \in L(V) \otimes A$ such that $[p] \neq [\varphi(p)]$ in $K_0^G(A)$. For example, let $A = \mathbf{C}$, and let G be any compact group having two distinct homomorphisms σ and τ to the unit circle S^1 in $\mathbf{C}$. Let V and W be $\mathbf{C}$ with the representation defined by σ and τ respectively. Then $L(V)$ and $L(W)$ are both isomorphic to $\mathbf{C}$ with the trivial action of G, so there is an equivariant isomorphism $\varphi : L(V) \longrightarrow L(W)$ such that $\varphi(1_V) = 1_W$. But $K_0^G(\mathbf{C}) = R(G)$, and $[1_V] = [V]$, which is not equal to $[1_W] = [W]$.

2.4.8 Remark. For nonunital A, the Grothendieck group of $S_G(A)$ is generally much smaller than $K_0^G(A)$. One must instead use the Grothendieck group of $S_G(A^+)$, and find the kernel of the map from it to $K_0^G(\mathbf{C})$ induced by the homomorphism $a + \lambda \cdot 1 \longrightarrow \lambda$. However, we will eventually prove that if A is a G-algebra with an increasing approximate identity of G-invariant projections, then the Grothendieck group of $S_G(A)$ *is* $K_0^G(A)$.

We devote the remainder of this section to several other ways of computing $S_G(A)$, which are sometimes more useful.

2.4.9 Definition. Let (G,A,α) be a unital G-Banach algebra. Then idempotents $p \in L(V)\otimes A$ and $q \in L(W)\otimes A$, where V and W are finite dimensional representation spaces of G, are said to be similar if there is a G-invariant invertible element $u \in L(V,W)\otimes A$ such that $upu^{-1} = q$. If A is in fact a C^*-algebra, and p and q are projections, then we say that they are unitarily equivalent if the element u above can be chosen to be unitary. (We will soon see that similar projections are necessarily unitarily equivalent.)

We also mention the C^*-version of Murray-von Neumann equivalence: projections $p \in L(V)\otimes A$ and $q \in L(W)\otimes A$ are Murray-von Neumann equivalent in the C^*-algebraic sense if there is a G-invariant element $u \in L(V,W)\otimes A$ such that $uu^* = q$ and $u^*u = p$. Again, we will soon see that this is actually the same as ordinary Murray-von Neumann equivalence of projections.

2.4.10 Lemma. Let (G,A,α) be a unital G-algebra. Then:

(1) Every G-invariant idempotent is similar to a G-invariant projection.

(2) If two G-invariant projections are similar, then they are unitarily equivalent.

Proof. (1) (Compare [53], theorem 26.) Let e be a G-invariant idempotent. Let $a = 1 + (e - e^*)(e - e^*)^*$, which is invertible in A. As in the proof of theorem 26 in [53], one can check that $p = ee^*a^{-1}$ is a projection such that $ep = p$ and $pe = e$; we omit the details. Also p is clearly G-invariant. Let $x = 1 - p + e$, which is a G-invariant element of A; the element $1 + p - e$ is easily shown to be an inverse for x. Furthermore, $xex^{-1} = p$. Thus, e and p are similar.

(2) (Compare [53], theorem 27.) Let p and q be G-invariant projections in A, and assume that there is a G-invariant element x such that $q = xpx^{-1}$. Let $u = x(x^*x)^{-1/2}$. Then an easy calculation shows that u is a G-invariant unitary. Since x^*x commutes with p (easy to check), $(x^*x)^{-1/2}$ commutes with p, and therefore $upu^* = q$, as desired. Q.E.D.

2.4.11 Proposition. (1) If (G,A,α) is a unital G-Banach algebra, then $S_G(A)$ is equal to the set of equivalence classes in $P_G(A)$ for the equivalence relation generated by similarity and the relation $p \sim p \oplus 0$ for any idempotent p and the zero element of any $L(V)\otimes A$.

(2) If (G,A,α) is a G-algebra, then $S_G(A)$ is canonically isomorphic to the set of equivalence classes of projections for the C^* version of Murray-von Neumann equivalence.

(3) If (G,A,α) is a unital G-algebra, then $S_G(A)$ is canonically isomorphic to the set of equivalence classes of projections for the equivalence relation generated by unitary equivalence and the relation $p \sim p \oplus 0$ for any projection p and the zero element of any $L(V)\otimes A$.

Proof. (1) Let $\sim$ be the equivalence relation generated by similarity and $p \sim p \oplus 0$. We have to show that G-invariant idempotents p and q are Murray-von Neumann equivalent if and only if $p \sim q$.

Assume $p \sim q$. Since the equivalence relation $\sim$ is generated by similarity and $p \sim p \oplus 0$, we need only show that similarity implies Murray-von Neumann equivalence, and that p is Murray-von Neumann equivalent to $p \oplus 0$. The second fact was proved in the course of lemma 2.4.3. For the first, let p be similar to q so that $p = xqx^{-1}$. (If $p \in L(V)\otimes A$ and $q \in L(W)\otimes A$, then $x \in L(W,V)\otimes A$.) Then with $a = pxq$ and $b = qx^{-1}p$, we have $ab = p$ and $ba = q$. So p and q are Murray-von Neumann equivalent.

For the other direction, let $p \in L(V)\otimes A$ and $q \in L(W)\otimes A$ be Murray-von Neumann equivalent. Then there are G-invariant elements $x \in L(V,W)\otimes A$ and $y \in L(W,V)\otimes A$ such that $yx = p$ and $xy = q$. Then $xp = xyx = qx$ and similarly $py = yq$. Let

$$z = \begin{pmatrix} x & 1-q \\ 1-p & -y \end{pmatrix},$$

which is a G-invariant element of $L(V\oplus W, W\oplus V)\otimes A$, and has the inverse

$$z^{-1} = \begin{pmatrix} y & 1-p \\ 1-q & -x \end{pmatrix}.$$

It easy to check that $z(p\oplus 0)z^{-1} = xpy \oplus 0 = q \oplus 0$. Therefore $p \sim p\oplus 0 \sim q \oplus 0 \sim q$, as desired.

(2) To prove this statement, we must prove that in a G-algebra, every G-invariant idempotent is Murray-von Neumann equivalent to a G-invariant projection, and that if two G-invariant projections are Murray-von Neumann equivalent, then they are Murray-von Neumann equivalent in the C^*-algebra sense. To do the first part, let $e \in L(V)\otimes A$ be a G-invariant idempotent. By part (1) of the previous lemma, there is a G-invariant projection p and a G-invariant invertible element x, both in $(L(V)\otimes A)^+$, such that $xex^{-1} = p$. Let $u = pxe$ and $v = ex^{-1}p$. Then $p, u, v \in L(V)\otimes A$ and satisfy $uv = p$ and $vu = e$. So p and e are Murray-von Neumann equivalent, as desired.

To do the second part, let $p \in L(V)\otimes A$ and $q \in L(W)\otimes A$ be Murray-von Neumann equivalent G-invariant projections. Thus there are G-invariant elements $x \in L(V,W)\otimes A$ and $y \in L(W,V)\otimes A$ such that $yx = p$ and $xy = q$. Let a be the element of $L(V\oplus W)\otimes A^+$ defined by

$$a = \begin{pmatrix} 1-p & -y \\ x & 1-q \end{pmatrix}.$$

(which is obtained from the matrix z in the proof of (1) by interchanging its rows). Then just as in the proof of (1), we have $a\,(p \oplus 0)\,a^{-1} = (0 \oplus q)$. By part (2) of the previous lemma, $p \oplus 0$ and $0 \oplus q$ are unitarily equivalent in $L\,(V \oplus W) \otimes A^{+}$, that is, there exists a G-invariant unitary u such that $u\,(p \oplus 0)u^{*} = (0 \oplus q)$. Setting $v = u\,(p \oplus 0)$, we obtain $v \in L\,(V \oplus W) \otimes A$, $v^{*}v = p \oplus 0$, and $vv^{*} = 0 \oplus q$. Since the obvious Murray-von Neumann equivalences of p with $p \oplus 0$ and q with $0 \oplus q$ are also equivalences in the C^{*}-algebra sense (because p and q are selfadjoint), we find that p is Murray-von Neumann equivalent to q in the C^{*}-algebra sense.

(3) Combine the previous lemma and part (1) of this proposition. Q.E.D.

It would perhaps be appropriate to compute some examples of equivariant K-theory here, besides the trivial one $K_0^G(\mathbf{C}) = R(G)$. Unfortunately, equivariant K-theory is in general very hard to compute, and we do not yet have enough tools at our disposal to compute many interesting examples. We will not actually take up space in this chapter with examples; rather, we refer to some examples computed in later chapters, specifically, to examples 4.1.7, 4.2.3, and 4.4.12, and proposition 9.2.2. The first two of these are calculated using Julg's theorem (section 6), and we do not explicitly write down the equivariant K-theory there, although we do the hard part of the calculation. The third one depends on corollary 1.2.3 and on some deep facts about ordinary K-theory. Proposition 9.2.2 covers a number of specific cases, and depends on the results of sections 5 and 7 (but not on Julg's theorem).

2.5. Direct Limits

In this section, we compute the equivariant K-theory of a direct limit of G-algebras. We restrict ourselves to C^{*}-algebras for most of this section, since we make essential use of the fact that homomorphisms have norm 1. (It is not clear that direct limits exist in the category of Banach algebras.) The key lemma for computing the K-theory of direct limits, lemma 2.5.3, says that if two projections are sufficiently close, then they are equivalent. This lemma is one of the important reasons for introducing the construction of K-theory in terms of projections, and it will be used frequently in later chapters. At the end of this section we give, as another application of this lemma, a short proof of the homotopy invariance of equivariant K-theory.

We start with:

2.5.1 Proposition. (Compare [11], beginning of section 3.) Direct limits exist in the category of G-algebras for any fixed group G.

Proof (sketch). Let $(G, A_\lambda, \alpha_\lambda)_{\lambda \in D}$ be a directed system of G-algebras with equivariant maps $\varphi_{\lambda\mu} : A_\lambda \longrightarrow A_\mu$ for $\lambda < \mu$. Let A_0 be the algebraic direct $\varinjlim A_\lambda$, which is a *-algebra in an obvious way, and carries an obvious action of G. Let φ_λ^0 be the canonical (equivariant) homomorphism from A_λ to A_0. We recall that A_0 is the union of the images of the φ_λ for $\lambda \in D$, and that $\varphi_\lambda^0(a)=0$ if and only if there is $\mu > \lambda$ such that $\varphi_{\lambda\mu}(a)=0$.

Let $a \in A_0$. Then $a = \varphi_\lambda^0(b)$ for some λ and some $b \in A_\lambda$. Define $\| a \| = \lim_\mu \| \varphi_{\lambda\mu}(b) \|$; the limit exists because $\| \varphi_{\lambda\mu}(b) \|$ is a decreasing net of nonnegative real numbers. This limit is easily seen not to depend on the choice of λ and b. Then $\| \quad \|$ is a C^*-algebra seminorm such that the G-action is isometric. Let I be the ideal of elements of norm 0, and let A be the completion of A_0 / I, which is a G-algebra. It is easy to check that A satisfies the appropriate universal property for $\varinjlim A_\lambda$ in the category of G-algebras, using the maps $\varphi_\lambda : A_\lambda \longrightarrow A$ obtained by composing φ_λ^0 with the canonical map from A_0 to A Q.E.D.

2.5.2 Remark. Direct limits of exact sequences are easily seen to be exact. One can also check that direct limits commute with crossed products, but we do not need this.

The next proposition shows that direct limits commute with K_0^G. First, we need the lemma referred to at the beginning of this section. For later use, we also prove the Banach algebra formulation. We return to our assumption that G is compact.

2.5.3 Lemma. (1) Let p and q be G-invariant idempotents in a unital G-Banach algebra. If $\| p - q \| < (\| p \| + \| q \|)^{-1}$ then p and q are similar.

(2) Let p and q be G-invariant projections in a unital G-algebra. If $\| p - q \| < \frac{1}{2}$, then p and q are unitarily equivalent.

Proof. (1) Let $v = pq + (1-p)(1-q)$, which is G-invariant. Then

$$\| 1 - v \| = \| p(p-q) + (p-q)q \| < 1,$$

so v is invertible. It is easily seen that $pv = vq$, whence $p = vqv^{-1}$.

(2) If p and q are projections then $\| p \| = \| q \| = 1$. Hence the element v in the proof of (1) is invertible for $\| p - q \| < \frac{1}{2}$, and satisfies $vqv^{-1} = p$. Now by lemma 2.4.10 (2), p and q are unitarily equivalent. Q.E.D.

2.5.4 Proposition. (Compare [11], proposition 5.1.) Let $(G, A_\lambda, \alpha_\lambda)_{\lambda \in D}$ be a directed system of G-algebras. Then there is a natural isomorphism $K_0^G(\varinjlim A_\lambda) \simeq \varinjlim K_0^G(A_\lambda)$.

Proof. We first consider the case where all A_λ and all $\varphi_{\lambda\mu} : A_\lambda \longrightarrow A_\mu$ are unital. Let $A = \varinjlim A_\lambda$, and let $\varphi_\lambda : A_\lambda \longrightarrow A$ be the canonical map. Then clearly $(\varphi_\lambda)_* = (\varphi_{\lambda\mu})_* \cdot (\varphi_\mu)_*$. By the universal property of direct limits, there is therefore a homomorphism L from $\varinjlim K_0^G(A_\lambda)$ to $K_0^G(A)$. We must then show that every $\eta \in K_0^G(A)$ has the form $(\varphi_\lambda)_*(\eta_0)$ for some λ (that is, L is surjective), and that if

$\eta \in K_0^G(A_\lambda)$ with $(\varphi_\lambda)_*(\eta) = 0$, then $(\varphi_{\lambda\mu})_*(\eta) = 0$ for some μ (that is, L is injective). From the projection description of $K_0^G(A)$ (corollary 2.4.5), using the description of $S_G(A)$ given in lemma 2.4.11 (3), it is sufficient to prove the following: If $p \in A$ is a G-invariant projection, then there is $\lambda \in D$ and a G-invariant projection $q \in A_\lambda$ such that $\varphi_\lambda(q)$ is unitarily equivalent to p, and if $p, q \in A_\lambda$ are G-invariant projections such that $\varphi_\lambda(p)$ and $\varphi_\lambda(q)$ are unitarily equivalent, then $\varphi_{\lambda\mu}(p)$ and $\varphi_{\lambda\mu}(q)$ are unitarily equivalent for some $\mu > \lambda$. (We need only consider projections in A_λ or A, because we can replace A_λ and A by $L(V)\otimes A_\lambda$ and $L(V)\otimes A$ if necessary, since $\varinjlim (L(V)\otimes A_\lambda)$ is canonically isomorphic to $L(V)\otimes A$.)

Let $p \in A$ be a G-invariant projection. Then there is $\lambda_1 \in D$ and $a \in A_{\lambda_1}$ such that $\| \varphi_{\lambda_1}(a) - p \| < \frac{1}{4}$. Replacing a by $\frac{1}{2}(a + a^*)$ and averaging it over G, we may assume a is selfadjoint and G-invariant. The element $\varphi_{\lambda_1}(\frac{1}{2} - a)$ is invertible, with inverse $c \in A$. Choose λ_2 such that there is $d \in A_{\lambda_2}$ with

$$\| \varphi_{\lambda_2}(d) - c \| < \| \tfrac{1}{2} - a \|^{-1}.$$

It follows that

$$\| \varphi_{\lambda_2}(d)\varphi_{\lambda_1}(\tfrac{1}{2} - a) - 1 \| < 1 \quad \text{and} \quad \| \varphi_{\lambda_1}(\tfrac{1}{2} - a)\varphi_{\lambda_2}(d) - 1 \| < 1.$$

Therefore there is $\lambda \geq \lambda_1, \lambda_2$ such that

$$\| \varphi_{\lambda_2\lambda}(d)\varphi_{\lambda_1\lambda}(\tfrac{1}{2} - a) - 1 \| < 1 \quad \text{and} \quad \| \varphi_{\lambda_1\lambda}(\tfrac{1}{2} - a)\varphi_{\lambda_2\lambda}(d) - 1 \| < 1.$$

Then $\varphi_{\lambda_1\lambda}(\frac{1}{2} - a)$ is invertible, so that $\frac{1}{2}$ is not in the spectrum $sp(\varphi_{\lambda_1\lambda}(a))$. Let f be the characteristic function of $[\frac{1}{2}, \infty)$, which is continuous on $sp(\varphi_{\lambda_1\lambda}(a))$, and let $q = f(\varphi_{\lambda_1\lambda}(a))$, which is a G-invariant projection in A_λ. We have $\varphi_\lambda(q) = f(\varphi_{\lambda_1}(a))$ and $\| f(\varphi_{\lambda_1}(a)) - p \| < \frac{1}{2}$, so $\varphi_\lambda(q)$ is unitarily equivalent to p by lemma 2.5.3 (2), as needed.

For the other part, let p and q be G-invariant projections in A_λ such that $\varphi_\lambda(p)$ and $\varphi_\lambda(q)$ are unitarily equivalent. Let $u \in A$ be a G-invariant unitary such that $u\varphi_\lambda(p)u^* = \varphi_\lambda(q)$. Find $\mu_1 \geq \lambda$ such that there is $c \in A_{\mu_1}$ with $\| \varphi_{\mu_1}(c) - u \| < \frac{1}{12}$; by averaging over G we may assume that c is G-invariant. By an argument similar to one used in the previous paragraph, there is $\mu_2 \geq \mu_1$ such that $d = \varphi_{\mu_1\mu_2}(c)$ is invertible. Let $v = d(d^*d)^{-1/2}$, which is unitary and G-invariant. A calculation shows $\| \varphi_{\mu_2}(v) - u \| < \frac{1}{4}$, whence

$$\| \varphi_{\mu_2}(v)\varphi_\lambda(p)\varphi_{\mu_2}(v)^* - \varphi_\lambda(q) \| < \tfrac{1}{2}.$$

Therefore there is $\mu \geq \mu_2$ such that

$$\| \varphi_{\mu_2\mu}(v)\varphi_{\lambda\mu}(p)\varphi_{\mu_2\mu}(v)^* - \varphi_{\lambda\mu}(q) \| < \tfrac{1}{2}.$$

By lemma 2.5.3 (2), $\varphi_{\lambda\mu}(q)$ and $\varphi_{\mu_2\mu}(v)\varphi_{\lambda\mu}(p)\varphi_{\mu_2\mu}(v)^*$ are unitarily equivalent. Hence $\varphi_{\lambda\mu}(q)$ and $\varphi_{\lambda\mu}(p)$ are unitarily equivalent, as desired.

Now consider the case where not all the algebras or not all of the maps are unital. Then $A^+ \simeq \varinjlim A_\lambda^+$ with maps $\varphi_{\lambda\mu}^+$. Let $r_\lambda : A_\lambda^+ \longrightarrow \mathbf{C}$ and $r : A^+ \longrightarrow \mathbf{C}$ be the usual maps. Then under the isomorphism $K_0^G(A^+) \simeq \varinjlim K_0^G(A_\lambda^+)$ we obtain $Ker\,(r_*) \simeq \varinjlim Ker\,(r_\lambda)_*$, that is, $K_0^G(A) \simeq \varinjlim K_0^G(A_\lambda)$ as desired. Q.E.D.

2.5.5 Corollary. (Compare [51], remark after 3.3.) Let A be a G-algebra having an increasing approximate identity consisting of G-invariant projections $\{e_n\}_{n\in D}$. Then $K_0^G(A)$ is isomorphic to the Grothendieck group of $S_G(A)$; moreover, every class in $S_G(A)$ is represented by a projection p in some $L(V)\otimes A$ such that $p \le 1_V \otimes e_n$ for some n.

Proof. Evidently $A = \varinjlim e_n A e_n$; therefore $K_0^G(A) \simeq \varinjlim K_0^G(e_n A e_n)$. So every element of $K_0^G(A)$ has the form $[p]-[q]$ where $p \in L(V)\otimes e_m A e_m$ and $q \in L(W)\otimes e_n A e_n$. For the last statement, let $p \in L(V)\otimes A$ be a G-invariant projection. From the proof of the proposition, it is clear that p is unitarily equivalent in $L(V)\otimes A^+$ to a projection $q \in L(V)\otimes (e_n A e_n)^+$, but evidently q must actually be in $L(V)\otimes e_n A e_n$. Q.E.D.

2.5.6 Remark. The hypothesis in this corollary that the approximate identity of projections be increasing is unnecessary, at least for separable C^*-algebras. In fact, if A is a separable G-algebra which has an approximate identity of G-invariant projections, then one can show that the approximate identity can be chosen so that it is also increasing. However, we don't need this refinement, so we don't prove it.

We will also need the Banach algebra version of corollary 2.5.5 when dealing with algebras of the form $L^1(G, A)$. Since we do not need the equivariant version, we will state the result only for Banach algebras B with no group action. We will therefore write $S(B)$ rather than $S_G(B)$.

2.5.7 Proposition. Let B be a Banach algebra with an increasing net of idempotents $\{e_n\}_{n\in D}$ such that $\overline{\bigcup_{n\in D} e_n B e_n} = B$. Then $K_0(B)$ is the Grothendieck group of $S(B)$, and every element of $S(B)$ is represented by an idempotent which is dominated by some $1_V \otimes e_n$.

Proof. We will use the method of proof of proposition 2.5.4 and corollary 2.5.5, only indicating how the proofs need to be modified for the case at hand. First, observe that

$$\overline{\bigcup_{n\in D} (e_n B e_n)^+} = B^+ . \qquad (*)$$

The argument of corollary 2.5.5 then shows that it is enough to prove that $K_0(B^+) = \varinjlim K_0((e_n B e_n)^+)$. The argument in the first paragraph of the proof of proposition 2.5.4 now shows that it is sufficient to prove that every idempotent in B^+

is similar to one in $(e_n Be_n)^+$ for some n, and that two idempotents in $(e_n Be_n)^+$ which are similar in B^+ are also similar in $(e_m Be_m)^+$ for some $m \geq n$. The first of these statements is an immediate consequence of (*), lemma 2.5.3 (1), and the following lemma, which replaces the second paragraph of the proof of proposition 2.5.4. The second statement is obtained from a corresponding but much simpler modification of the third paragraph of that proof; we omit the details. Q.E.D.

2.5.8 Lemma. Let B be a Banach algebra, let $p \in B$ be an idempotent, and let f be the function on $\mathbf{C}$ defined by $f(x+iy)=0$ for $x<\frac{1}{2}$ and $f(x+iy)=1$ for $x>\frac{1}{2}$. Then there is $\varepsilon>0$ such that for every $a \in B$ with $\| p-a \| < \varepsilon$, the spectrum $sp(a)$ is contained in the domain of f, and the idempotent $f(a)$ (defined using the holomorphic functional calculus) satisfies $\| p-f(a) \| < (\| p \| + \| f(a) \|)^{-1}$.

Proof. Let Γ be the contour in $\mathbf{C}$ consisting of the circles $|z|=\frac{1}{4}$ and $|z-1|=\frac{1}{4}$. Let

$$M = sup\{ \| (z-p)^{-1} \| : z \in \Gamma \},$$

and let $\varepsilon = M^{-2}(2\| p \| + 1)^{-1}$. Observe that this ε satisfies $\varepsilon < \frac{1}{4}$ and $\varepsilon < \frac{1}{2}M^{-1}$. Let $\| p-a \| < \varepsilon$. Then

$$sp(a) \subset \{ z \in \mathbf{C} \colon |z| < \tfrac{1}{4} \text{ or } |z-1| < \tfrac{1}{4} \}.$$

Therefore f is holomorphic on a neighborhood of $sp(a)$, and

$$f(a) = \frac{1}{2\pi i}\int_\Gamma f(z)(z-a)^{-1}dz.$$

Since $f(p)=p$, we have

$$\| p-f(a) \| \leq \frac{1}{2\pi} l(\Gamma) \, \| f \|_\infty \, sup\{ \| (z-p)^{-1} - (z-a)^{-1} \| : z \in \Gamma \},$$

where $l(\Gamma)$ is the length of Γ. The estimate

$$\| (b+h)^{-1} - b^{-1} \| \leq 2\| b^{-1} \|^2 \| h \| \quad \text{for} \quad \| h \| \leq \tfrac{1}{2} \| b^{-1} \|^{-1},$$

applied to $b=z-p$ and $h=p-a$, together with the fact that

$$\| p-a \| < M^{-2}(2\| p \| + 1)^{-1} < \tfrac{1}{2}M^{-1},$$

and the fact that $l(\Gamma)=\pi$, now shows that

$$\| p-f(a) \| < (2\| p \| + 1)^{-1} < (\| p \| + \| f(a) \|)^{-1},$$

as desired. Q.E.D.

The representation of classes in $K_0^G(A)$ by projections also yields a short proof of homotopy invariance of K_0^G.

2.5.9 Definition. Let (G,A,α) and (G,B,β) be G-Banach algebras. Then equivariant homomorphisms $\psi_0, \psi_1 : A \longrightarrow B$ are called G-homotopic if there are equivariant homomorphisms $\varphi_t : A \longrightarrow B$ for $t \in [0,1]$, varying continuously in the sense that $t \longrightarrow \varphi_t(a)$ is continuous for $a \in A$, such that $\varphi_0 = \psi_0$ and $\varphi_1 = \psi_1$. (The conditions on φ_t are exactly the conditions that the formula $\varphi(a)(t) = \varphi_t(a)$ define an equivariant homomorphism from A to $B \otimes C([0,1])$, where G acts trivially on $[0,1]$. The proof of this equivalence uses the uniform boundedness principle.) If A and B are C^*-algebras, then the homotopy above is called a G-homotopy of G-algebra maps if in addition all the φ_t are $*$-homomorphisms.

2.5.10 Proposition. Let $\psi_0, \psi_1 : A \longrightarrow B$ be G-homotopic homomorphisms of G-Banach algebras. Then $(\psi_0)_*$ and $(\psi_1)_* : K_0^G(A) \longrightarrow K_0^G(B)$ are equal.

Proof. It is obviously sufficient to prove this when everything is unital. Let (φ_t) be a G-homotopy from ψ_0 to ψ_1. Let $p \in L(V) \otimes A$ be a G-invariant idempotent. Since $[0,1]$ is compact and $t \longrightarrow (id_{L(V)} \otimes \varphi_t)(p)$ is continuous, there is a constant M such that $\| (id_{L(V)} \otimes \varphi_t)(p) \| \leq M$ for $t \in [0,1]$, and there are $t_0, \ldots, t_n \in [0,1]$, with $t_0 = 0$ and $t_n = 1$, such that

$$\| (id_{L(V)} \otimes \varphi_{t_i})(p) - (id_{L(V)} \otimes \varphi_{t_{i+1}})(p) \| < \frac{1}{2M} \qquad (i = 0, \ldots, n-1).$$

By lemma 2.5.3 (1), the idempotents $(id_{L(V)} \otimes \varphi_{t_i})(p)$ and $(id_{L(V)} \otimes \varphi_{t_{i+1}})(p)$ are similar, for $i = 0, \ldots, n-1$. Therefore $(\varphi_0)_*([p]) = (\varphi_1)_*([p])$, as desired. Q.E.D.

2.6. Julg's Theorem

In this section, we prove the theorem of Julg [51] relating the equivariant K-theory of G-algebras to the ordinary K-theory of crossed products. (See also [39] for commutative C^*-algebras.)

2.6.1 Julg's Theorem ([51]). Let (G,A,α) be a unital G-algebra. Then there is a natural isomorphism $K_0^G(A) \simeq K_0(C^*(G,A,\alpha))$.

The proof is divided into two steps: a straightforward one to show $K_0^G(A) \simeq K_0(L^1(G,A,\alpha))$, using the integrated form of a G-action on a module, and a technical one to show $K_0(L^1(G,A,\alpha)) \simeq K_0(C^*(G,A,\alpha))$. The first of these is the following lemma.

2.6.2 Lemma. There is a natural isomorphism $K_0^G(A) \simeq K_0(L^1(G,A,\alpha))$.

Proof. Let (p_n) be the increasing net of idempotents in $L^1(G,A,\alpha)$ constructed in lemma 2.2.4. By proposition 2.5.7, $K_0(L^1(G,A,\alpha))$ is isomorphic to the Grothendieck group of $S(L^1(G,A,\alpha))$. We may in fact consider idempotents dominated by some

$1 \otimes p_n$. Then the correspondence between modules and idempotents in the proof of theorem 2.4.4 shows that $K_0(L^1(G,A,\alpha))$ is the Grothendieck group of the semigroup of isomorphism classes of $L^1(G,A,\alpha)$-modules of the form $e\,[\mathbf{C}^k \otimes L^1(G,A,\alpha)]$ for idempotents $e \in M_k \otimes L^1(G,A,\alpha)$ such that $e \leq 1 \otimes p_n$ for some n. (Recall that M_k is the algebra of $k \times k$ matrices with entries in $\mathbf{C}$.)

Let (G,E,λ) be a finitely generated projective (G,A,α)-module. Let $\Phi(E)$ be E with the $L^1(G,A,\alpha)$-module structure defined in lemma 2.2.3. We want to prove that Φ defines an isomorphism on K-theory.

The first, and longest, step is to show that $\Phi(E)$ is a module of the correct form. By lemma 2.2.5, E is a direct summand in some $V \otimes A$, where V is a finite dimensional representation space of G. Since Φ preserves direct sums, it suffices to show that $\Phi(V \otimes A)$ has the correct form. Since V is a direct sum of irreducible representation spaces of G, it is in fact sufficient to show this for irreducible V. Let χ be the normalized character of the obvious representation of G on $\widetilde{V}$, the complex conjugate of V, and let e be a minimal idempotent in $\chi L^1(G)\chi$. We may choose e such that $e^* = e$. We regard e as an element of $L^1(G,A,\alpha)$, and we claim that $\Phi(V \otimes A) \simeq eL^1(G,A,\alpha)$, which clearly has the correct form.

From standard results on compact groups, $L^1(G)e \simeq \widetilde{V}$ as left $L^1(G)$-modules (using the integrated form of the action of G on V). Therefore we may assume that $V = (L^1(G)e)^{\sim}$. Define $\varphi : (L^1(G)e)^{\sim} \otimes A \longrightarrow e\,L^1(G,A,\alpha)$ by $\varphi(f \otimes a) = f^* a$, where the product is evaluated as the product of an element $f^* \in L^1(G,A,\alpha)$ and a multiplier a of $L^1(G,A,\alpha)$, as in [79], 7.6.2 and 7.6.3. That is, $\varphi(f \otimes a)(h) = \overline{f(h^{-1})}\,\alpha_h(a)$. Since $fe = f$, we have $ef^* = f^*$, whence $\varphi(f \otimes a) \in e\,L^1(G,A,\alpha)$ as desired. Furthermore, φ is a vector space homomorphism. (The complex conjugate on $(L^1(G)e)^{\sim}$ cancels the conjugate linearity of $f \longrightarrow f^*$).

We prove that φ is an $L^1(G,A,\alpha)$-module homomorphism. To this end, let $a \in A$, $f \in (L^1(G)e)^{\sim}$, $b \in L^1(G,A,\alpha)$, and let π be the left regular representation of G on $L^2(G)$. Then

$$\begin{aligned} \varphi((f \otimes a)b)(g) &= \varphi\left[\int_G \pi_{h^{-1}}(f) \otimes \alpha_{h^{-1}}(ab(h))\,dh\right](g) \\ &= \int_G \overline{f(hg^{-1})}\,\alpha_{gh^{-1}}(ab(h))\,dh = (\varphi(f \otimes a)b)(g), \end{aligned}$$

by a change of variables, as desired.

To construct an inverse for φ, we proceed as follows: let $\xi_1, \ldots, \xi_n$ be an orthonormal basis for $L^1(G)e$, regarded as a subset of $L^2(G)$, and let $m_{ij}(g) = \langle \pi_g \xi_i, \xi_j \rangle$ for $a \leq i,j \leq n$. Then m_{ij} is a continuous function on G, and one has, from the theory of representation of compact groups (see for example [29], 15.2.5), the following properties:

$$m_{ij}\, m_{kl} = \delta_{jk}\, m_{il} \quad \text{in } L^1(G), \text{ where } \delta_{jk} = \begin{cases} 1 & j = k \\ 0 & j \neq k, \end{cases}$$

$$< m_{ij}, m_{kl} > = \delta_{ik}\, \delta_{jl}\, n \quad \text{in } L^2(G),$$

and

$$\sum_1^n m_{ii} = \chi, \text{ and } m_{ii} \text{ is a minimal idempotent.}$$

We may assume that $m_{11} = e$. Then the inverse to φ is given by the formula

$$\psi(a) = \frac{1}{n}\sum_1^n m_{i1} \otimes a_i, \quad \text{for } a \in e\, L^1(G,A,\alpha),$$

where

$$a_i = \int_G m_{i1}(h^{-1})\, \alpha_{h^{-1}}(a(h))\, dh .$$

We want to show that $\psi = \varphi^{-1}$, and we do the easy part first. The module $(L^1(G)e)^{\sim} \otimes A$ is spanned as a vector space by the elements $m_{i1} \otimes a$ for $a \in A$ and $1 \le i \le n$, and if $x = \varphi(m_{i1} \otimes a)$, we have

$$x_j = \int_G m_{j1}(h^{-1})\, \overline{m_{i1}(h^{-1})}\, a dh = n\,\delta_{ij}\, a$$

by the orthogonality relations. So $\psi \circ \varphi$ is the identity.

For the other direction, we need the following formula for $a \in L^1(G,A,\alpha)$:

$$\int_G e(h)\, \alpha_h(a(h^{-1}g))\, dh \tag{*}$$
$$= \frac{1}{n}\sum_{i=1}^n m_{1i}(g) \int_G m_{i1}(g^{-1}h)\, \alpha_h(a(h^{-1}g))\, dh .$$

To prove this, notice that it is sufficient to prove it for a of the form $a(g) = f(g)x^*(g)$, where f runs through a set of continuous functions which span a dense subspace of $L^1(G)$ and $x \in L^1(G,A,\alpha)$ is a constant function $x(g) = x_0^*$ for $g \in G$. The equation to be verified then becomes

$$\left[\int_G e(h) f(h^{-1}g)\, dh\right] \alpha_g(x_0)$$
$$= \frac{1}{n}\sum_{i=1}^n \left[\int_G m_{i1}(g^{-1}h)\, \overline{f^*(g^{-1}h)}\, dh\right] m_{1i}(g)\, \alpha_g(x_0).$$

We may now drop $\alpha_g(x_0)$ and rewrite this equation as

$$ef = \frac{1}{n}\sum_{i=1}^n < m_{i1}, f^* > m_{1i} .$$

If $f = m_{1j}$ then $f^* = m_{j1}$ and both sides are equal to m_{1j}. If $f = m_{kj}$ for some $k \ne 1$, then both sides are zero (recall that $e = m_{11}$). Finally, if f is a continuous function in $L^2(G)$ which is orthogonal to χ, then again both sides are zero. We have now considered enough functions to span $L^1(G)$, so the formula (*) has been proved.

We now have, for $a \in eL^1(G,A)$,

$$\frac{1}{n}\sum_1^n \varphi(m_{i1}\otimes a_i)(g)$$

$$= \frac{1}{n}\sum_1^n m_{1i}(g)\int_G m_{i1}(h^{-1})\,\alpha_{gh^{-1}}(a(h))\,dh = ea(g) = e(g),$$

where the second last step follows from a change of variables and (*). Thus, $\varphi\circ\psi$ is the identity. We have now succeeded in proving that

$$\Phi((L^1(G)e)^{\sim}\otimes A) \simeq e\,L^1(G,A,\alpha),$$

and therefore that $\Phi(E)$ is a module of the correct form for any E.

It is now obvious that Φ is well defined on isomorphism classes, and that it therefore defines a semigroup homomorphism. We next show that Φ is a surjective on isomorphism classes of modules. Let E be an $L^1(G,A,\alpha)$-module of the form $e[\mathbf{C}^m\otimes L^1(G,A,\alpha)]$. We may assume e is dominated by $1\otimes p_n$ for some n. Then E is a direct summand of $\mathbf{C}^m\otimes(p_n L^1(G,A,\alpha))$. Using the fact that the elements of G and of A define multipliers of $L^1(G,A,\alpha)$ (see [79], 7.6.2 and 7.6.3 again), we can make E and $\mathbf{C}^m\otimes(p_n L^1(G,A,\alpha))$ into (G,A,α)-modules, and E is a direct summand of $\mathbf{C}^m\otimes p_n L^1(G,A,\alpha)$ as (G,A,α)-modules. Write $p_n = \sum_1^l \chi_j$, the χ_j being normalized characters of irreducible representations of G, and let V_j be the corresponding representation space for $j=1,\dots,l$. Then from the previous argument it follows that $\mathbf{C}^m\otimes(p_n L^1(G,A,\alpha)) \simeq \bigoplus_{j=1}^{l}(\tilde{V}_j)^m\otimes A$ as (G,A,α)-modules. Therefore E is finitely generated and projective.

To see that Φ is injective on isomorphism classes, we need only observe that the argument in the previous paragraph using multipliers recovers the (G,A,α)-module structure on E for any $\Phi(E)$.

We have thus shown that Φ defines an isomorphism between the semigroup of finitely generated projective (G,A,α)-modules and $S(L^1(G,A,\alpha))$. Therefore it defines an isomorphism $K_0^G(A)\simeq K_0(L^1(G,A,\alpha))$. Q.E.D.

The proof of the isomorphism $K_0(L^1(G,A,\alpha))\simeq K_0(C^*(G,A,\alpha))$ rests on the following technical lemma, and the construction of a subalgebra I such that the lemma can be applied with B equal to either $L^1(G,A,\alpha)$ or $C^*(G,A,\alpha)$. The lemma basically asserts that the inclusion from I into B is an isomorphism on K-theory, although we don't actually consider the K-theory of algebras which are not Banach algebras.

2.6.3 Lemma. Let A be a Banach algebra. Let I be a dense hereditary subalgebra of A, that is, a dense subalgebra such that $IAI\subset I$. Then for any $n>0$, we have:

(1) For every idempotent $p\in M_n(A)$ there is an idempotent $q\in M_n(I)$ which is Murray-von Neumann equivalent to p.

(2) If p and q are idempotents in $M_n(I)$ which are Murray-von Neumann equivalent in $M_n(A)$, then they are so in $M_n(I)$.

Proof. We first observe that by replacing A with $M_n(A)$ and I with $M_n(I)$, we may consider only idempotents in A and I.

(1) Let $p \in A$ be an idempotent. Let f be the function given by $f(x+iy)=0$ for $x<½$ and $f(x+iy)=1$ for $x>½$. By lemma 2.5.8, there is $\varepsilon>0$ such that whenever $\| p-a \| < \varepsilon$, we have $\| p-f(a) \| < (\| p \| + \| f(a) \|)^{-1}$. By assumption there is $a \in I$ such that $\| p-a \| < \varepsilon$. Then p is similar to $f(a)$ in A^+ by lemma 2.5.3(1), whence p is Murray-von Neumann equivalent to $f(a)$. It remains only to show that $f(a) \in I$. Now $f(z)=zg(z)z$, where $g(x+iy)=0$ for $x<½$ and $g(x+iy)=(x+iy)^{-2}$ for $x>½$. Consequently, $f(a)=ag(a)a \in IAI \subset I$, as required.

(2) Let p and q be idempotents in I, and suppose that there are $u, v \in A$ such that $uv=p$ and $vu=q$. Then puq and qvp are in I because $IAI \subset I$, and

$$(puq)(qvp) = p(uv)^3 p = p^5 = p,$$

while similarly $(qvp)(puq)=q$. So p and q are Murray-von-Neumann equivalent in I.

Q.E.D.

Proof of theorem 2.6.1. We only have to prove that the inclusion of $L^1(G,A,\alpha)$ in $C^*(G,A,\alpha)$ induces an isomorphism on K-theory. Define an A-valued inner product in the space $C(G,A)$ of continuous functions from G to A by

$$<x,y> = \int_G \alpha_g^{-1}(y(g)^* x(g))\, dg .$$

If we set $(xb)(g)=x(g)\alpha_g(b)$ for $x \in C(G,A)$ and $b \in A$, then $C(G,A)$ is a right A-module and $<xb,y> = <x,y>b$ for $x,y \in C(G,A)$ and $b \in B$. In fact, $C(G,A)$ is a pre-Hilbert B-module in the sense of [76], definition 2.1, and the formula $\| x \|_2 = \| <x,x> \|^{½}$ defines a norm on $C(G,A)$ by [76], proposition 2.3 or [84], proposition 2.10. We claim that if $a, x \in C(G,A)$, then $\| ax \|_2 \le \| a \| \, \| x \|_2$, where the product is evaluated in $L^1(G,A,\alpha)$ and $\| \ \|$ is the norm for $C^*(G,A,\alpha)$. This is actually a special case of the results in section 2 of [37], and it can also be derived by showing that $<x,y> = p(y^*x)$ where $p : C(G,A) \longrightarrow A$ is a generalized conditional expectation (evaluation at the identity in G) in the sense of definition 4.12 of [84]. However, we prefer to give a direct proof here.

To prove the claim, let φ be any state on A. For $x,y \in C(G,A)$, set $<x,y>_\varphi = \varphi(<x,y>)$, so that $< \cdot , \cdot >_\varphi$ is an ordinary complex valued inner product on $C(G,A)$. Let H denote the Hilbert space obtained by forming the Hausdorff completion of $C(G,A)$ for the associated norm $\| \ \|_\varphi$. Define a representation σ of A on $C(G,A)$ by $(\sigma(b)x)(g) = bx(g)$ for $b \in A$ and $x \in C(G,A)$. Then

$$\| \sigma(b)x \|_\varphi^2 = \varphi\left[\int_G \alpha_g^{-1}(x(g)^* b^* bx(g))\, dg\right] \le \| b \|^2 \| x \|_\varphi^2,$$

since

$$\alpha_g^{-1}(x(g)^* b^* bx(g)) \le \| b \|^2 \alpha_g^{-1}(x(g)^* x(g))$$

as elements of A. It is now easily seen that σ extends to a representation of A on H, just as in the GNS construction. We further define a representation u of G on H by $(u_h x)(g) = \alpha_h(x(h^{-1}g))$ for $g, h \in G$ and $x \in C(G,A)$. This formula does define a unitary representation on H, because

$$<u_h x, u_h y>_\varphi = \varphi\left[\int_G \alpha_{g^{-1}h}(y(h^{-1}g)^* x(h^{-1}g))\, dg\right] = <x,y>_\varphi .$$

It is immediate to check that the pair (u,σ) is in fact a covariant representation of (G,A,α), and that the associated representation π of $C^*(G,A,\alpha)$ on H is given by $\pi(a)x = ax$ for $a, x \in C(G,A)$, where the multiplication is as in $L^1(G,A,\alpha)$. Consequently $\| ax \|_\varphi^2 \le \| a \|^2 \| x \|_\varphi^2$, that is, $\varphi(<ax, ax>) \le \| a \|^2 \varphi(<x,x>)$. Since this holds for all states φ, we obtain $\| <ax, ax> \| \le \| a \|^2 \| <x,x> \|$, which is just $\| ax \|_2 \le \| a \| \, \| x \|_2$. This proves the claim.

Let J be the part of the completion of $C(G,A)$ with respect to $\| \ \|_2$ which lies in $C^*(G,A)$. That is, J is the set of all $x \in C^*(G,A)$ such that there is a sequence (x_n) in $C(G,A)$ such that $x_n \longrightarrow x$ for $\| \ \|$ and (x_n) is a Cauchy sequence for $\| \ \|_2$. The inequality $\| ax \|_2 \le \| a \| \, \| x \|_2$ implies that J is a left ideal in $C^*(G,A)$. Indeed, if $a_n, x_n \in C(G,A)$ (which is dense in $C^*(G,A)$) are such that $a_n \longrightarrow a$ and $x_n \longrightarrow x$ for $\| \ \|$ and (x_n) is Cauchy for $\| \ \|_2$, then $a_n x_n \longrightarrow ax$ for $\| \ \|$ and $(a_n x_n)$ is Cauchy for $\| \ \|_2$.

Let $I = J^* J$ (the linear span of the products $a^* b$ for $a, b \in J$), which is a dense hereditary subalgebra of $C^*(G,A,\alpha)$. Let $x, y \in C(G,A)$, and define $\rho_g(y)$ by $\rho_g(y)(h) = y(hg)$ for $g, h \in G$. Then

$$\begin{aligned} x^* y(g) &= \int_G \alpha_h(x(h^{-1})^*)\, \alpha_h(y(h^{-1}g))\, dh \\ &= \int_G \alpha_{h^{-1}}(x(h)^* \rho_g(y)(h))\, dh = <x, \rho_g(y)> . \end{aligned}$$

Now $\| <x, \rho_g(y)> \|_2 \le \| x \|_2 \| \rho_g(y) \|_2$ by the analog of the Cauchy-Schwartz inequality for C^*-algebra valued inner products ([84], proposition 2.9, or [76], proposition 2.3). Furthermore $\| \rho_g(y) \|_2 = \| y \|_2$ by a change of variables. Consequently $\| y^* x \|_\infty \le \| x \|_2 \| y \|_2$, where $\| \ \|_\infty$ is the sup norm on $C(G,A)$. Therefore if x_n, $y_n \in C(G,A)$ with $x_n \longrightarrow x$ and $y_n \longrightarrow y$ in $\| \ \|_2$ for $x, y \in J$, we find that $y_n^* x_n \longrightarrow y^* x$ uniformly, so that $y^* x \in C(G,A)$. It follows that $I = J^* J \subset C(G,A) \subset L^1(G,A,\alpha)$.

We have shown that I satisfies the hypotheses of lemma 2.6.3 relative to its inclusion in $L^1(G,A,\alpha)$ and relative to its inclusion in $C^*(G,A,\alpha)$. Two applications of this lemma therefore show that the conclusion holds for the inclusion of $L^1(G,A,\alpha)$ in $C^*(G,A,\alpha)$, so that $S(L^1(G,A,\alpha)) \longrightarrow S(C^*(G,A,\alpha))$ is an isomorphism of semigroups. By lemma 2.2.4, there is an increasing net $\{e_n\}_{n \in D}$ of idempotents in $L^1(G)$ such that $\bigcup_{n \in D} e_n L^1(G,A,\alpha) e_n$ is dense in $L^1(G,A,\alpha)$ and $\bigcup_{n \in D} e_n C^*(G,A,\alpha) e_n$ is dense in $C^*(G,A,\alpha)$. Therefore proposition 2.5.7 implies that $K_0(L^1(G,A,\alpha)) \longrightarrow K_0(C^*(G,A,\alpha))$ is an isomorphism.

Naturality is obvious. Q.E.D.

2.7. The $R(G)$-Module Structure on $K_0(C^*(G,A,\alpha))$

Unfortunately, Julg's theorem only gives us an isomorphism $K_0^G(A) \simeq K_0(C^*(G,A,\alpha))$ as abelian groups. We can, and do, give $K_0(C^*(G,A,\alpha))$ the unique $R(G)$-module structure which makes this isomorphism an $R(G)$-module homomorphism. However, in order to apply Julg's theorem, we need to be able to compute this structure in terms of the crossed product. The idea which enables us to do this is exterior equivalence of actions.

Before defining exterior equivalence, we recall a few facts about multiplier algebras (see [79], section 3.12). If A is a C^*-algebra, then a multiplier ("double centralizer" in [79]) of A is a pair (r, l) of linear maps from A to A such that, for $x, y \in A$, we have $r(xy) = xr(y)$, $l(xy) = l(x)y$, and $r(x)y = xl(y)$. (Thus r and l behave like the right and left multiplications by some element.) Multipliers are necessarily bounded ([79], 3.12.2), and the set of all of them is a unital C^*-algebra $M(A)$ with appropriate operations ([79], 3.12.4). $M(A)$ is called the multiplier algebra of A, and A is an ideal in it, when we identify $a \in A$ with the multiplier (r_a, l_a) defined by $r_a(x) = xa$ and $l_a(x) = ax$. If A has a unit then $M(A) = A$.

If $\alpha: G \longrightarrow Aut(A)$ is a continuous action, then each automorphism α_g extends to an automorphism of $M(A)$, which will usually also be denoted α_g. Unfortunately the resulting action $G \longrightarrow Aut(M(A))$ is unlikely to be continuous for the norm topology on $M(A)$. The best we can do is say that it is continuous for the strict topology, which is the topology defined by the seminorms on $M(A)$ given by $x \longrightarrow \|ax\| + \|xa\|$, for $a \in A$ ([18], definition 3.4).

We now define exterior equivalence. For the definition and the example following it, we let G be an arbitrary topological group.

2.7.1 Definition. (See [79], 8.11.3.) Let A be a C^*-algebra, and let $\alpha, \beta: G \longrightarrow Aut(A)$ be two continuous actions of a group G on A. (Unlike in [79], we do not require that G be abelian.) Then α and β are said to be exterior equivalent if there is a function $g \longrightarrow u_g$ from G into the unitary group of $M(A)$ such that:

(1) $u_{gh} = u_g\alpha_g(u_h)$ for $g, h \in G$.

(2) $\beta_g(a) = u_g\alpha_g(a)u_g^*$ for $a \in A$ and $g \in G$.

(3) $g \longrightarrow u_g a$ is continuous as a function from G to A for all $a \in A$ (that is, $g \longrightarrow u_g$ is strictly continuous).

The function $g \longrightarrow u_g$ is called a one-cocycle, or simply a cocycle.

2.7.2 Example. Let A be any C^*-algebra, let α be the trivial action of a group G on A, and let β be an inner action of G on A. Here we say an action β is inner if it has the form $\beta_g(a) = w_g a w_g^*$, where $g \longrightarrow w_g$ is a strictly continuous homomorphism from G into the unitary group of $M(A)$. (The requirement that w be strictly continuous is exactly what is needed to ensure that β is continuous.) Then α and β are exterior equivalent, with $u_g = w_g$.

2.7.3 Warning. An inner action is not the same thing as an action by inner automorphisms. (An inner automorphism of a C^*-algebra A is of course an automorphism of the form $a \longrightarrow uau^*$ for some unitary $u \in M(A)$.) There are actions $\alpha : G \longrightarrow Aut(A)$ such that each α_g is inner, $\alpha_g(a) = u_g a u_g^*$, but the unitaries u_g implementing the action cannot be chosen such that $u_g u_h = u_{gh}$ for $g, h \in G$. (The simplest such action is studied in example 4.2.3.)

If continuous actions α and β of a compact (or even locally compact) group G on a C^*-algebra A are exterior equivalent, then it is easy to see that $L^1(G,A,\alpha) \simeq L^1(G,A,\beta)$, and therefore that $C^*(G,A,\alpha) \simeq C^*(G,A,\beta)$. (We actually do this calculation in the course of the proof of theorem 2.8.3.) It is therefore very tempting to conclude that $K_0^G(A)$ is the same for either action, and this is in fact true. But without knowing how to compute the $R(G)$-module structure on $K_0(C^*(G,A,\alpha))$ in terms of the crossed products, we can't prove that the isomorphism we obtain is an $R(G)$-module homomorphism. Nevertheless, we need to know that certain exterior equivalent actions yield the same equivariant K-theory in order to find a useful way to compute the $R(G)$-module structure on $K_0(C^*(G,A,\alpha))$. Fortunately, there is a different proof that works for enough cases to enable us to proceed. We consider algebras with approximate identities of projections so as to be able to use it on the compact operators as well as on the algebras of the form $L(V) \otimes A$ for A unital.

2.7.4 Proposition. Let A be a C^*-algebra, and let $\alpha, \beta : G \longrightarrow Aut(A)$ be exterior equivalent actions of a compact group G on A. Suppose A has an increasing approximate identity of projections which are invariant for both actions. Then $K_0^G(A)$ for the action α is naturally isomorphic as an $R(G)$-module to $K_0^G(A)$ for the action β.

Proof. Let $g \longrightarrow u_g$ be as in definition 2.7.1. We consider the unital case first. Let (G,E,λ) be a finitely generated projective (G,A,α)-module. Define an action μ of G on E by $\mu_g(e) = \lambda_g(e) u_g^*$ for $e \in E$ and $g \in G$. Since

$$\mu_g(ea) = \lambda_g(e)\alpha_g(a)u_g^* = \mu_g(e)\beta_g(a),$$

we find that (G,E,μ) is a (G,A,β)-module, and it is of course still finitely generated and projective. This correspondence clearly defines an isomorphism of $R(G)$-modules between the groups $K_0^G(A)$ for the two actions.

For the general case, let $(p_n)_{n \in D}$ be the approximate identity of the hypothesis. Then also $u_g p_n u_g^* = p_n$ for all g and n, and $g \longrightarrow p_n u_g p_n$ defines an exterior equivalence between the restricted actions $g \longrightarrow \alpha_g|_{p_n A p_n}$ and $g \longrightarrow \beta_g|_{p_n A p_n}$. Since $A = \varinjlim p_n A p_n$, we obtain $K_0^G(A) = \varinjlim K_0^G(p_n A p_n)$ relative to both actions, by proposition 2.5.4. Therefore the unital case implies that $K_0^G(A)$ is the same for both actions.

Q.E.D.

If H is any Hilbert space, we let $K(H)$ denote the C^*-algebra of compact operators on H. Let $g \longrightarrow u_g$ be a continuous (in the strong operator topology) unitary representation of an arbitrary topological group G on H. Then there is a continuous inner action α of G on $K(H)$ defined by $\alpha_g(a) = u_g a u_g^*$ for $a \in K(H)$ and $g \in G$. (Recall that the multiplier algebra of $K(H)$ is the C^*-algebra $L(H)$ of all bounded operators on H.) Whenever H is a Hilbert space carrying a unitary representation of a group G, the algebra $K(H)$ will be assumed to carry this action of G. If A is any G-algebra, then we give $K(H) \otimes A$ the diagonal action of G.

Returning to our convention that all groups are compact, we prove the following special case of the stability of equivariant K-theory.

2.7.5 Lemma. Let (G,A,α) be a unital G-algebra, and let u be a unitary representation of G on a Hilbert space H. Then there is a natural isomorphism $K_0^G(K(H) \otimes A) \simeq K_0^G(A)$. If $p \in K(H)$ is any G-invariant projection, and if $\varphi: A \longrightarrow K(H) \otimes A$ is defined by $\varphi(a) = p \otimes a$, then this isomorphism identifies φ_* with multiplication by the class $[pH] \in R(G)$.

Proof. Since G is compact, u is a direct sum of finite dimensional representations, so $K(H)$ has an increasing approximate identity consisting of G-invariant projections. Therefore so does $K(H) \otimes A$. Furthermore, the action on $K(H)$ is exterior equivalent to the trivial action ι of G on $K(H)$, using the cocycle $g \longrightarrow u_g$ (compare example 2.7.2). Therefore the diagonal action on $K(H) \otimes A$ is exterior equivalent to the action $\iota \otimes \alpha$, using the cocycle $g \longrightarrow u_g \otimes 1$. By the previous proposition, we may thus assume for the first part of the proof that the representation of G on H is trivial. The proof now proceeds just as for ordinary K-theory; we carry it out anyway.

Let r be any one-dimensional projection in $K(H)$, and define $\varphi: A \longrightarrow K(H) \otimes A$ by $\varphi(a) = r \otimes a$. Since any two one-dimensional projections in $K(H)$ are connected by a continuous path of projections, the corresponding maps φ are G-homotopic. By proposition 2.5.10, φ_* therefore does not depend on which projection is chosen. To prove that φ_* is an isomorphism, we use the fact (corollary 2.5.5) that $K_0^G(K(H) \otimes A)$ can be computed in terms of G-invariant projections in various algebras $L(V) \otimes K(H) \otimes A$, which we rewrite as $K(H) \otimes L(V) \otimes A$.

Let $p_i \in L(V_i) \otimes A$ for $i = 1,2$ be G-invariant projections such that their images in $K(H) \otimes L(V_i) \otimes A$ are Murray-von Neumann equivalent. Replacing V_1 and V_2 by $V = V_1 \oplus V_2$, we may assume that p_1 and p_2 are in the same $L(V) \otimes A$. To simplify the notation, let $B = L(V) \otimes A$, and let φ also denote the map $a \longrightarrow r \otimes a$ from B to $K(H) \otimes B$. Since $\varphi(p_1)$ and $\varphi(p_2)$ are Murray-von Neumann equivalent in $K(H) \otimes B$, and lie in $(r \otimes 1)(K(H) \otimes B)(r \otimes 1)$, they are Murray-von Neumann equivalent in $(r \otimes 1)(K(H) \otimes B)(r \otimes 1)$. (Choose u and v such that $uv = \varphi(p_1)$ and $vu = \varphi(p_2)$. Set

$a=\varphi(p_1)u\varphi(p_2)$ and $b=\varphi(p_2)v\,\varphi(p_1)$. Then $ab=\varphi(p_1)$ and $ba=\varphi(p_2)$. This calculation was done in the proof of lemma 2.6.3 (2), so we don't repeat it here.) Since φ defines an isomorphism of B with $(r\otimes 1)(K(H)\otimes B)(r\otimes 1)$, we conclude that p_1 and p_2 are Murray-von Neumann equivalent in B. It follows from this argument that φ_* is injective.

To see surjectivity, let $p\in K(H)\otimes L(V)\otimes A$ be a G-invariant projection, and write again $B=L(V)\otimes A$. By the last statement in corollary 2.5.5, we can take p to be dominated by $e\otimes 1$ for some projection $e\in K(H)$ without changing its equivalence class. Then p is easily seen to be equivalent to a projection q in $K(H)\otimes L(eH)\otimes B$ such that q is dominated by $r\otimes e\otimes 1$. Since q is in the range of the map from $L(eH)\otimes L(V)\otimes A$ to $K(H)\otimes L(eH)\otimes L(V)\otimes A$ induced by φ, it follows that $[p]$ is in the range of φ_*. Thus, φ_* is surjective, and the first part of the lemma is proved.

We now turn to the second part of the lemma. Again let u be a unitary representation of G on a Hilbert space H, and now let $p\in K(H)$ be a G-invariant projection. Define $\varphi: A\longrightarrow K(H)\otimes A$ by $\varphi(a)=p\otimes a$. Let ε be the exterior equivalence isomorphism from $K_0^G(K(H)\otimes A)$ to $K_0^G(K(H_0)\otimes A)$, where H_0 is the Hilbert space H but with the trivial action of G. Let r be a one-dimensional projection in $K(H_0)$, and define $\psi: A\longrightarrow K(H_0)\otimes A$ by $\psi(a)=r\otimes a$. The proof of the first part of the lemma showed that ψ_* is an isomorphism; the assertion of the second part is that $\psi_*^{-1}\circ\varepsilon\circ\varphi_*$, which is a homomorphism from $K_0^G(A)$ to $K_0^G(A)$, is multiplication by the element $[pH]\in R(G)$.

Let (G,E,λ) be a finitely generated projective (G,A,α)-module. Write $E=q\,(L(V)\otimes A)$ for some G-invariant projection q and some finite dimensional representation space V of G. Then

$$\varphi_*([E])=[(p\otimes q)(K(H)\otimes L(V)\otimes A)]=[pK(H)\otimes E].$$

Here $pK(H)\otimes E$ is a $K(H)\otimes A$-module in the obvious way, the action of G on E is just λ, and the action of G on $pK(H)$ is defined by $a\longrightarrow u_g a u_g^*$. (This makes sense because p is G-invariant.) An inspection of the proof of proposition 2.7.4 (where ε is defined) shows that $\varepsilon([pK(H)\otimes E])$ is represented by the finitely generated projective $(G,K(H_0)\otimes A)$-module $(G,pK(H)\otimes E,\,\mu\otimes\lambda)$, where μ is the action of G on $pK(H)$ defined by $\mu_g(a)=u_g a$ (rather than $u_g a u_g^*$).

The final step is to apply ψ_*^{-1}. We will actually apply ψ_* to $[pH\otimes E]$ and show that it is represented by the module $(G,pK(H)\otimes E,\,\mu\otimes\lambda)$ obtained above. It is clear that $\psi_*([pH\otimes E])$ is represented by $(G,rK(H_0)\otimes pH\otimes E,\,\nu\otimes\lambda)$, where ν is the action of G on $rK(H_0)\otimes pH$ defined by $\nu_g(b\otimes\xi)=b\otimes u_g\xi$. (Compare this with the calculation of $\varphi_*([E])$.) It is clearly sufficient to prove that

$$(G,pK(H),\mu)\simeq(G,rK(H_0)\otimes pH,\nu)$$

as $(G,K(H_0))$-modules (recall that G acts trivially on $K(H_0)$.) Before carrying out the proof, let us consider a special case. Let u be irreducible, so that in particular $n=dim(H)<\infty$, and let $p=1$. Then $rK(H_0)\simeq\mathbf{C}^n$, with the trivial action of G. The desired isomorphism then says that the representation of G on $L(H)$ defined by left multiplication by u_g is the direct sum of n copies of u, which is a standard fact in the theory of representations of compact groups. In the general case, however, we cannot appeal directly to this fact, because we must ensure that the isomorphism preserves the $K(H_0)$-module structures. Therefore, we let $\xi_0 \in H_0$ be a unit vector in the range of the one-dimensional projection r, and define $\sigma : rK(H_0) \otimes pH \longrightarrow pK(H)$ by $\sigma(b \otimes \xi)(\eta) = <\eta, b^* \xi_0> \xi$. The map σ is easily seen to be bijective (in particular, its image really is in $pK(H)$), and the following computations show that it is a $(G, K(H_0))$-module homomorphism:

$$u_g \, \sigma(b \otimes \xi)(\eta) = u_g <\eta, b^* \xi_0> \xi = \sigma(b \otimes u_g \xi)(\eta),$$

and

$$\sigma(b \otimes \xi)(c\eta) = <c\eta, b^* \xi_0> \xi = \sigma(bc \otimes \xi)(\eta),$$

for $g \in G$, $b \in rK(H_0)$, $c \in K(H_0)$, $\xi \in pH$, and $\eta \in H$. Q.E.D.

2.7.6 Warning. This proposition need not hold if the action of G on $K(H)$ comes from a projective unitary representation of G on H rather than an honest unitary representation. (See example 4.2.3.)

We now have a method for computing the $R(G)$-module structure on $K_0(C^*(G,A,\alpha))$ for a unital G-algebra (G,A,α). Specifically, let V be a representation space of G of finite dimension n. Define $\varphi : A \longrightarrow L(V) \otimes A$ by $\varphi(a) = 1 \otimes a$, and let $\overline{\varphi} : C^*(G,A) \longrightarrow C^*(G, L(V) \otimes A)$ be the induced map on crossed products. Let $\varepsilon : C^*(G, L(V) \otimes A) \longrightarrow C^*(G, M_n \otimes A)$ be the isomorphism of crossed products induced by the canonical exterior equivalence of $(G, L(V) \otimes A)$ with $(G, M_n \otimes A)$, as in the proof of the first part of lemma 2.7.5. (Here M_n has the trivial G-action.) Finally, let $\sigma : K_0(C^*(G, M_n \otimes A)) \longrightarrow K_0(C^*(G,A))$ be the stability isomorphism in ordinary K-theory, where of course we identify $C^*(G, M_n \otimes A)$ with $M_n \otimes C^*(G,A)$. Then the composite $\sigma \cdot \varepsilon \cdot \varphi_*$ is multiplication by $[V]$ on $K_0(C^*(G,A))$.

There are two problems that must be overcome. The first is that this method can so far only be applied to unital G-algebras. The next lemma provides the means for getting around this difficulty. A more serious objection is that the map ε does not come from a map of G-algebras. This means, for example, that one cannot immediately conclude that multiplication by $[V]$ commutes with certain natural maps between the ordinary K-groups of crossed products, such as connecting homomorphisms. The solution to this is to use the space $V \oplus \mathbf{C}$ (where G acts trivially on $\mathbf{C}$), instead of V. The result we obtain (theorem 2.7.9) is not as pretty as the procedure sketched above, but all the maps that appear do come from maps of G-algebras.

2.7.7 Lemma. Let (G,A,α) be a G-algebra. Then there is a natural exact sequence

$$0 \longrightarrow K_0^G(A) \longrightarrow K_0^G(A^+) \longrightarrow R(G) \longrightarrow 0.$$

Proof. Of course we identify $R(G)$ with $K_0^G(\mathbf{C})$. The first map is injective and the sequence is exact in the middle, by definition for nonunital A, and by the comments accompanying the definition for unital A. The second map is surjective since the map $A^+ \longrightarrow \mathbf{C}$ has an equivariant right inverse. Q.E.D.

2.7.8 Definition. Let (G,A,α) be a G-algebra. We define an $R(G)$-module structure on $K_0(C^*(G,A,\alpha))$ as follows. If V is any finite dimensional representation space of G, let $V\oplus\mathbf{C}$ be its direct sum with a one-dimensional trivial representation space. Let $p_V, p_{\mathbf{C}} \in L(V\oplus\mathbf{C})$ be the projections on V and $\mathbf{C}$ respectively. Define $\varphi_V, \varphi_{\mathbf{C}}: A \longrightarrow L(V\oplus\mathbf{C})\otimes A$ by $\varphi_V(a) = p_V \otimes a$ and $\varphi_{\mathbf{C}}(a) = p_{\mathbf{C}}\otimes a$, and let $\bar{\varphi}_V$ and $\bar{\varphi}_{\mathbf{C}}$ be the corresponding maps on the crossed products by G. Then for $\eta \in K_0(C^*(G,A,\alpha))$, we define $[V]\eta$ to be $(\bar{\varphi}_{\mathbf{C}})_*^{-1}\cdot(\bar{\varphi}_V)_*(\eta)$.

One should of course check that $(\bar{\varphi}_{\mathbf{C}})_*$ in invertible, and that with this definition, one has $([V][W]) = [V]([W]\eta)$ and $([V]+[W])\eta = [V]\eta + [W]\eta$ for $\eta\in K_0(C^*(G,A,\alpha))$ and for finite dimensional representation spaces V and W of G. These facts are all immediate consequences of the proof of the following theorem.

2.7.9 Theorem. Let (G,A,α) be an arbitrary G-algebra. Then there is natural isomorphism of $R(G)$-modules $K_0^G(A) \simeq K_0(C^*(G,A,\alpha))$, where the $R(G)$-module structure on $K_0(C^*(G,A,\alpha))$ is given by the previous definition.

Proof. We first consider the unital case. Let V be a finite dimensional representation space of G, and let φ_V, $\varphi_{\mathbf{C}}$, $\bar{\varphi}_V$, and $\bar{\varphi}_{\mathbf{C}}$ be as in the previous definition. Then by naturality in Julg's theorem (2.6.1), there is a commutative diagram in which the vertical arrows are the isomorphisms of 2.6.1:

$$\begin{array}{ccccc}
K_0^G(A) & \xrightarrow{(\varphi_V)_*} & K_0^G(L(V\oplus\mathbf{C})\otimes A) & \xleftarrow{(\varphi_{\mathbf{C}})_*} & K_0^G(A) \\
\downarrow & & \downarrow & & \downarrow \\
K_0(C^*(G,A)) & \xrightarrow{(\bar{\varphi}_V)_*} & K_0(C^*(G,L(V\oplus\mathbf{C})\otimes A) & \xleftarrow{(\bar{\varphi}_{\mathbf{C}})_*} & K_0(C^*(G,A))
\end{array}$$

By lemma 2.7.5, $(\varphi_{\mathbf{C}})_*$ is invertible and $(\varphi_{\mathbf{C}})_*^{-1}\cdot(\varphi_V)_*$ is multiplication by $[V]$ on $K_0^G(A)$. Since $(\varphi_{\mathbf{C}})_*$ is invertible and the vertical arrows are isomorphisms, $(\bar{\varphi}_{\mathbf{C}})_*$ is also invertible. By definition, $(\bar{\varphi}_{\mathbf{C}})_*^{-1}\cdot(\bar{\varphi}_V)_*$ is multiplication by $[V]$ on $K_0(C^*(G,A))$. The commutativity of the diagram now implies that $K_0^G(A) \longrightarrow K_0(C^*(G,A))$ commutes with multiplication by $[V]$. This map is therefore an $R(G)$-module homomorphism.

Now consider the nonunital case. It is easily seen that the sequence of crossed products

$$0 \longrightarrow C^*(G,A) \longrightarrow C^*(G,A^+) \longrightarrow C^*(G) \longrightarrow 0$$

is split exact, where $C^*(G) = C^*(G,\mathbf{C})$. (Actually, we will soon prove that taking crossed products preserves exactness in general. See lemma 2.8.2.) Therefore there is a (split) exact sequence

$$0 \longrightarrow K_0(C^*(G,A)) \longrightarrow K_0(C^*(G,A^+)) \longrightarrow K_0(C^*(G)) \longrightarrow 0,$$

obtained from the six term exact sequence for ordinary K-theory. The commutative diagram

$$\begin{array}{ccccccccc} 0 & \longrightarrow & K_0(C^*(G,A)) & \longrightarrow & K_0(C^*(G,A^+)) & \longrightarrow & K_0(C^*(G)) & \longrightarrow & 0 \\ & & \downarrow (\overline{\varphi}_{\mathbf{C}})_* & & \downarrow (\overline{\varphi}_{\mathbf{C}})_* & & \downarrow (\overline{\varphi}_{\mathbf{C}})_* & & \\ 0 & \longrightarrow & K_0(C^*(G,A)) & \longrightarrow & K_0(C^*(G,A^+)) & \longrightarrow & K_0(C^*(G)) & \longrightarrow & 0 \end{array}$$

now shows that $(\overline{\varphi}_{\mathbf{C}})_*$ is invertible even for nonunital A. The required homomorphism λ from $K_0^G(A)$ to $K_0(C^*(G,A))$ is the unique map (necessarily an isomorphism of abelian groups) making the following diagram commutative:

$$\begin{array}{ccccccccc} 0 & \longrightarrow & K_0^G(A) & \longrightarrow & K_0^G(A^+) & \longrightarrow & K_0^G(\mathbf{C}) & \longrightarrow & 0 \\ & & \lambda \downarrow & & \downarrow & & \downarrow & & \\ 0 & \longrightarrow & K_0(C^*(G,A)) & \longrightarrow & K_0(C^*(G,A^+)) & \longrightarrow & K_0(C^*(G)) & \longrightarrow & 0. \end{array}$$

Here, the middle and right vertical arrows are the isomorphisms of Julg's theorem. Another diagram chase now shows that λ commutes with the multiplications by $[V]$ for finite dimensional representation spaces V of G, and is therefore an $R(G)$-module isomorphism.

Q.E.D.

If G is abelian, the procedure in the discussion before the last proposition takes a more useful form, and we close this section by presenting it. We first recall the definition of the dual action on a crossed product $C^*(G,A,\alpha)$, where G is a locally compact abelian group (see [79], 7.8.3). For each element τ of the Pontrjagin dual $\widehat{G}$ of G, there is an automorphism $\widehat{\alpha}_\tau$ of $C^*(G,A,\alpha)$ defined on $C_c(G,A)$ by $\widehat{\alpha}_\tau(a)(g) = \tau(g)a(g)$ for $a \in C_c(G,A)$ and $g \in G$. The homomorphism $\widehat{\alpha}: \widehat{G} \longrightarrow Aut\,(C^*(G,A,\alpha))$ is then called the dual action of α; it is a continuous action.

If G is a compact abelian group, then $R(G)$ can be canonically identified with the ring $\mathbf{Z}[\widehat{G}]$ of all formal finite linear combinations of elements of $\widehat{G}$ with integer coefficients. (Multiplication is determined by the requirement that $(n\sigma)(m\tau) = (nm)(\sigma\tau)$ for $n, m \in \mathbf{Z}$ and $\sigma, \tau \in \widehat{G}$). It follows that an $R(G)$-module structure on an abelian group is determined by the multiplications by the elements of $\widehat{G}$. With this in mind, and returning to our convention that G is a compact group, we have:

2.7.10 Proposition. Let G be abelian, and let (G,A,α) be a G-algebra. Then for $\tau \in \widehat{G}$ and $\eta \in K_0(C^*(G,A,\alpha))$, we have $\tau\eta = (\widehat{\alpha}_\tau)_*(\eta)$.

Proof. First consider the case in which A is unital, and let (G,E,λ) be a finitely generated projective (G,A,α)-module. Let Φ be the functor assigning to a (G,A,α)-module the corresponding $L^1(G,A,\alpha)$-module (lemma 2.2.3). Then $\Phi(E)$ is just E with the module structure given by

$$ea = \int_G \lambda_g^{-1}(e)\,\alpha_g^{-1}(a(g))\,dg$$

for $e \in E$ and $a \in L^1(G,A,\alpha)$, and $(\hat{\alpha}_\tau)_* \Phi(E)$ is E with the module structure

$$ea = \int_G \lambda_g^{-1}(e)\,\alpha_g^{-1}((\hat{\alpha}_\tau^{-1}(a))(g))dg\,. \tag{*}$$

Also τE is E with G-action $(\tau\lambda)_g(e) = \tau(g)\lambda_g(e)$, so $\Phi(\tau E)$ is E with the module structure

$$ea = \int_G \tau(g)^{-1}\lambda_g^{-1}(e)\,\alpha_g^{-1}(a(g))\,dg\,. \tag{**}$$

Since (*) and (**) are the same, we have $\Phi(\tau E) \simeq (\hat{\alpha}_\tau)_* \Phi(E)$. It follows that multiplication by τ is $(\hat{\alpha}_\tau)_*$. The nonunital case follows by naturality and the usual diagram chase using lemma 2.7.7. Q.E.D.

2.8. The Homology Theory K_*^G.

2.8.1 Definition. Let (G,A,α) be a G-algebra. We define $K_1^G(A) = K_0^G(SA)$, where SA is the suspension $C_0((0,1))\otimes A$ of A, and G acts trivially on $(0,1)$. Then K_1^G is also a functor from G-algebras to $R(G)$-modules. We write $K_*^G(A)$ for the direct sum $K_0^G(A) \oplus K_1^G(A)$. Homomorphisms of K_*^G are assumed to preserve the grading, that is, to map K_0^G to K_0^G and K_1^G to K_1^G.

We are going to prove that K_*^G is a periodic homology theory on G-algebras by using Julg's theorem to reduce to the case of ordinary K-theory. We therefore need to following lemma:

2.8.2 Lemma. Let

$$0 \longrightarrow I \xrightarrow{\ i\ } A \xrightarrow{\ \varphi\ } B \longrightarrow 0$$

be an equivariant exact sequence of G-algebras, where G is any locally compact group. Then the sequence of crossed products

$$0 \longrightarrow C^*(G,I) \xrightarrow{\ j\ } C^*(G,A) \xrightarrow{\ \psi\ } C^*(G,B) \longrightarrow 0$$

is exact.

Proof. Recall ([79], 7.6.4) that there is a one-to-one correspondence between representations of a crossed product $C^*(G,A,\alpha)$ and covariant representations of the G-algebra (G,A,α), assigning to a covariant representation its integrated form. Let α be the action of G on A, regard I as an ideal in A, and let β be the action of G on B.

Injectivity of j: Let a be a nonzero element of $C^*(G,I,\alpha)$. Then there is a representation π of $C^*(G,I,\alpha)$ such that $\pi(a) \neq 0$. Let (u, σ) be the corresponding covariant representation of (G,I,α). By [29], 2.10.4, σ extends uniquely to a representation σ_0 of A on the same Hilbert space. Since, for $g \in G$, the representation $a \longrightarrow u_g^* \sigma_0(\alpha_g(a))u_g$ also extends σ_0, it must be equal to σ_0. Hence (u, σ_0) is a covariant representation of (G,A,α). Let π_0 be the integrated form of (u, σ_0). Then it is clear that $\pi_0(j(a)) = \pi(a)$, and hence $j(a) \neq 0$.

The composite $\psi \cdot j$ is zero: obvious.

Ker $\psi \subset C^*(G,I,\alpha)$: Let $b \in C^*(G,A,\alpha)$, and suppose that $b \notin C^*(G,I,\alpha)$. Then there is a representation π of $C^*(G,A,\alpha)$ which annihilates $C^*(G,I,\alpha)$ but not b, since $C^*(G,I,\alpha)$ is clearly a closed ideal in $C^*(G,A,\alpha)$. Furthermore, π is the integrated form of a covariant representation (u, σ) of (G,A,α), and from the construction of σ in the proof of [79], 7.6.4, it follows that $\sigma|_I = 0$. Therefore σ defines a representation of $A/I \simeq B$, so that π defines a representation of $C^*(G,B,\beta)$. Since $\pi(b) \neq 0$, it follows that $b \notin$ *Ker* ψ.

Surjectivity of ψ: Let π be a faithful representation of $C^*(G,B,\beta)$. Then π is the integrated form of a covariant representation (u,σ) of (G,B,β). Clearly $(u, \sigma \cdot \varphi)$ is a covariant representation of (G,A,α), and its integrated form is exactly $\pi \cdot \psi$. Since φ is surjective, the integrated form of $(u, \sigma \cdot \varphi)$ has the same image as the integrated form of (u,σ). Thus π and $\pi \cdot \psi$ have the same image. Since π is injective, ψ must be surjective.

Q.E.D.

We now return to our convention that all groups are compact.

2.8.3 Theorem. The functors K_0^G and K_1^G define a homology theory of period two on the category of G-algebras and equivariant homomorphisms, with values in the category of $R(G)$-modules. That is:

(1) Exactness: If

$$0 \longrightarrow I \longrightarrow A \longrightarrow B \longrightarrow 0$$

is an equivariant exact sequence of G-algebras, then there is a natural six term ("long") exact sequence of $R(G)$-modules

$$\begin{array}{ccccc} K_0^G(I) & \longrightarrow & K_0^G(A) & \longrightarrow & K_0^G(B) \\ \uparrow & & & & \downarrow \\ K_1^G(B) & \longleftarrow & K_1^G(A) & \longleftarrow & K_1^G(I). \end{array}$$

(2) Bott periodicity: There is a natural isomorphism $K_*^G(S^2 A) \simeq K_*^G(A)$, where $S^2 A$ is the second suspension.

(3) Homotopy invariance: If $\varphi, \psi : A \longrightarrow B$ are G-homotopic maps of G-algebras, then $\varphi_*, \psi_* : K_*^G(A) \longrightarrow K_*^G(B)$ are equal.

Furthermore, one has:

(4) Stability: $K_*^G(K(H)\otimes A)\simeq K_*^G(A)$ if the action of G on $K(H)$ is given by a unitary representation of G on the Hilbert space H. If the action of G on $K(H)$ is trivial, then the isomorphism is induced by the map from A to $K(H)\otimes A$ defined by $a \longrightarrow p\otimes a$ for any one-dimensional projection $p \in K(H)$.

(5) Exterior invariance: $K_*^G(A)$ is unchanged if the action of G on A is replaced by an exterior equivalent one.

(6) Continuity: If $(A_\lambda)_{\lambda\in D}$ is a directed system of G-algebras, then $K_*^G(\varinjlim A_\lambda)\simeq \varinjlim K_*^G(A_\lambda)$. In particular, if $\bigoplus_{i\in I} A_i$ denotes the direct sum (c_0 sum) of G-algebras A_i for $i\in I$, then $K_*^G(\bigoplus_{i\in I} A_i)\simeq \bigoplus_{i\in I} K_*^G(A_i)$.

(7) Natural isomorphism with ordinary K-theory of crossed products: There is a natural isomorphism of $R(G)$-modules $K_*^G(A)\simeq K_*(C^*(G,A))$, where the $R(G)$-module structure on $K_*(C^*(G,A))$ is given by definition 2.7.8. (Notice that

$$K_1(C^*(G,A))\simeq K_0(S\,C^*(G,A))\simeq K_0(C^*(G,SA))$$

is the K-theory of a crossed product, so that definition 2.7.8 applies). This isomorphism commutes with the connecting homomorphisms. If G is abelian, then the isomorphism $K_*^G(A)\simeq K_*(C^*(G,A,\alpha))$ identifies multiplication by the class $[\tau]$ of $\tau\in\widehat{G}$ with $(\widehat{\alpha}_\tau)_*$.

Proof. Statement (7), except for the parts about connecting homomorphisms and abelian groups, follows from theorem 2.7.9. We then define the connecting homomorphisms ∂_i^G for $i=0,1$ for the equivariant exact sequence

$$0\longrightarrow I\longrightarrow A\longrightarrow B\longrightarrow 0$$

to be the unique maps making the following diagram commute, where the maps ∂_i are the usual connecting homomorphisms for the exact sequence of crossed products (see the previous lemma):

$$\begin{array}{ccc} K_i^G(B) & \xrightarrow{\partial_i^G} & K_{1-i}^G(I) \\ \downarrow\simeq & & \downarrow\simeq \\ K_i(C^*(G,B)) & \xrightarrow{\partial_i} & K_{1-i}(C^*(G,I)). \end{array}$$

To show that ∂_i^G is an $R(G)$-module homomorphism, it is enough to show that ∂_i is. Let V be a finite dimensional representation space of G, let $p,q\in L(V\oplus\mathbf{C})$ be the projections on V and $\mathbf{C}$ respectively, and let $\varphi_B,\psi_B : B\longrightarrow L(V\oplus\mathbf{C})\otimes B$ be the maps $\varphi_B(b)=p\otimes b$ and $\psi_B(b)=q\otimes b$, and similarly for B replaced by I. Then the naturality of the connecting homomorphism in ordinary K-theory implies that the following diagram commutes:

$$\begin{array}{ccccc} K_i(C^*(G,B)) & \xrightarrow{(\varphi_B)_*} & K_i(C^*(G,L(V\oplus \mathbf{C})\otimes B)) & \xleftarrow{(\psi_B)_*} & K_i(C^*(G,B)). \\ \downarrow \partial_i & & \downarrow \partial_i & & \downarrow \partial_i \\ K_{1-i}(C^*(G,I)) & \xrightarrow{(\varphi_I)_*} & K_{1-i}(C^*(G,L(V\oplus \mathbf{C})\otimes I)) & \xleftarrow{(\psi_I)_*} & K_{1-i}(C^*(G,I)). \end{array}$$

Thus, $\partial_i \cdot (\psi_B)_*^{-1} \cdot (\varphi_B)_* = (\psi_I)_*^{-1} \cdot (\varphi_I)_* \cdot \partial_i$, so that ∂_i commutes with multiplication by $[V]$ and is thus an $R(G)$-module homomorphism.

We have defined the connecting homomorphisms for K_*^G in such a way that (7) holds, and we have also proved that all the maps in the sequence in (1) are $R(G)$-module homomorphisms. They are clearly all natural, and the sequence is exact because the corresponding sequence in (7) is exact by lemma 2.8.2 and exactness for ordinary K-theory. The part about abelian groups in (7) follows from proposition 2.7.10, since if α is the action on A and β is the action on SA, then $\hat{\beta}_\tau$ is just the suspension of $\hat{\alpha}_\tau$ under the isomorphism $C^*(G,SA) \simeq S\,C^*(G,A)$. So (1) and (7) have been proved.

(3) Let $\varphi, \psi : A \longrightarrow B$ be G-homotopic. Then φ^+ and $\psi^+ : A^+ \longrightarrow B^+$ are also G-homotopic, and therefore by proposition 2.5.10 they induce the same map $K_0^G(A^+) \longrightarrow K_0^G(B^+)$. By restriction to $K_0^G(A)$, we find that φ and ψ induce the same map $K_0^G(A) \longrightarrow K_0^G(B)$. Now the suspensions $S\varphi, S\psi : SA \longrightarrow SB$ are also G-homotopic, and $K_1^G(A) = K_0^G(SA)$ (similarly for B). This, together with the result already obtained, implies that φ_* and ψ_* are equal on K_1^G as well.

(2) We define the cone CA over A to be $CA = C_0((0,1])\otimes A$, where G acts trivially on the first factor. Notice that $SA = C_0((0,1))\otimes A$ is embedded in it as an ideal, and that the quotient is just A. Furthermore, CA is equivariantly contractible, that is, the identity map id_{CA} is G-homotopic to the zero map. (Regard $C_0((0,1])$ as the set of functions f on $[0,1]$ such that $f(0)=0$, set $f_t(s)=f(ts)$ for $s,t\in[0,1]$, and define $\varphi_t : CA \longrightarrow CA$ by $\varphi_t(f\otimes a)=f_t\otimes a$. Then $\varphi_1 = id_{CA}$ and $\varphi_0 = 0$.) It follows from (3) that $K_*^G(CA)=0$, and the exact sequence of (1), applied to the short exact sequence

$$0 \longrightarrow SA \longrightarrow CA \longrightarrow A \longrightarrow 0,$$

now yields a natural isomorphism $K_1^G(SA) \simeq K_0^G(A)$. Since $K_1^G(SA) = K_0^G(S^2A)$ by definition, (2) is proved.

(4) This follows by extending the stability lemma 2.7.5 in a manner analogous to the way the homotopy lemma was extended in the proof of (3).

(6) That $K_*^G(\varinjlim A_\lambda) \simeq \varinjlim K_*^G(A_\lambda)$ is proved analogously to (3), using the fact that $\varinjlim SA_\lambda \simeq S(\varinjlim A_\lambda)$. To verify the assertion about direct sums, we first observe that it holds for finite direct sums. Indeed, the formula $K_i^G(A\oplus B) \simeq K_i^G(A)\oplus K_i^G(B)$ can easily be proved directly for $i=0$ and by taking suspensions for $i=1$, or one can apply the usual reasoning and the exact sequence of (1) to the split exact sequence

$$0 \longrightarrow A \xrightarrow{i_1} A \oplus B \underset{p_2}{\overset{i_2}{\rightleftarrows}} B \longrightarrow 0.$$

That K_*^G commutes with finite direct sums now follows by induction. Now let F denote the set of all finite subsets of the arbitrary index set I, ordered by inclusion. For $S \in F$, set $A_S = \bigoplus_{i \in S} A_i$. Then one easily verifies that, with the obvious maps,

$$\bigoplus_{i \in I} A_i \simeq \varinjlim_{S \in F} A_S,$$

whence

$$K_*^G(\bigoplus_{i \in I} A_i) \simeq \varinjlim_{S \in F} K_*^G(A_S) \simeq \varinjlim_{S \in F} \left[\bigoplus_{i \in S} K_*^G(A_i)\right] \simeq \bigoplus_{i \in I} K_*^G(A_i),$$

as desired.

(5) We are now in a position to use the argument outlined in the discussion preceding proposition 2.7.3. (Note, however, that the argument used in the proof of 2.7.3 does not work, because exterior equivalent actions $\alpha, \beta : G \longrightarrow Aut(A)$ do not usually extend to exterior equivalent actions on A^+.)

By assumption, there are unitaries u_g in the multiplier algebra of A such that $\beta_g(a) = u_g \alpha_g(a) u_g^*$ and $g \longrightarrow u_g a$ is continuous from G to A for every $a \in A$. Define $\varphi : L^1(G,A,\alpha) \longrightarrow L^1(G,A,\beta)$ by $\varphi(a)(g) = a(g) u_g^*$, for $a \in L^1(G,A,\alpha)$ and $g \in G$. Then φ is isometric and bijective; furthermore, it is easily verified that it is a *-isomorphism of Banach algebras. Therefore it defines an isomorphism of crossed products, also called φ, from $C^*(G,A,\alpha)$ to $C^*(G,A,\beta)$. (So far G could have been any locally compact group.) Since now obviously $K_*(C^*(G,A,\alpha)) \simeq K_*(C^*(G,A,\beta))$, we have the desired isomorphism. It remains only to check that it is an $R(G)$-module homomorphism. The proof is similar to the proof that the connecting homomorphisms are $R(G)$-module homomorphisms, and we omit it. (Notice, however, that for abelian G it is much easier: a direct calculation shows that φ intertwines the dual actions on the crossed products.) Q.E.D.

2.8.4 Corollary. Let $K_G^1(X) = K_1^G(C_0(X))$ for locally compact G-spaces X. Then the functors K_G^0 and K_G^1 define a continuous periodic cohomology theory of period two on the category of locally compact G-spaces and proper equivariant maps, with values in the category of $R(G)$-modules. In particular, if X is a locally compact G-space and U is an open G-invariant subset, then there is a long exact sequence of $R(G)$-modules:

$$\begin{array}{ccccc} K_G^0(U) & \longrightarrow & K_G^0(X) & \longrightarrow & K_G^0(X-U) \\ \uparrow & & & & \downarrow \\ K_G^1(X-U) & \longleftarrow & K_G^1(X) & \longleftarrow & K_G^1(U). \end{array}$$

Furthermore, if X is compact then $K_G^1(X)$ is a module over the unital ring $K_G^0(X)$, compatibly with the $R(G)$-module structures given above.

Proof. Everything is obvious from the preceding theorem, except for the final statement. That $K_G^0(X)$ is a unital $R(G)$-algebra is shown in proposition 2.1.4, so we need only make $K_G^1(X)$ into a $K_G^0(X)$-module and check that the $R(G)$-module structures match.

In the proof of lemma 1.2.1 (which was an elementary consequence of the parts of the present corollary which have already been proved), it was shown that $K_G^1(X)$ is isomorphic to the kernel of the map $\varphi^*: K_G^0(S^1 \times X) \longrightarrow K_G^0(X)$, where φ identifies X with $\{z\} \times X \subset S^1 \times X$ for a certain fixed $z \in S^1$. Now $K_G^0(S^1 \times X)$ is a unital ring. The unital ring homomorphism $\pi^*: K_G^0(X) \longrightarrow K_G^0(S^1 \times X)$ induced by the projection $\pi: S^1 \times X \longrightarrow X$ therefore makes $K_G^0(S^1 \times X)$ a $K_G^0(X)$-algebra. Using the fact that $\varphi^* \circ \pi^* = (\pi \circ \varphi)^*$ is the identity on $K_G^0(X)$, it is now easily shown that φ^* is a $K_G^0(X)$-algebra homomorphism. Therefore its kernel, which is $K_G^1(X)$, is a $K_G^0(X)$-module, as desired. The $R(G)$-module structures are clearly compatible. Q.E.D.

2.8.5 Remark. Using theorem 2.8.3, one can very quickly compute the equivariant K-theory of the trivial G-algebra (G,A,α), where A is any C^*-algebra and $\alpha_g(a) = a$ for all $g \in G$ and $a \in A$. It is just $K_*^G(A) \simeq R(G) \otimes K_*(A)$. Indeed, $C^*(G,A,\alpha) \simeq C^*(G) \otimes A$, and since $C^*(G)$ is a direct sum of full matrix algebras we obtain

$$K_*(C^*(G,A,\alpha)) \simeq K_0(C^*(G)) \otimes K_*(A) \simeq R(G) \otimes K_*(A).$$

The formula for the $R(G)$-module structure can now be easily used to show that this is an isomorphism of $R(G)$-modules as well. (Of course, multiplication by an element r of $R(G)$ on $R(G) \otimes K_*(A)$ is given by $r\,(s \otimes \eta) = rs \otimes \eta$.)

For later use, we state and prove the following explicit form of the Bott periodicity isomorphism.

2.8.6 Proposition. Let A be a unital G-algebra, and let X be the space $(0,1) \times (0,1)$, with the trivial G-action, so that $S^2A = C_0(X) \otimes A$. Then the Bott periodicity isomorphism $s_A^G: K_0(A) \longrightarrow K_0(S^2A)$ is given by

$$[p] \longrightarrow [e \otimes p] - [1 \otimes p], \qquad (*)$$

for G-invariant projections p in algebras $L(V) \otimes A$, where 1 is the identity in $C_0(X)^+$ and e is a certain rank 1 projection in $M_2(C_0(X)^+)$ which does not depend on A.

Proof. This is known for the non-equivariant case, but we lack an adequate reference for it so we will give some justification. (For spaces, this is found for example in [2], theorem 2.4.9.) Let us first, however, show how to deduce the equivariant version from the non-equivariant version. First, notice that if Bott periodicity is given by (*) for unital C^*-algebras, then the naturality of (*) and corollary 2.5.5 show that Bott periodicity is also given by (*) for algebras having an increasing approximate identity of projections. In particular this applies to $C^*(G,A)$. Now define $t: K_0^G(A) \longrightarrow K_0^G(S^2A)$ by $t([p]) = [e \otimes p] - [1 \otimes p]$ for $[p] \in S_G(A)$.

Consider the diagram

$$\begin{array}{ccc} K_0^G(A) & \xrightarrow{\quad\quad\quad\quad t \quad\quad\quad\quad} & K_0^G(S^2A) \\ \downarrow\simeq & & \downarrow\simeq \\ K_0(C^*(G,A)) \xrightarrow{s_{C^*(G,A)}} & K_0(S^2C^*(G,A)) \xrightarrow{\simeq} & K_0(C^*(G,S^2A)) \end{array}.$$

Since $s_{C^*(G,A)}$ has the form (*), it is easily seen that this diagram commutes. But s_A^G was defined to be the unique map which, when put in place of t, makes the diagram commute. So $s_A^G = t$, as desired.

We now prove the result in the non-equivariant case. We have to compute the image under the isomorphism $K_0(A) \simeq K_0(S^2A)$ of the class of a projection $p \in M_n(A)$. We follow [105], sections 8 and 9. (As pointed out earlier, the restriction in [105] that the algebras be commutative is unnecessary.) The first stage of the isomorphism is $K_0(A) \simeq K_1(SA)$. We identify SA^+ as the set of continuous functions f from $[0,1]$ into A such that $f(0) = f(1)$ and is a multiple of the identity. Then by [105], 9.2, the image of $[p]$ in $K_1(SA)$ is represented by $u = 1 - p + zp$, where z is the function $z(t) = e^{2\pi i t}$. The second stage is the isomorphism $K_1(SA) \simeq K_0(S^2A)$, which is the connecting homomorphism associated with the exact sequence

$$0 \longrightarrow S(SA) \longrightarrow C(SA) \longrightarrow SA \longrightarrow 0.$$

Following the procedure for computing the connecting homomorphism ([105], 8.1), we first choose a unitary $v \in M_{2n}(C(SA)^+)$ such that the image of v in SA is $\begin{pmatrix} u & 0 \\ 0 & u^* \end{pmatrix}$.

Our choice is

$$v = \begin{pmatrix} u & 0 \\ 0 & 1 \end{pmatrix} \begin{pmatrix} s & c \\ -c & s \end{pmatrix} \begin{pmatrix} 1 & 0 \\ 0 & u^* \end{pmatrix} \begin{pmatrix} s & -c \\ c & s \end{pmatrix},$$

where $c, s : [0,1] \longrightarrow \mathbf{C}$ are $c(\lambda) = \cos(\frac{\pi}{2}\lambda)$ and $s(\lambda) = \sin(\frac{\pi}{2}\lambda)$. Then the image of $[u]$ in $K_0(S^2A)$ is the difference

$$\left[v \begin{pmatrix} 1 & 0 \\ 0 & 0 \end{pmatrix} v^* \right] - \left[\begin{pmatrix} 1 & 0 \\ 0 & 0 \end{pmatrix} \right].$$

If we compute this and rewrite things in terms of functions from $[0,1]\times[0,1]$ into $M_n(A)$ which are constant on the boundary of $[0,1]\times[0,1]$, we obtain

$$v \begin{pmatrix} 1 & 0 \\ 0 & 0 \end{pmatrix} v^* = \begin{pmatrix} 1 - p + (c^4 + s^4 + s^2c^2(z + \bar{z}))p & sc\,(z-1)(zs^2 + c^2)p \\ sc\,(\bar{z}-1)(\bar{z}s^2 + c^2)p & (sc)^2(z-1)(\bar{z}-1)p \end{pmatrix},$$

where $c(t,\lambda) = \cos(\frac{\pi}{2}\lambda)$, $s(t,\lambda) = \sin(\frac{\pi}{2}\lambda)$, and $z(t,\lambda) = e^{2\pi i t}$. If we set

$$e = \begin{bmatrix} c^4+s^4+s^2c^2(z+\bar{z}) & sc\,(z-1)(zs^2+c^2) \\ sc\,(\bar{z}-1)(\bar{z}s^2+c^2) & (sc)^2(z-1)(\bar{z}-1) \end{bmatrix} \in M_2(C_0(X)^+),$$

then clearly

$$\left[v\begin{pmatrix}1&0\\0&0\end{pmatrix}v^*\right]-\left[\begin{pmatrix}1&0\\0&0\end{pmatrix}\right]=\left[\begin{pmatrix}1-p&0\\0&0\end{pmatrix}\right]+[e\otimes p]-\left[\begin{pmatrix}1&0\\0&0\end{pmatrix}\right]=[e\otimes p]-[1\otimes p],$$

as desired.

Q.E.D.

2.8.7 Remark. Just as in [3] for the case of locally compact spaces, there is actually a much stronger form of Bott periodicity, as follows: Let G be a compact Lie group, and let V be a complex representation space of G, of finite dimension n. If the representation is trivial, then $C_0(V)\otimes A \simeq S^{2n} A$ for any G-algebra A, so, from what we have proved, $K_0^G(C_0(V)\otimes A)\simeq K_0^G(A)$. But this is actually true for any unitary representation on V. There are two ways of approaching this. One is to replace the families of elliptic operators indexed by a compact space X used in [3] by the equivariant analog of the elliptic operators over C^*-algebras in [68]. The other approach, which requires ordinary elliptic operators, also uses Kasparov's KK-theory and is found in [55]. We will have something to say about it in the next chapter. We do not, however, use this result.

If A is a unital C^*-algebra then $K_1(A)$ can also be constructed in terms of unitaries in the algebras $M_n(A)$. (In [105], $K_1(A)$ is defined in terms of invertible elements in $M_n(A)$ for commutative unital Banach algebras.) We will need the analogous construction of $K_1^G(A)$ in chapter 6, so we give it here. Unfortunately it is somewhat more complicated.

First, we need some notation. Let $R_+(G)$ be the set of unitary equivalence classes of finite dimensional representations of G. (Thus, $R_+(G)$ is the subset of $R(G)$ consisting of the elements which are the classes of representations.) Then $R_+(G)$ is a directed set for the ordering $V\le W$ if V is equivalent to a direct summand of W. (We will confuse representation spaces V with their classes $[V]$ in $R_+(G)$ for the purposes of this construction.) If A is a unital G-algebra and $V\in R_+(G)$, we let $U^G(V,A)$ be the group of G-invariant unitary elements in $L(V)\otimes A$, and we let $U_0^G(V,A)$ be the connected component of the identity in $U^G(V,A)$. Let $\bar{U}^G(V,A)=U^G(V,A)/\,U_0^G(V,A)$. Finally, if (G,A,α) is any G-algebra, then we denote by A^G the C^*-subalgebra of A consisting of the fixed points for the action of G on A. That is,

$$A^G = \{a\in A : \alpha_g(a) = a \text{ for all } g\in G\}.$$

Let $V, W\in R_+(G)$ with $V\le W$. Then $W\simeq V\oplus V_0$ for some $V_0\in R_+(G)$. We define an embedding $U^G(V,A)\longrightarrow U^G(W,A)$ by $u\longrightarrow u\oplus 1_{L(V_0)\otimes A}$. This map is clearly a group homomorphism, and it defines a group homomorphism $i_{V,W}\colon \bar{U}^G(V,A)\longrightarrow \bar{U}^G(W,A)$. We now claim that $i_{V,W}$ does not depend on the choice of the isomorphism $W\simeq V\oplus V_0$.

Indeed, V_0 is determined up to unitary equivalence by V and W. If r_0 and r_1 are any two G-invariant unitary operators from $V \oplus V_0$ to W, then $r_1^* r_0$ is a G-invariant unitary in $L(W)$. Since $L(W)^G$ is finite dimensional, its unitary group is connected. A homotopy from $r_1^* r_0$ to 1 can clearly be used to produce a homotopy $t \longrightarrow r_t$ for $t \in [0,1]$ of G-invariant unitaries from r_0 to r_1. The two possible maps from $U^G(V,A)$ to $U^G(W,A)$ are given by $u \longrightarrow r_t(u \oplus 1_{L(V_0)\otimes A})r_t^*$ for $t = 0,1$. These are obviously homotopic, and therefore define the same map from $\overline{U}^G(V,A)$ to $\overline{U}^G(W,A)$. Thus, the claim has been proved. It is now clear that if V, W, $X \in R_+(G)$ with $V \leq W \leq X$, then $i_{W,X} \cdot i_{V,W} = i_{V,X}$. We have therefore constructed a directed system of groups.

2.8.8 Theorem. Let (G,A,α) be a G-algebra. Then there is a natural isomorphism
$$K_1^G(A) \simeq \varinjlim_{V \in R_+(G)} \overline{U}^G(V,A^+).$$

We point out that $\varinjlim \overline{U}^G(V,A^+)$ is indeed a functor on G-algebras: if $\varphi : A \longrightarrow B$ is equivariant, then φ^+ defines, for every $V \in R_+(G)$, a group homomorphism from $U^G(V,A^+)$ to $U^G(V,B^+)$ which maps $U_0^G(V,A^+)$ into $U_0^G(V,B^+)$. Therefore φ^+ defines a homomorphism from $\overline{U}^G(V,A^+)$ to $\overline{U}^G(V,B^+)$. These homomorphisms clearly commute with the maps $i_{V,W}$. Thus, the direct limit is a functor.

We also point out that, if A is unital, then $\varinjlim \overline{U}^G(V,A^+) = \varinjlim \overline{U}^G(V,A)$. Indeed, in this case $A^+ = A \oplus \mathbf{C}$, whence $\overline{U}^G(V,A^+) = \overline{U}^G(V,A) \oplus \overline{U}^G(V,\mathbf{C})$. But $\overline{U}^G(V,\mathbf{C}) = \{1\}$ because $U^G(V,\mathbf{C})$, being the unitary group of the finite dimensional C^*-algebra $L(V)^G$, is connected.

A complete proof of this theorem seems to require the methods of section 8 of [105], adapted to G-algebras and using unitaries rather than invertible elements. This proof would be fairly long, and would use a fair amount of the material that has been avoided by deriving periodicity from Julg's theorem. Since we only use this result for finite abelian groups, we will only prove it in that case, and the proof is much simpler. First, we need the following special case, which is well known but for which I have been unable to find a reference.

2.8.9 Lemma. If A is a unital C^*-algebra, then theorem 2.8.8 holds for the trivial group $G = \{e\}$, with A^+ replaced by A.

Proof. Of course, here $R_+(G) = \mathbf{Z}^+$, the set of positive integers, with its usual ordering. Also, $U^G(V,A)$ is just the unitary group of the C^*-algebra $M_n(A)$, where $n = \dim V$. We will write it as $U_n(A)$, and its connected component of the identity as $U_n^0(A)$. Then we must show that $K_1(A) \simeq \varinjlim U_n(A)/U_n^0(A)$. Now $K_1(A) \simeq \varinjlim GL_n(A)/GL_n^0(A)$ (see [105], section 5, or [26], section 2), where $GL_n(A)$ is the group of invertible elements in $M_n(A)$, and $GL_n^0(A)$ is its connected component of the identity. The maps are defined analogously to the maps for the groups $U_n(A)/U_n^0(A)$. It is therefore enough to show that $U_n^0(A) = GL_n^0(A) \cap U_n(A)$ and that

$U_n(A) \cdot GL_n^0(A) = GL_n(A)$. (This will enable one to use the standard isomorphism theorems of groups to produce a canonical isomorphism $U_n(A)/U_n^0(A) \simeq GL_n(A)/GL_n^0(A)$.) In both cases, only the inclusion $\supset$ is nontrivial. For the first relation, let x_t be a continuous path of invertible elements with $x_0 = 1$ and x_1 unitary. Then $u_t = x_t(x_t^* x_t)^{-1/2}$ is a continuous path of unitaries with $u_0 = 1$ and $u_1 = x_1$. For the second relation, let $x \in GL_n(A)$. Then $x = x(x^*x)^{-1/2}(x^*x)^{1/2}$, where $x(x^*x)^{-1/2} \in U_n(A)$, and $(x^*x)^{1/2} \in GL_n^0(A)$ because of the continuous path $t \longrightarrow (x^*x)^{1/2 t}$ connecting it to 1 in $GL_n(A)$. Q.E.D.

Proof of theorem 2.8.8, in case G is finite abelian. The direct limit is unchanged if $R_+(G)$ is replaced by a cofinal subset. We choose the set S of all V_0^n for $n \in \mathbf{Z}^+$, where V_0 is the space of the regular representation of G, and V_0^n is the direct sum of n copies of V_0. (That S is cofinal in $R_+(G)$ follows from the fact that every irreducible representation of G occurs in V_0 with positive multiplicity.) Now $L(V_0^n)\otimes A^+ \simeq L(V_0)\otimes M_n(A^+)$. By Takai duality ([79], 7.9.3), there is an equivariant isomorphism

$$C^*(\hat{G}, C^*(G, M_n(A^+))) \simeq L(V_0)\otimes M_n(A^+),$$

where the double crossed product carries the double dual action of $(\hat{G})\hat{} = G$. It is easily seen that the set of fixed points for this action of G is $C^*(G, M_n(A^+))$, which is isomorphic to $M_n(C^*(G,A^+))$. Therefore there is a canonical isomorphism

$$U_n(C^*(G,A^+)) \simeq U_n((L(V_0)\otimes M_n(A^+))^G) = U^G(V_0^n, A^+).$$

Consequently,

$$K_1^G(A^+) \simeq K_1(C^*(G,A^+)) \simeq \varinjlim_n U_n(C^*(G,A^+))/U_n^0(C^*(G,A^+))$$

$$\simeq \varinjlim_n \bar{U}^G(V_0^n, A^+) \simeq \varinjlim_{V \in R_+(G)} \bar{U}^G(V, A^+).$$

It remains only to prove that $K_1^G(A^+) \simeq K_1^G(A)$. This follows immediately from consideration of the split equivariant exact sequence

$$0 \longrightarrow A \longrightarrow A^+ \longrightarrow \mathbf{C} \longrightarrow 0,$$

together with the fact that $K_1^G(\mathbf{C}) = 0$. Q.E.D.

For later use, we give the following description of the connecting homomorphisms in the long exact sequence of 2.8.3 (1). We note that this description can be combined with the proof of proposition 2.8.6 to give an alternate proof that the connecting homomorphisms are $R(G)$-module homomorphisms.

2.8.10 Lemma. The connecting homomorphisms associated to an equivariant short exact sequence are composites of suspension isomorphisms, homomorphisms induced by equivariant homomorphisms of G-algebras, and their inverses.

Proof. If $\varphi : A \longrightarrow B$ is an equivariant map of G-algebras, we define its mapping cone $C\varphi$ to be the G-algebra given by

$$C\varphi = \{ (a, f) \in A \oplus CB : \varphi(a) = f(1) \},$$

where $CB = C_0((0,1]) \otimes B$ as in the proof of theorem 2.8.3 (2), and we regard CB as an algebra of continuous functions from $(0,1]$ to B. This definition is essentially the same as definition 2.1 of [96], except that we use $(0,1]$ rather than $[0,1)$. (The change is required for consistency with our earlier definition of CB, which was taken from [105], section 8.)

Let

$$0 \longrightarrow I \xrightarrow{\varphi} A \xrightarrow{\pi} B \longrightarrow 0$$

be an equivariant exact sequence of G-algebras, and consider the following diagram:

$$\begin{array}{ccccccccc} 0 \longrightarrow & SB & \xrightarrow{\psi} & CB & \xrightarrow{ev_1} & B & \longrightarrow 0 \\ & \uparrow \lambda & & \uparrow C(\pi) & & \uparrow = & \\ 0 \longrightarrow & C\varphi & \xrightarrow{p_2} & CA & \xrightarrow{\kappa} & B & \longrightarrow 0 \\ & \downarrow p_1 & & \downarrow ev_1 & & \downarrow = & \\ 0 \longrightarrow & I & \xrightarrow{\varphi} & A & \xrightarrow{\pi} & B & \longrightarrow 0 . \end{array} \qquad (*)$$

The maps are defined as follows: ev_1 is evaluation of functions at the point $1 \in (0,1]$; ψ is the obvious inclusion of SB in CB; p_1 and p_2 are the projections on the first and second summands of $C\varphi$; $\lambda : C\varphi \longrightarrow SB$ is $\lambda(j, f) = \pi \cdot f$, which is in SB because $f(1) \in I$ implies that $\pi(f(1)) = 0$; $C(\pi)$ is the obvious map coming from π; and finally $\kappa : CA \longrightarrow B$ is defined by $\kappa(f) = \pi(f(1))$. All these maps are equivariant, and routine calculations show that the diagram commutes and has exact rows.

The homomorphism λ is surjective and satisfies

$$Ker(\lambda) = \{ (j, f) \in C\varphi : f(t) \in I \text{ for all } t \} \simeq CI .$$

Furthermore, CI is equivariantly contractible, so $K_*^G(CI) = 0$. The long exact sequence

$$\cdots \longrightarrow K_i^G(CI) \longrightarrow K_i^G(C\varphi) \xrightarrow{\lambda_*} K_i^G(SB) \longrightarrow K_{1-i}^G(CI) \longrightarrow \cdots$$

therefore implies that λ_* is an isomorphism.

Apply K_*^G to the diagram (*), and let ∂_j be the connecting homomorphism for the j-th row. Then we obtain the following diagram, which commutes since connecting homomorphisms are natural:

$$\begin{array}{ccc} K_i^G(B) & \xrightarrow{\partial_1} & K_{1-i}^G(SB) \\ =\uparrow & & \uparrow \lambda_* \\ K_i^G(B) & \xrightarrow{\partial_2} & K_{1-i}^G(C\varphi) \\ =\downarrow & & \downarrow (p_1)_* \\ K_i^G(B) & \xrightarrow{\partial_3} & K_{1-i}^G(I). \end{array}$$

The map ∂_1 is a suspension isomorphism (see the proof of theorem 2.8.3 (2)), and λ_* is invertible. Therefore we obtain $\partial_3 = (p_1)_* \cdot (\lambda_*)^{-1} \cdot \partial_1$, which has the desired form.

Q.E.D.

2.9. The Equivariant K-Theory of Hereditary Subalgebras and Twisted Products

In this section, we treat two topics which will be needed later. The first is the equivariant K-theory of G-invariant hereditary subalgebras, and the second is the C^* analog of the space $G \times_H X$, where X is an H-space and H is a closed subgroup of G. As usual, we assume G is compact.

Recall that a closed subalgebra B of a C^*-algebra A is said to be hereditary if whenever $0 \le a \le b$ and $b \in B$ then $a \in B$. (We assume here that hereditary subalgebras are closed, unlike in section 6, where a certain dense hereditary subalgebra played a crucial role.) From [79], 1.5.2, it follows that a hereditary subalgebra is characterized by the property $\overline{BAB} = B$ (the notation means the closed linear span of all products $b_1 a b_2$, for $a \in A$ and $b_1, b_2 \in B$). Indeed, if B is a hereditary subalgebra, then $B = L \cap L^*$ for a left ideal L, and we have

$$B = \overline{B^3} \subset \overline{L^*AL} \subset \overline{L^*L} \subset L \cap L^* = B,$$

where the first equality follows from the existence of an approximate identity in B. On the other hand, a closed subalgebra B satisfies $\overline{BAB} = \overline{AB} \cap (\overline{AB})^*$ (the inclusion $\supset$ follows from the existence of an approximate identity in B), so that if $B = \overline{BAB}$ then B is hereditary.

We further recall ([15]) that a hereditary subalgebra B of A is called full if it is not contained in any proper two-sided ideal of A.

2.9.1 Proposition. Let (G,A,α) be a separable G-algebra, and let B be a G-invariant full hereditary subalgebra of A. Let $i : B \longrightarrow A$ be the inclusion. Then $i_* : K_*^G(B) \longrightarrow K_*^G(A)$ is an isomorphism.

Proof. This is true without the group, by proposition 1.2 of [77], or by a slight modification of proposition 2.10 of [15] so as to apply to K-theory rather than *Ext*. In view of the isomorphism $K_*^G(A) \simeq K_*(C^*(G,A))$, it is therefore sufficient to show that $C^*(G,B)$ is a full hereditary subalgebra in $C^*(G,A)$.

To show it is hereditary, the remarks before the proposition actually imply that it is sufficient to prove

$$C^*(G,B)\, C^*(G,A)\, C^*(G,B) \subset C^*(G,B),$$

since the reverse inclusion of the closures always holds. In fact, we need only consider the L^1 algebras. So let $a, b \in L^1(G,B)$ and let $x \in L^1(G,A)$. Then

$$(axb)(g) = \int_{G\times G} a(h)\, \alpha_h(x(k))\, \alpha_{hk}(b(k^{-1}h^{-1}g))\, dh\, dk\,,$$

and $axb \in L^1(G,B)$ since the integrand is in $\overline{BAB} \subset B$ for all values of g.

Now let $I \subset C^*(G,A)$ be an ideal containing $C^*(G,B)$. Then I is also an ideal in the multiplier algebra of $C^*(G,A)$. Let $a \in A$. Then a can be regarded as a multiplier of $C^*(G,A)$ ([79], 7.6.3), so that for $f \in L^1(G,B)$, the functions $g \longrightarrow af(g)$ and $g \longrightarrow f(g)\alpha_g(a)$ are in I. We want to show that the function $g \longrightarrow f(g)a$ is also in I.

Let $f \in C(G,B)$, and let $\varepsilon > 0$. Let $U_1, \ldots, U_n$ be an open cover of G such that $\| \alpha_g^{-1}(a) - \alpha_h^{-1}(a) \| < \varepsilon$ whenever g and h are both in the same U_k. Let $\varphi_1, \ldots, \varphi_n$ be a partition of unity on G, subordinate to $U_1, \ldots, U_n$. Let $a_k = \alpha_{g_k}^{-1}(a)$ for some $g_k \in U_k$. Then the function

$$g \longrightarrow \sum_{k=1}^{n} (\varphi_k f)(g)\, \alpha_g(a_k)$$

is in I, and for all $g \in G$,

$$\| f(g)a - \sum_{k=1}^{n} \varphi_k(g) f(g)\, \alpha_g(a_k) \| \le \sum_{k=1}^{n} \varphi_k(g)\, \| f(g) \|\ \| a - \alpha_{gg_k^{-1}}(a) \|$$

$$\le \| f \|_\infty \sum_{k=1}^{n} \varphi_k(g)\, \| \alpha_g^{-1}(a) - \alpha_{g_k}^{-1}(a) \| .$$

Since $\| \alpha_g^{-1}(a) - \alpha_{g_k}^{-1}(a) \| < \varepsilon$ whenever $\varphi_k(g) \neq 0$, and $\sum_{k=1}^{n} \varphi_k(g) = 1$, we find that this expression is less than $\varepsilon \| f \|_\infty$ for all $g \in G$. Since the sup norm on $C(G,A)$ dominates the norm it inherits as a subset of $C^*(G,A)$, and since I is a closed ideal in $C^*(G,A)$, it follows that the function $g \longrightarrow f(g)a$ is in I, as desired.

As a special case of what we have just proved, if $f \in C(G,B)$ has the form $f(g) = f_0(g)b$, for some $f_0 \in C(G)$ and some $b \in B$, we find that the function $g \longrightarrow f(g)a_1 b a_2$ is in I for every $a_1, a_2 \in A$. Since $\overline{ABA} = A$ (B being full), such functions span a sup norm dense subset of $C(G,A)$. Since $C(G,A)$ is dense in $C^*(G,A)$, we conclude that $I = C^*(G,A)$, and so $C^*(G,B)$ is full as desired. Q.E.D.

The other topic in this section is twisted products, to which we now turn. If G is a compact group, H is a closed subgroup of G, and X is an H-space, then we can define an action of H on $G\times X$ by $h\cdot(g,x)=(gh^{-1},hx)$ for $h\in G$. Then the twisted product of G and X over H, which we denote by $G\times_H X$, is the orbit space $(G\times X)/H$. It becomes a G-space in a natural way when, with $[g,x]$ denoting the image of (g,x) in $(G\times X)/H$, we define $g_0\cdot[g,x]=[g_0g,x]$ for $g,g_0\in G$ and $x\in X$. One reason for the usefulness of this construction is that $K_G^*(G\times_H X)$ is isomorphic to $K_H^*(X)$. (See for example [98], page 132, example (iii).) The analogous construction for C^*-algebras is given in the following definition. Before stating it, let us recall that if $\alpha: G\longrightarrow Aut(A)$ is an action of an arbitrary group G on a C^*-algebra A, then the set of fixed points for the action is denoted by A^G. It is a C^*-subalgebra of A, and it is unital if A is. Returning to our standing assumption that G is compact, we easily see that if X is a locally compact G-space then $C_0(X)^G$ is naturally isomorphic to $C_0(X/G)$. (In the general case, $C_0(X)^G$ will be the set of continuous functions vanishing at infinity on the locally compact but not necessarily Hausdorff space X/G.)

2.9.2 Definition ([56], section 5, definition 2). Let H be a closed subgroup of G, and let (H,A,α) be an H-algebra. Then the twisted product of G and A is the G-algebra $(G,C(G\times_H A),\beta)$ constructed as follows. Define an action $\rho: H\longrightarrow Aut(C(G,A))$ by $\rho_h(f)(g)=\alpha_h(f(gh))$ for $f\in C(G,A)$, $g\in G$, and $h\in H$. Then $C(G\times_H A)$ is defined to be $C(G,A)^H$, and the action β is defined by $\beta_g(f)(g_0)=f(g^{-1}g_0)$ for $g,g_0\in G$ and $f\in C(G,A)^H$. (One readily checks that $\beta_g(f)$ is again in $C(G,A)^H$.)

2.9.3 Remark. A routine calculation now shows that if X is an H-space then $C(G\times_H C_0(X))$ is naturally isomorphic to $C_0(G\times_H X)$ as G-algebras.

For the following proposition, we regard an $R(H)$-module (where H is a closed subgroup of G) as an $R(G)$-module via the restriction map $R(G)\longrightarrow R(H)$, which is a ring homomorphism defined by regarding a representation space of G as a representation space of its subgroup H.

2.9.4 Proposition. Let H be a closed subgroup of G, and let (H,A,α) be an H-algebra. Then there is a natural isomorphism of $R(G)$-modules $K_*^G(C(G\times_H A))\simeq K_*^H(A)$.

Proof. For the usual reasons, it is sufficient to prove this for K_0 and for A unital. Define $\varphi: C(G\times_H A)\longrightarrow A$ by $\varphi(f)=f(e)$, where e is the identity in G.

Let V be a representation space of G, and let V_H be the same space regarded as a representation space of H. We claim that $id_{L(V)}\otimes\varphi$ maps $(L(V)\otimes C(G\times_H A))^G$ into $(L(V_H)\otimes A)^H$. To prove the claim, let λ be the action of G on $L(V)$. Also regard elements of $L(V)\otimes C(G\times_H A)$ as functions from G into $L(V)\otimes A$; a function $f: G\longrightarrow L(V)\otimes A$ is in $L(V)\otimes C(G\times_H A)$ exactly when $f(g)=(id_{L(V)}\otimes\alpha_h)(f(gh))$ for $g\in G$ and $h\in H$. If f is G-invariant, then in particular,

$$f(e)=(\lambda_h\otimes id_A)(f(h^{-1}))=(\lambda_h\otimes\alpha_h)(f(e))$$

for $h \in H$, and $f(e) \in (L(V_H)\otimes A)^H$, as desired.

By considering equivariant K-theory in terms of invariant projections, it follows immediately that φ defines a group homomorphism φ_* from $K_0^G(C(G\times_H A))$ to $K_0^H(A)$. Furthermore, if W is a finite dimensional representation space of G, then $[W]\cdot[p]=[1_{L(W)}\otimes p]$, and $\varphi(1_{L(W)}\otimes p)$ is obviously equal to $1_{L(W_H)}\otimes p$. Therefore φ_* is an $R(G)$-module homomorphism.

To show that φ_* is bijective, we are going to need the following fact: every finite dimensional representation space W of H is a direct summand of a space V_H for some finite dimensional representation space V of G. It will be convenient to work in terms of representations, so let w be the representation of H on W, which we may clearly take to be irreducible. Let λ_H be the left regular representation of H on $L^2(H)$, and let λ_G be the left regular representation of G on $L^2(G)$. Then w is equivalent to a subrepresentation of λ_H, and $\lambda_G|_H$ is a multiple (infinite unless H has finite index) of λ_H. Write λ_G as a direct sum of irreducible, and hence finite dimensional, subrepresentations λ_i. Since w is irreducible and equivalent to a subrepresentation of $\lambda_G|_H$, it is equivalent to a subrepresentation of some $\lambda_i|_H$, and we take V to be the space of λ_i.

Now let V be a finite dimensional representation space of G, let λ be the action of G on $L(V)$, and define a homomorphism

$$\psi: (L(V_H)\otimes A)^H \longrightarrow (L(V)\otimes C(G\times_H A))^G$$

by $\psi(a)(g)=(\lambda_g\otimes id_A)(a)$. We check that $\psi(a)\in L(V)\otimes C(G\times_H A)$: for $h\in H$ and $g\in G$, we have

$$(id_{L(V)}\otimes\alpha_h)(\psi(a)(gh)) = (\lambda_{gh}\otimes\alpha_h)(a) = (\lambda_g\otimes id_A)(a) = \psi(a)(g),$$

since $(\lambda_h\otimes\alpha_h)(a)=a$. Also, we check that $\psi(a)$ is fixed under the action of G: for $g, g_0\in G$, we have

$$\begin{aligned}(\lambda_g\otimes\beta_g)(\psi(a))(g_0) &= (\lambda_g\otimes id_A)(\psi(a)(g^{-1}g_0))\\ &= (\lambda_{g_0}\otimes id_A)(a) = \psi(a)(g_0).\end{aligned}$$

We now observe that ψ is an inverse for φ on invariant elements: $\varphi(\psi(a)) = (\lambda_e\otimes id_A)(a)=a$, for $a\in(L(V_H)\otimes A)^H$, and if $f\in(L(V)\otimes C(G\times_H A))^G$, then

$$\begin{aligned}\psi\cdot\varphi(f)(g) &= (\lambda_g\otimes id_A)(f(e))\\ &= (\lambda_g\otimes id_{C(G\times_H A)})(f)(e) = (id_{L(V)}\otimes\beta_g^{-1})(f)(e) = f(g).\end{aligned}$$

Here the third step follows from the G-invariance of f and the last step follows from the fact that $f\in C(C\times_H A)$.

Now it is easy to prove that φ_* is bijective. Let $p\in L(W)\otimes A$ be an H-invariant projection. Replacing W by a larger space, we may assume $W=V_H$ for some V. Then $[p]=\varphi_*([\psi(p)])$. Similarly, let $p, q\in L(V)\otimes C(G\times_H A)$ be two G-invariant projections such that $\varphi(p)$ and $\varphi(q)$ are unitarily equivalent. Then there is an H-invariant unitary $u \in L(V_H) \otimes A$ such that $u\varphi(p)u^* = \varphi(q)$. Applying ψ to both sides of this equation yields $\psi(u)p\,\psi(u)^* = q$, so p and q are unitarily equivalent. Thus φ_* is an isomorphism as required. Q.E.D.

2.9.5 Remark. One could presumably also approach this as follows: The methods of [88] should yield a strong Morita equivalence between $C^*(H,A)$ and $C^*(G, C(G \times_H A))$. (Situation 4 of [88] is exactly the case in which A is commutative.) If A and G are separable, then it follows from [17] that the algebras are stably isomorphic and hence have the same K-theory. However, it is not clear how to trace the $R(G)$-module structure in this approach.

We will need to be able to recognize a G-algebra of the form $C(G \times_H A)$ in certain situations, so we also prove the following proposition, which deals with an especially simple case:

2.9.6 Proposition. Let H be a closed subgroup of G of finite index, and let R be a set of left coset representatives for H in G. Let (G,A,α) be a G-algebra containing an H-invariant ideal I such that

$$A = \bigoplus_{g \in R} \alpha_g[I].$$

(That is, α has a system of imprimitivity based on G/H.) Then A is equivariantly isomorphic to $C(G \times_H I)$.

Proof. One of the ideals $\alpha_g[I]$ is just I. Let π be the projection from A onto I associated with the direct sum decomposition. Define $\varphi : A \longrightarrow C(G \times_H I)$ by $\varphi(a)(g) = \pi(\alpha_g^{-1}(a))$. Let $g \in G$ and $h \in H$. Then

$$\alpha_h(\pi(\alpha_{gh}^{-1}(a))) = \pi(\alpha_g^{-1}(a)),$$

since π commutes with α_h. This proves that the image of φ really is in $C(G \times_H I)$. That φ is a homomorphism is now obvious. We will prove that φ is the required isomorphism.

Injectivity of φ: obvious.

Surjectivity of φ: Let $f \in C(G \times_H I)$, and set $a = \sum_{g \in R} \alpha_g(f(g))$. Let $k \in G$, and let $g \in R$ be the unique element with $k^{-1}g \in H$. Then

$$\varphi(a)(k) = \pi(\alpha_{k^{-1}g}(f(g)) = \alpha_{k^{-1}g}(f(g)) = f(k).$$

The first step is justified by the fact that π is zero on the other terms, while the last step follows from $f \in C(G \times_H I)$. Therefore $\varphi(a) = f$, and φ is surjective.

Equivariance of φ: Let β be the action of G on $C(G \times_H I)$. Then we have, for $g, k \in G$,

$$\varphi(\alpha_g(a))(k) = \pi(\alpha_{k^{-1}g}(a)) = \beta_g(\varphi(a))(k),$$

as desired. Q.E.D.

Although proposition 2.9.4 is the last result about equivariant K-theory that we prove in this chapter, we should at least mention a few other results. In Chapter 5 we prove some results about the relation between $K_*^G(A)$ and $K_*^H(A)$, where H is a subgroup of a finite group G. In Chapter 6 we examine a pairing

$$K_*^G(A) \times K_*^G(B) \longrightarrow K_*^G(A \otimes B)$$

in connection with proving Künneth theorems for certain localized theories $K_*^G(\ )_P$. Finally, we neither use nor prove the following Mayer-Vietoris sequence (compare [96], theorem 4.5), which is nevertheless worthy of mention: if

$$\begin{array}{ccc} A & \longrightarrow & B \\ \downarrow & & \downarrow \\ C & \longrightarrow & D \end{array}$$

is an equivariant pullback diagram of G-algebras, and one of the two maps to D is surjective, then there is a long exact sequence

$$\begin{array}{ccccc} K_0^G(A) & \longrightarrow & K_0^G(B) \oplus K_0^G(C) & \longrightarrow & K_0^G(D) \\ \uparrow & & & & \downarrow \\ K_1^G(D) & \longleftarrow & K_1^G(B) \oplus K_1^G(C) & \longleftarrow & K_1^G(A). \end{array}$$

In fact, most of [96] applies equally well to the equivariant situation.

Chapter 3

Introduction to Equivariant KK-Theory

In this chapter, we define the equivariant KK-theory of C^*-algebras, first introduced by Kasparov in [55]. Kasparov's KK-theory is a functor of two variables which simultaneously generalizes K-theory and the Brown-Douglas-Fillmore Ext-theory ([16]). While Kasparov's definition is a generalization of Atiyah's construction in [4] using elliptic operators, we follow the approach of Cuntz in [25], where KK-theory is obtained as generalization of K-theory. We prefer Cuntz's approach because we find his construction more convenient for our applications of KK-theory. The nature of the product is also intuitively much clearer in his approach. We fill in many of the details which have been omitted in [25], and also make the essentially trivial modifications needed to handle the equivariant case, which is not considered there.

There are essentially no examples in this chapter. We do show that $KK_G^*(\mathbf{C},A)$ is isomorphic to $K_*^G(A)$, and homotopy invariance, stability, Bott periodicity, and exact sequences enable one to do somewhat more. Indeed, with the aid of several further general theorems proved there, we will be able to give in chapter 9 a general procedure for computing equivariant KK-theory for AF algebras, provided that the group actions satisfy a certain technical condition. If G is trivial then under certain circumstances there are universal coefficient theorems which express $KK^*(A,B)$ in terms of $K_*(A)$ and $K_*(B)$ ([91], [92]). However, $KK_G^*(A,B)$ is very hard to compute in general. In fact, in the uses we make of KK-theory we do not actually compute a single nontrivial example of a KK-group — we merely show that certain localizations $KK_G^*(A,B)_P$ at certain prime ideals P in $R(G)$ are or are not zero.

Section 1 of this chapter contains some necessary preliminaries on the algebra $K(H)\otimes A$ and lifting of derivations. We also introduce the G-multiplier algebra, consisting of the elements in the multiplier algebra on which G acts continuously. It replaces the multiplier algebra in situations involving group actions. In section 2, we define the basic objects of KK-theory, namely prequasihomomorphisms and quasihomomorphisms. These should be regarded as formal differences of homomorphisms. We then define the $R(G)$-modules $KK_G(A,B)$.

Section 3 is devoted to extendible quasihomomorphisms. These are needed for the definition of the product in KK-theory, and the main result of this section is a rather technical proposition to the effect that sufficiently many extendible quasihomomorphisms exist. In section 4, we then take up the definition of the product. We also give a second construction of it, which is more useful for computations but is unsatisfactory for proving such results as the fact that the product is well defined. Using this second construction, we prove what is perhaps the central theorem of KK-theory, the associativity of the product. In the next section, we give a more general form of the product (considered in [55] but not in [25]). We than use it, together with an adaptation of a trick of Atiyah (section 1 of [3]), to prove stability and Bott periodicity for KK-theory.

In the last section, we define the generalized homology and cohomology theory of two variables, KK_G^*, and give its basic properties and its relationship to equivariant K-theory. We do not give proofs for the long exact sequences, but instead refer to the papers of Kasparov ([55]) and Cuntz and Skandalis ([27]).

Throughout this chapter, G is a compact group. In all results pertaining to equivariant KK-theory, we assume that G is second countable. (This assumption is needed to prove that $KK_G^0(\mathbf{C}, A) \simeq K_0^G(A)$.) We consider only G-algebras; G-Banach algebras will not be mentioned. From the third section on, we assume that all G-algebras are separable.

3.1. Preliminaries

We will need to make extensive use of multiplier algebras in this chapter. As indicated in the discussion at the beginning of section 2.7, the G-action on the multiplier algebra of a G-algebra need not be continuous. We are therefore led to the following definition.

3.1.1 Definition. Let (G,A,α) be a G-algebra. We define the G-multiplier algebra of A to be

$$M_G(A) = \{a \in M(A) : g \longrightarrow \alpha_g(a) \text{ is continuous for the norm topology on } M(A)\}.$$

It is easily seen that $M_G(A)$ is a norm closed G-invariant *-subalgebra of $M(A)$, containing the unit of $M(A)$, and that the action of G on it is continuous. Of course, if G is discrete then $M_G(A) = M(A)$.

3.1.2 Proposition. (Compare [25], proposition 1.1.) Let I be a G-invariant ideal in a G-algebra (G,A,α), and let $\varphi : I \longrightarrow B$ be a surjective equivariant homomorphism to another G-algebra (G,B,β). Then φ extends uniquely to an equivariant homomorphism $\Phi : A \longrightarrow M_G(B)$. If I is essential in A (that is, has a nonzero intersection with every other nonzero ideal of A), and φ is an isomorphism, then Φ is injective.

Proof. If $a \in A$, we let $\Phi(a)$ be the multiplier (ρ, λ) defined by $\rho(\varphi(x)) = \varphi(xa)$ and $\lambda(\varphi(x)) = \varphi(ax)$. Since $Ker\,\varphi$, being an ideal in I, is also an ideal in A, it follows that ρ and λ are well defined. It is easy to verify that Φ is a homomorphism. Since Φ must then be continuous, it follows that its image must actually be in $M_G(B)$, rather than just $M(B)$.

For the second assertion, let $x \in A$, $x \neq 0$. Then $\overline{AxAI} = \overline{AxA} \cap I \neq 0$, whence $xI \neq 0$ (because $xAI \subset xI$). The injectivity of φ now implies that $\Phi(x) \neq 0$. Q.E.D.

3.1.3 Corollary. Let (G,A,α) be a G-algebra, and let B be a G-invariant C^*-subalgebra of $M_G(A)$ containing A. Then $M_G(B)$ can be canonically identified with a subalgebra of $M_G(A)$.

Proof. We first show that A is an essential ideal in B. Let J be an ideal in B, with $J \cap A = 0$. Then $JA = 0$, and since J is a subset of $M(A)$ it follows that $J = 0$. Now we apply the proposition to the identity map from A to A. Q.E.D.

3.1.4 Lemma. (Compare [79], 3.12.12.) Let (G,A,α) be a G-algebra, let $D \subset A$ be a G-invariant subalgebra, and for $i,j = 1,2$, let R_{ij} be a G-invariant subset of A. Suppose that $D \subset \overline{DR_{11}D} = R_{11}$, $\overline{DR_{12}} = R_{12}$, $\overline{R_{21}D} = R_{21}$, and that the subset B of $M_2(A)$ consisting of all 2×2 matrices (x_{ij}), where $x_{ij} \in R_{ij}$, is a C^*-algebra. Then the inclusion map $D \longrightarrow B$ given by $a \longrightarrow \begin{pmatrix} a & 0 \\ 0 & 0 \end{pmatrix}$ extends to an equivariant inclusion $\varphi : M_G(D) \longrightarrow M_G(B)$, such that

$$\varphi(a)\begin{pmatrix} x_{11} & x_{12} \\ x_{21} & x_{22} \end{pmatrix} = \varphi(a)\begin{pmatrix} x_{11} & x_{12} \\ 0 & 0 \end{pmatrix} \quad \text{and} \quad \begin{pmatrix} x_{11} & x_{12} \\ x_{12} & x_{22} \end{pmatrix}\varphi(a) = \begin{pmatrix} x_{11} & 0 \\ x_{21} & 0 \end{pmatrix}\varphi(a),$$

for $x = (x_{ij}) \in B$ and $a \in M_G(D)$.

Proof. We follow [79], 3.12.12, and regard $M(D)$ as a subset of the second dual D'' of D, and $M(B)$ as a subset of B''. Then we certainly have an injection $\varphi : M(D) \longrightarrow B''$, given by $\varphi(a) = \begin{pmatrix} a & 0 \\ 0 & 0 \end{pmatrix}$, where B'' is regarded as a subset of $M_2(A'')$. Let $a \in M(D)$, let $x = (x_{ij}) \in B$, and let (e_λ) be an approximate identity for D. Then the hypotheses imply that $e_\lambda x_{11} \longrightarrow x_{11}$, $x_{11}e_\lambda \longrightarrow x_{11}$, $e_\lambda x_{12} \longrightarrow x_{12}$, and $x_{21}e_\lambda \longrightarrow x_{21}$. Therefore

$$\varphi(a)x = \begin{pmatrix} ax_{11} & ax_{12} \\ 0 & 0 \end{pmatrix} = \lim_\lambda \begin{pmatrix} ae_\lambda x_{11} & ae_\lambda x_{12} \\ 0 & 0 \end{pmatrix},$$

which is an element of B because $ae_\lambda \in D$. Similarly $x\varphi(a) = \begin{pmatrix} x_{11}a & 0 \\ x_{12}a & 0 \end{pmatrix} \in B$. We have shown that $\varphi[M(D)] \subset M(B)$ and it follows from the continuity of φ, together with the obvious fact that it is equivariant, that $\varphi[M_G(D)] \subset M_G(B)$. Q.E.D.

3.1.5 Convention. From now through the end of this chapter, K will be the algebra of compact operators on a separable Hilbert space, carrying the G-action described as follows. Let l^2 be the Hilbert space of square summable sequences, with the trivial representation of G. Let $L^2(G)$, taken with respect to Haar measure on G, carry the left regular representation λ, and let $H = l^2 \otimes L^2(G)$ carry the diagonal representation $u_g(\xi \otimes \eta) = \xi \otimes \lambda_g(\eta)$. Finally, let $K = K(H)$ carry the inner G-action $\kappa_g(a) = u_g a u_g^*$. For any G-algebra (G,A,α), the algebra $K \otimes A$ will always be assumed to carry the diagonal action, given by $(\kappa \otimes \alpha)_g(k \otimes a) = \kappa_g(k) \otimes \alpha_g(a)$.

3.1.6 Remark. The important thing about the representation u is that every irreducible representation of G occurs infinitely often as a direct summand of u. It is thus a universal representation of G on a separable Hilbert space, in the sense that any representation of G on a separable Hilbert space is equivalent to a subrepresentation of u. In particular, if V is any separable Hilbert space with a representation of G, then $V\oplus H$ is isomorphic to H as a representation space of G. Furthermore, $V\otimes H$ is also isomorphic to H. (Proof: it is sufficient to show that every irreducible representation of G occurs infinitely often in $V\otimes H$. Since $H\simeq L^2(G)\otimes L^2(G)\otimes l^2$, it is enough to show that the trivial one-dimensional representation occurs in $V\otimes L^2(G)$. Let π be any irreducible representation of G occurring in V. Then the complex conjugate representation $\bar{\pi}$ of π occurs in $L^2(G)$. Therefore $\pi\otimes\bar{\pi}$ occurs in $V\otimes L^2(G)$. Since the trivial one-dimensional representation is a subrepresentation of $\pi\otimes\bar{\pi}$ (with multiplicity 1), it occurs in $V\otimes L^2(G)$.)

As consequences of these facts about H, we find that there are spatial isomorphisms $M_2(K)\simeq K$ and in fact $K(V)\otimes K\simeq K$. (Here we call an isomorphism spatial if it is implemented by a G-invariant unitary on the underlying spaces.) We will in this chapter always assume that isomorphisms as above are spatial. However, there is nothing canonical about these isomorphisms, so we will need part (1) of the following lemma. Parts (2) and (3) will also be needed later.

3.1.7 Lemma. (1) Any two spatial automorphisms of K are G-homotopic through a path of spatial automorphisms.

(2) Let H be the Hilbert space in 3.1.5. Then there is a path $t\longrightarrow v_t$ of G-invariant isometries in $L(H)$ such that $t\longrightarrow v_t$ is continuous for the *-strong operator topology, $v_0=1$, $v_1v_1^*[H]$ and $(1-v_1v_1^*)[H]$ are both isomorphic to H, and $v_tv_t^*\geq v_1v_1^*$ for all t.

(3) The composite $\eta: K\longrightarrow M_2(K)\simeq K$, where the first map is $a\longrightarrow\begin{bmatrix} a & 0\\ 0 & 0\end{bmatrix}$, is G-homotopic to the identity map on K.

Proof. (1) It is obviously sufficient to prove that the group of G-invariant unitaries in $L(H)$ is connected, where H is the Hilbert space in 3.1.5. So let $u\in L(H)$ be a G-invariant unitary. Choose as usual a bounded Borel branch f of the log function on $sp(u)$, and set $u_t=\exp(tf(u))$ for $t\in[0,1]$. Obviously u_t is G-invariant, and is a path connecting u to 1.

(2) We first do this for the case that G is the one element group. In that case, we can take $H=L^2([0,1])$. Define v_t by

$$v_t(f)(s)=\begin{cases}(1+t)^{1/2}f((1+t)s) & s\leq(1+t)^{-1}\\ 0 & s>(1+t)^{-1}.\end{cases}$$

Then $t\longrightarrow v_t$ satisfies all the required conditions. To do the general case, we simply use $v_t\otimes 1$, where 1 is the identity operator on $L^2(G)$.

(3) Let v_t be as in part (2), and let $u: H\longrightarrow H\oplus H$ be a G-invariant unitary which identifies $v_1v_1^*[H]$ with $H\oplus 0$ and $(1-v_1v_1^*)[H]$ with $0\oplus H$. Then η is the composite

$$K(H)\xrightarrow{\simeq}K(H\oplus 0)\longrightarrow K(H\oplus H)\xrightarrow{\simeq}K(H),$$

when the middle map comes from the inclusion of $H \oplus 0$ in $H \oplus H$ and the last map is the spatial isomorphism implemented by u. Now set $\eta_t(a) = v_t^* \eta(a) v_t$. Since $v_t v_t^* \geq v_1 v_1^*$, we conclude that η_t is actually a homomorphism. The *-strong operator continuity of v_t now implies that $t \longrightarrow \eta_t$ is a homotopy. We have $\eta_0 = \eta$, while η_1 is the spatial automorphism implemented by the unitary $v_1^* u^* w$, where w is the inclusion of H in $H \oplus H$ on the first summand. By part (1), η_1 is homotopic to the identity. Therefore so is η.

Q.E.D.

Finally, we will need the equivariant version of Pedersen's derivation lifting theorem, 1.3 in [25].

3.1.8 Theorem. Let $\pi: B \longrightarrow A$ be a surjective equivariant homomorphism of separable G-algebras (G,A,α) and (G,B,β). Let δ be an equivariant derivation of A, that is, $\delta: A \longrightarrow A$ is a bounded equivariant linear map such that $\delta(xy) = \delta(x)y + x\delta(y)$ and $\delta(x^*) = \delta(x)^*$ for all $x, y \in A$. Then there is an equivariant derivation d of B such that $\pi \circ d = \delta \circ \pi$.

Proof. By [79], 8.6.15, there is a derivation d_0 such that $\pi \circ d_0 = \delta \circ \pi$. Define d by

$$d(x) = \int_G (\alpha_g^{-1} \circ d_0 \circ \alpha_g)(x)\, dg .$$

Then d is evidently a derivation satisfying $\alpha_g \circ d = d \circ \alpha_g$, and we have $\pi \circ d = \delta \circ \pi$ because δ and π are equivariant.

Q.E.D.

3.2. The Definition of $KK_G(A,B)$

If B is a unital G-algebra, then $K_0^G(B)$ is the set of differences $[p] - [q]$ of equivalence classes of G-invariant projections in various algebras $L(V) \otimes B$ for finite dimensional representation spaces V of G. Equivalently, one can consider the set of differences of equivalence classes of G-invariant projections in $K \otimes B$. If B is not unital, then in general there are not enough projections in $K \otimes B$, and one must use projections in $K \otimes B^+$ instead. Here one must further require that the p and q be equivalent modulo $K \otimes B$; it is easy to see that we can in fact require that $p - q \in K \otimes B$. (Of course, $p - q$ is usually not a projection.)

We can view a projection $p \in K \otimes B^+$ as a homomorphism from $\mathbf{C}$ to $K \otimes B^+$, defined by $z \longrightarrow zp$. This suggests that one consider "differences of equivalence classes" of equivariant homomorphisms, say φ and ψ, from an arbitrary G-algebra A into $K \otimes B^+$, subject to the condition that $\varphi(a) - \psi(a) \in K \otimes B$. Usually there are not enough of these (for example, take $B = \mathbf{C}$), so we consider instead equivariant homomorphisms into $M_G(K \otimes B)$. For technical reasons, we make the following slightly different definition. It can, however, be shown that each object satisfying the conditions of this definition is equivalent, in an appropriate sense, to a pair

(φ,ψ) of equivariant homomorphisms from A to $M_G(K\otimes B)$ such that $\varphi(a)-\psi(a)\in K\otimes B$ for $a\in A$. (See proposition 3.6.7.)

3.2.1 Definition. ([25], 2.1.) Let A and B be C^*-algebras. A prequasihomomorphism from A to B is a triple (φ,ψ,μ), where φ and ψ are homomorphisms from A to a C^*-algebra E, μ is a homomorphism from an ideal J of E to B, and $\varphi(a)-\psi(a)\in J$ for all $a\in A$. If (G,A,α) and (G,B,β) are G-algebras, then (φ,ψ,μ) as above is called an equivariant prequasihomomorphism if there is a continuous action of G on E such that J is a G-invariant ideal in E and the homomorphisms φ, ψ, and μ are all equivariant.

Following [25], we will write

$$(\varphi,\psi):\ A\longrightarrow E\ \triangleright J\xrightarrow{\ \mu\ }B$$

in this situation.

3.2.2 Definitions. An equivariant homomorphism $\eta: A\longrightarrow B$ can be regarded as an equivariant prequasihomomorphism $(\eta,0,id_B)$ with $E=J=B$. If A_0 and B_0 are also G-algebras, $\eta: A_0\longrightarrow A$ and $\lambda: B\longrightarrow B_0$ are equivariant homomorphisms, and $\Phi=(\varphi,\psi,\mu)$ is an equivariant prequasihomomorphism as above, then we define equivariant prequasihomomorphisms $\Phi\eta$ and $\lambda\Phi$ by $\Phi\eta=(\varphi\circ\eta,\psi\circ\eta,\mu)$ and $\lambda\Phi=(\varphi,\psi,\lambda\circ\mu)$. Also, we say that two equivariant prequasihomomorphisms

$$(\varphi_i,\psi_i): A\longrightarrow E_i\triangleright J_i\xrightarrow{\ \mu_i\ }G$$

are isomorphic if there is an equivariant isomorphism $\lambda: E_1\longrightarrow E_2$ such that $\lambda[J_1]=J_2$, $\lambda\circ\varphi_1=\varphi_2$, $\lambda\circ\psi_1=\psi_2$, and $\mu_2\circ\lambda=\mu_1$. We will generally identify isomorphic prequasihomomorphisms.

Now let $\Phi=(\varphi,\psi,\mu)$ be a prequasihomomorphism. We define the associated maps $D_\Phi: A\longrightarrow B$ and $Q_\Phi: A\times A\longrightarrow B$ by the formulas

$$D_\Phi(a)=\mu(\varphi(a)-\psi(a)) \quad\text{and}\quad Q_\Phi(a,b)=\mu[\psi(a)(\varphi(b)-\psi(b))].$$

They are continuous, D_Φ is linear, and Q_Φ is bilinear. If Φ is equivariant, then D_Φ and Q_Φ are as well. (Q_Φ being equivariant means that $Q_\Phi(\alpha_g(a),\alpha_g(b))=\beta_g(Q_\Phi(a,b))$.) They contain all the relevant information about Φ.

3.2.3 Definition. ([25], 2.2.) A quasihomomorphism from A to B is a prequasihomomorphism

$$(\varphi,\psi):\ A\longrightarrow E\triangleright J\xrightarrow{\ \mu\ }B$$

such that:

(1) E is generated as a C^*-algebra by the images of φ and ψ.

(2) J is generated as an ideal in E by $\{\varphi(a)-\psi(a): a\in A\}$.

(3) J is an essential ideal in E.

(4) μ is injective.

If (φ, ψ, μ) is a prequasihomomorphism such that μ is injective, we will often drop μ from the notation.

3.2.4 Proposition. ([25], 2.3.) Let (G,A,α) and (G,B,β) be G-algebras. Then:

(1) If Φ_1 and Φ_2 are equivariant quasihomomorphisms from A to B such that $D_{\Phi_1} = D_{\Phi_2}$ and $Q_{\Phi_1} = Q_{\Phi_2}$, then there is a unique isomorphism from Φ_1 to Φ_2.

(2) If Φ is a quasihomomorphism from A to B such that D_Φ and Q_Φ are equivariant, then Φ is equivariant.

(3) For every prequasihomomorphism Φ_0 from A to B there is a quasihomomorphism Φ from A to B such that $D_\Phi = D_{\Phi_0}$ and $Q_\Phi = Q_{\Phi_0}$. If Φ_0 is equivariant then so is Φ.

Proof. (1) Let Φ_i be the quasihomomorphism

$$\Phi_i = (\varphi_i, \psi_i) : A \longrightarrow E_i \triangleright J_i \longrightarrow B .$$

Then the image of J_i in B must be the subalgebra of B generated by the images of D_{Φ_i} and Q_{Φ_i}, whence $J_1 \simeq J_2$. Since J_i is an essential ideal in E_i, we may identify E_i with a subalgebra of $M_G(J_1)$ by proposition 3.1.2. Since $Q_{\Phi_1} = Q_{\Phi_2}$, we obtain $\psi_1(a)x = \psi_2(a)x$ for all $a \in A$ and $x \in J$. Therefore, $\psi_1 = \psi_2$ as maps from A to $M_G(J_1)$. Since $D_{\Phi_1} = D_{\Phi_2}$, we also obtain $\varphi_1 = \varphi_2$. Thus, under the identifications made, $\Phi_1 = \Phi_2$, and the isomorphism identifying them is unique because the embeddings of E_1 and E_2 in $M_G(J)$ are unique.

(2) Let

$$\Phi = (\varphi, \psi) : A \longrightarrow E \triangleright J \longrightarrow B .$$

Let $g \in G$, and set $\Psi = \beta_g \Phi \alpha_g^{-1}$, which is again a quasihomomorphism. By assumption, $D_\Phi = D_\Psi$ and $Q_\Phi = Q_\Psi$. Therefore by (1), applied to the case in which G is trivial, there is a unique isomorphism from Φ to Ψ, given by $\eta_g : E \longrightarrow E$. By uniqueness we must have $\eta_g \circ \eta_h = \eta_{gh}$, and we therefore have a (perhaps discontinuous) action of G on E such that J is G-invariant and φ and ψ are equivariant. Now for $a \in A$ the maps $g \longrightarrow \varphi(\alpha_g(a))$ and $g \longrightarrow \psi(\alpha_g(a))$ are continuous. Since φ and ψ are equivariant, the action of G is continuous on their images. These images generate E as a C^*-algebra, and a standard argument now shows that $g \longrightarrow \eta_g(x)$ is continuous for all $x \in E$. Therefore Φ is equivariant.

(3) Let

$$\Phi_0 = (\varphi_0, \psi_0) : A \longrightarrow E_0 \triangleright J_0 \overset{\mu_0}{\longrightarrow} B .$$

Let J be the C^*-subalgebra of B generated by the images of D_{Φ_0} and Q_{Φ_0}, and let μ be its inclusion in B. Let E_1 be the C^*-subalgebra of E generated by the images of φ_0 and ψ_0, and let J_1 be the C^*-subalgebra of E generated by all elements $\varphi_0(a) - \psi_0(a)$

and $\psi_0(a)(\varphi_0(b)-\psi_0(b))$ for $a, b \in A$. Then obviously $\mu_0[J_1]=J$. Furthermore, J_1 is an ideal in E_1, since the relation $\psi_0(a)J_1 \subset J_1$ is obvious, the relation $\varphi_0(a)J_1 \subset J_1$ follows because $\varphi_0(a)-\psi_0(a) \in J_1$, and J_1 is then also preserved under right multiplication by $\psi_0(a)$ and $\varphi_0(a)$ because it is selfadjoint. Therefore there is by proposition 3.1.2 a homomorphism $\gamma: E_1 \longrightarrow M(J)$ extending $\mu|_{J_1}$. Let E be the C^*-subalgebra of $M(J)$ generated by the images of $\varphi=\gamma\cdot\varphi_0$ and $\psi=\gamma\cdot\psi_0$. Then (φ,ψ,μ) is obviously a quasihomomorphism (J is an essential ideal in E because E is a subalgebra of $M(J)$), and $D_\Phi = D_{\Phi_0}$, $Q_\Phi = Q_{\Phi_0}$.

If Φ_0 is equivariant, then so are D_{Φ_0} and Q_{Φ_0}, whence Φ is equivariant by part (2). Uniqueness of Φ up to isomorphism follows from part (1). Q.E.D.

From now until the end of the chapter, we will use the following conventions: All (pre-) quasihomomorphisms will be assumed to be equivariant unless otherwise stated, and we will abbreviate the words prequasihomomorphism to "p.q.h." and quasihomomorphism to "q.h.". If Φ is a p.q.h. from A to B, then the q.h. obtained from Φ by part (3) of the previous proposition will be called the q.h. defined by Φ. If Φ is a q.h. from A to B, and $\eta: A_0 \longrightarrow A$ and $\lambda: B \longrightarrow B_0$ are equivariant homomorphisms, then we will by abuse of notation write $\Phi\eta$ and $\lambda\Phi$ for the q.h.'s defined by the p.q.h.'s $\Phi\eta$ and $\lambda\Phi$.

3.2.5 Definition. Let A and B be G-algebras. Two q.h.'s Φ_0 and Φ_1 from A to B are called G-homotopic if there is a family $\{\Psi_t\}$ for $t \in [0,1]$ of q.h.'s from A to B such that $\Psi_0=\Phi_0$, $\Psi_1=\Phi_1$, and the maps $t \longrightarrow D_{\Psi_t}(a)$ and $t \longrightarrow Q_{\Psi_t}(a,b)$ are continuous for all $a, b \in A$.

3.2.6 Proposition. ([25], proposition 2.4.) Two q.h.'s Φ_0 and Φ_1 as above are G-homotopic if and only if there is a q.h. Φ from A to $B\otimes C([0,1])$ such that $ev_t\,\Phi=\Phi_t$ for $t=0,1$. Here G acts trivially on $[0,1]$, and $ev_t : B\otimes C([0,1]) \longrightarrow B$ is evaluation at t, where $B\otimes C([0,1])$ is regarded as the algebra of continuous functions from $[0,1]$ to B.

Proof. If Φ exists, then $t \longrightarrow ev_t\Phi$ is obviously the required homotopy. So assume that $t \longrightarrow \Phi_t$ is a G-homotopy, where

$$\Phi_t = (\varphi_t,\psi_t): A \longrightarrow E_t \rhd J_t \longrightarrow B .$$

Let $F = \prod_{t\in[0,1]} E_t$, and define $\varphi,\psi: A \longrightarrow F$ by $\varphi(a)_t = \varphi_t(a)$ and $\psi(a)_t = \psi_t(a)$. Let E be the C^*-subalgebra of F generated by the images of φ and ψ. Let S be the C^*-subalgebra of E generated by the elements $\varphi(a)-\psi(a)$ for $a \in A$, and let J be the ideal in E generated by S. The continuity of $t \longrightarrow D_{\Phi_t}(a)$ implies that S can be regarded as a subalgebra of $B\otimes C([0,1])$, and the continuity of $t \longrightarrow Q_{\Phi_t}(a,b)$ further implies that J can be regarded as a subalgebra of $B\otimes C([0,1])$. It is now clear that

$$\Phi = (\varphi, \psi) : A \longrightarrow E \triangleright J \longrightarrow B \otimes C([0,1])$$

is a q.h. with $ev_t \Phi = \Phi_t$ for $t = 0, 1$, as desired. Q.E.D.

We will also need to consider homotopies of p.q.h.'s. The following definition is appropriate.

3.2.7 Definition. Let Φ_0 and Φ_1 be p.q.h.'s from A to B. A G-homotopy from Φ_0 to Φ_1 consists of a family of p.q.h.'s

$$\Psi_t = (\varphi_t, \psi_t) : A \longrightarrow E_t \triangleright J_t \xrightarrow{\mu_t} B$$

for $t \in [0,1]$, together with G-invariant subalgebras

$$E \subset \prod_{t \in [0,1]} E_t \quad \text{and} \quad J \subset \prod_{t \in [0,1]} J_t ,$$

such that the following conditions are satisfied:

(1) $\Psi_0 = \Phi_0$ and $\Psi_1 = \Phi_1$.

(2) There is a p.q.h.

$$\Psi = (\varphi, \psi) : A \longrightarrow E \triangleright J \xrightarrow{\mu} B \otimes C([0,1]),$$

where $\varphi(a)_t = \varphi_t(a)$, $\psi(a)_t = \psi_t(a)$, and $\mu(j)(t) = \mu_t(j_t)$ for $a \in A$, $j \in J$, and $t \in [0,1]$. Here $B \otimes C([0,1])$ is identified with the set of continuous functions from $[0,1]$ to B, and G acts trivially on $[0,1]$. (The requirement that Ψ be a p.q.h. of course includes such requirements as the ranges of φ and ψ being contained in E and the continuity of $t \longrightarrow \mu_t(j_t)$ for $j \in J$.)

(3) The projections from E to E_t and from J to J_t are surjective for every t.

It might be more appropriate to require that E and J be algebras over $C([0,1])$, or even that they be the sets of continuous sections of bundles of C^*-algebras over $[0,1]$ with fibers E_t and J_t respectively. (See [49] for definitions.) However, we do not need the added restrictions.

3.2.8 Example. Let A and B be G-algebras, and for $t \in [0,1]$ let

$$\Psi_t = (\varphi_t, \psi_t) : A \longrightarrow E_0 \triangleright J_0 \xrightarrow{\mu_t} B$$

be p.q.h.'s from A to B. If $t \longrightarrow \varphi_t$, $t \longrightarrow \psi_t$, and $t \longrightarrow \mu_t$ are all G-homotopies of homomorphisms, then we obtain an obvious homotopy from Ψ_0 to Ψ_1 by setting $E = E_0 \otimes C([0,1])$ and $J = J_0 \otimes C([0,1])$, regarded as the algebras of continuous functions from $[0,1]$ to E_0 and J_0 respectively.

From now on, we will assume that all homotopies are G-homotopies. We summarize some of the properties of homotopies of p.q.h.'s in the following lemma.

3.2.9 Lemma. (1) Homotopy of p.q.h.'s is an equivalence relation.

(2) If two p.q.h.'s are homotopic, then the q.h.'s they define are homotopic as q.h.'s (definition 3.2.5).

(3) A p.q.h. is homotopic to the q.h. it defines.

Proof. (1) The only property which is not immediately obvious is transitivity. So let

$$\Phi_t = (\varphi_t, \psi_t) : A \longrightarrow E_t \triangleright J_t \xrightarrow{\mu_t} B$$

for $t \in [0,2]$ be p.q.h.'s defining homotopies from Φ_0 to Φ_1 and Φ_1 to Φ_2, with appropriate subalgebras (as in definition 3.2.7)

$$F_0 \subset \prod_{t\in[0,1]} E_t \quad \text{and} \quad F_1 \subset \prod_{t\in[1,2]} E_t ,$$

and with corresponding ideals $I_0 \subset F_0$ and $I_1 \subset F_1$. Let

$$\pi_0 : \prod_{t\in[0,1]} E_t \longrightarrow E_1 \quad \text{and} \quad \pi_1 : \prod_{t\in[1,2]} E_t \longrightarrow E_1$$

be the projections. Further let

$$F = \{(a,b) \in F_0 \oplus F_1 : \pi_0(a) = \pi_1(b)\}$$

and

$$I = \{(a,b) \in I_0 \oplus I_1 : \pi_0(a) = \pi_1(b)\}.$$

Then F can be identified with an obvious subalgebra of $\prod_{t\in[0,2]} E_t$, and similarly for I. Clearly we have a p.q.h.

$$(\varphi, \psi) : A \longrightarrow F \triangleright I \longrightarrow B \otimes C([0,2]),$$

and we have thus shown that Φ_0 is homotopic to Φ_2.

(2) Adopt the notation of definition 3.2.7. Using proposition 3.2.6, it is clear the q.h. defined by Ψ yields a homotopy of q.h.'s from the q.h. defined by Ψ_0 to the q.h. defined by Ψ_1.

(3) Let

$$\Phi_0 = (\varphi_0, \psi_0) : A \longrightarrow E_0 \triangleright J_0 \xrightarrow{\mu_0} B$$

be a p.q.h. Let E, E_1, J, J_1, γ, μ, φ, and ψ have the same meanings as in the proof of proposition 3.2.4 (3). Then

$$\Phi = (\varphi, \psi) : A \longrightarrow E \triangleright J \xrightarrow{\mu} B$$

is the q.h. defined by Φ_0, and

$$\Phi_1 = (\varphi_0, \psi_0) : A \longrightarrow E_1 \triangleright J_1 \xrightarrow{\mu_0 | J_1} B$$

is a p.q.h. with $E_1 \subset E_0$ and $J_1 \subset J_0$.

Define a homotopy from Φ_0 to Φ_1 by setting $\Phi_t = \Phi_0$ for $0 \le t < 1$. The required subalgebras of $\prod E_t$ and $\prod J_t$ are given by

$$F_0 = \{ f \in E_0 \otimes C([0,1]) : f(1) \in E_1 \}$$

and

$$I_0 = \{ f \in J_0 \otimes C([0,1]) : f(1) \in J_1 \}.$$

Here $E_0 \otimes C([0,1])$ and $J_0 \otimes C([0,1])$ are regarded as algebras of continuous functions on $[0,1]$. Clearly

$$A \longrightarrow F_0 \triangleright I_0 \longrightarrow B \otimes C([0,1]),$$

with the obvious maps, is a p.q.h. So Φ_0 is homotopic to Φ_1.

Now construct a homotopy from Φ_1 to Φ by setting $\Phi_t = \Phi$ for $1 < t \le 2$. We set

$$F_1 = \{ (a, f) \in E_1 \oplus E \otimes C([1,2]) : \gamma(a) = f(1) \}$$

and

$$I_1 = \{ (a, f) \in J_1 \oplus J \otimes C([1,2]) : \mu_0(a) = \mu(f(1)) \}.$$

Note that F_1 can be viewed as a subalgebra of $\prod_{t \in [1,2]} E_t$ via the map λ defined by

$$\lambda(a, f)_t = \begin{cases} a & t = 1 \\ f(t) & 1 < t \le 2. \end{cases}$$

Similarly, I_1 can be viewed as a subalgebra of $\prod_{t \in [1,2]} J_t$. This shows that Φ_1 is homotopic to Φ. By part (1), we conclude that Φ_0 is homotopic to Φ. Q.E.D.

3.2.10 Definition. Let (G, A, α) and (G, B, β) be G-algebras. We define $KK_G(A, B)$ to be the set of G-homotopy classes of q.h.'s from A to $K \otimes B$. (Recall our convention about the action of G on K.) We write $[\Phi]$ for the G-homotopy class of a q.h. Φ, and also for the class of the q.h. defined by a p.q.h. Φ.

For $i = 1, 2$, let

$$\Phi_i = (\varphi_i, \psi_i) : A \longrightarrow E_i \triangleright J_i \longrightarrow K \otimes B$$

be a q.h. (or a p.q.h.). Then the direct sum of Φ_1 and Φ_2 is the p.q.h.

$$\Phi_1 \oplus \Phi_2 = (\varphi_1 \oplus \varphi_2, \psi_1 \oplus \psi_2) : A \longrightarrow E_1 \oplus E_2 \triangleright J_1 \oplus J_2 \longrightarrow (K \otimes B) \oplus (K \otimes B) \longrightarrow K \otimes B,$$

where the last map comes from $K \oplus K \longrightarrow M_2(K) \simeq K$. We then define addition in $KK_G(A, B)$ by $[\Phi_1] + [\Phi_2] = [\Phi_1 \oplus \Phi_2]$. If V is a finite dimensional representation space of G, we further define $[V][\Phi_1]$ to be the class of the q.h.

$$(\varphi_1, \psi_1): A \longrightarrow E_1 \triangleright J_1 \longrightarrow K\otimes B \stackrel{\lambda}{\longrightarrow} L(V)\otimes K\otimes B \simeq K\otimes B,$$

where $\lambda(x) = 1\otimes x$ and the last isomorphism is as in remark 3.1.6.

Finally, let A_0 and B_0 be two more G-algebras, and let $\lambda: A_0 \longrightarrow A$ and $\eta: B \longrightarrow B_0$ be equivariant homomorphisms. We then define, for any q.h. Φ from A to B, an element $\lambda^*([\Phi]) \in KK_G(A_0, B)$ by $\lambda^*([\Phi]) = [\Phi\lambda]$, and an element $\eta_*([\Phi]) \in KK_G(A, B_0)$ by $\eta_*([\Phi]) = [(id_K\otimes\eta)\Phi]$.

3.2.11 Remark. It follows from lemma 3.2.9 (3) that $KK_G(A, B)$ could equally well have been defined as the set of G-homotopy classes of p.q.h.'s, and that addition, multiplication by elements of $R(G)$, and the maps λ^* and η_* could also have be defined in terms of p.q.h.'s.

3.2.12 Theorem. With the operations in definition 3.2.10, $KK_G(A,B)$ is a functor of two variables, contravariant in the first and covariant in the second, from G-algebras to $R(G)$-modules. It is homotopy invariant in both variables.

Proof. The first step is to show that the operations in the definition are well defined. If in the definition of addition either Φ_1 or Φ_2 is replaced by a G-homotopic q.h., then clearly the direct sum of Φ_1 and Φ_2 is also replaced by a G-homotopic q.h. Furthermore, if any other choice is made for the isomorphism $M_2(K) \simeq K$, lemma 3.1.7 (1) implies that the resulting direct sums are homotopic. Thus, addition is well defined on homotopy classes. Similarly, if any other choice of the isomorphism $L(V)\otimes K \simeq K$ is made, the new quasihomomorphism is homotopic to the old one, and multiplication by $[V]$ is well defined.

Addition is easily seen to be commutative, since conjugating by a path of unitaries from $\begin{pmatrix}1&0\\0&1\end{pmatrix}$ to $\begin{pmatrix}0&1\\1&0\end{pmatrix}$ in M_2 shows that the maps $(a, b) \longrightarrow \begin{pmatrix}a&0\\0&b\end{pmatrix}$ and $(a, b) \longrightarrow \begin{pmatrix}b&0\\0&a\end{pmatrix}$ from $K\oplus K$ to $M_2(K)$ are G-homotopic, so that the p.q.h.'s $\Phi_1\oplus\Phi_2$ and $\Phi_2\oplus\Phi_1$ are G-homotopic. The zero element is the zero q.h., with $E = J = 0$; to show that $\Phi_1\oplus 0$ is G-homotopic to Φ_1, we only need a homotopy from the composite $K \longrightarrow K\oplus K \longrightarrow M_2(K) \simeq K$ to the identity on K. The existence of this homotopy is given in lemma 3.1.7 (3). The inverse of the q.h.

$$\Phi = (\varphi, \psi): A \longrightarrow E \triangleright J \longrightarrow K\otimes B$$

is (ψ, φ). To prove this, let Ψ_t be the homotopy of p.q.h.'s

$$A \longrightarrow M_2(E) \triangleright M_2(J) \longrightarrow M_2(K\otimes B) \simeq K\otimes B$$

given by

$$\Psi_t = (\varphi\oplus\psi, (u_t\otimes 1)(\psi\oplus\varphi)(u_t\otimes 1)^*),$$

where

$$u_t = \begin{bmatrix} \cos t & \sin t \\ -\sin t & \cos t \end{bmatrix}$$

for $t \in \left[0, \frac{\pi}{2}\right]$ is a path of unitaries in M_2 from $\begin{pmatrix} 1 & 0 \\ 0 & 1 \end{pmatrix}$ to $\begin{bmatrix} 0 & 1 \\ -1 & 0 \end{bmatrix}$. It is easy to check that

$$\varphi(a) \oplus \psi(a) - (u_t \otimes 1)(\psi(a) \oplus \varphi(a))(u_t \otimes 1)^* \in M_2(J)$$

for $a \in A$, given that $\varphi(a) - \psi(a) \in J$. Thus, Ψ_t really is a p.q.h. Clearly Ψ_0 defines the same p.q.h. as $(\varphi, \psi) \oplus (\psi, \varphi)$, and

$$\Psi_1 = (\varphi \oplus \psi, \varphi \oplus \psi) : A \longrightarrow M_2(E) \rhd M_2(J) \longrightarrow K \otimes B$$

defines the zero q.h. Thus, $-[\varphi, \psi] = [\psi, \varphi]$ as claimed.

We can now define multiplication by an arbitrary element $[V] - [W]$ of $R(G)$ by $([V] - [W])[\Phi] = [V][\Phi] - [W][\Phi]$, and it is trivial to check that $KK_G(A,B)$ is an $R(G)$-module with the operation.

Finally, we consider functoriality. Let $\lambda : A_0 \longrightarrow A$ and $\eta : B \longrightarrow B_0$ be equivariant homomorphisms. Then we easily see that λ^* is a well defined map from $KK_G(A,B)$ to $KK_G(A_0, B)$, which is an $R(G)$-module homomorphism and depends only on the G-homotopy class of λ; we omit the details. Similarly, $\eta_* : KK_G(A,B) \longrightarrow KK_G(A,B_0)$ is an $R(G)$-module homomorphism depending only on the G-homotopy class of η. Q.E.D.

3.2.13 Definition. Let $\varphi : A \longrightarrow B$ be an equivariant homomorphism of G-algebras. Then the element $[\varphi]$ of $KK_G(A,B)$ defined by φ is the class of the p.q.h.

$$(\varphi, 0) : A \longrightarrow B \rhd B \longrightarrow K \otimes B,$$

where the map $B \longrightarrow K \otimes B$ is $b \longrightarrow p \otimes b$ for a G-invariant projection $p \in K$ such that, with H being the Hilbert space on which K acts, pH is a one-dimensional subspace on which G acts trivially. The triviality of the action is important: there is a G-homotopy class of one-dimensional projections in K for every unitary equivalence class of one-dimensional representations of G, and we are choosing a specific one of them. If q is an arbitrary G-invariant projection in K, then the class of the p.q.h. obtained by replacing p with q is $[qH] \cdot [\varphi]$, which is usually different from $[\varphi]$.

3.3. Extendible Quasihomomorphisms

Toward the end of this section, we will need to assume that certain C^*-algebras are separable. To prove associativity of the product, in the next section, we will need to impose more separability conditions. To simplify matters, we therefore assume from now on that all G-algebras occurring as domains or codomains of q.h.'s and p.q.h.'s are separable. (Of course, we will still have to consider multiplier algebras, which in general are not separable.)

Let

$$(\varphi,\psi): A \longrightarrow F \rhd I \longrightarrow K\otimes B \quad \text{and} \quad (\sigma,\tau): B \longrightarrow E \rhd J \longrightarrow K\otimes C$$

be q.h.'s. If we think of (φ,ψ) as the "difference of equivalence classes" $[\varphi]-[\psi]$, and think of (σ,τ) as $[\sigma]-[\tau]$, then we would expect to be able to define their product $([\varphi]-[\psi])([\sigma]-[\tau])$ to be $[\sigma_0\varphi]-[\tau_0\varphi]-[\sigma_0\psi]+[\tau_0\psi]$, where $\sigma_0 = id_K\otimes\sigma$ and τ_0 is defined similarly. Unfortunately, the composites $\sigma_0\varphi$ etc. need not be defined. To get around this problem, we need to be able to extend the q.h. (σ,τ) as in the following definition.

3.3.1 Definition. Let

$$\Phi = (\varphi,\psi): A \longrightarrow E \rhd J \xrightarrow{\mu} B$$

be a p.q.h., where A is a G-invariant ideal in a larger G-algebra A_0 (which need not be separable). Then we say Φ is extendible to A_0 if there are equivariant homomorphisms $\varphi_0,\psi_0: A_0 \longrightarrow M_G(E)$ extending φ and ψ, such that, if (e_λ) is an approximate identity for A which is quasicentral for A_0 ([79], 3.12.13), then

$$\varphi_0(a)x = \lim_\lambda \varphi(ae_\lambda)x \quad \text{and} \quad \psi_0(a)x = \lim_\lambda \psi(ae_\lambda)x\,, \tag{*}$$

for $a\in A_0$ and $x \in E$. (Note that quasicentral approximate identities always exist, by [79], 3.12.14.) The triple (φ_0,ψ_0,μ) will be called an extended (to A_0) p.q.h.

It is clear from the definition that if a p.q.h. is extendible, then its extension is unique, being determined by the formulas (*).

3.3.2 Lemma. If a p.q.h. from A to B is extendible to A_0, then so is the q.h. it defines.

Proof. Let

$$\Phi_0 = (\varphi_0,\psi_0): A \longrightarrow E_0 \rhd J_0 \xrightarrow{\mu_0} B$$

be a p.q.h., and let E, E_1, J, J_1, γ, μ, φ, and ψ have the same meanings that they did in the construction of the q.h. Φ defined by Φ_0 in the proof of proposition 3.2.4 (3). That is, E_1 is the subalgebra of E_0 generated by the images of φ and ψ, J_1 is the ideal in E_1 generated by $\{\varphi(a)-\psi(a): a\in A\}$, $J=\mu_0[J_1]$ and μ is the inclusion of J in B, and E is the image of E_1 in $M_G(J)$ under the natural map γ. Also $\varphi=\gamma\cdot\varphi_0$ and $\psi=\gamma\cdot\psi_0$.

Now let $\bar\varphi_0,\bar\psi_0: A_0 \longrightarrow M_G(E_0)$ be an extension of Φ_0 to A_0. Let D_1 be the subalgebra of $M_G(E_0)$ generated by the images of $\bar\varphi_0$ and $\bar\psi_0$. Then the formulas (*) in the definition of extendibility imply that E_1 is an ideal in D_1. Therefore J_1 is also an ideal in D_1, and there is an equivariant map $\bar\gamma: D_1 \longrightarrow M_G(J)$ which extends γ.

We now define $\overline{\varphi}, \overline{\psi}: A_0 \longrightarrow M_G(J)$ by $\overline{\varphi} = \overline{\gamma} \cdot \overline{\varphi}_0$ and $\overline{\psi} = \overline{\gamma} \cdot \overline{\psi}_0$. By corollary 3.1.3, $M_G(E)$ is a subalgebra of $M_G(J)$. The formulas (*) imply that $\overline{\varphi}(a)x$ and $\overline{\psi}(a)x$ are in E for $a \in A_0$ and $x \in E$, so that the images of $\overline{\varphi}$ and $\overline{\psi}$ are actually in $M_G(E)$. It is now trivial to check that $\overline{\varphi}$ and $\overline{\psi}$ define an extension to A_0 of the q.h.

$$\Phi = (\varphi, \psi): A \longrightarrow E \rhd J \longrightarrow B$$

defined by Φ_0. Q.E.D.

We will refer to $(\overline{\varphi}, \overline{\psi}, \mu)$ as the extended q.h. defined by the extended p.q.h. $(\overline{\varphi}_0, \overline{\psi}_0, \mu_0)$.

We now want to define a homotopy for extended p.q.h.'s and q.h.'s, in a way that extends the definitions of the previous section. The definition for extended q.h.'s is clear.

3.3.3 Definition. A homotopy of extended (to A_0) q.h.'s from A to B is an extended (to A_0) q.h. Φ from A to $B \otimes C([0,1])$ such that $ev_0\Phi$ and $ev_1\Phi$ are the given extended q.h.'s.

Before defining homotopy for extended p.q.h.'s, recall that a homotopy of p.q.h.'s consists of a family of p.q.h.'s

$$\Phi_t = (\varphi_t, \psi_t): A \longrightarrow E_t \rhd J_t \xrightarrow{\mu_t} B,$$

together with a p.q.h.

$$\Phi = (\varphi, \psi): A \longrightarrow E \rhd J \xrightarrow{\mu} B \otimes C([0,1]),$$

where

$$E \subset \prod_{s \in [0,1]} E_s \quad \text{and} \quad J \subset \prod_{s \in [0,1]} J_s,$$

and φ, ψ, and μ are defined in the obvious way. It is required that the restriction π_t to E of the projection $\prod E_s \longrightarrow E_t$ be surjective for all $t \in [0,1]$, and that $\pi_t[J] = J_t$. By proposition 3.1.2, π_t extends to an equivariant homomorphism, also called π_t, from $M_G(E)$ to $M_G(E_t)$. The homomorphisms π_t define a homomorphism

$$\pi: M_G(E) \longrightarrow \prod_{s \in [0,1]} M_G(E_s),$$

which is easily seen to be injective. We can thus identify $M_G(E)$ with a subalgebra of $\prod_s M_G(E_s)$.

We are now ready for:

3.3.4 Definition. Let A and B be (separable) G-algebras, and let A_0 be a not necessarily separable G-algebra which contains A as an ideal. A homotopy of extended (to A_0) p.q.h.'s from A to B is a family $\Phi_t = (\varphi_t, \psi_t, \mu_t)$ of extended p.q.h.'s from A to

B, together with an extended p.q.h. (φ, ψ, μ) from A to $B \otimes C([0,1])$, such that:

(1) The family of p.q.h.'s

$$(\varphi_t|_A, \psi_t|_A): A \longrightarrow E_t \rhd J_t \xrightarrow{\mu_t} B,$$

together with the p.q.h.

$$(\varphi|_A, \psi|_A): A \longrightarrow E \rhd J \xrightarrow{\mu} B \otimes C([0,1]),$$

is a homotopy of p.q.h.'s, and

(2) For all $a \in A_0$, we have $\varphi(a)_t = \varphi_t(a)$ and $\psi(a)_t = \psi_t(a)$ using the identification above of $M_G(E)$ with a subalgebra of $\prod_t M_G(E_t)$.

We point out that a homotopy of extended q.h.'s given by an extended q.h. (φ, ψ, μ) from A to $B \otimes C([0,1])$ yields a homotopy of the given extended q.h.'s as p.q.h.'s. One need only set $\Phi_t = (\varphi, \psi, ev_t \cdot \mu)$, where ev_t is evaluation of functions from $[0,1]$ to B at t, and then replace each Φ_t by the extended q.h. it defines. Alternatively, one can apply parts (1) and (3) of the next lemma, which is the analog of lemma 3.2.9.

3.3.5 Lemma. (1) Homotopy of extended p.q.h.'s is an equivalence relation.

(2) If two extended p.q.h.'s are homotopic, then the extended q.h.'s they define are homotopic as extended q.h.'s.

(3) An extended p.q.h. is homotopic to the extended q.h. it defines.

Proof. The proofs of parts (1) and (2) require only minor additions to the proof of the corresponding parts of lemma 3.2.9. We therefore prove only part (3). Adopt the notation of the proof of lemma 3.3.2. Let

$$\Phi_1 = (\varphi_0, \psi_0): A \longrightarrow E_1 \rhd J_1 \xrightarrow{\mu_0|_{J_1}} B.$$

Then Φ_1 is an extendible p.q.h., with the extension given by $\eta \cdot \bar{\varphi}_0, \eta \cdot \bar{\psi}_0: A_0 \longrightarrow M_G(E_1)$, where η is the map from D_1 to $M_G(E_1)$ obtained from proposition 3.1.2 and the fact that E_1 is an ideal in D_1. Note that there is also a canonical map $\omega: M_G(E_1) \longrightarrow M_G(E)$, and we can identify $M_G(E)$ with a subalgebra of $M_G(J)$ by corollary 3.1.3. We then have $\bar{\gamma} = \omega \cdot \eta$.

Recall from the proof of lemma 3.2.9 (3) that we defined a homotopy from Φ_0 to Φ_1 by setting $\Phi_t = \Phi_0$ for $0 \le t < 1$, with the required subalgebras of $\prod E_t$ and $\prod J_t$ given by

$$F_0 = \{f \in E_0 \otimes C([0,1]) : f(1) \in E_1\}$$

and

$$I_0 = \{f \in J_0 \otimes C([0,1]) : f(1) \in J_1\}.$$

Define $\sigma_0: A_0 \longrightarrow M_G(F_0)$ by $\sigma_0(a)(t) = \bar{\varphi}_0(a)$ for $0 \le t < 1$ and $\sigma_0(a)(1) = \eta(\bar{\varphi}_0(a))$. It is clear that $\sigma_0(a)$ really is in $M_G(F_0)$. Now let (e_λ) be an approximate identity in A which is quasicentral for A_0, and let $a \in A_0$ and $x \in F_0$. Then for $s, t \in [0,1]$, we have

$$\| \bar{\varphi}_0(ae_\lambda)x(s) - \bar{\varphi}_0(ae_\lambda)x(t) \| \le \| a \| \; \| x(s) - x(t) \|$$

for all λ (since $\|e_\lambda\| \le 1$), and

$$\| \bar{\varphi}_0(a)x(s) - \bar{\varphi}_0(a)x(t) \| \le \| a \| \; \| x(s) - x(t) \| .$$

Therefore the functions $\bar{\varphi}_0(ae_\lambda)x$ and $\bar{\varphi}_0(a)x$ are equicontinuous. By assumption, $\bar{\varphi}_0(ae_\lambda)x(t) \longrightarrow \bar{\varphi}_0(a)x(t)$ for every $t \in [0,1]$, and equicontinuity together with the compactness of $[0,1]$ implies that the convergence is in fact uniform. Since $\bar{\varphi}_0(b)y = \eta(\bar{\varphi}_0(b))y$ for all $b \in A_0$ and $y \in E_1$, we have in fact shown that $\sigma_0(ae_\lambda)x \longrightarrow \sigma_0(a)x$. Replacing $\bar{\varphi}_0$ by $\bar{\psi}_0$, we define τ_0 analogously and conclude that $\tau_0(ae_\lambda)x \longrightarrow \tau_0(a)x$ also. So (σ_0, τ_0) defines an extension to A_0 of the p.q.h.

$$A \longrightarrow F_0 \quad I_0 \longrightarrow B$$

(with the obvious maps). Therefore the extended p.q.h.'s $(\bar{\varphi}_0, \bar{\psi}_0, \mu_0)$ and $(\eta \cdot \bar{\varphi}_0, \eta \cdot \bar{\psi}_0, \mu_0|_{J_1})$ are homotopic.

A similar argument, starting from the homotopy from Φ_1 to Φ given in the proof of lemma 3.2.9 (3), also shows that the extended p.q.h.'s $(\eta \cdot \bar{\varphi}_0, \eta \cdot \bar{\psi}_0, \mu_0|_{J_1})$ and $(\bar{\varphi}, \bar{\psi}, \mu)$ are homotopic. Therefore $(\bar{\varphi}_0, \bar{\psi}_0, \mu_0)$ is homotopic to $(\bar{\varphi}, \bar{\psi}, \mu)$. Q.E.D.

3.3.6 Proposition. (Compare [25], proposition 3.1.) Let

$$\Phi = (\varphi, \psi): A \longrightarrow E \rhd J \longrightarrow K \otimes B$$

be a q.h. Let A_0 be a separable G-algebra which contains A as an ideal. Then there is a q.h. Σ from A to B which is extendible to A_0 and homotopic to Φ. The extension of Σ to A_0 is unique up to homotopy of q.h.'s extended to A_0.

Proof. Let $R = \varphi[A]$, let $S = \psi[A]$, and let Z be the C^*-subalgebra of $M_2(E)$ given by

$$Z = \begin{bmatrix} \overline{R+RJR} & \overline{RS+RJS} \\ \overline{SR+SJR} & \overline{S+SJS} \end{bmatrix}.$$

That is, Z consists of all 2×2 matrices whose ij entries are in the subsets of E appearing in the ij place in the above matrix. It is clear that Z is a separable closed selfadjoint G-invariant subspace of $M_2(E)$, but we must show that $Z^2 \subset Z$. We exploit the fact that J is an ideal in E. Then the only nontrivial parts are to show that $RSR \subset R + RJR$ and $SRS \subset S + SJS$. We do only one of them. Let $a, b, c \in A$. Then

$$\varphi(a)\psi(b)\varphi(c) = \varphi(abc) + \varphi(a)(\psi(b) - \varphi(b))\varphi(c) \in R + RJR,$$

as desired.

Since $\varphi: A \longrightarrow R$ is surjective, proposition 3.1.2 provides an equivariant homomorphism from A_0 to $M_G(R)$. By lemma 3.1.4, the map from R to Z given by $r \longrightarrow \begin{bmatrix} r & 0 \\ 0 & 0 \end{bmatrix}$ extends to a homomorphism from $M_G(R)$ to $M_G(Z)$. (The various conditions $D \subset \overline{DR_{11}D} = R_{11}$ etc. of the lemma are easily seen to be satisfied, by using an approximate identity for R.) We therefore obtain by composition an equivariant homomorphism $\sigma_0 : A_0 \longrightarrow M_G(Z)$ extending the map from A to Z given by $a \longrightarrow \begin{bmatrix} \varphi(a) & 0 \\ 0 & 0 \end{bmatrix}$. A similar argument yields an equivariant homomorphism $\gamma : A_0 \longrightarrow M_G(Z)$ extending the map from A to Z given by $a \longrightarrow \begin{bmatrix} 0 & 0 \\ 0 & \psi(a) \end{bmatrix}$.

Let $I = Z \cap M_2(J)$. Since $\varphi(a) - \psi(a) \in J$ for all $a \in A$, it follows that φ and ψ define the same map from A to E/J. Since Φ is a q.h. (and not just a p.q.h.), we have $R + J = S + J = E$, whence $\overline{RS+J} = \overline{(R+J)(S+J)} = E$, and similarly $\overline{SR+J} = E$. Therefore $Z + M_2(J) = M_2(E)$, and it follows that Z/I is canonically isomorphic to $M_2(E/J)$. Let π be the quotient map.

Let $u = \begin{bmatrix} 0 & 1 \\ 1 & 0 \end{bmatrix}$, and let δ_0 be a derivation of M_2 such that $\exp(\delta_0) = \mathrm{ad}(u)$. (Here $\mathrm{ad}(u)(a) = uau^*$. The derivation δ_0 can be taken to be of the form $\delta_0(a) = da - ad$, where d is chosen so that $\exp(d) = u$. See [79], 8.6.1.) Then $\delta_0 \otimes id_{E/J}$ is easily seen to be an equivariant derivation of $M_2(E/J)$, so that by theorem 3.1.8 there is an equivariant derivation δ of Z which lifts δ_0, that is, we have $\pi \circ \delta = \delta_0 \circ \pi$. Let $\eta = \exp(\delta)$, and extend η to an equivariant automorphism, still called η, of $M_G(Z)$. Set $\tau_0 = \eta \circ \gamma$, which is an equivariant homomorphism from A_0 to $M_G(Z)$. Set $\sigma = \sigma_0|_A$ and $\tau = \tau_0|_A$. Define $\mu : I \longrightarrow K \otimes B$ to be the composite

$$I \longrightarrow M_2(J) \longrightarrow M_2(K \otimes B) \simeq K \otimes B .$$

Then we claim that

$$\Sigma = (\sigma, \tau) : A \longrightarrow Z \triangleright I \xrightarrow{\mu} K \otimes B$$

is a p.q.h. which is extendible to A_0 and homotopic to Φ. To obtain the first statement of the proposition, it will then be sufficient to replace Σ by the q.h. it defines, which is also extendible to A_0 by lemma 3.3.2.

We first show that Σ is a p.q.h. Let $\pi_0 : E \longrightarrow E/J$ be the quotient map. Then for $a \in A$ we have

$$\pi(\sigma(a) - \tau(a)) = \pi\left[\begin{bmatrix} \varphi(a) & 0 \\ 0 & 0 \end{bmatrix} - \eta\begin{bmatrix} 0 & 0 \\ 0 & \psi(a) \end{bmatrix}\right]$$

$$= \begin{bmatrix} \pi_0 \circ \varphi(a) & 0 \\ 0 & 0 \end{bmatrix} - \begin{bmatrix} 0 & 1 \\ 1 & 0 \end{bmatrix}\begin{bmatrix} 0 & 0 \\ 0 & \pi_0 \circ \psi(a) \end{bmatrix}\begin{bmatrix} 0 & 1 \\ 1 & 0 \end{bmatrix}^* = 0,$$

using the fact that $\pi \circ \eta = \exp(\pi \circ \delta) = \exp(\delta_0 \otimes id_{E/J}) \circ \pi$. Thus, $\sigma(a) - \tau(a) \in I$, and Σ is a p.q.h.

Next we show that Σ is extendible. The extensions σ_0 and τ_0 from A_0 to $M_G(Z)$ are already given, so we need only check the condition (*). Let (e_λ) be an approximate identity in A, let $a \in M_G(A)$, and let $x = \begin{bmatrix} x_{11} & x_{12} \\ x_{21} & x_{22} \end{bmatrix} \in Z$. Then (using the last part of the statement of lemma 3.1.4), we have

$$\lim_{\lambda} \sigma(ae_\lambda)x = \sigma_0(a) \lim_{\lambda} \sigma(e_\lambda)x$$

$$= \sigma_0(a) \lim_{\lambda} \begin{pmatrix} \sigma(e_\lambda)x_{11} & \sigma(e_\lambda)x_{12} \\ 0 & 0 \end{pmatrix} = \sigma_0(a) \begin{pmatrix} x_{11} & x_{12} \\ 0 & 0 \end{pmatrix} = \sigma_0(a)x .$$

Here the second to the last step follows from the fact that $\sigma(e_\lambda)$ is an approximate identity for R, and the relations $x_{11} \in \overline{R+RJR}$ and $x_{12} \in \overline{RS+RJS}$. The calculation for τ_0 is similar (the presence of the automorphism η requires only minor changes).

Finally we must show that Σ is homotopic to Φ. We replace Σ by the p.q.h.

$$\Sigma_0 = (\sigma,\tau) : A \longrightarrow M_2(E) \triangleright M_2(J) \longrightarrow K \otimes B$$

obtained by replacing Z by the algebra $M_2(E)$ which contains it, and using the fact that the map $I \longrightarrow K \otimes B$ has a canonical extension to $M_2(J)$. Since Σ_0 defines the same q.h. as Σ, we need only prove that Σ_0 is homotopic to Φ. We next replace Φ by the p.q.h.

$$\Phi_0 = (0 \oplus \varphi, 0 \oplus \psi) : A \longrightarrow M_2(E) \triangleright M_2(J) \longrightarrow K \otimes B ,$$

where the map from $M_2(J)$ to $K \otimes B$ is the same as the one used in the definition of Σ. The proof that the zero q.h. is the zero element of $KK_G(A,B)$ (in the proof of theorem 3.2.12) implies that the q.h. defined by Φ_0 is homotopic to Φ. Therefore it suffices to find a homotopy of p.q.h.'s from Φ_0 to Σ_0.

Let $\rho_t = \exp(t\delta_0 \otimes id_E)$, which is an automorphism of $M_2(E)$ for $t \in [0,1]$. Let $\eta_t = \exp(t\delta)$, which is an automorphism of Z for $t \in [0,1]$; we regard η_t as a homomorphism from Z to $M_2(E)$. Then set

$$\varphi_t = \rho_t \circ \begin{pmatrix} 0 & 0 \\ 0 & \varphi \end{pmatrix} \quad \text{and} \quad \psi_t = \eta_t \circ \begin{pmatrix} 0 & 0 \\ 0 & \psi \end{pmatrix} .$$

(To see that the second definition makes sense, note that $\begin{pmatrix} 0 & 0 \\ 0 & \psi(a) \end{pmatrix} \in Z$ for $a \in A$). The only nontrivial calculation needed to show that $t \longrightarrow (\varphi_t, \psi_t)$ defines a homotopy is showing that $\varphi_t(a) - \psi_t(a) \in M_2(J)$ for $a \in A$. This, however, follows immediately from the observation that, modulo $M_2(J)$, the automorphisms ρ_t and η_t are both $\exp(t\delta_0 \otimes id_{E/J})$. (Recall that $Z/(M_2(J) \cap Z) = M_2(E/J)$.) We have thus proved the existence part of the proposition.

We now turn to the uniqueness part. First, let us assume that Φ is already extendible to A_0, with maps $\varphi_0, \psi_0 : A_0 \longrightarrow M_G(E)$. Let $R_0 = \varphi_0[A_0]$ and let $S_0 = \psi_0[A_0]$, which are subalgebras of $M_G(E)$. Let E_0 be the C^*-subalgebra of $M_G(E)$ generated by R_0 and S_0, and let J_0 be the ideal in E_0 generated by all $\varphi_0(a) - \psi_0(a)$ for $a \in A_0$. Finally, let

$$Z_0 = \begin{pmatrix} \overline{R_0 + R_0 J_0 R_0} & \overline{R_0 S_0 + R_0 J_0 S_0} \\ \overline{S_0 R_0 + S_0 J_0 R_0} & \overline{S_0 + S_0 J_0 S_0} \end{pmatrix}.$$

(Thus, R_0, S_0, etc. are the analogs of R, S, etc. obtained by using φ_0, ψ_0, and A_0 instead of φ, ψ, and A.) It is obvious that R is an ideal in R_0 and S is an ideal in S_0. We now claim that E is an ideal in E_0 and Z is an ideal in Z_0. Using the condition (*) in definition 3.3.1, we obtain $S_0 R \subset \overline{SR}$, and $R_0 S \subset \overline{RS}$. (Sample computation: if $a \in A_0$, $b \in A$, and (e_λ) is an approximate identity for A which is quasicentral for A_0, then the condition (*) implies that

$$\psi_0(a)\varphi(b) = \lim_\lambda \psi(ae_\lambda)\varphi(b) \in \overline{SR},$$

whence $S_0 R \subset \overline{SR}$.) It follows that $E_0 E \subset E$. The proof that $EE_0 \subset E$ is similar.

We next show that $RJ_0 \subset \overline{RJ}$. Let $j = \varphi_0(a) - \psi_0(a)$ for some $a \in A_0$. Since R_0 and S_0 generate E_0, it is enough to show that the subsets RjE_0, RR_0jE_0, RS_0jE_0, $RS_0R_0jE_0$, etc. are contained in $\overline{RJ}$. We do the last of these:

$$RS_0R_0jE_0 \subset \overline{RSR_0jE_0} \subset \overline{RSRjE_0} \subset \overline{RSRJE_0} \subset \overline{RJ}.$$

The last inclusion follows from the fact that J is an ideal in E and E is an ideal in E_0, whence J is an ideal in E_0. The second last inclusion follows from the relation $Rj \subset \overline{RJ}$, which is proved in the same way that $S_0R \subset \overline{SR}$ was proved. Similar arguments enable us to show that $R_0 J \subset \overline{RJ}$, and also to prove the analogous inclusions with R replaced by S and with the factors in the other order. With these facts in hand, it is straightforward to prove that Z is an ideal in Z_0.

Just as before, let $I_0 = Z_0 \cap M_2(J_0)$. Then Z_0/I_0 is canonically isomorphic to $M_2(E_0/J_0)$. Since A_0 is separable, Z_0 is also, and there is by theorem 3.1.8 a derivation d of Z_0 lifting the derivation $\delta_0 \otimes id_{E_0/J_0}$ of $M_2(E_0/J_0)$. Since Z is an ideal in Z_0, it follows that $d[Z] \subset Z$. The methods of the previous paragraph show that $\overline{I_0 Z} = \overline{IZ}$, whence $I_0 \cap Z = I$. It is now clear that $d|_Z$ is a lift of $\delta_0 \otimes id_{E/J}$, and thus would have been a possible choice for the derivation δ used above. But then $t\delta + (1-t)d|_Z$ is also a derivation of Z which lifts $\delta_0 \otimes id_{E/J}$, and substituting it for δ in the construction of the extendible p.q.h. Σ yields a homotopy of extended p.q.h.'s from the one constructed using δ to the one constructed using $d|_Z$. (The one tricky point is to verify condition (*) of definition 3.3.1 for the appropriate extended p.q.h. from A to $K \otimes B \otimes C([0,1])$. This requires an equicontinuity argument similar to the one in the proof of lemma 3.3.4 (3).) We may therefore assume that $d|_Z = \delta$.

We now show how to extend the homotopy constructed previously to a homotopy of extendible p.q.h.'s. First, we observe that Φ_0 is also extendible, and is homotopic to Φ as an extended p.q.h. Next, define an automorphism ρ_t of $M_2(E_0)$ by $\rho_t = \exp(t\delta_0 \otimes id_{E_0})$. Then ρ_t is the extension of the previous ρ_t to a larger algebra. Also define an automorphism η_t of Z_0 by $\eta_t = \exp(td)$, again an extension of the previous η_t, since $d|_Z = \delta$. We regard η_t as a map from Z_0 to $M_2(E_0)$ (which contains

Z_0). Let

$$\varphi_t^0 = \rho_t \cdot \begin{bmatrix} 0 & 0 \\ 0 & \varphi_0 \end{bmatrix} \quad \text{and} \quad \psi_t^0 = \eta_t \cdot \begin{bmatrix} 0 & 0 \\ 0 & \psi_0 \end{bmatrix}.$$

Then $\varphi_t^0|_A = \varphi_t$ and $\psi_t^0|_A = \psi_t$.

We want to show that $t \longrightarrow (\varphi_t^0, \psi_t^0)$ defines a homotopy of extended p.q.h.'s. It is sufficient to prove that if $x \in M_2(E) \otimes C([0,1])$, (e_λ) is an approximate identity for A which is quasicentral for A_0, and $a \in A_0$, then $\varphi_t^0(ae_\lambda)x(t) \longrightarrow \varphi_t^0(a)x(t)$ and $\psi_t^0(ae_\lambda)x(t) \longrightarrow \psi_t^0(a)x(t)$, both uniformly in $t \in [0,1]$. Since the proof for ψ_t^0 is harder, we omit the proof for φ_t^0. The following argument is essentially due to Nigel Higson.

First, observe that $f_\lambda = \begin{bmatrix} \varphi(e_\lambda) & 0 \\ 0 & \psi(e_\lambda) \end{bmatrix}$ is an approximate identity for Z. Now set $u_\lambda = f_\lambda \otimes 1$, which is an approximate identity for $Z \otimes C([0,1])$. It is immediate from the definition of Z_0 and the fact that (φ_0, ψ_0) is an extension of (φ, ψ) that $\lim_\lambda z f_\lambda x = zx$ for all $z \in Z_0$ and $x \in M_2(E)$. Consequently $\lim_\lambda z u_\lambda x = zx$ for all $z \in Z_0 \otimes C([0,1])$ and $x \in M_2(E) \otimes C([0,1])$.

Since (e_λ) is quasicentral for A_0, we have $\| e_\lambda a - a e_\lambda \| \longrightarrow 0$. Define $\psi^0 : A_0 \longrightarrow M_2(E_0) \otimes C([0,1])$ by $\psi^0(a)(t) = \psi_t^0(a)$. We must show that for $x \in M_2(E) \otimes C([0,1])$, we have $\lim_\lambda \psi^0(ae_\lambda)x = \psi^0(a)x$, and by quasicentrality it is sufficient to show that $\lim_\lambda \psi^0(e_\lambda a)x = \psi^0(a)x$. Define

$$\eta : Z \otimes C([0,1]) \longrightarrow M_2(E) \otimes C([0,1])$$

by $\eta(z)(t) = \eta_t(z(t))$. Then η is actually an automorphism of $Z \otimes C([0,1])$, so that $(\eta(u_\lambda))$ is also an approximate identity for $Z \otimes C([0,1])$. Furthermore, $\eta(u_\lambda)\psi^0(a) = \psi^0(e_\lambda a)$ for $a \in A_0$. Therefore we have, for $a \in A_0$, $x \in M_2(E) \otimes C([0,1])$, and any fixed λ_0,

$$\begin{aligned} \| \psi^0(e_\lambda a)x - \psi^0(a)x \| \le{} & \| \eta(u_\lambda) \| \; \| \psi^0(a)x - \psi^0(a)u_{\lambda_0}x \| \\ & + \| \eta(u_\lambda)\psi^0(a)u_{\lambda_0} - \psi^0(a)u_{\lambda_0} \| \; \| x \| \\ & + \| \psi^0(a)u_{\lambda_0}x - \psi^0(a)x \| . \end{aligned}$$

The result of the previous paragraph implies that λ_0 can be chosen such that $\| \psi^0(a)u_{\lambda_0}x - \psi^0(a)x \| < \frac{\varepsilon}{4}$. Since $\psi^0(a)u_{\lambda_0} \in Z$, we have

$$\| \eta(u_\lambda)\psi^0(a)u_{\lambda_0} - \psi^0(a)u_{\lambda_0} \| < \frac{\varepsilon}{2\|x\|}$$

for all sufficiently large λ. Since $\| \eta(u_\lambda) \| \le 1$ for all λ, we obtain $\| \psi^0(e_\lambda a)x - \psi^0(a)x \| < \varepsilon$ for all sufficiently large λ. Thus, $\psi^0(e_\lambda a)x \longrightarrow \psi^0(a)x$, as desired.

We have now proved that we have a homotopy of extended p.q.h.'s. In particular, we find that Σ_0 is extendible by taking $t = 1$. Clearly Σ_0 and Σ define the same extended q.h., and by lemma 3.3.5 (2), we conclude that the extended q.h. defined by Σ is homotopic to the extended q.h. (φ_0, ψ_0).

Now let Φ be any q.h., and let Σ_0 and Σ_1 be two extended q.h.'s which are homotopic to Φ as ordinary q.h.'s. Then Σ_0 and Σ_1 are homotopic as ordinary q.h.'s. Let Σ be a homotopy, which is a q.h. from A to $B \otimes C([0,1])$, and let T be the extended q.h. constructed from it as in the existence part of the proof. The previous argument shows Σ_0 and $ev_0 T$, and Σ_1 and $ev_1 T$, are homotopic pairs of extended q.h.'s, while T is a homotopy of extended q.h.'s from $ev_0 T$ to $ev_1 T$. Therefore Σ_0 and Σ_1 are homotopic, and uniqueness is proved. Q.E.D.

We point out that separability of A_0 was used only in the uniqueness part. Also, the requirement that the approximate identities in the definition of extendibility be quasicentral was used at only one point, namely in the construction of the homotopy of extended p.q.h.'s in the uniqueness part of the proof.

We now obtain the analogous result for p.q.h.'s.

3.3.7 Corollary. Let Φ be a p.q.h. from A to $K \otimes B$, and let A_0 be a separable G-algebra which contains A as an ideal. Then there is a q.h. Σ from A to $K \otimes B$ which is homotopic to Φ and extendible to A_0. The extension of Σ is unique up to homotopy of p.q.h.'s extended to A_0.

Proof. Let Φ_0 be the q.h. defined by Φ, and let Σ be the extendible q.h. homotopic to Φ_0 whose existence was shown in the previous proposition. Since Φ_0 is homotopic to Φ by lemma 3.2.9 (3), it follows that Σ is also homotopic to Φ. The existence part is thus proved. For the uniqueness part, let T be any extendible p.q.h. which is homotopic to Φ. Then lemma 3.3.5 (3) implies that the extension of T to A_0 is homotopic to the extended q.h. T_0 that it defines, and the previous proposition now implies that T_0 is homotopic to the extension of Σ. This proves uniqueness. Q.E.D.

3.3.8 Remark. In the next section, we are going to need to require the intermediate algebras E and J of a p.q.h.

$$\Phi = (\varphi, \psi) : A \longrightarrow E \triangleright J \longrightarrow B$$

to be separable. We therefore point out that the previous corollary remains true if only p.q.h.'s with separable intermediate algebras are considered, because all the constructions done in its proof preserve this property.

3.4. The Kasparov Product

Our attempt to define the product now runs into a new problem. Let

$$\Phi_0 = (\varphi, \psi) : A_0 \longrightarrow E \rhd J \longrightarrow K \otimes A_1$$

be a q.h., and let Φ_1 be a q.h. from A_1 to $K \otimes A_2$. Let

$$\Sigma = (\sigma, \tau) : J \longrightarrow F \rhd I \longrightarrow K \otimes A_2$$

be a q.h. which is homotopic to the restriction to J of the obvious p.q.h. $id_K \otimes \Phi_1$ from $K \otimes A_1$ to $K \otimes A_2$, and which is extendible to E. Call the extended maps also σ and τ. Then we are hoping to define a p.q.h., to be thought of intuitively as $[\sigma\varphi] - [\tau\varphi] - [\sigma\psi] + [\tau\psi]$, which represents the product $[\Phi_0][\Phi_1]$. A good candidate would seem to be the direct sum $(\sigma\varphi, \tau\varphi) \oplus (\tau\psi, \sigma\psi)$. Unfortunately it may not be true that $\sigma\varphi(a) - \tau\varphi(a) \in I$ for $a \in A_0$, since $\varphi(a)$ need not be in J. A similar problem occurs if we try to use $(\sigma\varphi, \sigma\psi) \oplus (\tau\psi, \tau\varphi)$ instead. Therefore, in the following definition, we need to introduce a certain automorphism which "twists" the second component $\tau\varphi \oplus \sigma\psi$ in such a way that we do in fact obtain a p.q.h. Then we must prove that an appropriate automorphism actually exists — a nontrivial fact.

3.4.1 Definition. Let A_0, A_1, and A_2 be G-algebras, and let

$$\Phi_0 = (\varphi, \psi) : A_0 \longrightarrow E \rhd J \xrightarrow{\mu} K \otimes A_1,$$

and

$$\Phi_1 = (\varphi_1, \psi_1) : A_1 \longrightarrow E_1 \rhd J_1 \xrightarrow{\mu_1} K \otimes A_2$$

be p.q.h.'s such that the intermediate algebras E and E_1 are separable. To define their product, we proceed as follows: Replace Φ_1 by

$$id_K \otimes \Phi_1 = (id \otimes \varphi_1, id \otimes \psi_1) : K \otimes A_1 \longrightarrow K \otimes E_1 \rhd K \otimes J_1 \xrightarrow{id \otimes \mu_1} K \otimes K \otimes A_2 \simeq K \otimes A_2 .$$

Next, let

$$\Sigma = (\sigma, \tau) : J \longrightarrow F \rhd I \xrightarrow{\nu} K \otimes A_2$$

be a p.q.h. which is extendible to E and homotopic to the composite $(id_K \otimes \Phi_1)\mu$, which is a p.q.h. from J to $K \otimes A_2$. We also call the extended maps σ and τ. We next define a subalgebra Q of $M_2(M_G(F))$ and find an equivariant derivation δ of Q such that, with $\gamma = \exp(\delta)$, the expression

$$\left(\begin{bmatrix}\sigma\varphi & 0\\ 0 & \tau\psi\end{bmatrix}, \gamma\begin{bmatrix}\tau\varphi & 0\\ 0 & \sigma\psi\end{bmatrix}\right) : A_0 \longrightarrow M_2(M_G(F)) \rhd M_2(I) \xrightarrow{\nu} M_2(K\otimes A_2) \simeq K\otimes A_2 \quad (*)$$

is defined and is a p.q.h. We then set $[\Phi_0][\Phi_1]$ equal to the class in $KK_G(A_0, A_2)$ of this p.q.h. (We write the product in the same order as Cuntz and Kasparov do, which is the reverse of the usual convention for the composition of ordinary functions.)

In order to define Q, we first define equivariant functions (not homomorphisms!) from A_0 to $M_G(F)$ as follows:

$$D_1(a) = \sigma\varphi(a) - \tau\varphi(a),$$
$$D_2(a) = \sigma\varphi(a) - \sigma\psi(a),$$
$$D_3(a) = \sigma\psi(a) - \tau\psi(a),$$

and

$$D_4(a) = \tau\varphi(a) - \tau\psi(a).$$

We then let Q be the C^*-subalgebra of $M_2(M_G(F))$ generated by all 2×2 matrices whose entries are in the ranges of some D_i, and by the matrices of the form $\begin{bmatrix}\sigma\varphi(a) & 0\\ 0 & \sigma\varphi(a)\end{bmatrix}$ for $a \in A_0$. The derivation δ is any derivation of Q satisfying the following condition: for $a \in A_0$ and $(a_{ij}) \in M_2(A_0)$, we have

$$\left.\begin{array}{l}\delta\left[(D_1(a_{ij}))\right] - \dfrac{\pi i}{2}\left(\begin{bmatrix}0 & 1\\ 1 & 0\end{bmatrix}(D_1(a_{ij})) - (D_1(a_{ij}))\begin{bmatrix}0 & 1\\ 1 & 0\end{bmatrix}\right) \in M_2(I),\\[2ex] \delta\left[(D_2(a_{ij}))\right] \in M_2(I),\\[2ex] \text{and}\\[2ex] \delta\begin{bmatrix}\sigma\varphi(a) & 0\\ 0 & \sigma\varphi(a)\end{bmatrix} \in M_2(I).\end{array}\right\} \quad (**)$$

We now want to show that this definition makes sense. This is the purpose of the next three propositions, which show that (*) really does define a p.q.h., that a derivation δ satisfying (**) exists, and that the class in $KK_G(A_0, A_2)$ of the p.q.h. (*) depends only on the homotopy classes of Φ_0 and Φ_1, and not on the choices of Σ and δ.

3.4.2 Proposition. Under the hypotheses of definition 3.4.1, if δ is a derivation satisfying (**) and $\gamma = \exp(\delta)$, then the formula (*) defines a p.q.h.

Proof. Since the exponential of an inner derivation $a \longrightarrow da - ad$ is the inner automorphism $a \longrightarrow \exp(d)a\exp(d)^{-1}$, and since $\exp\left(\frac{\pi i}{2}\begin{bmatrix}0 & 1\\ 1 & 0\end{bmatrix}\right) = \begin{bmatrix}0 & 1\\ 1 & 0\end{bmatrix}$, the relation (**) implies the following for the automorphism $\gamma = \exp(\delta)$: for $a \in A_0$ and $(a_{ij}) \in M_2(A_0)$,

$$\gamma\left[(D_1(a_{ij}))\right] - \begin{bmatrix} 0 & 1 \\ 1 & 0 \end{bmatrix} (D_1(a_{ij})) \begin{bmatrix} 0 & 1 \\ 1 & 0 \end{bmatrix} \in M_2(I),$$

$$\gamma\left[(D_2(a_{ij}))\right] - (D_2(a_{ij})) \in M_2(I),$$

and

$$\gamma\begin{bmatrix} \sigma\varphi(a) & 0 \\ 0 & \sigma\varphi(a) \end{bmatrix} - \begin{bmatrix} \sigma\varphi(a) & 0 \\ 0 & \sigma\varphi(a) \end{bmatrix} \in M_2(I).$$

(To prove this, we work in $Q/(Q \cap M_2(I))$, and observe that the derivation δ preserves the ideal $Q \cap M_2(I)$.) In order to prove that, for $a \in A_0$,

$$\begin{bmatrix} \sigma\varphi(a) & 0 \\ 0 & \tau\psi(a) \end{bmatrix} - \gamma\begin{bmatrix} \tau\varphi(a) & 0 \\ 0 & \sigma\psi(a) \end{bmatrix} \in M_2(I),$$

we rewrite the entries as follows:

$$\tau\psi(a) = \sigma\varphi(a) - D_1(a) - D_4(a),$$

$$\tau\varphi(a) = \sigma\varphi(a) - D_1(a),$$

and

$$\sigma\psi(a) = \sigma\varphi(a) - D_2(a).$$

A straightforward calculation, using the results of the beginning of the proof and the easily verified fact that $D_2(a) - D_4(a) \in I$ for $a \in A_0$, now proves the desired result. Q.E.D.

3.4.3 Proposition. Under the hypotheses of definition 3.4.1, there exists an equivariant derivation δ satisfying (*).

Proof. We retain the notation of definition 3.4.1. Further, let P be the image of Q in $M_2(M_G(F)/I)$, and let R be the C^*-subalgebra of $M_G(F)/I$ generated by the images of all $D_i(a)$ and $\sigma\varphi(a)$ for $a \in A_0$. Then $P \subset M_2(R)$. For $i = 1, 2$, let T_i be the ideal in R generated by the images of the $D_i(a)$ for $a \in A_0$, and let Y be the C^*-algebra consisting of the images of $\sigma\varphi(a)$ in $M_G(F)/I$ for $a \in A_0$. We are going to construct an equivariant derivation δ_0 of P which is zero on $M_2(T_2)$ and on the algebra

$$Z = \left\{ \begin{bmatrix} y & 0 \\ 0 & y \end{bmatrix} : y \in Y \right\},$$

but which acts like

$$x \longrightarrow \frac{\pi i}{2}\left[\begin{bmatrix} 0 & 1 \\ 1 & 0 \end{bmatrix} x - x \begin{bmatrix} 0 & 1 \\ 1 & 0 \end{bmatrix}\right]$$

on $M_2(T_1)$. Then δ will be an equivariant lifting of δ_0 to Q. In order to construct δ_0, we will need several relations between the $M_2(T_i)$ and Z.

We start out by proving the following facts about the maps D_i: if $a, b, x \in A_0$, then

(1) $D_i(a) - D_j(a) \in I$ if i and j have the same parity (that is, both odd or both even);

(2) $D_i(a)D_j(b) \in I$ if i and j have opposite parity; and

(3) $D_i(a)\sigma\varphi(x)D_j(b) \in I$ if i and j have opposite parity.

The first one is easy: $D_1(a) - D_3(a) = D_2(a) - D_4(a) = (\sigma - \tau)(\varphi - \psi)(a)$, which is in I. For the second one, we first consider the case $i = 1$ and $j = 2$. Then

$$D_1(a)D_2(b) = (\sigma - \tau)\left[\varphi(a)(\varphi(b) - \psi(b))\right] - \tau(\varphi(a))(D_1(b) - D_3(b)),$$

and both terms on the right are in I because I and J are ideals. Next, the relation $D_2(a)D_1(b) \in I$ follows by taking adjoints, since $D_i(a)^* = D_i(a^*)$ for all a. The other combinations now follow by applications of (1), for example

$$D_3(a)D_2(b) = D_1(a)D_2(b) - (D_1(a) - D_3(a))D_2(b) \in I,$$

since I is an ideal.

Finally, the third expression satisfies, for $i = 1$ and $j = 2$,

$$\begin{aligned} D_1(a)\sigma(\varphi(x))D_2(b) = {} & (\sigma - \tau)\left[\varphi(a)\varphi(x)(\varphi(b) - \psi(b))\right] \\ & - \tau(\varphi(ax))(D_1(b) - D_3(b)) \\ & + \tau(\varphi(a))(\tau - \sigma)(\varphi(x))\sigma(\varphi(b) - \psi(b)). \end{aligned}$$

(One checks that, after everything is multiplied out, all the unwanted terms cancel out.) The first two terms are clearly in I. To show that the last one is, let $w = \varphi(x)$, let $s = \sigma(\varphi(b) - \psi(b))$, and let (e_λ) be an approximate identity for J which is quasicentral for E. Then $w \in E$, $s \in F$, and condition (*) in the definition of extendibility implies that

$$(\tau(w) - \sigma(w))s = \lim_\lambda (\tau(we_\lambda) - \sigma(we_\lambda))s.$$

Since $we_\lambda \in J$, it follows that $\tau(we_\lambda) - \sigma(we_\lambda) \in I$, whence $(\tau(w) - \sigma(w))s \in I$, and we have shown that $D_1(a)\sigma(\varphi(x))D_2(b) \in I$. Now the other combinations are in I by the same reasoning as used for (2).

We use these facts to show that $T_1T_2 = 0$. From the definition of T_i, it is sufficient to show that if $a, b \in A_0$ and r, s, and t are finite products of elements of $M_G(F)$ of the form $D_i(x)$ and $\sigma\varphi(x)$ for $x \in A_0$, then $rD_1(a)sD_2(b)t \in I$. Since I is an ideal, it is enough to show that $D_1(a)sD_2(b) \in I$. Write

$$s = s_0 D_{i_1}(x_1)s_1 \cdot \cdots \cdot s_{n-1}D_{i_n}(x_n)s_n,$$

where each s_j is either 1 or $\sigma\varphi(y_j)$, and $x_j, y_j \in A_0$. Then in the product $D_1(a)sD_2(b)$ there must be a factor of the form $D_{i_k}(x_k)s_k D_{i_{k+1}}(x_{k+1})$, where i_k and i_{k+1} have opposite parity. (We set $i_0 = 1$, $i_{n+1} = 2$, $x_0 = a$, and $x_{n+1} = b$.) If $s_k = 1$ this expression is in I by (2), and if $s_k = \sigma\varphi(y_k)$ this expression is in I by (3). So $D_1(a)sD_2(b) \in I$, and it follows that $T_1T_2 = 0$. Since T_1 and T_2 are ideals in R, it follows that $T_1 \cap T_2 = 0$.

We now want to show that $M_2(T_1) + M_2(T_2) + Z = P$. First, this sum is a C^*-algebra by [79], 1.5.8, since $M_2(T_1)$ and $M_2(T_2)$ are ideals in $M_2(R)$. Furthermore, $M_2(T_1) + M_2(T_2) + Z$ contains all 2×2 matrices with entries of the form $\pi_0(D_1(a))$ or $\pi_0(D_2(a))$, where $\pi_0 : M_G(F) \longrightarrow M_G(F)/I$ is the quotient map. Therefore it also contains all 2×2 matrices with entries of the form $\pi_0(D_3(a))$ or $\pi_0(D_4(a))$, since $D_1(a) - D_3(a)$ and $D_2(a) - D_4(a)$ are in I. Finally, $M_2(T_1) + M_2(T_2) + Z$ contains all matrices $\begin{bmatrix} \pi_0\sigma\varphi(a) & 0 \\ 0 & \pi_0\sigma\varphi(a) \end{bmatrix}$, since they are in Z. Thus $P \subset M_2(T_1) + M_2(T_2) + Z$.

We now have to show the opposite inclusion. That $Z \subset P$ is obvious. The proofs of the inclusions $M_2(T_i) \subset P$ are similar, so we do only one of them, namely the one for $i = 1$. The algebra T_1 is the C^*-subalgebra of R generated by all $\pi_0(D_1(a))$ and $\pi_0(\sigma\varphi(a)D_1(b))$, for $a, b \in A_0$. (We need not consider multiplications by elements of the image of D_3 because of relation (1) above, and by elements of the images of D_2 and D_4 because of relations (2) and (3).) Let $a, b \in A_0$. Then $\begin{bmatrix} \pi_0 D_1(a) & 0 \\ 0 & 0 \end{bmatrix}$ is obviously in P, and

$$\begin{bmatrix} \pi_0(\sigma\varphi(a)D_1(b)) & 0 \\ 0 & 0 \end{bmatrix} = \begin{bmatrix} \pi_0\sigma\varphi(a) & 0 \\ 0 & \pi_0\sigma\varphi(a) \end{bmatrix} \begin{bmatrix} \pi_0 D_1(a) & 0 \\ 0 & 0 \end{bmatrix}$$

is also in P. Similar arguments apply to the cases in which $\pi_0 D_1(a)$ and $\pi_0(\sigma\varphi(a)D_1(b))$ appear in the other three corners, and so $M_2(T_1) \subset P$ as desired. We have now shown that $P = M_2(T_1) + M_2(T_2) + Z$.

Let d be the element of M_2 defined by $d = \frac{\pi i}{2}\begin{bmatrix} 0 & 1 \\ 1 & 0 \end{bmatrix}$. Define a derivation δ_0 of P by $\delta_0(s_1 + s_2 + z) = ds_1 - s_1 d$ for $s_i \in M_2(T_i)$ and $z \in Z$. We must show that δ_0 is well defined. (It will then obviously be a derivation, and hence bounded by [79], 8.6.3.) So assume $s_1 + s_2 + z = r_1 + r_2 + y$, with $r_i \in M_2(T_i)$ and $y \in Z$. Then $y - z \in Z \cap M_2(T_1 + T_2)$. From the definition of Z, the element, $y - z$ must have the form $\begin{bmatrix} x & 0 \\ 0 & x \end{bmatrix}$ for some $x \in T_1 + T_2$. Write $x = t_1 + t_2$ for $t_i \in T_i$, so that

$$y - z = \begin{bmatrix} t_1 & 0 \\ 0 & t_1 \end{bmatrix} + \begin{bmatrix} t_2 & 0 \\ 0 & t_2 \end{bmatrix}.$$

Let $u_i = \begin{bmatrix} t_i & 0 \\ 0 & t_i \end{bmatrix}$, and rewrite this equation as $s_1 - r_1 + s_2 - r_2 = u_1 + u_2$. Since r_i, s_i, $u_i \in M_2(T_i)$ and $T_1 \cap T_2 = 0$, it follows that $s_1 - r_1 - u_1 = -s_2 + r_2 + u_2 = 0$. Therefore

$$\delta_0(r_1+r_2+y) = dr_1 - r_1 d = ds_1 - s_1 d - (du_1 - u_1 d)$$
$$= ds_1 - s_1 d = \delta_0(s_1+s_2+z),$$

where the third step follows from the relation $du_1 = u_1 d$. Consequently δ_0 is well defined. It is obviously equivariant. Let π be the projection from Q to P. Then by theorem 3.1.8 there is an equivariant derivation δ of Q such that $\pi\delta = \delta_0\pi$. (Recall that everything is separable.) Clearly δ satisfies the condition (**) of definition 3.4.1.

Q.E.D.

3.4.4 Remark. It is clear that a derivation δ of Q satisfies (**) of 3.4.1 if and only if it is a lifting of the derivation δ_0 of P constructed in the proof of the previous proposition.

3.4.5 Proposition. (Compare [25], theorem 3.3.) The product defined in 3.4.1 exists and is a well defined map from $KK_G(A_0, A_1) \times KK_G(A_1, A_2)$ to $KK_G(A_0, A_2)$.

Proof. Propositions 3.4.2 and 3.4.3 show that the procedure in definition 3.4.1 can be carried out. We must now show that the result is independent of the various choices made. First, suppose ε is a different derivation satisfying (**) of 3.4.1. Then $t\delta + (1-t)\varepsilon$ is also a derivation satisfying (**), and, with $\gamma_t = \exp(t\delta + (1-t)\varepsilon)$ for $t \in [0,1]$, we can define a p.q.h. from A_0 to A_2 using γ_t instead of γ. This clearly defines a homotopy of p.q.h.'s, so the homotopy class of the product is independent of the choice of δ.

Next, we show that the product $[\Phi_0][\Phi_1]$ does not depend on the choice of the extendible p.q.h. Σ. Let Φ_0 and Φ_1 be as in definition 3.4.1, and let, for $i = 0, 1$,

$$\Sigma_i = (\sigma_i, \tau_i) : J \longrightarrow F_i \rhd I_i \xrightarrow{\nu_i} K \otimes A_2$$

be two p.q.h.'s which are homotopic to $(id_K \otimes \Phi_0)\mu$ and extendible to E. By corollary 3.3.7, there is a homotopy consisting of extendible p.q.h.'s

$$\Sigma_t = (\sigma_t, \tau_t) : J \longrightarrow F_t \rhd I_t \xrightarrow{\nu_t} K \otimes A_2$$

for $t \in [0,1]$, and an extendible p.q.h.

$$\Sigma = (\sigma, \tau) : J \longrightarrow F \rhd I \xrightarrow{\nu} K \otimes A_2 \otimes C([0,1]),$$

with $F \subset \prod_{s \in [0,1]} F_s$ and $I \subset \prod_{s \in [0,1]} I_s$. (We are of course assuming that F and I are separable.) Call the extensions of σ_t, τ_t, σ, and τ also σ_t, etc. Also let π_t be the projection from $M_2\left(\prod_s M_G(F_s)\right)$ to $M_2(M_G(F_t))$. Apply the construction in the definition of the product to obtain a derivation δ of the appropriate subalgebra Q of $M_2(M_G(F))$ such that, with $\gamma = \exp(\delta)$, we obtain a p.q.h.

$$\Omega = \left(\begin{bmatrix} \sigma\varphi & 0 \\ 0 & \tau\psi \end{bmatrix}, \gamma\cdot \begin{bmatrix} \tau\varphi & 0 \\ 0 & \sigma\psi \end{bmatrix} \right) : A_0 \longrightarrow M_2(M_G(F)) \triangleright M_2(I)$$

$$\xrightarrow{\nu} M_2(K \otimes A_2 \otimes C([0,1])) \simeq K \otimes A_2 \otimes C([0,1]).$$

Now $M_2(M_G(F))$ is in a natural way a subalgebra of $M_2\left(\prod_s M_G(F_s)\right)$. Consequently π_t defines a map, also called π_t, from Q to $M_2(M_G(F_t))$, and one sees that its image is exactly the subalgebra Q_t of $M_2(M_G(F_t))$ analogous to Q. By standard properties of derivations, there is a derivation δ_t of Q_t such that $\pi_t \circ \delta = \delta_t \circ \pi_t$, and this derivation is easily seen to satisfy (**) of definition 3.4.1. Therefore, with $\gamma_t = \exp(\delta_t)$,

$$\Omega_t = \left(\begin{bmatrix} \sigma_t\varphi & 0 \\ 0 & \tau_t\psi \end{bmatrix}, \gamma_t \circ \begin{bmatrix} \tau_t\varphi & 0 \\ 0 & \sigma_t\psi \end{bmatrix} \right) : A_0 \longrightarrow M_2(M_G(F_t)) \triangleright M_2(I_t)$$

$$\xrightarrow{\nu_t} M_2(K \otimes A_2) \simeq K \otimes A_2$$

is a p.q.h., and the p.q.h.'s Ω_t, together with Ω, define a homotopy from Ω_0 to Ω_1. But Ω_0 and Ω_1 define the product of $[\Phi_0]$ and $[\Phi_1]$, computed using the extendible p.q.h.'s Σ_0 and Σ_1 respectively.

We now see that the product depends only on the homotopy class of Φ_1. Indeed, replacing Φ_1 by a homotopic p.q.h. merely results in replacing Σ by a homotopic extendible p.q.h.

Finally, we consider the effect of replacing Φ_0 by a homotopic p.q.h. Let $t \longrightarrow \Lambda_t$, together with Λ, define a homotopy of p.q.h.'s from A_0 to $K \otimes A_1$. Write

$$\Lambda_t = (\lambda_t, \eta_t) : A_0 \longrightarrow D_t \triangleright L_t \xrightarrow{\varepsilon_t} K \otimes A_1$$

and

$$\Lambda = (\lambda, \eta) : A_0 \longrightarrow D \triangleright L \xrightarrow{\varepsilon} K \otimes A_1 \otimes C([0,1]).$$

Let π_t be the projection map from D to D_t. We are assuming that D_t is separable for every t, and we may therefore assume that D is also separable. To see this, we replace the given homotopy by a new one consisting of three stages: a homotopy from Λ_0 to the q.h. it defines, the homotopy of q.h.'s defined by the given homotopy, and finally a homotopy from the q.h. defined by Λ_1 to Λ_1. The algebra corresponding to D in the middle stage is separable because A_0 and A_1 are, while the separability of the corresponding algebras in the first and last stages is clear from the construction of the homotopies given in the proof of lemma 3.2.9 (3).

We now want to construct a p.q.h. which is homotopic to $(id_K \otimes \Phi_1 \otimes id_{C([0,1])})\,\varepsilon$ and extendible to D, and which is related in an appropriate manner to the p.q.h.'s

$(id_K \otimes \Phi_1)\varepsilon_t$ for $t \in [0,1]$. We must therefore pay some attention to the details of the construction of extendible p.q.h.'s. Let

$$\Sigma_t = (\sigma_t, \tau_t) : L_t \longrightarrow F_t \triangleright I_t \overset{\nu_t}{\longrightarrow} K \otimes A_2$$

be the q.h. defined by $(id_K \otimes \Phi_1)\varepsilon_t$. Let F be the C^*-algebra generated by the images of the obvious maps $\sigma, \tau : L \longrightarrow \prod_{t \in [0,1]} F_t$, and let $I \subset \prod_{t \in [0,1]} I_t$ be the ideal in F generated by the differences $\sigma(x) - \tau(x)$ for $x \in L$. Then we have a p.q.h.

$$\Sigma = (\sigma, \tau) : L \longrightarrow F \triangleright I \overset{\nu}{\longrightarrow} K \otimes A_2 \otimes C([0,1]),$$

which is in fact a q.h. (There are only two nontrivial points to check, namely that the image of ν is actually in $K \otimes A_2 \otimes C([0,1])$ and that I is essential in F. The proofs of these facts are, however, not difficult.) Furthermore, the obvious projection maps from F to F_t and from I to I_t are surjective for all $t \in [0,1]$.

We now construct an extendible p.q.h. from Σ according to the procedure in the proof of proposition 3.3.6. It is given by

$$\bar{\Sigma} = (\bar{\sigma}, \bar{\tau}) : L \longrightarrow Z \triangleright Z \cap M_2(I) \longrightarrow M_2(K \otimes A_2 \otimes C([0,1])) \longrightarrow (K \otimes A_2 \otimes C([0,1])),$$

where Z is a certain subalgebra of $M_2(F)$, $\bar{\sigma} = \sigma \oplus 0$, and $\bar{\tau} = \exp(\delta) \cdot (0 \oplus \tau)$ for an appropriate derivation δ of Z. (There is no need to pass to the q.h. defined by $\bar{\Sigma}$.) Let Z_t be the subalgebra of F_t obtained by carrying out the same construction for Σ_t. It is clear from the definition of Z in the proof of proposition 3.3.6 that the obvious projection κ_t from Z to $M_2(F_t)$ has image exactly equal to Z_t. Therefore Z can be identified with a subalgebra of $\prod_{t \in [0,1]} Z_t$. Furthermore, there is a unique derivation δ_t of Z_t such that $\kappa_t \cdot \delta = \delta_t \cdot \kappa_t$, and δ_t can be used in the construction of an extendible p.q.h. $\bar{\Sigma}_t$ from Σ_t. We obtain

$$\bar{\Sigma}_t = (\sigma_t \oplus 0, \exp(\delta_t) \cdot (0 \oplus \tau_t)) : L_t \longrightarrow Z_t \triangleright Z_t \cap M_2(I_t) \longrightarrow K \otimes A_2.$$

We now proceed with the construction of the product of Λ and $\bar{\Sigma}$ according to definition 3.4.1. Using the same kind of argument as in the previous paragraph, we obtain a p.q.h. Ω representing the product of Λ and $\bar{\Sigma}$, and p.q.h.'s Ω_t representing the product of Λ_t and $\bar{\Sigma}_t$ for $t \in [0,1]$, such that each intermediate algebra for Ω is a subalgebra of the product of the corresponding algebras for Ω_t in such a way that $t \longrightarrow \Omega_t$ and Ω constitute a homotopy of p.q.h.'s. So Ω_0 is homotopic to Ω_1. But $[\Omega_t] = [\Lambda_t][\Phi_1]$, so we have shown that the product only depends on the homotopy class of the first factor.
Q.E.D.

3.4.6 Proposition. Let A, A_0, B, and B_0 be G-algebras, and let $\lambda : A_0 \longrightarrow A$ and $\eta : B \longrightarrow B_0$ be equivariant homomorphisms. Let Φ be a p.q.h. from A to B. Then $\lambda^*([\Phi]) = [\lambda][\Phi]$ and $\eta_*([\Phi]) = [\Phi][\eta]$, where $[\lambda]$ and $[\eta]$ are the classes in KK-theory defined by λ and η as in definition 3.2.13.

Proof. Write

$$\Phi = (\varphi, \psi): A \longrightarrow E \triangleright J \xrightarrow{\mu} K\otimes B .$$

Replacing Φ if necessary by a homotopic p.q.h. (such as the q.h. it defines), we may assume that E and J are separable. (Clearly none of the expressions in the proposition is changed by such a replacement.)

We first compute $[\lambda][\Phi]$. Represent $[\lambda]$ by the p.q.h.

$$(\lambda, 0): A_0 \longrightarrow A \triangleright A \longrightarrow K\otimes A ,$$

where the map from A to $K\otimes A$ is $a \longrightarrow p\otimes a$ for a projection p on a one dimensional subspace, of the standard Hilbert space H, on which G acts trivially. The restriction of $id_K\otimes\Phi$ to A is clearly homotopic to Φ, and Φ is trivially extendible to A. If D_1 is as in the construction of the product, then for $a \in A_0$ we have

$$D_1(a) = \varphi(\lambda(a)) - \psi(\lambda(a)) \in J .$$

The derivation δ may therefore be taken to be zero, and the product $[\lambda][\Phi]$ is represented by the p.q.h.

$$\left(\begin{bmatrix} \varphi\circ\lambda & 0 \\ 0 & 0 \end{bmatrix}, \begin{bmatrix} \psi\circ\lambda & 0 \\ 0 & 0 \end{bmatrix} \right): A_0 \longrightarrow M_2(E) \triangleright M_2(J) \longrightarrow M_2(K\otimes B) \simeq K\otimes B .$$

This p.q.h. defines, up to homotopy, the same q.h. as

$$\Phi\lambda = (\varphi\circ\lambda, \psi\circ\lambda): A_0 \longrightarrow E \triangleright J \longrightarrow K\otimes B ,$$

whence $\lambda^*([\Phi]) = [\Phi\lambda] = [\lambda][\Phi]$.

We now compute $[\Phi][\eta]$. Without loss of generality, we may assume that Φ is in fact a q.h. We then need a p.q.h. which is homotopic to $((id_K\otimes\eta)\circ\mu, 0, id_{K\otimes B_0})$ and extendible to E. We choose

$$\Sigma = (id_J, 0): J \longrightarrow J \triangleright J \xrightarrow{(id_K\otimes\eta)\circ\mu} K\otimes B_0 .$$

(To see that Σ is homotopic to $((id_K\otimes\eta)\circ\mu, 0, id_{K\otimes B_0})$, observe that both define the same q.h.) Observe that the canonical inclusion of E in $M_G(J)$ defines an extension of Σ to E, since the approximate identity condition is trivially satisfied. A calculation similar to that of the other case shows that the algebra Q in the definition of the product is $M_2(E)$, and we may take δ to be the inner derivation defined by $\frac{\pi i}{2}\begin{bmatrix} 0 & 1 \\ 1 & 0 \end{bmatrix}$, so that γ is conjugation by $\begin{bmatrix} 0 & 1 \\ 1 & 0 \end{bmatrix}$. It is easily seen that the resulting product defines the same q.h. as $\eta\Phi$, as desired. Q.E.D.

3.4.7 Corollary. Let A, B, and C be G-algebras, and let $\varphi: A \longrightarrow B$ and $\psi: B \longrightarrow C$ be equivariant homomorphisms. Then $[\varphi][\psi] = [\psi\circ\varphi]$ as elements of $KK_G(A, C)$.

Proof. Immediate. Q.E.D.

3.4.8 Theorem. The Kasparov product is an $R(G)$-bilinear map from $KK_G(A_0, A_1) \times KK_G(A_1, A_2)$ to $KK_G(A_0, A_2)$.

Proof. The proof of additivity in the second variable is trivial. Additivity in the first variable requires the following construction. Let

$$\Phi_i = (\varphi_i, \psi_i) : A_0 \longrightarrow E_i \rhd J_i \xrightarrow{\mu_i} K \otimes A_1$$

for $i = 1, 2$ be p.q.h.'s, and let Ψ be a p.q.h. from A_1 to A_2. Let

$$\Sigma_i = (\sigma_i, \tau_i) : J_i \longrightarrow F_i \rhd I_i \xrightarrow{\nu_i} K \otimes A_2$$

be a p.q.h. which is homotopic to $(id_K \otimes \Psi)\mu_i$ and extendible to E_i. Then the p.q.h. $\Sigma = \lambda(\Sigma_1 \oplus \Sigma_2)$, where λ is the composite

$$(K \otimes A_2) \oplus (K \otimes A_2) \longrightarrow M_2(K \otimes A_2) \simeq K \otimes A_2,$$

is homotopic to $(id_K \otimes \Psi)(\mu_1 \oplus \mu_2)$ and extendible to $E_1 \oplus E_2$, and can therefore be used in the construction of the product $([\Phi_1] + [\Phi_2])[\Psi]$. The proof of additivity in the first variable is now obvious.

The proof of the relation

$$[V]([\Phi][\Psi]) = ([V][\Phi])[\Psi] = [\Phi]([V][\Psi]),$$

for a finite dimensional representation space V of G, is similar and is omitted.

Q.E.D.

We point out that this theorem, combined with proposition 3.4.6, provides an alternate proof that the maps on KK-theory defined by homomorphisms of the algebras are $R(G)$-linear.

We now prove a sequence of lemmas leading up to the proof that the product is associative. We will also find a much more convenient way of computing the product of a class in $KK_G(A, B)$ and a class in $KK_G(B, C)$. First, a definition.

3.4.9 Definition. Let

$$(\varphi, \psi) : A \longrightarrow E \rhd J \xrightarrow{\mu} B$$

be a p.q.h. It is called split if μ is injective, and if there is an equivariant homomorphism $\pi : E \longrightarrow A$ with kernel J such that $\pi\varphi = id_A$. (That is,

$$0 \longrightarrow J \longrightarrow E \underset{\pi}{\overset{\varphi}{\rightleftarrows}} A \longrightarrow 0$$

is an equivariantly split exact sequence.)

3.4.10 Lemma. ([25], lemma 4.2.) For any q.h. Φ from A to B there is split p.q.h. from A to B which defines Φ.

Proof. Write

$$\Phi = (\varphi, \psi) : A \longrightarrow E \triangleright J \longrightarrow B .$$

Let φ_0 be the homomorphism from A to $E \oplus A$ given by $\varphi_0 = \varphi \oplus id_A$, that is, $\varphi_0(a) = (\varphi(a), a)$ for $a \in A$, and let $\psi_0 = \psi \oplus id_A$. Let E_0 be the C^*-subalgebra of $E \oplus A$ generated by the images of φ_0 and ψ_0, and let π be the restriction to E_0 of the projection $E \oplus A \longrightarrow A$. Then it is easy to see that

$$(\varphi_0, \psi_0) : A \longrightarrow E_0 \triangleright J \longrightarrow B ,$$

with the splitting π, is the required p.q.h. Q.E.D.

3.4.11 Lemma. ([25], proposition 4.3.) Let

$$0 \longrightarrow J \longrightarrow E \underset{\pi}{\overset{\lambda}{\leftrightarrows}} A \longrightarrow 0$$

be an equivariantly split exact sequence of G-algebras (recall that according to our standing convention, everything is separable). If B is another G-algebra, then every p.q.h. from J to $K \otimes B$ is homotopic to the restriction of a p.q.h. from E to $K \otimes B$.

Proof. Let

$$\Phi = (\varphi, \psi) : J \longrightarrow F \triangleright I \overset{\mu}{\longrightarrow} B$$

be a p.q.h. from J to B. By corollary 3.3.7, we may assume that (φ, ψ, μ) is extendible to E, that is, that φ and ψ are the restrictions of maps, which we again call φ and ψ, from E to $M_G(F)$, satisfying the condition (*) in definition 3.3.1. (Note that this is not the same as saying that (φ, ψ, μ) defines a p.q.h. from E to B!) Furthermore, there is a p.q.h.

$$(id_E, \lambda\pi) : E \longrightarrow E \triangleright J .$$

We now construct the product of $(id_E, \lambda\pi)$ and (φ, ψ, μ) according to definition 3.4.1. Let Q be the subalgebra of $M_2(M_G(F))$ defined there, and let $D_2(a) = \varphi(a) - \varphi\lambda\pi(a)$ for $a \in E$, as in that definition. Let δ be a derivation of Q satisfying (**) of definition 3.4.1; such a derivation exists by proposition 3.4.3. In particular, we have

$$\delta \begin{bmatrix} D_2(a) & 0 \\ 0 & 0 \end{bmatrix} \in M_2(I) \qquad (*)$$

for $a \in E$. Furthermore, with $\gamma = \exp(\delta)$,

$$\Psi = \left(\begin{pmatrix} \varphi & 0 \\ 0 & \psi\lambda\pi \end{pmatrix}, \gamma\cdot \begin{pmatrix} \psi & 0 \\ 0 & \varphi\lambda\pi \end{pmatrix} \right) : E \longrightarrow M_2(M_G(F)) \rhd M_2(I) \longrightarrow K\otimes B$$

is a p.q.h. The relation (*), combined with $\pi(a)=0$ and $\varphi(a)-\psi(a)\in I$ for $a\in J$, yields after a short calculation

$$\begin{pmatrix} \varphi(a) & 0 \\ 0 & 0 \end{pmatrix} - \exp(t\delta)\cdot \begin{pmatrix} \psi(a) & 0 \\ 0 & 0 \end{pmatrix} \in M_2(I)$$

for $a\in J$ and $t\in[0,1]$. (Recall that a derivation automatically preserves ideals.) Therefore

$$t \longrightarrow \Sigma_t = \left(\begin{pmatrix} \varphi & 0 \\ 0 & 0 \end{pmatrix}, \exp(t\delta)\cdot \begin{pmatrix} \psi & 0 \\ 0 & 0 \end{pmatrix} \right)$$

defines a homotopy from $\Sigma_1 = \Psi|_J$ to

$$\Sigma_0 = \left(\begin{pmatrix} \varphi & 0 \\ 0 & 0 \end{pmatrix}, \begin{pmatrix} \psi & 0 \\ 0 & 0 \end{pmatrix} \right) : J \longrightarrow M_2(M_G(F)) \rhd M_2(I) \longrightarrow K\otimes B .$$

Since Σ_0 is homotopic to Ψ, we are done. Q.E.D.

3.4.12 Lemma. (See [25], lemma 4.1.) Let

$$\Phi = (\varphi,\psi) : A_0 \longrightarrow E \rhd J \xrightarrow{\mu} K\otimes A_1$$

be a p.q.h., and let Ψ be a p.q.h. from A_1 to $K\otimes A_2$. Suppose that $(id_K\otimes\Psi)\mu$ is homotopic to the restriction to J of a p.q.h.

$$\Sigma = (\sigma,\tau) : E \longrightarrow F \rhd I \longrightarrow K\otimes A_2 .$$

Then the product $[\Phi][\Psi]$ is represented by the p.q.h.

$$\Omega = (\sigma\varphi\oplus\tau\psi, \tau\varphi\oplus\sigma\psi) : A_0 \longrightarrow M_2(F) \rhd M_2(I) \longrightarrow M_2(K\otimes A_2) \simeq K\otimes A_2 .$$

Proof. Let Φ_0 be the p.q.h. from A_0 to $K\otimes E$ given by

$$\Phi_0 = (\varphi,\psi) : A_0 \longrightarrow E \rhd J \longrightarrow E \longrightarrow K\otimes E ,$$

where the map from J to E is just the inclusion. Let μ_0 be the map from J to $K\otimes E$. Then $(id_K\otimes\Psi)\mu$ is homotopic to $(id_K\otimes\Sigma)\mu_0$. The proof of homotopy invariance of the product in its second variable therefore shows that $[\Phi][\Psi]=[\Phi_0][\Sigma]$. Now Φ_0 is homotopic to

$$(\varphi,\psi) : A_0 \longrightarrow E \rhd E \longrightarrow K\otimes E ,$$

which is in turn homotopic to

$$(\varphi\oplus 0, \psi\oplus 0) : A_0 \longrightarrow M_2(E) \rhd M_2(E) \longrightarrow K\otimes E .$$

Let $t \longrightarrow u_t$ be a continuous path of unitaries in M_2 with $u_0 = 1$ and $u_1 = \begin{pmatrix} 0 & 1 \\ 1 & 0 \end{pmatrix}$. Replacing $\psi \oplus 0$ by $u_t(\psi \oplus 0)u_t^*$, we see that this last p.q.h. is homotopic to

$$(\varphi \oplus 0, 0 \oplus \psi): A_0 \longrightarrow M_2(E) \rhd M_2(E) \longrightarrow K \otimes E.$$

Homotopy invariance of the product in the first variable now implies that

$$[\Phi_0][\Sigma] = [(\varphi \oplus 0, 0 \oplus \psi)][\Sigma] = ([\varphi] - [\psi])[\Sigma].$$

Combining bilinearity of the product and proposition 3.4.6, we obtain $([\varphi] - [\psi])[\Sigma] = [\Omega]$. So $[\Phi][\Psi] = [\Omega]$, as desired. Q.E.D.

It is tempting to try to prove this lemma directly from the definition of the product, since the derivation δ can be chosen to be zero. However, Σ cannot be used in the definition of the product, since the hypotheses do not imply that Σ is an extension to E of $\Sigma|_J$. (The approximate identity condition, (*) in definition 3.3.1, need not hold.)

We now have enough information to state an alternate method of computing the product, and to prove that it is associative.

3.4.13 Corollary. Let A_0, A_1, and A_2 be G-algebras, let Φ be a q.h. from A_0 to $K \otimes A_1$, and let Ψ be a q.h. from A_1 to $K \otimes A_2$. Then the product $[\Phi][\Psi] \in KK_G(A_0, A_2)$ can be computed as follows:

(1) Find a split p.q.h.

$$\Phi_0 = (\varphi, \psi): A_0 \longrightarrow E \rhd J \xrightarrow{\mu} K \otimes A_1,$$

which defines the q.h. Φ.

(2) Find a p.q.h.

$$\Psi_0 = (\sigma, \tau): E \longrightarrow F \rhd I \longrightarrow K \otimes A_2$$

such that $\Psi_0|_J$ is homotopic to $(id_K \otimes \Psi)|_J$.

(3) Then $[\Phi][\Psi]$ is the class of the p.q.h.

$$(\sigma\varphi \oplus \tau\psi, \tau\varphi \oplus \sigma\psi): A_0 \longrightarrow M_2(F) \rhd M_2(I) \longrightarrow M_2(K \otimes A_2) \simeq K \otimes A_2.$$

Proof. Step (1) can be carried out by lemma 3.4.10, step (2) can be carried out by lemma 3.4.11, and the class of the p.q.h. in (3) is equal to $[\Phi][\Psi]$ by lemma 3.4.12. Q.E.D.

One may wonder why we don't *define* the product $[\Phi][\Psi]$ by the construction in this result. The answer is that, even though many of the properties of the product would be easier to prove, we would really have only shifted the difficulties. Indeed, the proof of lemma 3.4.11 requires the machinery of extendible q.h.'s and also the derivation used in the definition of the product, and we would still need to use the uniqueness of the extendible q.h. to prove that the product is well defined.

3.4.14 Theorem. ([25], theorem 4.4.) The product is associative. That is, if A_i is a G-algebra for $i = 0, 1, 2, 3$, and Φ_i is a q.h. from A_i to $K \otimes A_{i+1}$ for $i = 0, 1, 2$, then $[\Phi_0]([\Phi_1][\Phi_2]) = ([\Phi_0][\Phi_1])[\Phi_2]$.

Proof. Represent $[\Phi_0]$ by a split p.q.h. (lemma 3.4.10)

$$(\varphi_0, \psi_0) : A_0 \longrightarrow E_0 \rhd J_0 \longrightarrow K \otimes A_1 .$$

By lemma 3.4.11, there is a p.q.h.

$$(\varphi_1, \psi_1) : E_0 \longrightarrow E_1 \rhd J_1 \longrightarrow K \otimes A_2$$

whose restriction to J_0 defines a q.h. homotopic to the one defined by the restriction of $id_K \otimes \Phi_1$ to J_0. By lemma 3.4.10, we may assume that (φ_1, ψ_1) is also split. Now another application of lemma 3.4.11 yields a p.q.h.

$$\Psi = (\varphi_2, \psi_2) : E_1 \longrightarrow E_2 \rhd J_2 \longrightarrow K \otimes A_3$$

whose restriction to J_1 defines a q.h. homotopic to the one defined by the restriction of $id_K \otimes \Phi_2$ to J_1. Then lemma 3.4.12 implies that $[\Phi_0][\Phi_1] = [\Sigma]$, where

$$\Sigma = (\varphi_1\varphi_0 \oplus \psi_1\psi_0, \psi_1\varphi_0 \oplus \varphi_1\psi_0) : A_0 \longrightarrow M_2(E_1) \rhd M_2(J_1) \longrightarrow M_2(K \otimes A_2) \simeq K \otimes A_2 .$$

The triple product $([\Phi_0][\Phi_1])[\Phi_2]$ is now represented by the p.q.h.

$$\Omega = ((\varphi_2\varphi_1\varphi_0 \oplus \varphi_2\psi_1\psi_0) \oplus (\psi_2\psi_1\varphi_0 \oplus \psi_2\varphi_1\psi_0), (\psi_2\varphi_1\varphi_0 \oplus \psi_2\psi_1\psi_0) \oplus (\varphi_2\psi_1\varphi_0 \oplus \varphi_2\varphi_1\psi_0)) :$$
$$A_0 \longrightarrow M_4(E_2) \rhd M_4(J_2) \longrightarrow M_4(K \otimes A_3) \simeq K \otimes A_3 .$$

(This expression is obtained by using lemma 3.4.12 on Σ and $id_{M_2} \otimes \Psi$.) If we compute the other product by the same method, we obtain

$$((\varphi_2\varphi_1\varphi_0 \oplus \psi_2\psi_1\varphi_0) \oplus (\psi_2\varphi_1\psi_0 \oplus \varphi_2\psi_1\psi_0), (\psi_2\varphi_1\varphi_0 \oplus \varphi_2\psi_1\varphi_0) \oplus (\varphi_2\varphi_1\psi_0 \oplus \psi_2\psi_1\psi_0)) :$$
$$A_0 \longrightarrow M_4(E_2) \rhd M_4(J_2) \longrightarrow M_4(K \otimes A_3) \simeq K \otimes A_3 .$$

Since this one differs from Ω only in the order of the summands (the last three must be permuted cyclically), it is homotopic to Ω, and we have shown that $([\Phi_0][\Phi_1])[\Phi_2] = [\Phi_0]([\Phi_1][\Phi_2])$. Q.E.D.

3.4.15 Corollary. If A is a G-algebra, then $KK_G(A, A)$ is a unital $R(G)$-algebra.

Proof. Everything has been proved already except the assertion that $KK_G(A, A)$ has an identity. But clearly $[id_A]$ is a identity for it. Q.E.D.

3.4.16 Notation. We write 1_A for the identity of $KK_G(A, A)$.

3.5. The General Form of the Product. Stability and Periodicity

In this section, we construct the general form of the product, which is a bilinear map

$$KK_G(A_1, B_1 \otimes D) \times KK_G(D \otimes A_2, B_2) \longrightarrow KK_G(A_1 \otimes A_2, B_1 \otimes B_2).$$

We then apply it to prove that KK_G is stable and satisfies Bott periodicity. Passing to the general case from the case we have already considered (in which $B_1 = A_2 = \mathbf{C}$) is actually very easy, with the help of the map in the following definition.

3.5.1 Definition. (Compare [55], section 4, definition 4.) Let A, B, and D be G-algebras (separable, as always in this chapter). We define a map

$$\tau_D : KK_G(A, B) \longrightarrow KK_G(A \otimes D, B \otimes D)$$

as follows: if

$$\Phi = (\varphi, \psi) : A \longrightarrow E \triangleright J \xrightarrow{\mu} K \otimes B$$

is a p.q.h., then $\tau_D([\Phi])$ is the class of the p.q.h.

$$\Phi \otimes id_D = (\varphi \otimes id_D, \psi \otimes id_D) : A \otimes D \longrightarrow E \otimes D \triangleright J \otimes D \xrightarrow{\mu \otimes id_D} K \otimes B \otimes D.$$

(All tensor products appearing here are minimal, or injective, C^*-tensor products — see [104], definition IV.4.8.) We will, when convenient, regard τ_D as a map from $KK_G(A, B)$ to $KK_G(D \otimes A, D \otimes B)$.

3.5.2 Proposition. Let A, B, C, D, and E be G-algebras. Then the maps defined above are well defined and satisfy the following properties:

(1) $\tau_D : KK_G(A, B) \longrightarrow KK_G(A \otimes D,\ B \otimes D)$ is an $R(G)$-module homomorphism.

(2) Multiplicativity: If $\eta \in KK_G(A, B)$ and $\lambda \in KK_G(B, C)$, then $\tau_D(\eta\lambda) = \tau_D(\eta)\tau_D(\lambda)$.

(3) $\tau_D(1_A) = 1_{A \otimes D}$.

(4) If we identify $D \otimes E$ with $E \otimes D$ via the obvious isomorphism, then $\tau_D \cdot \tau_E = \tau_E \cdot \tau_D = \tau_{D \otimes E}$.

Proof. Everything is obvious except for (2), which follows immediately from corollary 3.4.13. Q.E.D.

3.5.3 Definition. (Compare [55], section 4, theorem 4.) Let A_1, A_2, B_1, B_2, and D be G-algebras. Define a pairing $(\eta, \lambda) \longrightarrow \eta\lambda$ (or $\eta \otimes_D \lambda$ when necessary to avoid confusion) from $KK_G(A_1, B_1 \otimes D) \times KK_G(D \otimes A_2, B_2)$ to $KK_G(A_1 \otimes A_2,\ B_1 \otimes B_2)$ as follows: $\eta\lambda = \tau_{A_2}(\eta)\tau_{B_1}(\lambda)$, where the product on the right is the product of $\tau_{A_2}(\eta) \in KK_G(A_1 \otimes A_2, B_1 \otimes D \otimes A_2)$ and $\tau_{B_1}(\lambda) \in KK_G(B_1 \otimes D \otimes A_2,\ B_1 \otimes B_2)$, as defined earlier.

We state the basic properties of the product as a theorem.

3.5.4 Theorem ([55], section 4, theorem 4). (1) The product defined above is $R(G)$-bilinear.

(2) Associativity: If $\eta \in KK_G(A_1, B_1 \otimes D)$, $\lambda \in KK_G(D \otimes A_2, B_2 \otimes E)$, and $\mu \in KK_G(E \otimes A_3, B_3)$, then $(\eta\lambda)\mu = \eta(\lambda\mu)$ as elements of $KK_G(A_1 \otimes A_2 \otimes A_3, B_1 \otimes B_2 \otimes B_3)$.

(3) If $\eta \in KK_G(A_1, B_1 \otimes D_1 \otimes D)$ and $\lambda \in KK_G(D \otimes D_2 \otimes A_2, B_2)$, then $\tau_{D_2}(\eta) \otimes_{D_1 \otimes D \otimes D_2} \tau_{D_1}(\lambda) = \eta \otimes_D \lambda$ as elements of $KK_G(A_1 \otimes D_2 \otimes A_2, B_1 \otimes D_1 \otimes B_2)$.

(4) Multiplicitivity of τ: If $\eta \in KK_G(A_1, B_1 \otimes D)$ and $\lambda \in KK_G(D \otimes A_2, B_2)$, then $\tau_E(\eta \otimes_D \lambda) = \tau_E(\eta) \otimes_{D \otimes E} \tau_E(\lambda)$.

Proof. (1) is obvious. (2) follows immediately from the associativity of the product as defined earlier, theorem 3.4.14. (3) holds by definition, and (4) follows from the similar statement in proposition 3.5.2 (part (2)). Q.E.D.

3.5.5 Proposition. Let A_1, A_2, B_1, and B_2 be G-algebras. Then the product

$$KK_G(A_1, B_1) \times KK_G(A_2, B_2) \longrightarrow KK_G(A_1 \otimes A_2, B_1 \otimes B_2)$$

is commutative in the following sense: if $\eta_1 \in KK_G(A_1, B_1)$ and $\eta_2 \in KK_G(A_2, B_2)$, then the identifications $A_1 \otimes A_2 \simeq A_2 \otimes A_1$ and $B_1 \otimes B_2 \simeq B_2 \otimes B_1$ by the standard isomorphisms identify $\eta_1 \otimes \eta_2 \in KK_G(A_1 \otimes A_2, B_1 \otimes B_2)$ with $\eta_2 \otimes \eta_1 \in KK_G(A_2 \otimes A_1, B_2 \otimes B_1)$.

Proof. Let η_i be represented by the q.h.

$$\Phi_i = (\varphi_i, \psi_i) : A_i \longrightarrow E_i \triangleright J_i \longrightarrow K \otimes B_i .$$

Also, let f represent the flip isomorphisms $A_1 \otimes A_2 \longrightarrow A_2 \otimes A_1$, $B_1 \otimes B_2 \longrightarrow B_2 \otimes B_1$, $J_1 \otimes J_2 \longrightarrow J_2 \otimes J_1$, and also $M_G(J_1 \otimes J_2) \longrightarrow M_G(J_2 \otimes J_1)$. Then we must show that $[f^{-1}]([\Phi_1] \otimes [\Phi_2])[f] = [\Phi_2] \otimes [\Phi_1]$. By definition, $[\Phi_1] \otimes [\Phi_2]$ is the product $[\Phi_1 \otimes id_{A_1}][id_{B_1} \otimes \Phi_2]$, which we now compute. We first find a p.q.h. which is homotopic to the restriction of $id_K \otimes id_{B_1} \otimes \Phi_2$ to $J_1 \otimes A_2$ and extendible to $E_1 \otimes A_2$. We choose the p.q.h.

$$(id_{J_1} \otimes \varphi_2, id_{J_1} \otimes \psi_2) : J_1 \otimes A_2 \longrightarrow J_1 \otimes E_2 \triangleright J_1 \otimes J_2$$
$$\longrightarrow (K \otimes B_1) \otimes (K \otimes B_2) \simeq K \otimes B_1 \otimes B_2 .$$

The required extension to $E_1 \otimes A_2$ is $(i \otimes \varphi_2, i \otimes \psi_2)$, where i is the inclusion of E_1 in $M_G(J_1)$. (Notice that $M_G(J_1) \otimes E_2$ can be identified with a subalgebra of $M_G(J_1 \otimes E_2)$, by proposition 3.1.2). The pair $(i \otimes \varphi_2, i \otimes \psi_2)$ is in fact an extension, in the sense of definition 3.3.1, because the homomorphism i satisfies, for $a \in E_1$, $x \in J_1$, and (e_λ) an approximate identity for J_1, the identity

$$i(a)x = \lim_\lambda i(a)e_\lambda x = \lim_\lambda i(ae_\lambda)x .$$

(We can ignore the second factor here.)

We must next choose a derivation δ of the algebra Q in that definition 3.4.1 such that the analog of (**) of definition holds. Let $d = \frac{\pi i}{2}\begin{pmatrix}0&1\\1&0\end{pmatrix} \in M_2(\mathbf{C})$; then the condition becomes:

$$\left.\begin{array}{l}\delta(x)-(dx-xd)\in M_2(J_1\otimes J_2),\\ \qquad \text{if } x \text{ is a matrix with entries in the range of } \varphi_1\otimes(\varphi_2-\psi_2);\\ \delta(x)\in M_2(J_1\otimes J_2),\\ \qquad \text{if } x \text{ is a matrix with entries in the range of } (\varphi_1-\psi_1)\otimes\varphi_2; \text{ and}\\ \delta\begin{bmatrix}\varphi_1\otimes\varphi_2(a) & 0\\ 0 & \varphi_1\otimes\varphi_2(a)\end{bmatrix}\in M_2(J_1\otimes J_1) \text{ for } a\in A_1\otimes A_2.\end{array}\right\} \quad (*)$$

(Q is the C^*-subalgebra of $M_2(M_G(J_1\otimes J_2))$ generated by the elements considered in (*).) Let $\gamma=\exp(\delta)$. Then the product $[\Phi_1\otimes id_{A_2}][id_{B_1}\otimes\Phi_2]$ is represented by the p.q.h.

$$\Sigma = \left[\begin{bmatrix}\varphi_1\otimes\varphi_2 & 0\\ 0 & \psi_1\otimes\psi_2\end{bmatrix}, \gamma\cdot\begin{bmatrix}\varphi_1\otimes\psi_2 & 0\\ 0 & \psi_1\otimes\varphi_2\end{bmatrix}\right] : A_1\otimes A_2 \longrightarrow M_2(M_G(J_1\otimes J_2))$$

$$\triangleright\ M_2(J_1\otimes J_2)\longrightarrow M_2(K\otimes B_1\otimes B_2)\simeq K\otimes B_1\otimes B_2 .$$

A similar calculation shows that $\eta_2\otimes\eta_1$ is represented by the p.q.h.

$$\Sigma_0 = \left[\begin{bmatrix}\varphi_2\otimes\varphi_1 & 0\\ 0 & \psi_2\otimes\psi_1\end{bmatrix}, \gamma_0\cdot\begin{bmatrix}\varphi_2\otimes\psi_1 & 0\\ 0 & \psi_2\otimes\varphi_1\end{bmatrix}\right] : A_2\otimes A_1 \longrightarrow M_2(M_G(J_2\otimes J_1))$$

$$\triangleright\ M_2(J_2\otimes J_1)\longrightarrow M_2(K\otimes B_2\otimes B_1)\simeq K\otimes B_2\otimes B_1 .$$

Here, $\gamma_0=\exp(\delta_0)$, and δ_0 is a derivation of the corresponding algebra Q_0 such that the analog of (*) holds, namely

$$\left.\begin{array}{l}\delta_0(x)-(dx-xd)\in M_2(J_2\otimes J_1),\\ \qquad \text{if } x \text{ is a matrix with entries in the range of } \varphi_2\otimes(\varphi_1-\psi_1);\\ \delta_0(x)\in M_2(J_2\otimes J_1),\\ \qquad \text{if } x \text{ is a matrix with entries in the range of } (\varphi_2-\psi_2)\otimes\varphi_1; \text{ and}\\ \delta_0\begin{bmatrix}\varphi_2\otimes\varphi_1(a) & 0\\ 0 & \varphi_2\otimes\varphi_1(a)\end{bmatrix}\in M_2(J_2\otimes J_1) \text{ for } a\in A_2\otimes A_2.\end{array}\right\} \quad (**)$$

We also write f for the isomorphism $Q\simeq Q_0$ induced by the canonical isomorphism $J_1\otimes J_2\simeq J_2\otimes J_1$. We then choose for δ_0 the derivation

$$\delta_0(x) = dx - xd - f\cdot\delta\cdot f^{-1}(x),$$

where δ is the derivation satisfying (*) chosen above. Then δ_0 does in fact satisfy (**). (Notice that d commutes with the elements

$$\begin{bmatrix} \varphi_2\otimes\varphi_1(a) & 0 \\ 0 & \varphi_2\otimes\varphi_1(a) \end{bmatrix}$$

for $a \in A_2\otimes A_1$. Also, if $x \in Q_0$ then $dx - xd \in Q_0$, so δ_0 really is a derivation of Q_0.)

We are now going to show how to choose δ such that the derivations $x \longrightarrow dx - xd$ and $x \longrightarrow f\cdot\delta\cdot f^{-1}(x)$ of Q_0 commute. Recall from the proof of proposition 3.4.3 that δ is an equivariant lifting of a certain derivation $\bar{\delta}$ on $P = Q/(Q \cap M_2(J_1\otimes J_1))$, where $\bar{\delta}$ is uniquely determined by (*). Let ε be the derivation of Q defined by $\varepsilon(x) = dx - xd$, and let $\bar{\varepsilon}$ be the corresponding derivation of P. It is clear from (*) that $\bar{\varepsilon}$ commutes with $\bar{\delta}$. Therefore $\exp(t\bar{\varepsilon})$ commutes with $\bar{\delta}$ for all $t \in \mathbf{R}$. Now one easily checks that $\exp(t\varepsilon)$ commutes with the action of G on Q. Furthermore, $\exp(4\varepsilon)$ is the identity on Q. Let $H = G \times \mathbf{R}/4\mathbf{Z}$. Then H is a compact group acting on Q by automorphisms in the obvious way, and by theorem 3.1.8, we can choose δ to be H-equivariant. This means that δ is G-equivariant and commutes with $\exp(t\varepsilon)$ for all $t \in \mathbf{R}$. If we differentiate the equation

$$\exp(t\varepsilon)\delta = \delta\exp(t\varepsilon)$$

with respect to t, and set $t = 0$, we find $\varepsilon\delta = \delta\varepsilon$. Now $f\varepsilon f^{-1}(x) = dx - xd$, where d is now regarded as an element of $M_2(M_G(J_2\otimes J_1))$, and clearly $f\varepsilon f^{-1}$ commutes with $f\delta f^{-1}$, as desired.

It follows that $\exp(\delta_0) = \exp(f\delta f^{-1})\exp(f\varepsilon f^{-1})$. Since $\exp(f\varepsilon f^{-1})(x) = uxu^*$ where $u = \exp(d) = \begin{bmatrix} 0 & 1 \\ 1 & 0 \end{bmatrix}$, and $\exp(f\delta f^{-1}) = f\gamma f^{-1}$, it follows that

$$\gamma_0\cdot\begin{bmatrix} \varphi_2\otimes\psi_1 & 0 \\ 0 & \psi_2\otimes\varphi_1 \end{bmatrix} = f\gamma f^{-1}\cdot\begin{bmatrix} \psi_2\otimes\varphi_1 & 0 \\ 0 & \varphi_2\otimes\psi_1 \end{bmatrix}.$$

It is now clear that $\Sigma_0 = f\Sigma f^{-1}$, which is the desired result. Q.E.D.

3.5.6 Definition. (See [26], definition preceding proposition 5.3.) Let A and B be G-algebras, and let $\eta \in KK_G(A,B)$. We say η is invertible if there is $\lambda \in KK_G(B,A)$ such that $\eta\lambda = 1_A$ and $\lambda\eta = 1_B$. If $KK_G(A,B)$ contains an invertible element, we will say that A and B are generalized G-homotopy equivalent.

The next proposition shows that KK-theory cannot distinguish generalized homotopy equivalent G-algebras.

3.5.7 Proposition. Let R and S be generalized G-homotopy equivalent G-algebras. Then for any G-algebras A and B, there are natural isomorphisms

$$KK_G(R\otimes A, B) \simeq KK_G(S\otimes A, B) \quad \text{and} \quad KK_G(A, R\otimes B) \simeq KK_G(A, S\otimes B).$$

Proof. Let $\lambda \in KK_G(R,S)$ and $\mu \in KK_G(S,R)$ satisfy $\lambda\mu = 1_R$ and $\mu\lambda = 1_S$. Define

$$k : KK_G(R\otimes A, B) \longrightarrow KK_G(S\otimes A, B) \quad \text{and} \quad l : KK_G(S\otimes A, B) \longrightarrow KK_G(R\otimes A, B)$$

by $k(x) = \tau_A(\mu)x$ and $l(x) = \tau_A(\lambda)x$. Then for $x \in KK_G(R\otimes A, B)$, we have

$$l \cdot k(x) = \tau_A(\lambda)\tau_A(\mu)x = \tau_A(1_R)x = 1_{R\otimes A} \cdot x = x .$$

Similarly $k \cdot l$ is the identity on $KK_G(S\otimes A, B)$. The isomorphisms are obviously natural.

The isomorphism $KK_G(A, R\otimes B) \simeq KK_G(A, S\otimes B)$ is proved similarly. Q.E.D.

We point out that generalized homotopy equivalent C^*-algebras are indistinguishable for arbitrary stable, half-exact, and homotopy invariant functors on C^*-algebras. (See [26], especially proposition 5.3, for details.)

We are now going to prove stability and Bott periodicity for KK-theory by constructing generalized G-homotopy equivalences between $\mathbf{C}$ and the algebras $K(V)$ for a separable Hilbert space V carrying a unitary representation of G, and $C_0(V)$ for a finite dimensional representation space V of G. The following lemma, based on a trick in section 1 of [3], will be useful.

3.5.8 Lemma. Let A be a G-algebra. Suppose there are $\lambda \in KK_G(\mathbf{C}, A)$ and $\mu \in KK_G(A, \mathbf{C})$ such that $\lambda\mu = 1_{\mathbf{C}}$. Suppose further that there is an automorphism φ of A such that the maps $A\otimes A \longrightarrow A\otimes A$ defined by $a\otimes b \longrightarrow a\otimes\varphi(b)$ and $a\otimes b \longrightarrow b\otimes a$ are G-homotopic. Then $\mu\lambda = 1_A$.

Proof. We claim that the following identity holds, where B is an arbitrary G-algebra:

$$r\otimes s = s\otimes([\varphi]\otimes_A r) \quad \text{for} \quad r \in KK_G(A, \mathbf{C}) \quad \text{and} \quad s \in KK_G(A, B). \qquad (*)$$

To prove it, let $f : A\otimes A \longrightarrow A\otimes A$ by the flip automorphism $f(a\otimes b) = b\otimes a$. By proposition 3.5.5, $r\otimes s = f^*(s\otimes r)$. (We identify $\mathbf{C}\otimes B$ and $B\otimes\mathbf{C}$ with B, and under this identification, the flip from $\mathbf{C}\otimes B$ to $B\otimes\mathbf{C}$ becomes id_B.) Since f is homotopic to $id_A\otimes\varphi$, we obtain

$$f^*(s\otimes r) = s\otimes\varphi^*(r) = s\otimes([\varphi]\otimes_A r),$$

as desired.

Set $\tilde{\mu} = [\varphi]\otimes_A\mu$. We want to prove that $\tilde{\mu}\otimes\lambda = 1_A$. By proposition 3.5.5, $\tilde{\mu}\otimes\lambda = \lambda\otimes\tilde{\mu}$. The product $\lambda\otimes\tilde{\mu}$ is by definition the product $\tau_A(\lambda)\otimes_{A\otimes A}\tau_A(\tilde{\mu})$, where $\tau_A(\lambda)$ is taken to be $\lambda\otimes 1_A$ and $\tau_A(\tilde{\mu})$ is taken to be $1_A\otimes\tilde{\mu}$. Now $1_A\otimes\tilde{\mu} = \mu\otimes 1_A$ by (*) and the definition of $\tilde{\mu}$. Therefore

$$\lambda\otimes\tilde{\mu} = (\lambda\otimes 1_A)\otimes_{A\otimes A}(\mu\otimes 1_A) = \tau_A(\lambda)\otimes_{A\otimes A}\tau_A(\mu) = \tau_A(\lambda\otimes_A\mu) = \tau_A(1_{\mathbf{C}}) = 1_A ,$$

as desired. (What has happened here is that the flip automorphism f has been used to convert a product of the form $(r\otimes 1_A)\otimes_{A\otimes A}(1_A\otimes s)$, which is $r\otimes s$, into one of the form $(r\otimes 1_A)\otimes_{A\otimes A}(s\otimes 1_A)$, which is $(r\otimes_A s)\otimes 1_A$, for a different s.) We thus have $\tilde{\mu}\otimes\lambda = 1_A$. Therefore, by associativity,

$$\mu = 1_A\otimes_A\mu = \tilde{\mu}\otimes\lambda\otimes_A\mu = \tilde{\mu}\otimes 1_{\mathbf{C}} = \tilde{\mu} .$$

So $\mu\otimes\lambda = 1_A$. Q.E.D.

3.5.9 Remark. Under the hypotheses of the previous lemma, we can now conclude, as in [3], that $[\varphi] = 1_A$. Indeed, $[\varphi] = [\varphi] \otimes_A \mu \otimes \lambda = \tilde{\mu} \otimes \lambda = 1_A$.

3.5.10 Theorem. (Stability; see [55], section 5, theorem 1.) Let V be any separable Hilbert space with a unitary representation of G. Then $K(V)$ is generalized G-homotopy equivalent to $\mathbf{C}$. In particular, there are natural isomorphisms

$$KK_G(K(V) \otimes A, B) \simeq KK_G(A, B) \simeq KK_G(A, K(V) \otimes B).$$

Proof. Let H be the universal Hilbert representation space of G, that is, $H = l^2 \otimes L^2(G)$. (See 3.1.5.) Let V_0 be any irreducible subspace of V, and let H_0 be a subspace of H such that H_0 is isomorphic to the complex conjugate representation space $\widetilde{V}_0$ of V_0. Now $\widetilde{V}_0 \otimes V_0$ has a one-dimensional invariant subspace on which the action of G is trivial. Let W be the corresponding subspace of $H_0 \otimes V_0$, and regard W as a subspace of $H \otimes V$. Let $p \in K(H) \otimes K(V)$ be the projection on W. (We identify $K(H) \otimes K(V)$ with $K(H \otimes V)$.) We define $\lambda \in KK_G(\mathbf{C}, K(V))$ to be the class of the homomorphism $z \longrightarrow zp$ from $\mathbf{C}$ to $K(H) \otimes K(V)$. We further let $\mu \in KK_G(K(V), \mathbf{C})$ be the class defined by the composite $K(V) \longrightarrow K(H) \otimes K(V) \simeq K(H)$, where the first map is $a \longrightarrow q \otimes a$ for some one-dimensional G-invariant projection q such that qH carries the trivial representation of G. It follows from corollary 3.4.7 that $\lambda\mu$ is the class defined by the composite of the maps defining λ and μ. This composite is a map from $\mathbf{C}$ to $K(H)$ sending 1 to a projection on a one-dimensional trivial representation space in H. (Recall that by convention the isomorphism $K(H) \otimes K(V) \simeq K(H)$ is assumed to be spatial. See remark 3.1.6.) Therefore $\lambda\mu = 1_{\mathbf{C}}$.

In order to be able to apply the previous lemma, it remains to verify the condition concerning the flip automorphism of $K(V) \otimes K(V)$. (This means, of course, the map $a \otimes b \longrightarrow b \otimes a$.) There is a canonical isomorphism $K(V) \otimes K(V) \simeq K(V \otimes V)$. Using it, we can identify the flip on $K(V) \otimes K(V)$ with the spatial automorphism of $K(V \otimes V)$ obtained from the flip on $V \otimes V$. By lemma 3.1.7 (1), the flip on $K(V) \otimes K(V)$ is therefore G-homotopic to the identity. The previous lemma now implies that $\mu\lambda = 1_{K(V)}$, so that λ is invertible. Q.E.D.

3.5.11 Theorem (Bott periodicity; see [55], section 5, theorem 7). Let V be a finite dimensional Hilbert space, with a unitary representation of G. Then $C_0(V)$ is generalized G-homotopy equivalent to $\mathbf{C}$. In particular, there are natural isomorphisms

$$KK_G(C_0(V) \otimes A, B) \simeq KK_G(A, B) \simeq KK_G(A, C_0(V) \otimes B).$$

The proof of this theorem is somewhat messy, depending on, among other things, elliptic operators. The length of what follows is partly due to the fact that we are filling in some of what was left out in [55] and partly due to the fact that quasihomomorphisms are less well adapted to the proof of this theorem than Kasparov's approach. Of course, we want to construct $\lambda \in KK_G(\mathbf{C}, C_0(V))$ and $\mu \in KK_G(C_0(V), \mathbf{C})$ such that $\lambda\mu = 1_{\mathbf{C}}$ and $\mu\lambda = 1_{C_0(V)}$. We break the construction up into several parts.

3.5.12 Lemma. Let $\pi: A \longrightarrow B$ be a surjective map of unital G-algebras. Let V_1 and V_2 be finite dimensional representation spaces of G, and let $p_i \in L(V_i)\otimes A$ be G-invariant projections such that the element $[p_1]-[p_2]$ of $K_0^G(A)$ is the image of an element of $K_0^G(Ker\,\pi)$. Then there is a finite dimensional representation space W of G which contains V_1 and V_2 as subspaces, a G-invariant projection $e \in L(W)$ such that eW is orthogonal to V_1 and V_2, and a G-invariant unitary $v \in L(W)\otimes A$ such that

$$\pi(v(p_1+e\otimes 1)v^* - (p_2+e\otimes 1)) = 0 .$$

(Here we identify $p_i \in L(V_i)\otimes A$ with its image in $L(W)\otimes A$).

Proof. By the long exact sequence in equivariant K-theory (theorem 2.8.3 (1)), we have $\pi_*([p_1]) - \pi_*([p_2]) = 0$. Write π also for the maps $id_{L(V)}\otimes\pi: L(V)\otimes A \longrightarrow L(V)\otimes B$ for various representation spaces V. Then corollary 2.4.5 and proposition 2.4.11 (3) imply that there is a finite dimensional representation space W_1 of G and a G-invariant projection $q \in L(W_1)\otimes A$, such that the projections $r_1 = \pi(p_1)\oplus 0\oplus q$ and $r_2 = 0\oplus\pi(p_2)\oplus q$ are unitarily equivalent in $L(V_1\oplus V_2\oplus W_1)\otimes B$. By increasing the size of q and W_1, we may assume that q has the form $e_0\otimes 1$ for some G-invariant projection $e_0 \in L(W_1)$. (See proposition 2.2.5 and theorem 2.4.4.) Let u be a G-invariant unitary implementing the equivalence, thus $ur_1u^* = r_2$. Let $W_2 = V_1\oplus V_2\oplus W_1$, and let $W = W_2\oplus W_2$. Then in $L(W)\otimes B$ we have

$$(u\oplus u^*)(r_1\oplus 0)(u\oplus u^*)^* = r_2\oplus 0 .$$

Now [105], 5.4, applied to the Banach algebra $(L(W_2)\otimes B)^G$, implies that $u\oplus u^*$ is in the connected component of the identity in the group of invertible elements of $M_2((L(W_2)\otimes B)^G) = (L(W)\otimes B)^G$. Therefore by [105], 4.8, there is an invertible element $v \in (L(W)\otimes A)^G$ such that $\pi(v) = u\oplus u^*$. Replacing v by $(vv^*)^{-1/2}v$, we may assume that v is unitary. Let $e = 0\oplus 0\oplus e_0\oplus 0$, identify p_1 with $p_1\oplus 0\oplus 0\oplus 0$, and identify p_2 with $0\oplus p_2\oplus 0\oplus 0$, where in each case the first summand is in $L(V_1)$ or $L(V_1)\otimes A$, the second in $L(V_2)$ or $L(V_2)\otimes A$, the third in $L(W_1)$ or $L(W_1)\otimes A$, and the fourth in $L(W_2)$ or $L(W_2)\otimes A$. Then $\pi(p_1+e\otimes 1) = r_1\oplus 0$ and $\pi(p_2+e\otimes 1) = r_2\oplus 0$, whence

$$\pi(v(p_1+e\otimes 1)v^* - (p_2+e\otimes 1)) = 0 ,$$

as desired. Q.E.D.

3.5.13 Definition of λ. (Compare [3], section 4.) Let P be the projective space associated with the space $V\oplus \mathbf{C}$, where $\mathbf{C}$ carries the trivial action of G. Then P is a compact G-space, and V can be canonically identified, via the map $v \longrightarrow (v,1)$, with an open subset of P. Let H_0 be the standard line bundle on P, whose fiber over $x \in P$ is the line in $V\oplus\mathbf{C}$ represented by x. Let H_0^{-i} denote the $-i$-th tensor power of H_0, which is the i-th tensor power of the dual bundle to H_0. Further let $\Lambda^i(V^*)$ be the i-th exterior power of the dual space V^* of V. Let

$$E_0 = \bigoplus_{i \text{ even}} H_0^{-i} \otimes \Lambda^i (V^*) \quad \text{and} \quad E_1 = \bigoplus_{i \text{ odd}} H_0^{-i} \otimes \Lambda^i (V^*). \tag{*}$$

It is shown in [3], section 4, and [2], end of section 2.6, that the element $[E_0] - [E_1]$ of $K_G^0(P)$ is the image of an element $\lambda_V^* \in K_G^0(V)$ under the inclusion of V in P. Let $p_0 \in L(V_0) \otimes C(P)$ and $p_1 \in L(V_1) \otimes C(P)$ be G-invariant projections such that $p_i [V_i \otimes C(P)]$ is the space $\Gamma(E_i)$ of continuous sections of E_i for $i = 0, 1$. (These exist by combining theorem 2.4.4 with Swan's theorem, theorem 2.3.1.) Let π be the projection from $C(P)$ to $C(P)/C_0(V)$. Then by the previous lemma, there is a representation space W containing V_0 and V_1, a G-invariant projection $e \in L(W) \otimes C(P)$ which is orthogonal to p_0 and p_1 and of the form $e = e_0 \otimes 1$, and a G-invariant unitary $u \in L(W) \otimes C(P)$, such that

$$u(p_0 + e)u^* - (p_1 + e) \in L(W) \otimes C_0(V).$$

We now define a p.q.h. Φ from $\mathbf{C}$ to $K \otimes C_0(V)$ by

$$\Phi = (\varphi, \psi) : \mathbf{C} \longrightarrow L(W) \otimes C(P) \rhd L(W) \otimes C_0(V) \longrightarrow K \otimes C_0(V),$$

where $\varphi(z) = z(u(p_0 + e)u^*)$ and $\psi(z) = z(p_1 + e)$ for $z \in \mathbf{C}$, and the last map is induced by an inclusion of W in the universal separable representation space of G. Then $\lambda = [\Phi]$.

We note for future reference the following facts:

$$up_0u^*[W \otimes C(P)] \simeq \Gamma(E_0),$$

$$p_1[W \otimes C(P)] \simeq \Gamma(E_1),$$

and

$$ueu^*[W \otimes C(P)] \simeq e[W \otimes C(P)] \simeq \Gamma(e_0 W \times P),$$

where $e_0 W \times P$ is the trivial bundle over P with fiber $e_0 W$.

3.5.14 Facts about pseudodifferential operators. Here, we recall some facts which will be needed for the construction of an inverse for λ. We work over P, but the statements hold for any compact manifold without boundary. Throughout this collection of results, E, F_1, and F_2 are smooth (that is, C^∞) vector bundles on P. We let $\Gamma^\infty(E)$ be the space of smooth sections of E. The following material can be found in [75], and also in sections 5 and 6 of [7], where the equivariant case is also treated.

(1) Associated with E there is a "discrete Sobolev chain" of Hilbert spaces $H^s(E)$ for $s \in \mathbf{Z}$, with $H^s(E) \subset H^{s-1}(E)$ for $s \in \mathbf{Z}$. (This inclusion is not isometric!) We have $\Gamma^\infty(E) \subset H^s(E)$ as a dense subspace for all s. The inner product on $H^s(E)$ is determined in addition by a hermitian metric on E and a strictly positive smooth measure on P. However, the space $H^s(E)$ is independent of these choices. In order to ensure that G acts on $H^s(E)$ by a unitary representation, it is sufficient to choose the metric and the measure to be G-invariant. In particular, $H^0(E)$ is the space of L^2 sections of E for the given measure and metric. Consequently, if $\xi \in H^0(E)$ and $f \in C(P)$, then $f\xi \in H^0(E)$. Thus, multiplication by functions defines a

representation of $C(P)$ on $H^0(E)$ for any E. We will usually write f for the operator of multiplication by f.

Reference: [75], chapter 9. We ignore the spaces $H^s(E)$ for $s \notin \mathbf{Z}$. The statements about multiplication by functions are obvious consequences of the other statements.

(2) A linear map t from $\Gamma^\infty(F_1)$ to $\Gamma^\infty(F_2)$ is an operator of order k if it extends to a continuous linear map from $H^s(F_1)$ to $H^{s-k}(F_2)$ for each $s \in \mathbf{Z}$. If t is a pseudodifferential operator ($t \in Int_k(F_1, F_2)$ in [75]), then it is determined up to a compact perturbation by its symbol. If it is elliptic, then it is a Fredholm operator from $H^s(F_1)$ to $H^{s-k}(F_2)$, and its kernel as a map from $H^s(F_1)$ to $H^{s-k}(F_2)$ is contained in $\Gamma^\infty(F_1)$, and is the same for every $s \in \mathbf{Z}$. Its Fredholm index is also independent of s. If t is a pseudodifferential operator of order 0, and f stands for multiplication by $f \in C(P)$ on $H^0(F_i)$ for $i = 1, 2$, then $tf - ft$ is compact.

Reference: [75], chapter 11, where we again ignore the spaces $H^s(E)$ for $s \notin \mathbf{Z}$. The last assertion follows from the fact that tf and ft are pseudodifferential operators with the same symbol.

3.5.15 Definition of μ. (Compare [3], section 4.) Let Ω_0 denote the smooth vector bundle on P such that $\Gamma^\infty(\Omega_0)$ is the direct sum of the spaces of smooth differential forms of type $(0, 2k)$ as k ranges over the nonnegative integers. Let Ω_1 be the corresponding bundle for forms of type $(0, 2k+1)$. Then there is an elliptic G-invariant first order differential operator $\bar\partial + \bar\partial^*$ from $\Gamma^\infty(\Omega_0)$ to $\Gamma^\infty(\Omega_1)$ ([75], page 325, where $\bar\partial^*$ is written v), whose kernel is one-dimensional and whose index is 1 ([3], page 123). By (2) above, we have an operator $\bar\partial + \bar\partial^* : H^1(\Omega_0) \longrightarrow H^0(\Omega_1)$ which is surjective and has a one-dimensional kernel.

We further consider the operator

$$\Box = \bar\partial\bar\partial^* + \bar\partial^*\bar\partial : \Gamma^\infty(\Omega_0) \longrightarrow \Gamma^\infty(\Omega_0).$$

Gårding's lemma ([40], chapter 0, section 5) implies that $(1 + \Box)^{1/2}$ is an invertible operator from $H^1(\Omega_0)$ to $H^0(\Omega_0)$. We can therefore consider the G-invariant pseudodifferential operator of order 0 given by

$$d_0 = (\bar\partial + \bar\partial^*)(1 + \Box)^{-1/2} : H^0(\Omega_0) \longrightarrow H^0(\Omega_1),$$

which is a surjective Fredholm operator with one-dimensional kernel. Finally, set $d = (d_0 d_0^*)^{-1/2} d_0$, which is also a G-invariant pseudodifferential operator, and which is a surjective partial isometry from $H^0(\Omega_0)$ to $H^0(\Omega_1)$. Now we can define a p.q.h.

$$\Sigma = (\sigma, \tau) : C(P) \longrightarrow L(H^0(\Omega_0)) \rhd K(H^0(\Omega_0)) \longrightarrow K$$

by setting $\sigma(f) = f$ (recall that this is multiplication by f on $H^0(\Omega_0)$), and $\tau(f) = d^* f d$. We have $\sigma(f) - \tau(f) \in K(H^0(\Omega_0))$ by the last statement in (2) above. We define $\mu \in KK_G(C_0(V), \mathbf{C})$ to be the class $[\Sigma|_{C_0(V)}]$.

Before proving theorem 3.5.11, we will need several more lemmas.

3.5.16 Lemma. Let H be a Hilbert space with a unitary representation of G. Let U be the group of G-invariant unitary elements of $L(H)$ such that $1-u \in K(H)$. Then U is connected.

Proof. Let $u \in U$. Then $1-u \in K(H)$. Since H is a direct sum of finite dimensional G-invariant subspaces, there is a finite rank projection p on a sufficiently large G-invariant subspace that $\| (u-1)-p(u-1)p \| < 1$. Equivalently, $\| u-(1-p+pup) \| < 1$. Therefore, with $v_t = (1-t)u + t(1-p+pup)$, we have $\| u - v_t \| < 1$ for $t \in [0,1]$. Hence v_t is invertible. Clearly $1-v_t \in K(H)$.

In particular, $v_1 = 1-p+pup$ is invertible. Hence pup is invertible in $L(pH)$. Since $L(pH)^G$ is finite dimensional, its group of invertible elements is connected. Therefore there is a continuous path $t \longrightarrow w_t$ for $t \in [1,2]$ of invertible elements in $L(pH)^G$ such that $w_1 = pup$ and $w_2 = p$, which is the identity of $L(pH)$. Let $v_t = 1-p+w_t$ for $t \in [1,2]$. Then $t \longrightarrow v_t$ for $t \in [0,2]$ is a continuous path of G-invariant invertible elements in $K(H)^+$ such that $1-v_t \in K(H)$, $v_0 = u$, and $v_2 = 1$. It remains only to replace v_t by $u_t = v_t(v_t^* v_t)^{-1/2}$, which is a path in U with $u_0 = u$ and $u_2 = 1$.

Q.E.D.

3.5.17 Lemma. Let H_1 and H_2 be Hilbert spaces with actions of G. Let $p \in L(H_1)$ and $q \in L(H_2)$ be G-invariant projections, and let $c, d \in L(H_1, H_2)$ be G-invariant Fredholm partial isometries such that:

(1) $c - d$ is compact.

(2) $cp - qc$ is compact.

(3) $cc^* \geq q$ and $dd^* \geq q$.

Then the p.q.h.'s from $\mathbf{C}$ to $\mathbf{C}$ defined by the pairs (p, c^*qc) and (p, d^*qd) have the same class in $KK_G^0(\mathbf{C}, \mathbf{C})$.

Proof. Condition (3) guarantees that c^*qc and d^*qd are G-invariant projections in $L(H_1)$, and conditions (1) and (2) guarantee that $p - c^*qc$ and $p - d^*qd$ are compact. So the pairs of the lemma really do define p.q.h.'s from $\mathbf{C}$ to $\mathbf{C}$ (using the maps $z \longrightarrow zp$, $z \longrightarrow zc^*qc$, and $z \longrightarrow zd^*qd$ from $\mathbf{C}$ to $L(H_1)$.)

Let $V = c[H_1]^\perp$, and let $W = d[H_1]^\perp$. Without changing the classes of the p.q.h.'s of the lemma, or the conditions (1), (2), and (3), we may replace H_1 by $H_1 \oplus V \oplus W$, c by the operator given by $(\xi, v, w) \longrightarrow c(\xi) + v$, and d by the operator given by $(\xi, v, w) \longrightarrow d(\xi) + w$. Thus we may assume, in addition to the other hypotheses, that c and d are surjective. Since c and d are G-invariant Fredholm operators, and since $c - d$ is compact, their kernels are equivariantly isomorphic finite dimensional representation spaces of G. (See [7], page 519.) Therefore there is a G-invariant partial isometry w from $Ker\ d$ to $Ker\ c$. Then $c^*d + w$ is a G-invariant unitary in $L(H_1)$ such that $1 - (c^*d + w) \in K(H_1)$.

By the previous lemma, there is a continuous path $t \longrightarrow u_t$ of G-invariant unitaries in $L(H_1)$ such that $1-u_t \in K(H_1)$, $u_0=1$, and $u_1=c^*d+w$. We have $cu_0=c$, $cu_1=cc^*d+cw=d$, and cu_t is a G-invariant surjective partial isometry such that cu_t-c is compact. The p.q.h.'s defined by the pairs (p, c^*qc) and (p, d^*qd) are now easily seen to be homotopic via the p.q.h.'s defined by the pairs $(p, (cu_t)^*q(cu_t)$.)

Q.E.D.

3.5.18 Extensions of the operators $\bar\partial$ and d to tensor products. The operator $\bar\partial$ has a canonical extension $\bar\partial_E : \Gamma^\infty(E\otimes\Omega_0) \longrightarrow \Gamma^\infty(E\otimes\Omega_1)$ for any holomorphic vector bundle E on P (see [75], pages 324-325). If E is a G-vector bundle then $\bar\partial_E$ will be G-invariant. This extension yields canonical G-invariant extensions $\bar\partial_E^*$, $\Box_E$, $(d_0)_E$, and d_E of the operators $\bar\partial^*$, $\Box$, d_0, and d. Furthermore, the kernels of the operators $\bar\partial_E+\bar\partial_E^*$ and d_E as operators on the appropriate Sobolev spaces are clearly isomorphic as representation spaces of G. A similar statement holds for cokernels.

According to [3], page 123, we therefore have

$$Ker\ d_E \simeq \bigoplus_{k\geq 0} H^{2k}(\mathbf{O}(E)) \quad \text{and} \quad Coker\ d_E \simeq \bigoplus_{k\geq 0} H^{2k+1}(\mathbf{O}(E)). \qquad (*)$$

Here $\mathbf{O}(E)$ is the sheaf of germs of holomorphic sections of E, and $H^m(\mathbf{O}(E))$ is the m-th sheaf cohomology group of $\mathbf{O}(E)$. It is a representation space of G since E is a G-vector bundle. (Unfortunately, standard notation for cohomology and for Sobolev spaces conflicts. However, the meaning of the letter H should be clear from the context.) We now consider the special cases that we will meet.

First, suppose E is a trivial bundle $V_0\times P$. Then there is a canonical identification of $\Gamma^\infty(E\otimes\Omega_i)$ with $V_0\otimes\Gamma^\infty(\Omega_i)$, and an analogous identification for the Sobolev spaces. The operators $\bar\partial_E$ etc. are all just $1_V\otimes\bar\partial$ etc., and the sheaf cohomology in (*) is just the tensor product of V and the cohomology of the sheaf of germs of holomorphic cross sections of the one-dimensional trivial vector bundle. The computation of [3], page 123, now shows that $Ker\ d_E \simeq V$ and $Coker\ d_E=0$.

We also need to know what happens for the bundles E_0 and E_1 in the definition of λ. (See (*) of 3.5.13.) The computation for this case is carried out in [3], and the result is that d_{E_0} is surjective with kernel isomorphic to the trivial one-dimensional representation space $\mathbf{C}$ of G, while d_{E_1} is invertible.

Proof of theorem 3.5.11. We verify the hypotheses of lemma 3.5.8. There are two parts, the first of which is the evaluation of the product $\lambda\mu$. Let W be the space in the definition of λ. Consider the p.q.h.

$$\Sigma_0=(\sigma_0,\tau_0): L(W)\otimes C(P) \longrightarrow L(W\otimes H^0(\Omega_0)) \triangleright K(W\otimes H^0(\Omega_0)) \longrightarrow L(W)\otimes K \simeq K,$$

where $\sigma_0=id_{L(W)}\otimes\sigma$ and $\tau_0=id_{L(W)}\otimes\tau$. Observe that $\Sigma_0\,|_{L(W)\otimes C_0(V)}$ is isomorphic to $(id_K\otimes\Sigma)\,|_{L(W)\otimes C_0(V)}$. (Here, of course, W is identified with a subspace of the universal separable representation space of G, as in the definition of Φ.) Lemma 3.4.12 now implies that the product $\lambda\mu=[\Phi][\Sigma\,|_{C_0(V)}]$ is represented by the p.q.h.

$$(\sigma_0\varphi \oplus \tau_0\psi,\ \tau_0\varphi \oplus \sigma_0\psi) : \mathbf{C} \longrightarrow M_2 \otimes L(W \otimes H^0(\Omega_0))$$
$$\triangleright M_2 \otimes L(W \otimes H^0(\Omega_0)) \longrightarrow M_2 \otimes L(W) \otimes K \simeq K .$$

This p.q.h. is the sum of the p.q.h.'s $(\sigma_0\varphi, \tau_0\varphi)$ and $(\tau_0\psi, \sigma_0\psi)$. Recall that $\varphi(1) = up_0u^* + ueu^*$, where the corresponding bundles are isomorphic to E_0 and $e_0W \times P$. Let q_0 and q_1 be the images of up_0u^* in $L(W \otimes H^0(\Omega_0))$ and $L(W \otimes H^0(\Omega_1))$ respectively, and let r_0 and r_1 be the corresponding images of ueu^*. Then $\sigma_0\varphi(1) = q_0 + r_0$ and $\tau_0\varphi(1) = d_W^*(q_1 + r_1)d_W$. Here d_W is the canonical extension of d to tensor products with the trivial bundle $W \times P$.

Write $W \times P$ as a direct sum of bundles $W \times P \simeq E_0 \oplus (e_0W \times P) \oplus F_0$. By the remarks on the extension of $\bar\partial$, there are canonical extensions d_{E_0} and d_{W_0} of d to operators on tensor products by E_0 and by $e_0W \times P$. (We set $W_0 = e_0W$.) Furthermore, $d_{E_0}d_{E_0}^* = q_1$ and $d_{W_0}d_{W_0}^* = r_1$. It is shown in section 5 of [7] that the operator $\bar\partial + \bar\partial^*$ has *some* G-invariant extension to an operator from $\Gamma^\infty(F_0 \otimes \Omega_0)$ to $\Gamma^\infty(F_0 \otimes \Omega_1)$, even though there is no reason to believe that F_0 can be given a holomorphic structure. We then get an extension of d to a G-invariant Fredholm partial isometry

$$d_{F_0} : H^0(F_0 \otimes \Omega_0) \longrightarrow H^0(F_0 \otimes \Omega_1) .$$

Now let $\tilde d_W = d_{E_0} \oplus d_{W_0} \oplus d_{F_0}$. Then $\tilde d_W \tilde d_W^* \geq q_1 + r_1$, and $\tilde d_W - d_W$ is compact because d_W and $\tilde d_W$ are both extensions of the same operator to an operator from $H^0(W \otimes \Omega_0)$ to $H^0(W \otimes \Omega_1)$ and hence have the same symbol. The p.q.h. $(\sigma_0\varphi, \tau_0\varphi)$ is determined by the pair of projections $(q_0 + r_0,\ d_W^*(q_1 + r_1)d_W)$, which by lemma 3.5.17 defines the same class in KK-theory as the p.q.h. determined by the pair $(q_0 + r_0,\ \tilde d_W^*(q_1 + r_1)\tilde d_W)$. The computations in the remarks on extensions of $\bar\partial$ (3.5.18) show that $q_0 + r_0 - \tilde d_W^*(q_1 + r_1)\tilde d_W$ is the projection on a space isomorphic to $\mathbf{C} \oplus W_0$. Therefore the class in $KK_G^0(\mathbf{C}, \mathbf{C})$ of the p.q.h. $(\sigma_0\varphi, \tau_0\varphi)$ is equal to the class of the p.q.h. determined by the pair $(1, 0)$ of projections in $L(\mathbf{C} \oplus W_0)$.

A similar computation shows that the class of $(\tau_0\psi, \sigma_0\psi)$ is equal to the class of the p.q.h. determined by the pair $(0, 1)$ of projections in $L(W_0)$. It is now clear that the sum of these two classes, which is $\lambda\mu$, is just $1_{\mathbf{C}}$.

The other, and much easier, part of the hypotheses of lemma 3.5.8 is the existence of a homotopy from the flip on $C_0(V) \otimes C_0(V)$ to an automorphism of the form $id_{C_0(V)} \otimes \varphi$. We will work at the level of spaces. Then the homotopy is defined by

$$h_t(u, v) = ((\cos t)u + (\sin t)v,\ -(\sin t)u + (\cos t)v)$$

for $t \in [0, \frac{\pi}{2}]$ and $u, v \in V$. The automorphism φ is determined by $v \longrightarrow -v$. Q.E.D.

3.6. Exact Sequences and Relations with K-Theory

3.6.1 Definition. Let A and B be G-algebras (separable, as usual). We define $KK_G^0(A,B) = KK_G(A,B)$ and $KK_G^1(A,B) = KK_G^0(SA,B)$. (Here SA is the suspension of A introduced in section 2.8.) We further write $KK_G^*(A,B)$ for the direct sum $KK_G^0(A,B) \oplus KK_G^1(A,B)$.

3.6.2 Lemma. Let A and B be G-algebras. Then there is a natural isomorphism $KK_G^1(A,B) \simeq KK_G^0(A,SB)$.

Proof. We have to show $KK_G^0(SA,B) \simeq KK_G^0(A,SB)$. Using Bott periodicity (theorem 3.5.11), it is clearly enough to show $KK_G^0(SA,S^2B) \simeq KK_G^0(A,S^3B)$. (We have abbreviated SSB to S^2B, etc.) Setting $C = SB$ and using Bott periodicity again, it is enough to show that $KK_G^0(SA,SC) \simeq KK_G^0(S^2A,S^2C)$.

Consider the maps $\tau_i : KK_G^0(S^iA,S^iC) \longrightarrow KK_G^0(S^{i+1}A,S^{i+1}C)$ for $i = 0,1,2$, where we define $S^0A = A$. Each of them is really τ_S as in definition 3.5.1, with $S = C_0(\mathbf{R})$, but of course they are maps between different groups. We want to prove that τ_1 is an isomorphism; this will be done by showing that $\tau_1 \cdot \tau_0$ and $\tau_2 \cdot \tau_1$ are both isomorphisms.

Write S^2 for $S \otimes S$, and let $\lambda \in KK_G^0(\mathbf{C},S^2)$ and $\mu \in KK_G^0(S^2,\mathbf{C})$ be the elements, constructed in the proof of the periodicity theorem, which satisfy $\lambda \otimes_{S^2} \mu = 1_{\mathbf{C}}$ and $\mu \otimes \lambda = 1_{S^2}$. Then for $x \in KK_G^0(A,\mathbf{C})$, we have

$$\tau_1 \cdot \tau_0(x) = 1_{S^2} \otimes x = \mu \otimes \lambda \otimes x .$$

Since the multiplications by μ and by λ are both isomorphisms, it follows that $\tau_1 \cdot \tau_0$ is an isomorphism. A similar argument shows that $\tau_2 \cdot \tau_1$ is an isomorphism. Therefore τ_1 is an isomorphism. Q.E.D.

3.6.3 Corollary. There is a natural product

$$KK_G^i(A_1, B_1 \otimes D) \times KK_G^j(D \otimes A_2, B_2) \longrightarrow KK_G^{i+j}(A_1 \otimes A_2, B_1 \otimes B_2)$$

for G-algebras A_1, A_2, B_1, B_2, and D. The indices are taken mod 2. This product satisfies the analogs of the properties (1)-(4) of theorem 3.5.4. In particular, it is $R(G)$-bilinear and associative.

Proof. Use theorem 3.5.4, Bott periodicity, and the previous lemma. Q.E.D.

3.6.4 Corollary. Let A be a G-algebra. Then for any G-algebra B, the groups $KK_G^i(A,B)$ are left $KK_G^0(A,A)$-modules, and the groups $KK_G^i(B,A)$ are right $KK_G^0(A,A)$-modules.

Proof. Obvious. Q.E.D.

We will refer to Kasparov's paper [55] for the proofs of the existence of long exact sequences. We therefore need:

3.6.5 Theorem ([25], section 5). The groups $KK_G^i(A,B)$ defined here are canonically isomorphic to those defined in [55].

The proof requires several lemmas, and also the Stabilization Theorem. We therefore start by introducing Hilbert C^*-modules, as in [54], definition 1. (See also [76] and [84].) However, we deviate from [54] in that we assume that the C^*-algebra valued inner product is linear in the first variable and conjugate linear in the second, rather than the other way around. We write H_B for the "Hilbert space over B", that is, $H_B = H \otimes B$, where $H = l^2 \otimes L^2(G)$, and the inner product is defined by $< \xi \otimes b, \eta \otimes c > = < \xi, \eta > c^* b$. (Of course, H_B is really the completion of the algebraic tensor product of H and B for the norm defined by $< \cdot, \cdot >$.) If E_1 and E_2 are Hilbert B-modules, we let $L(E_1, E_2)$ be the set of bounded adjointable module homomorphisms from E_1 to E_2, with its natural G-action, and we let $K(E_1, E_2)$ be the closed subset of "compact" operators, as in [54], definitions 3 and 4. (Again, see also [76] and [84].) If $E_1 = E_2 = E$, we write $L(E)$ and $K(E)$. Then $L(E)$ and $K(E)$ are C^*-algebras, $L(E)$ is naturally isomorphic to $M(K(E))$ ([54], theorem 1; see also [37], lemma 16), and $K(H_B) \simeq K \otimes B$ ([54], lemma 4). Thus, $L(H_B) \simeq M(K \otimes B)$. The Stabilization Theorem ([54], theorem 2; [67], theorem 2.5) asserts that if E is a countably generated Hilbert B-module, then there is an equivariant isomorphism $H_B \oplus E \simeq H_B$.

3.6.6 Lemma. Let B be a G-algebra, and let E be a countably generated Hilbert B-module. Let $z \in L(H_B)$ be the inclusion of H_B in $H_B \oplus E$, followed by an isomorphism $\varphi : H_B \oplus E \longrightarrow H_B$. Then there exists a path $t \longrightarrow z_t$ for $t \in [0,1]$, such that:

(1) z_t is a G-invariant isometry in $L(H_B)$ for every $t \in [0,1]$ (that is, $z_t^* z_t = 1$).

(2) $z_0 = 1$ and $z_1 = z$.

(3) $z_t z_t^* \geq zz^*$ for all t.

(4) $t \longrightarrow z_t$ is continuous for the strict topology on $L(H_B) \simeq M(K \otimes B)$.

Proof. We first recall that if C is a unital G-algebra and $u \in C$ is a G-invariant unitary, then $u \oplus u^* \in M_2(C)$ is connected to the identity by a norm continuous path of G-invariant unitaries. (Use [105], 5.4, and the retraction $x \longrightarrow (xx^*)^{-1/2} x$ of the group of invertible elements onto the unitary group.) We now use a trick of Cuntz ([22], proof of lemma 3.2) to show that if $u \in L(H_B)$ is any G-invariant unitary, then $u \oplus 1_{H_B}$ is connected to the identity in $L(H_B \oplus H_B)$ by a norm continuous path of G-invariant unitaries. We write H_B^∞ for the l^2 direct sum of countably many copies of H_B and a^∞ the operator on H_B^∞ given by the direct sum of countably many copies of $a \in L(H_B)$. Now

$$H_B \oplus H_B \simeq H_B \oplus H_B^\infty \oplus H_B^\infty$$

by expanding the second summand, and this isomorphism identifies $u \oplus 1_{H_B}$ with $u \oplus 1_{H_B^\infty} \oplus 1_{H_B^\infty}$. By the remark at the beginning of this paragraph, this operator is connected by an appropriate path to $u \oplus u^\infty \oplus (u^*)^\infty = u^\infty \oplus (u^*)^\infty$, which in turn is connected by an appropriate path to the identity. This proves that $u \oplus 1_{H_B}$ is connected to $1_{H_B} \oplus 1_{H_B}$ by the desired type of path.

Let u be the composite

$$H_B \oplus H_B \xrightarrow{1 \oplus \varphi^*} H_B \oplus H_B \oplus E \simeq H_B \oplus E \oplus H_B \xrightarrow{\varphi \oplus 1} H_B \oplus H_B .$$

Then u is a unitary in $L(H_B \oplus H_B)$, and since $H_B \oplus H_B \simeq H_B$, we find that $u \oplus 1_{H_B}$ can be connected to the identity on $H_B \oplus H_B \oplus H_B$ by a norm continuous path $t \longrightarrow x_t$ for $t \in [0,1]$ of G-invariant unitaries x_t on $H_B \oplus H_B \oplus H_B$.

We next use the fact that H_B is just an appropriate completion of $H \otimes B$, where H is the Hilbert space in 3.1.5. Using the fact that $H \oplus H \simeq H$, we tensor the path of isometries v_t of lemma 3.1.7 (2) with the identity on B to obtain a path $t \longrightarrow w_t \in L(H_B, H_B \oplus H_B \oplus H_B)$ such that w_0 is unitary, each w_t is an isometry, w_1 is the inclusion of H_B into $H_B \oplus H_B \oplus H_B$ as the first summand, and $w_t w_t^* \geq w_1 w_1^*$ for all t. The *-strong continuity of $t \longrightarrow v_t$ translates into the following continuity condition on $t \longrightarrow w_t$: the maps $t \longrightarrow w_t \xi$ and $t \longrightarrow w_t^* \eta$ are continuous for all $\xi \in H_B$ and $\eta \in H_B \oplus H_B \oplus H_B$.

We are now in a position to construct the desired path. We let z_t be the composite

$$H_B \xrightarrow{a_t} H_B \oplus H_B \oplus H_B \xrightarrow{b_t} H_B \oplus H_B \oplus H_B \xrightarrow{c_t} H_B ,$$

where for $t \in [0,1]$,

$$a_t = w_0, \quad b_t = x_{1-t}, \quad c_t = w_0^*;$$

$$a_{1+t} = w_t, \quad b_{1+t} = u \oplus 1_{H_B}, \quad c_{1+t} = w_0^*;$$

and

$$a_{2+t} = w_1, \quad b_{2+t} = u \oplus 1_{H_B}, \quad c_{2+t} = w_t^* .$$

We observe that z_t is unitary for $0 \leq t \leq 1$ and is an isometry for $1 \leq t \leq 2$. For $2 \leq t \leq 3$, we have $c_t c_t^* \geq w_1 w_1^*$, which is the projection on $H_B \oplus 0 \oplus 0 \subset H_B \oplus H_B \oplus H_B$. Since $w_1[H_B] = H_B \oplus 0 \oplus 0$ and $(u \oplus 1_{H_B})[H_B \oplus 0 \oplus 0] \subset H_B \oplus 0 \oplus 0$, we see that the domain projection of c_t dominates the range projection of $b_t a_t$, so that $z_t = c_t b_t a_t$ is an isometry in this case also. Now clearly $z_0 = 1_{H_B}$ and it is easily seen that $z_3 = z$.

It remains to check the strict continuity of $t \longrightarrow z_t$. But it is clear that $t \longrightarrow z_t \xi$ and $t \longrightarrow z_t^* \xi$ are continuous for every $\xi \in H_B$, and this form of continuity is easily seen to be equivalent to strict continuity in the multiplier algebra. Q.E.D.

3.6.7 Proposition. Let A and B be G-algebras. Then $KK_G(A,B)$ is equal to the set of homotopy classes of p.q.h.'s of the form

$$(\varphi,\psi): A \longrightarrow M_G(K\otimes B) \rhd K\otimes B, \qquad (*)$$

where homotopies are required to be defined by p.q.h.'s of the form

$$((\varphi_t,\psi_t)): A \longrightarrow M_G(K\otimes B\otimes C([0,1])) \rhd K\otimes B\otimes C([0,1]). \qquad (**)$$

Proof. We must show that every q.h. is homotopic to the q.h. defined by a p.q.h. of the form (*), and that if two p.q.h.'s of the form (*) define homotopic q.h.'s, then the are homotopic via a homotopy given by a p.q.h. of the form (**). We do the first part first, but in slightly greater generality so as to be able to use it in the second part.

Let

$$\Phi_0 = (\varphi_0,\psi_0): A \longrightarrow E \rhd J \overset{\mu_0}{\longrightarrow} K\otimes B$$

be a p.q.h. such that μ_0 is injective. Thus, J may be regarded as a subalgebra of $K\otimes B$. Let

$$R = \begin{bmatrix} \overline{J(K\otimes B)J} & \overline{J(K\otimes B)} \\ \overline{(K\otimes B)J} & K\otimes B \end{bmatrix},$$

regarded as a subalgebra of $M_2(K\otimes B)$. (That is, R is the set of all 2×2 matrices whose entries are in the indicated subsets of $K\otimes B$.) We identify $K\otimes B$ with $K(H_B)$. Then $\overline{J(K\otimes B)J}$ is identified with $K(\overline{JH_B})$, and R itself is identified with $K(\overline{JH_B}\oplus H_B)$. The Stabilization Theorem implies that there is an equivariant isomorphism $u: \overline{JH_B}\oplus H_B \longrightarrow H_B$. Therefore there is an isomorphism $\omega_0: R \longrightarrow K\otimes B$ given by $\omega_0(a) = uau^*$, and ω_0 extends to an isomorphism $\omega: M_G(R) \longrightarrow M_G(K\otimes B)$.

Define $\sigma_0: J \longrightarrow R$ by $\sigma_0(a) = a\oplus 0$. By lemma 3.1.4, σ_0 extends to a map $\sigma: M_G(J) \longrightarrow M_G(R)$. Let $\gamma: E \longrightarrow M_G(J)$ be the map obtained via proposition 3.1.2 from the fact that J is an ideal in E. We then have a p.q.h.

$$\Phi = (\varphi,\psi): A \longrightarrow M_G(K\otimes B) \rhd K\otimes B,$$

where $\varphi = \omega\circ\sigma\circ\gamma\circ\varphi_0$ and $\psi = \omega\circ\sigma\circ\gamma\circ\psi_0$. This p.q.h. clearly defines the same q.h. as the p.q.h.

$$(\gamma\circ\varphi_0,\gamma\circ\psi_0): A \longrightarrow M_G(J) \rhd J \overset{\sigma_0}{\longrightarrow} R \overset{\omega_0}{\longrightarrow} K\otimes B.$$

Now σ_0 is homotopic to the map $\sigma_1: J \longrightarrow R$ defined by $\sigma_1(a) = 0\oplus a$. (A homotopy is given by $\sigma_t(a) = u_t\,(a\oplus 0)u_t^*$, where $t \longrightarrow u_t$ is a path of unitaries in $M_2(\mathbf{C})$ from 1 to $\begin{bmatrix} 0 & 1 \\ 1 & 0 \end{bmatrix}$.) Furthermore, the previous lemma provides a strictly continuous path $t \longrightarrow z_t$ of G-invariant isometries in $M_G(K\otimes B)$ such that the formula $\eta_t(a) = z_t^* u\sigma_1(a)u^* z_t$ defines a homotopy of homomorphisms from J to $K\otimes B$ with $\eta_0 = \omega_0\circ\sigma_1$ and $\eta_1 = \mu_0$. (The countably generated Hilbert module appearing there is taken to be $\overline{JH_B}$.) It follows that the q.h. defined by Φ is homotopic to the q.h. defined by Φ_0, as desired.

For the second statement, we first assume that in (*) we already have $E = M_G(K\otimes B)$, $J = K\otimes B$, and $\mu_0 = id_{K\otimes B}$, and we prove that the p.q.h. Φ constructed from Φ_0 by the procedure above is homotopic to Φ_0 in the appropriate sense. We have

$$\Phi = (\varphi, \psi) : A \longrightarrow M_2(M_G(K\otimes B)) \xrightarrow{\omega} M_G(K\otimes B) \triangleright K\otimes B,$$

where $\varphi(a) = \varphi_0(a)\oplus 0$ and $\psi(a) = \psi_0(a)\oplus 0$ for $a \in A$, and where ω is the isomorphism implemented by an equivariant isomorphism $u : H_B \oplus H_B \longrightarrow H_B$. Use the previous lemma to find a strictly continuous path $t \longrightarrow z_t$ of G-invariant isometries in $M_2(M_G(K\otimes B))$ such that $z_0 = 1$, z_1 is u followed by the inclusion of H_B in $H_B \oplus H_B$ as the first summand (so that $z_1 z_1^* = \begin{pmatrix} 1 & 0 \\ 0 & 0 \end{pmatrix}$), and $z_t z_t^* \geq \begin{pmatrix} 1 & 0 \\ 0 & 0 \end{pmatrix}$ for all t. Now identify $M_G(K\otimes B\otimes C([0,1]))$ with the algebra of bounded strictly continuous functions from $[0,1]$ to $M_G(K\otimes B)$. (See [1], corollary 3.4, for the case in which G is trivial; the general case follows immediately.) For $x \in M_G(K\otimes B)$, define $\nu(x)(t) = u z_t^*(x \oplus 0) z_t u^*$. Then ν is a homomorphism from $M_G(K\otimes B)$ to $M_G(K\otimes B\otimes C([0,1]))$, and the required homotopy is given by the p.q.h.

$$(\nu\circ\varphi_0, \nu\circ\psi_0) : A \longrightarrow M_G(K\otimes B\otimes C([0,1])) \triangleright K\otimes B\otimes C([0,1]).$$

We now do the general case. Let Φ_0 and Φ_1 be p.q.h.'s of the form (*), and suppose that the q.h.'s that they define are homotopic. Lemma 3.2.9 (3) provides a homotopy from Φ_0 to the q.h. it defines. In the case at hand, it is clear that the maps to $K\otimes B$ obtained from the intermediate p.q.h.'s are all injective. We glue this homotopy to the homotopy from the q.h. defined by Φ_0 to the q.h. defined by Φ_1, and then to the analogous homotopy from that q.h. to Φ_1 itself. We obtain a homotopy from Φ_0 to Φ_1 such that the corresponding p.q.h. from A to $K\otimes B\otimes C([0,1])$ has an injective map to $K\otimes B\otimes C([0,1])$. We can therefore apply the construction in the first part of the proof to obtain a homotopy from the p.q.h. constructed from Φ_0 to the p.q.h. constructed from Φ_1, defined by p.q.h. of the form

$$(\sigma, \tau) : A \longrightarrow M_G(K\otimes B\otimes C([0,1])) \triangleright K\otimes B\otimes C([0,1]).$$

But we have already shown that Φ_0 and Φ_1 are homotopic to the p.q.h.'s constructed from them, via a homotopy of this type. Therefore Φ_0 and Φ_1 are homotopic via a homotopy of this type. Q.E.D.

We now have a definition of our $KK_G(A,B)$ suitable for use in the proof of theorem 3.6.5. However, we must still obtain a suitable version of Kasparov's definition of $KK_G(A,B)$. We start by recalling his definition from section 4 of [55]:

3.6.8 Definition. Let A and B be G-algebras (separable, as usual). We denote by $\mathbf{E}(A,B)$ the set of all triples $(\varepsilon, \varphi, f)$, where:

(1) ε is a grading operator on H_B, that is, H_B is the direct sum of the $+1$ and -1 eigenspaces for ε,

(2) $\varphi : A \longrightarrow L(H_B)$ is an equivariant homomorphism of degree 0, that is, φ is the direct sum of a representation of A on the $+1$ eigenspace of ε and one on the -1 eigenspace of ε,

(3) $f \in L(H_B)$ is an operator of degree 1, that is, f interchanges the $+1$ and -1 eigenspaces of ε, and

(4) for all $a \in A$, the elements

$$\varphi(a)f - f\varphi(a), \quad (f^2 - 1)\varphi(a), \quad \text{and} \quad (f - f^*)\varphi(a) \tag{*}$$

are in $K(H_B)$.

A homotopy between two elements of $\mathbf{E}(A,B)$ is an element of $\mathbf{E}(A, B \otimes C([0,1]))$ whose restrictions at the endpoints of $[0,1]$ are the given elements of $\mathbf{E}(A,B)$.

An element of $\mathbf{E}(A,B)$ is called degenerate if the elements (*) are all zero.

The direct sum of two elements of $\mathbf{E}(A,B)$ is defined componentwise in the obvious way.

The Kasparov group of (A,B) is the set of homotopy classes of elements of $\mathbf{E}(A,B)$, modulo the classes of degenerate elements of $\mathbf{E}(A,B)$. We will temporarily denote it by $\overline{KK}_G(A,B)$.

3.6.9 Lemma. (Compare [100], remark 3.) Let $(\varepsilon, \varphi, f) \in \mathbf{E}(A,B)$, and let $(\varepsilon_0, \varphi_0, f_0) \in \mathbf{E}(A,B)$ be degenerate. Then there is a homotopy from $(\varepsilon, \varphi, f)$ to $(\varepsilon, \varphi, f) \oplus (\varepsilon_0, \varphi_0, f_0)$ whose restriction at $t \in [0,1]$ is $(\varepsilon, \varphi, f)$ for $t = 0$ and $(\varepsilon, \varphi, f) \oplus (\varepsilon_0, \varphi_0, f_0)$ for all other t.

Proof. Let

$$E = (H_B \otimes C([0,1])) \oplus (H_B \otimes C_0((0,1])),$$

completed in the appropriate manner (as a Hilbert $B \otimes C([0,1])$-module). Define

$$\bar{\varepsilon} = (\varepsilon \otimes 1) \oplus (\varepsilon_0 \otimes 1),$$

$$\bar{\varphi}(a) = (\varphi(a) \otimes 1) \oplus (\varphi_0(a) \otimes 1) \quad \text{for } a \in A,$$

and

$$\bar{f} = (f \otimes 1) \oplus (f_0 \otimes 1).$$

By the Stabilization Theorem, $E \simeq H_{B \otimes C([0,1])}$. It is now clear that $(\bar{\varepsilon}, \bar{\varphi}, \bar{f})$ is the required homotopy. Q.E.D.

3.6.10 Proposition. $\overline{KK}_G(A,B)$ is canonically isomorphic to the set of homotopy classes of triples (φ, ψ, u), where φ and ψ are equivariant representations of A in $L(H_B)$ and $u \in L(H_B)$ is a G-invariant unitary such that $\varphi(a) - u\psi(a)u^* \in K(H_B)$ for all $a \in A$. A homotopy is, of course, a triple as above in which B is replaced by $B \otimes C([0,1])$.

Proof. First, we observe that any two degenerate elements of $\mathbf{E}(A,B)$ are homotopic. Indeed, two applications of the previous lemma show that both are homotopic to their direct sum. Therefore, in the definition of $\overline{KK_G}(A,B)$, it is unnecessary to divide out by the classes of the degenerate elements.

Next, observe that a triple (φ,ψ,u) as above defines a canonical element of $\mathbf{E}(A,B)$, namely

$$\left(\begin{bmatrix}1 & 0\\ 0 & -1\end{bmatrix}, \begin{bmatrix}\varphi & 0\\ 0 & \psi\end{bmatrix}, \begin{bmatrix}0 & u\\ u^* & 0\end{bmatrix}\right). \tag{*}$$

(We are of course using the fact that $H_B \oplus H_B \simeq H_B$.) Our first objective is therefore to prove that every element of $\mathbf{E}(A,B)$ is homotopic to one of the form (*). This will be done in a number of stages.

Let $(\varepsilon,\varphi,f) \in \mathbf{E}(A,B)$. Let E_0 be the $+1$ eigenspace of ε and let E_1 be the -1 eigenspace. Let ε_0 be a grading operator such that both its eigenspaces are isomorphic to H_B. Then $(\varepsilon,\varphi,f) \oplus (\varepsilon_0, 0, 0)$ is homotopic to (ε,φ,f) by the previous lemma, and both eigenspaces of $\varepsilon \oplus \varepsilon_0$ are isomorphic to H_B by the Stabilization Theorem, applied to $E_0 \oplus H_B$ and $E_1 \oplus H_B$. We may therefore assume that this was the case for our original ε.

Next, consider the triple $(\bar\varepsilon,\bar\varphi,\bar f)$ defined as follows. Identify $L(H_{B\otimes C([0,\frac12])})$ with the set of bounded strictly continuous functions from $[0,\frac12]$ into $M(K\otimes B)$, using corollary 3.4 of [1]. Then define $\bar\varepsilon(t) = \varepsilon$ and $\bar\varphi(a)(t) = \varphi(a)$ for all $t \in [0,\frac12]$ and $a \in A$. Further set $\bar f(t) = (1-t)f + tf^*$. Some calculation then shows that $(\bar\varepsilon,\bar\varphi,\bar f) \in \mathbf{E}(A, B\otimes C([0,\frac12]))$. Consequently (ε,φ,f) is homotopic to $(\varepsilon,\varphi,\frac12(f+f^*))$. We may therefore assume from the beginning that $f^* = f$.

We are now going to show that we may take $\|f\| \le 1$. It is clear from the technique of the previous paragraph that it is sufficient to construct a continuous path of operators f_t such that $f_0 = f$, $\|f_1\| \le 1$, and $(\varepsilon,\varphi,f_t) \in \mathbf{E}(A,B)$ for all $t \in [0,1]$. Let $h_t : \mathbf{R} \longrightarrow \mathbf{R}$ be the function

$$h_t(s) = \begin{cases} s & |s| \le 1 \\ s - t(s-1) & s > 1 \\ s - t(s+1) & s < -1, \end{cases}$$

and set $f_t = h_t(f)$. Clearly $t \longrightarrow f_t$ is continuous, $f_0 = f$, and $\|f_1\| \le 1$. We must show that $(\varepsilon,\varphi,f_t) \in \mathbf{E}(A,B)$ for all t. We point out that, for fixed ε and φ, the set

$$\{x \in L(H_B) : (\varepsilon,\varphi,x) \in \mathbf{E}(A,B)\}$$

is norm closed in $L(H_B)$. It therefore suffices to prove that $(\varepsilon,\varphi,x_n) \in \mathbf{E}(A,B)$ for some sequence (x_n) converging to f_t. The function h_t can be approximated uniformly on the spectrum $sp(f)$ of f by real polynomials $p_n(s)$. Since h_t is an odd function, we may assume that $p_n(s)$ contains only odd powers of s, and since $h_t(1) = 1$ we may assume that the sum of the coefficients of $p_n(s)$ is 1. Set

$x_n = p_n(f)$.

Since f has degree 1, so does every odd power of f, and therefore so does x_n. The relation $\varphi(a)f - f\varphi(a) \in K(H_B)$ for all $a \in A$ clearly implies that $\varphi(a)p(f) - p(f)\varphi(a) \in K(H_B)$ for all $a \in A$, for any polynomial p at all. The relation $(x_n - x_n^*)\varphi(a) \in K(H_B)$ holds automatically since x_n is selfadjoint. It remains to prove that $(x_n^2 - 1)\varphi(a) \in K(H_B)$. From the relation $(f^2 - 1)\varphi(a) \in K(H_B)$, we conclude that $(f^{n+2} - f^n)\varphi(a) \in K(H_B)$ for every n, and hence that $(f^{2n} - 1)\varphi(a) \in K(H_B)$ for every n. Now $p_n(s)$ has the form $p_n(s) = sq(s^2)$ for some polynomial $q(s)$. Therefore $p_n(s)^2 = s^2 q(s^2)^2$ contains only even powers of s. Since the coefficient sum of $p_n(s)^2$ is still 1, we can now conclude that

$$(x_n^2 - 1)\varphi(a) = (p_n(f)^2 - 1)\varphi(a) \in K(H_B),$$

as desired. Thus, we have shown that $(\varepsilon, \varphi, x_n) \in \mathbf{E}(A,B)$.

We may now assume that $\|f\| \le 1$. We then write $\varphi = \sigma \oplus \tau$, where σ and τ are the representations of A on the $+1$ and -1 eigenspaces of ε. Relative to this decomposition, f has the form $\begin{bmatrix} 0 & x \\ x^* & 0 \end{bmatrix}$, where $x \in L(H_B)$ satisfies $\|x\| \le 1$. The relation $\varphi(a)f - f\varphi(a) \in K(H_B)$ for all $a \in A$ is equivalent to the relation $\sigma(a)x - x\tau(a) \in K(H_B)$ for all $a \in A$, and the relation $(f^2 - 1)\varphi(a) \in K(H_B)$ is equivalent to the relations $(xx^* - 1)\sigma(a) \in K(H_B)$ and $(x^*x - 1)\tau(a) \in K(H_B)$.

We have written $(\varepsilon, \varphi, f)$ as

$$\left(\begin{bmatrix} 1 & 0 \\ 0 & -1 \end{bmatrix}, \begin{bmatrix} \sigma & 0 \\ 0 & \tau \end{bmatrix}, \begin{bmatrix} 0 & x \\ x^* & 0 \end{bmatrix} \right),$$

so the only step remaining is to make x unitary. Now it follows from the previous lemma that $(\varepsilon, \varphi, f)$ is homotopic to

$$\left(\begin{bmatrix} 1 & 0 \\ 0 & -1 \end{bmatrix}, \begin{bmatrix} \sigma\oplus 0 & 0 \\ 0 & \tau\oplus 0 \end{bmatrix}, \begin{bmatrix} 0 & x\oplus 0 \\ x^*\oplus 0 & 0 \end{bmatrix} \right), \tag{**}$$

where the components of the triple are now really 4×4 matrices. Set

$$u_t = \begin{bmatrix} x & t(1-xx^*)^{1/2} \\ -t(1-x^*x)^{1/2} & tx^* \end{bmatrix}.$$

(Note that it is defined since $\|x\| \le 1$.) Then

$$t \longrightarrow \left(\begin{bmatrix} 1 & 0 \\ 0 & -1 \end{bmatrix}, \begin{bmatrix} \sigma\oplus 0 & 0 \\ 0 & \tau\oplus 0 \end{bmatrix}, \begin{bmatrix} 0 & u_t \\ u_t^* & 0 \end{bmatrix} \right)$$

defines a homotopy of elements of $\mathbf{E}(A,B)$, as can be shown by a computation. At $t = 0$, we obtain the triple (**), and at $t = 1$ we obtain a triple of the required form (*), since u_1 is unitary. This completes the first part of the proof.

We now have to show that if two elements of $\mathbf{E}(A,B)$ of the form (*) are homotopic, then they are homotopic via a homotopy of the form (*). This is done by applying the first part of the proof to the entire homotopy, and then using the previous lemma several times; exact details are omitted. Q.E.D.

Proof of theorem 3.6.5. We define a map from $KK_G^0(A,B)$ to $\overline{KK}_G(A,B)$ by assigning to the p.q.h.

$$(\varphi,\psi): A \longrightarrow M_G(K\otimes B) \triangleright K\otimes B$$

the triple $(\varphi,\psi,1)$. The inverse map assigns to the triple (φ,ψ,u) the p.q.h.

$$(\varphi,u\psi u^*): A \longrightarrow L(H_B) \triangleright K(H_B),$$

where $u\psi u^*$ is the homomorphism $a \longrightarrow u\psi(a)u^*$. Here we are using the isomorphisms $K(H_B)\simeq K\otimes B$ and $L(H_B)\simeq M(K\otimes B)$. Since φ and $u\psi u^*$ are continuous and equivariant, their images must actually lie in $M_G(K\otimes B)$, and so we really do obtain the correct sort of p.q.h. Both of the maps we have defined clearly send homotopies to homotopies, and are therefore well defined.

If (φ,ψ) is a p.q.h., then the p.q.h. assigned to the triple $(\varphi,\psi,1)$ is just (φ,ψ) again. To prove that our maps are inverses of each other, it is therefore sufficient to prove that a given triple (φ,ψ,u) is homotopic to the triple $(\varphi,u\psi u^*,1)$ obtained from the p.q.h. assigned to it. By lemma 3.6.9, (φ,ψ,u) is homotopic to $(\varphi\oplus 0,\psi\oplus 0,u\oplus u^*)$. Let $t \longrightarrow v_t$ be a path of unitaries such that $v_0=u\oplus u^*$ and $v_1=1$. We then have a homotopy

$$t \longrightarrow (\varphi\oplus 0,\, v_t^*(u\psi u^*\oplus 0)v_t,\, v_t)$$

from $(\varphi\oplus 0,\psi\oplus 0,u\oplus u^*)$ to $(\varphi\oplus 0,u\psi u^*\oplus 0,1)$. This last triple is homotopic to $(\varphi,u\psi u^*,1)$, again by lemma 3.6.9. We have thus shown that our $KK_G^0(A,B)$ is the same as the group defined by Kasparov.

The group $KK_G^1(A,B)$ as defined here is therefore also the same as Kasparov's $KK_G^1(A,B)$, since the groups defined here and those of [55] both satisfy $KK_G^1(A,B)\simeq KK_G^0(SA,B)$. Q.E.D.

As a consequence, we obtain:

3.6.11 Theorem ([55], section 7, theorems 2 and 3). Let A and B be G-algebras, both separable as usual, and assume in addition that A is nuclear. Let I be a G-invariant ideal in A and let J be a G-invariant ideal in B. Then there are natural 6 term exact sequences

$$\begin{array}{ccccc} KK_G^0(A,J) & \longrightarrow & KK_G^0(A,B) & \longrightarrow & KK_G^0(A,B/J) \\ \uparrow & & & & \downarrow \\ KK_G^1(A,B/J) & \longleftarrow & KK_G^1(A,B) & \longleftarrow & KK_G^1(A,J) \end{array}$$

and

$$\begin{array}{ccccc} KK_G^0(A/I,B) & \longrightarrow & KK_G^0(A,B) & \longrightarrow & KK_G^0(I,B) \\ \uparrow & & & & \downarrow \\ KK_G^1(I,B) & \longleftarrow & KK_G^1(A,B) & \longleftarrow & KK_G^1(A/I,B). \end{array}$$

For the proof, we refer to [55], where this is proved by identifying $KK_G^1(A,B)$ with the group $Ext_G(A,B)$ of equivariant extensions. Another, simpler, proof appears in [27], where there is also an indication of how this result can be proved in terms of quasihomomorphisms. (The hypotheses in [27] differ slightly from those in theorem 3.6.11, but they cover all the cases we will actually encounter.)

The relation between KK-theory and K-theory is given by the next theorem.

3.6.12 Theorem ([55], section 6, theorem 3). If B is any separable G-algebra, then there is a natural isomorphism F from $K_*^G(B)$ to $KK_G^*(\mathbf{C},B)$. If B is unital, and $p,q \in K\otimes B$ are G-invariant projections, then $F([p]-[q])$ is the class of the p.q.h. $(\varphi,\psi):\mathbf{C} \longrightarrow K\otimes B$, where $\varphi(\lambda)=\lambda p$ and $\psi(\lambda)=\lambda q$. If B is not unital, and $p,q \in K\otimes B^+$ are G-invariant projections such that $p-q \in K\otimes B$, then $F([p]-[q])$ is the class of the p.q.h.

$$(\varphi,\psi):\mathbf{C} \longrightarrow K\otimes B^+ \rhd K\otimes B,$$

where φ and ψ are defined as before.

The proof of this theorem requires the equivariant version of the Mišcenko-Fomenko theory [68] of Fredholm operators in $L(H_B)$. We will state the results; the proofs are essentially the same as in [68], except that the submodules L_n of H_B generated by the first n standard basis elements must be replaced by the submodules $V_n\otimes B$, where (V_n) is an increasing sequence of finite dimensional G-invariant subspaces of $l^2\otimes L^2(G)$ whose union is dense.

3.6.13 Facts about Fredholm operators in $L(H_B)$. Let B be a unital G-algebra. A G-invariant element $a \in L(H_B)$ is called Fredholm if there are G-invariant projections $p,q \in K\otimes B$ such that, relative to the direct sum decompositions $H_B = pH_B \oplus (1-p)H_B$ for the domain and $H_B = qH_B \oplus (1-q)H_B$ for the range, a is represented by a matrix $\begin{bmatrix} x & 0 \\ 0 & y \end{bmatrix}$ with y invertible. The index of a is $[p]-[q] \in K_0^G(B)$. This index is well defined ([68], definition in section 1). If $a-1 \in K\otimes B$, then a is automatically Fredholm and has index 0 ([68], lemma 2.2). It is easily seen from the proof of that lemma that if $a^*=a$ then we may take $p=q$ and that x and y are then selfadjoint.

3.6.14 Lemma. Let B be a unital G-algebra, and let $p,q \in M(K\otimes B)$ be G-invariant projections such that $p-q \in K\otimes B$. Assume that, regarded as operators on H_B, the projections p and q satisfy $pH_B \simeq qH_B \simeq H_B$. Then there is a G-invariant partial isometry $u \in M(K\otimes B)$ such that $uu^* \le p$, $u^*u \le q$, and the elements $p-uu^*$, $q-u^*u$, and $p-u$ are all in $K\otimes B$.

Proof. Let $v = pq + (1-p)(1-q)$. Then

$$v - 1 = p(q-p) + (p-q)q \in K \otimes B.$$

Consequently, $vv^* - 1 \in K \otimes B$. Furthermore, $vq = pv$, whence p commutes with vv^*. So $pvv^*p - p$, regarded as an operator on pH_B, is in $K(pH_B)$. Since $pH_B \simeq H_B$, we can apply the facts about Fredholm operators to find a G-invariant projection $p_0 \in L(pH_B)$, that is, $p_0 \in M(K \otimes B)$ with $p_0 \leq p$, such that $p - p_0 \in K \otimes B$ and, relative to the decomposition $pH_B = (p - p_0)H_B \oplus p_0 H_B$, the element pvv^*p is represented by a matrix $\begin{bmatrix} x & 0 \\ 0 & y \end{bmatrix}$ with y invertible and selfadjoint. Set $a = \begin{bmatrix} 0 & 0 \\ 0 & y^{-1/2} \end{bmatrix}$, and regard a as an element of $M(K \otimes B)$. Let $u = apv$. Then it is readily verified that $uu^* = p_0$. Therefore u^*u is a projection. One checks that $u^*uq = qu^*u = u^*u$, so that $u^*u \leq q$.

By construction, $p - uu^* = p - p_0 \in K \otimes B$. Also,

$$p - u = p - apv = (p - p_0) + p_0(1 - v) + a(p(vv^*)^{1/2}p - p)v \in K \otimes B.$$

(Notice that $(vv^*)^{1/2} - 1 \in K \otimes B$ because $vv^* - 1 \in K \otimes B$.) Since $p - q \in K \otimes B$, we conclude that also $q - u \in K \otimes B$, from which it follows that

$$q - u^*u = (q-u)^*q + u^*(q-u) \in K \otimes B,$$

as desired. Q.E.D.

Proof of theorem 3.6.12. The first step of the proof is to show that if B is a unital G-algebra, then the map $F: K_0^G(B) \longrightarrow KK_G^0(\mathbf{C}, B)$ is an isomorphism.

Recall that if $p, q \in K \otimes B$ are G-invariant projections, then $F([p]-[q])$ is the class of the p.q.h. from $\mathbf{C}$ to $K \otimes B$ given by the maps $\lambda \longrightarrow \lambda p$ and $\lambda \longrightarrow \lambda q$. Since $K \otimes B$ contains all algebras $L(V) \otimes B$ for finite dimensional representation spaces V of G, we see that F is defined on all elements of $K_0^G(B)$. To show that F is well defined, we must show that if $[p_0]-[q_0] = [p_1]-[q_1]$ for G-invariant projections $p_0, p_1, q_0, q_1 \in K \otimes B$, then the corresponding p.q.h.'s are homotopic. Using the fact that any p.q.h. of the form $(\varphi, \varphi): \mathbf{C} \longrightarrow K \otimes B$ is homotopic to the zero p.q.h., we may replace p_0 and q_0 by $p_0 \oplus q_1$ and $q_0 \oplus q_1$, and also p_1 and q_1 by $q_0 \oplus p_1$ and $q_0 \oplus q_1$. Thus, we may assume that $q_0 = q_1$, and hence that $[p_0] = [p_1]$. Now the p.q.h.'s defined by (p_i, q_i) are homotopic to the ones defined by $(p_i \oplus r, q_i \oplus r)$, for any G-invariant projection $r \in K \otimes B$. Choose r such that $p_0 \oplus r$ and $p_1 \oplus r$ are unitarily equivalent in $M_2((K \otimes B)^+)$. Let u be a G-invariant unitary such that $u(p_0 \oplus r)u^* = p_1 \oplus r$. Going to $M_4((K \otimes B)^+)$, with $e_0 = p_0 \oplus r \oplus 0$, $e_1 = p_1 \oplus r \oplus 0$, and $v = u \oplus u^*$, we have $ve_0v^* = e_1$, and there is a continuous path of G-invariant unitaries $t \longrightarrow v_t$ such that $v_0 = 1$ and $v_1 = v$. Set $f_t = v_t e_0 v_t^*$. Then $f_0 = e_0$, $f_1 = e_1$, and the p.q.h. defined by $(e_0, q_0 \oplus r \oplus 0)$ is homotopic to the one defined by $(e_1, q_0 \oplus r \oplus 0)$ via the p.q.h.'s defined by $(f_t, q_0 \oplus r \oplus 0)$ for $t \in [0,1]$. This shows that F is well defined. That F is an $R(G)$-module homomorphism is easily shown.

We now construct an inverse for F. It follows from proposition 3.6.7 that $KK_G^0(\mathbf{C}, B)$ is isomorphic to the group of homotopy classes of pairs (p, q) of G-

invariant projections in $M(K\otimes B)$ such that $p-q\in K\otimes B$. By simply forming the direct sum with the (trivial) pair $(1,1)$, where 1 is the identity in $M(K\otimes B)$, we may assume that p and q satisfy, when regarded as operators on H_B, the relation $pH_B\simeq qH_B\simeq H_B$. According to the previous lemma, there is a G-invariant partial isometry $u\in M(K\otimes B)$ such that $uu^*\leq p$, $u^*u\leq q$, and $p-uu^*$, $q-u^*u$, and $p-u$ are all in $K\otimes B$. Denoting the equivalence class of the pair (p,q) by $[p,q]$, we define $Z: KK_G^0(\mathbf{C},B)\longrightarrow K_0^G(B)$ by $Z([p,q])=[p-uu^*]-[q-u^*u]$.

We must first show that $[p-uu^*]-[q-u^*u]$ is independent of the choice of u. So let v be another G-invariant partial isometry satisfying the same conditions. Assume first that $vv^*=uu^*$. Set $w=u^*v$. Then $ww^*=u^*u$ and $w^*w=v^*v$. Furthermore, w is a G-invariant element of $L(qH_B)$ such that $q-w\in K(qH_B)$. By 3.6.13, w is a Fredholm operator on qH_B of index 0, so that $[q-ww^*]=[q-w^*w]$. So $Z([p,q])$ is not changed if u is replaced by v. The same argument clearly works if vv^* is only assumed to be unitarily equivalent to uu^*, via a unitary in $(p(K\otimes B)p)^+$.

Now assume that $vv^*\leq uu^*$. By the previous paragraph, we may replace v with the G-invariant partial isometry $w=vv^*u$, since $ww^*=vv^*$. Now $w^*w=u^*(vv^*)u\leq u^*u$, and the projections uu^*-ww^* and u^*u-w^*w are Murray-von Neumann equivalent in $K\otimes B$ via the partial isometry $u-w$. It follows that

$$[p-uu^*]-[q-u^*u]=[p-vv^*]-[q-v^*v]$$

for this choice of v also.

Finally, consider the case in which v is arbitrary. Since $p(K\otimes B)p\simeq K\otimes B$ (because $pH_B\simeq H_B$), both $p-uu^*$ and $p-vv^*$ are unitarily equivalent in $(p(K\otimes B)p)^+$ to projections dominated by a G-invariant projection r which, under the isomorphism above, is sent to some $e\otimes 1$. We may therefore assume that $p-uu^*$, $p-vv^*\leq r$, that is, $uu^*, vv^*\geq p-r$. Set $w=(p-r)u$. Then $ww^*=p-r$, $w^*w\leq q$. and the expressions $p-ww^*$, $q-w^*w$, and $p-w$ are all in $K\otimes B$. The previous paragraph therefore implies that

$$[p-uu^*]-[q-u^*u]=[p-ww^*]-[q-w^*w]=[p-vv^*]-[q-v^*v].$$

We have thus shown that $Z([p,q])$ does not depend on the choice of u.

To complete the proof that Z is well defined, we must show that if (p_0,q_0) is homotopic to (p_1,q_1), then $Z([p_0,q_0])=Z([p_1,q_1])$. This is done by applying the previous lemma to the entire homotopy from (p_0,q_0) to (p_1,q_1), and using the fact that homotopic projections are equivalent. (This is a special case of proposition 2.5.10.)

We now show that Z is the inverse of F. It is obvious, just by making appropriate choices, that $Z\cdot F$ is the identity on $K_0^G(B)$. For the composite in the other order, let (p,q) be a pair of G-invariant projections with $pH_B\simeq qH_B\simeq H_B$ and $p-q\in K\otimes B$. Then $F\cdot Z([p,q])$ is represented by the pair $(p-uu^*,q-u^*u)$, where u is chosen as in the previous lemma. It is easy to see that $((p-uu^*)\oplus uu^*, (q-u^*u)\oplus u^*u)$ is homotopic to $(p\oplus 0, q\oplus 0)$, so it is sufficient to prove that $[uu^*,u^*u]=0$ in $KK_G^0(\mathbf{C},B)$.

We will actually show that $(uu^*\oplus 0, u^*u\oplus 0)$ is homotopic to $(uu^*\oplus 0, uu^*\oplus 0)$. Set $a_t = tu + (1-t)u^*u$ for $t\in[0,1]$, and let

$$x_t = \begin{bmatrix} a_t & (1-a_t a_t^*)^{1/2} \\ -(1-a_t^* a_t)^{1/2} & a_t^* \end{bmatrix}.$$

(This makes sense since $\| a_t \| \leq 1$.) Then x_t is unitary for all t, and we have

$$x_0(u^*u\oplus 0)x_0^* = u^*u\oplus 0 \quad \text{and} \quad x_1(u^*u\oplus 0)x_1^* = uu^*\oplus 0.$$

Thus, $t \longrightarrow (uu^*\oplus 0, x_t(u^*u\oplus 0)x_t^*)$ defines the required homotopy, provided we show that

$$uu^*\oplus 0 - x_t(u^*u\oplus 0)x_t^* \in M_2(K\otimes B)$$

for all t. Now $a_t - p$, $uu^* - p$, and $u^*u - p$ are all in $K\otimes B$, so, with $v = \begin{bmatrix} p & 1-p \\ -(1-p) & p \end{bmatrix}$, we have $x_t - v \in M_2(K\otimes B)$. Therefore

$$(uu^*\oplus 0 - x_t(u^*u\oplus 0)x_t^*) - (p\oplus 0 - v(p\oplus 0)v^*) \in M_2(K\otimes B).$$

But the second term is zero. So we have proved the existence of the required homotopy, and therefore that $F\circ Z$ is the identity. We have now proved the case $K_0^G(B)\simeq KK_G^0(\mathbf{C},B)$ of the theorem, for unital algebras B.

If B is not unital, we consider the commutative diagram with exact rows:

$$\begin{array}{ccccccccc} 0 & \longrightarrow & K_0^G(B) & \longrightarrow & K_0^G(B^+) & \longrightarrow & K_0^G(\mathbf{C}) & \longrightarrow & 0 \\ & & & & \downarrow F & & \downarrow F & & \\ 0 & \longrightarrow & KK_G^0(\mathbf{C},B) & \longrightarrow & KK_G^0(\mathbf{C},B^+) & \longrightarrow & KK_G^0(\mathbf{C},\mathbf{C}) & \longrightarrow & 0. \end{array} \qquad (*)$$

Here the bottom row is obtained by applying theorem 3.6.11 to the split exact sequence

$$0\longrightarrow B\longrightarrow B^+\longrightarrow \mathbf{C}\longrightarrow 0.$$

In the diagram (*), both vertical arrows are isomorphisms by what we have already proved. Consequently there is an isomorphism from $K_0^G(B)$ to $KK_G^0(\mathbf{C},B)$ which makes the diagram commute. We omit the verification that it is given by the formula in the statement of the theorem.

That $K_1^G(B)$ is naturally isomorphic to $KK_G^1(\mathbf{C},B)$ for any B now follows by taking suspensions on both sides. Q.E.D.

We point out that the definition of the inverse can be given in a different way, using the difference map introduced by Lance in [63]. If

$$(\varphi,\psi): \mathbf{C}\longrightarrow E \rhd J \xrightarrow{\ \mu\ } K\otimes B$$

is any equivariant p.q.h., then the corresponding element of $K_0^G(B)$ is $\mu_*\circ(\varphi/\psi)_*([1])$, where $[1]$ is the class of the identity in $\mathbf{C}$, and $(\varphi/\psi)_*$ is the equivariant version of the map defined in section 1 of [63]. We do not use this definition in our proof because the construction of $(\varphi/\psi)_*$ takes a convenient form only on K_1, while our definition of KK-theory is convenient to use only on KK^0.

It should also be pointed out that a careful comparison of Bott periodicity for K-theory (proposition 2.8.6) and for KK-theory (theorem 3.5.11 and its proof) reveals that the periodicity isomorphism $K_0^G(A) \longrightarrow K_0^G(S^2A)$ is the negative of the isomorphism $KK_G^0(\mathbf{C}, A) \longrightarrow KK_G^0(\mathbf{C}, S^2A)$. This difference is of course easily fixed, but it is not important.

The product in KK-theory is a useful substitute in the noncommutative case for the ring structure on $K_0^G(C(X))$ for compact G-spaces X. We will use it in this way in later chapters, perhaps most dramatically in section 4.5. It is important to know the relation between the two products, which we state in the following proposition:

3.6.15 Proposition. Let X be a compact G-space, and let $\eta_1, \eta_2 \in K_0^G(C(X))$. Write $\eta_1\eta_2$ for the product of η_1 and η_2 in the ring structure of $K_0^G(C(X))$, and write $\eta_1 \otimes \eta_2$ for the element in $K_0^G(C(X) \otimes C(X))$ whose image in $KK_G^0(\mathbf{C}, C(X) \otimes C(X))$ is the product of the images of η_1 and η_2 in $KK_G^0(\mathbf{C}, C(X))$. Then $\eta_1\eta_2 = \Delta_*(\eta_1 \otimes \eta_2)$, where $\Delta: C(X) \otimes C(X) \longrightarrow C(X)$ is defined by $\Delta(f_1 \otimes f_2) = f_1 f_2$.

We point out that the canonical identification of $C(X) \otimes C(X)$ with $C(X \times X)$ identifies Δ with the map given by $\Delta(f)(x) = f(x, x)$ for $x \in X$.

Proof of proposition 3.6.15. It is sufficient to prove this for $\eta_i = [E_i]$, where E_1 and E_2 are G-vector bundles over X. Let $p_i \in L(V_i) \otimes C(X)$ be G-invariant projections, satisfying $p_i[V_i \otimes C(X)] \simeq \Gamma(E_i)$ for $i = 1, 2$. The corresponding p.q.h.'s are given by

$$\Phi_i = (\varphi_i, 0) : \mathbf{C} \longrightarrow L(V_i) \otimes C(X),$$

where $\varphi_i(1) = p_i$, and their product is represented by the p.q.h.

$$\Phi = (\varphi, 0) : \mathbf{C} \longrightarrow L(V_1 \otimes V_2) \otimes C(X) \otimes C(X),$$

where $\varphi(1) = p_1 \otimes p_2$. The vector bundle E corresponding to this p.q.h. satisfies

$$\Gamma(E) \simeq (p_1 \otimes p_2)[V_1 \otimes V_2 \otimes C(X \times X)],$$

and can obviously be taken to be the "external" tensor product of E_1 and E_2, whose fiber over (x, y) is $(E_1)_x \otimes (E_2)_y$. Now $\Delta_*([E])$ is the class of the pullback of E to X via the diagonal map $x \longrightarrow (x, x)$. This pullback is just the "internal" tensor product $E_1 \otimes E_2$ used in the definition of the ring structure on $K_G^0(X)$. Thus, we have proved that $\Delta_*([E_1] \otimes [E_2]) = [E_1][E_2]$, as desired. Q.E.D.

This is the last result on KK-theory in this chapter. However, we should at least mention several others which appear later. In chapter 5, we consider the relationship between $KK_G^*(A,B)$ and $KK_H^*(A,B)$ when H is a subgroup of the finite group G. In chapter 9, we consider the KK-theory of certain twisted products, and also discuss the additivity of KK-theory in both of its variables. We furthermore give a general procedure for computing the equivariant KK-theory for AF algebras, provided that the actions leave sufficiently many finite dimensional subalgebras invariant.

Chapter 4

Basic Properties of K-Freeness

This chapter develops the basic properties of K-freeness, total K-freeness, and KK-freeness. We begin the first section by defining the obvious generalization of condition (6) of theorem 1.1.1, which we call "locally discrete K-theory." (Recall that theorem 1.1.1 gave a number of conditions on the equivariant K-theory of a compact space which are equivalent to free action.) After several lemmas, we give two examples to show that this condition does not behave as well as one might hope. Therefore in section 2 we introduce the stronger and more useful notions of K-free and totally K-free actions. We show by example that total K-freeness is stronger than K-freeness, and we develop their elementary properties. It will turn out in later chapters that total K-freeness is a better generalization of freeness of actions on spaces, but that K-freeness is more closely related to the Connes spectrum. In section 3 we then prove our first substantial theorem: if (G,A,α) is a separable G-algebra such that G is finite and acts freely on *Prim* (A), then α is totally K-free.

Section 4 defines KK-free actions and develops their elementary properties. Several counterexamples are also included. There are two reasons for studying KK-freeness. For one thing, it is interesting in its own right. But it also implies K-freeness (and, as will be seen later, sometimes even total K-freeness), and hence properties of KK-freeness can sometimes be used to prove properties of K-freeness. This possibility is illustrated in the last two sections. In section 5 we define $(G,C_0(X))$-algebras, which are essentially G-algebras which are algebras over $C_0(X)$, such that the G-actions are compatible. We then prove that if (G,A,α) is a $(G,C(X))$-algebra, where X is compact and everything is separable, and if G acts freely on X, then α is KK-free. Section 6 uses this result to show that is (G,A) is a $(G,C_0(X))$-algebra, where now X is locally compact and everything is separable, and if G acts freely on X, then α is totally K-free. We end the chapter with a construction of a "free unitization" of a G-algebra and a related example.

Throughout this chapter, G is a compact Lie group. All ideals in C^*-algebras are assumed to be closed.

4.1. Actions Having Locally Discrete K-Theory

4.1.1 Definition. Let (G,A,α) be a G-algebra. We say the action α has locally discrete K-theory if the localizations $K_*^G(A)_P$ are zero for all prime ideals P of $R(G)$ not containing the augmentation ideal $I(G)$.

The terminology is motivated as follows: from proposition 4.1.3 below, it is easily seen that α has locally discrete K-theory if and only if every finitely generated submodule M of $K_*^G(A)$ is discrete in the $I(G)$-adic topology, that is, there is an integer n such that $I(G)^n M = 0$. The word "locally" thus means "for finitely generated submodules." This condition does not imply that $K_*^G(A)$ is discrete in the $I(G)$-adic topology; see example 4.4.12 below. Since we still have some use for the condition that $K_*^G(A)$ be $I(G)$-adically discrete, we also make the following definition:

4.1.2 Definition. Let (G,A,α) be a G-algebra. Then α has discrete K-theory if there is n such that $I(G)^n K_*^G(A) = 0$.

The following result will be very useful for checking that an action has locally discrete K-theory.

4.1.3 Proposition. Let (G,A,α) be a G-algebra. Then the following are equivalent:

(1) The action α has locally discrete K-theory.

(2) For all $\eta \in K_*^G(A)$ there is n such that $I(G)^n \eta = 0$.

If G is abelian, (1) and (2) are equivalent to

(3) For all $\eta \in K_*(C^*(G,A))$ and $\tau \in \widehat{G}$, there is n such that $((\widehat{\alpha}_\tau)_* - id)^n(\eta) = 0$.

Proof. (1)$\Longleftrightarrow$(2) follows immediately from lemma 1.2.2, since $R(G)$ is noetherian by [97], corollary 3.3.

Now let G be abelian. By theorem 2.8.3 (7), condition (3) is equivalent to the following: for all $\eta \in K_*^G(A)$ and $\tau \in \widehat{G}$, there is n such that $(\tau - 1)^n \cdot \eta = 0$. Since the elements $\tau - 1$ are all in $I(G)$, the implication (2)$\Longrightarrow$(3) is obvious. For the converse, note that the elements $\tau - 1$ for $\tau \in \widehat{G}$ generate $I(G)$, since G is abelian. Furthermore, since $R(G)$ is noetherian, $I(G)$ has a finite set $r_1, \dots, r_n$ of generators. Writing each r_k as $r_k = \sum_{\tau \in S_k} s_\tau(\tau - 1)$ for appropriate finite subsets S_k of $\widehat{G}$ and elements $s_\tau \in R(G)$ for $\tau \in S_k$, we see that $I(G)$ is generated by finitely many elements $\tau_1 - 1, \dots, \tau_N - 1$, where the τ_k are in $\widehat{G}$. Now let $\eta \in K_*^G(A)$, and find $m_1, \dots, m_N$ such that $(\tau_k - 1)^{m_k}\eta = 0$. One can now check that $I(G)^M \eta = 0$ with $M = 1 + \sum_{k=1}^{N}(m_k - 1)$. (When M linear combinations of $\tau_1 - 1, \dots, \tau_N - 1$ with coefficients in $R(G)$ are multipled together, each term must contain some $\tau_k - 1$ to a power of at least m_k.) We have thus proved that (3)$\Longrightarrow$(2). Q.E.D.

It follows immediately from theorem 1.1.1 that if X is compact, then an action on $C(X)$ has locally discrete K-theory if and only if the corresponding action on X is free. What this condition really has to do with freeness is perhaps best seen in the following special case. Assume that G is finite, let B be any unital C^*-algebra, and let $A = C(G) \otimes B$. Let τ be the action on $C(G)$ given by translation on G, that is, $\tau_g(f)(h) = f(g^{-1}h)$ for $g, h \in G$ and $f \in C(G)$, and define an action α of G on A by $\alpha_g = \tau_g \otimes id_B$ for $g \in G$. Let (G,E,λ) be a finitely generated projective (G,A,α)-module.

Corresponding to the direct sum decomposition $A = \bigoplus_{g \in G} B$, there is a direct sum decomposition $E = \bigoplus_{g \in G} E_g$, in which each E_g is a finitely generated projective B-module. Let F be the summand corresponding to the identity element of G. Then $(G, C(G) \otimes F, \tau \otimes id_F)$ is a finitely generated projective (G,A,α)-module in the obvious way, and it is isomorphic to (G,E,λ) via the map which takes an element $f \in C(G) \otimes F$, regarded as a function from G to F, to $\sum_{g \in G} \lambda_g (f(g))$. (Note that $\lambda_g(f(g)) \in E_g$ for $g \in G$.) The freeness of the action of G on itself has thus enabled us to "untwist" the action of G on E. In particular, if V is any finite dimensional representation space of G, then the argument above, applied to $V \otimes E$, shows that $V \otimes E \simeq C(G) \otimes V \otimes F$ as (G,A,α)-modules, where the action of $g \in G$ on $C(G) \otimes V \otimes F$ is $\tau_g \otimes id_V \otimes id_F$. Since the original representation of G on V has disappeared from this formula, we see that $[V \otimes E]$ depends only on the dimension of V. We have thus shown that $I(G)$ annihilates $[E]$. (Note that τ is really the left regular representation of G, so that this computation is the analog of the fact that tensoring with the left regular representation "absorbs", up to multiplicity, all other representations of G.)

In spite of this example, we should not expect $I(G)$ to annihilate $K_*^G(A)$ for all of the G-algebras that we want to regard as free. Indeed, if we have an equivariant exact sequence

$$0 \longrightarrow I \longrightarrow A \longrightarrow B \longrightarrow 0,$$

and if $I(G)$ annihilates $K_*^G(I)$ and $K_*^G(B)$, the best conclusion that we can draw is that $I(G)^2$ annihilates $K_*^G(A)$. This phenomenon already occurs in the commutative case, for example with $A = C(S^3)$, $B = C(S^1)$, and $G = S^1 \subset \mathbf{C}$, acting on $S^3 \subset \mathbf{C}^2$ and $S^1 \subset \mathbf{C}$ by the usual scalar multiplication. On the other hand, if in the exact sequence above the actions on I and B have locally discrete K-theory, then the action on A does also. In fact, we have the following result.

4.1.4 Lemma. Let

$$0 \longrightarrow I \longrightarrow A \longrightarrow B \longrightarrow 0$$

be an equivariant exact sequence of G-algebras. If two of the three actions have locally discrete K-theory, than so does the third.

Proof. There is a six term exact sequence

$$\begin{array}{ccccc} K_0^G(I) & \longrightarrow & K_0^G(A) & \longrightarrow & K_0^G(B) \\ \uparrow & & & & \downarrow \\ K_1^G(B) & \longleftarrow & K_1^G(A) & \longleftarrow & K_1^G(I). \end{array}$$

We localize it at a prime ideal $P \subset R(G)$ which does not contain $I(G)$, and obtain the exact sequence

$$\begin{array}{ccccc} K_0^G(I)_P & \longrightarrow & K_0^G(A)_P & \longrightarrow & K_0^G(B)_P \\ \uparrow & & & & \downarrow \\ K_1^G(B)_P & \longleftarrow & K_1^G(A)_P & \longleftarrow & K_1^G(I)_P . \end{array}$$

By hypothesis, two of the three actions have locally discrete K-theory. Therefore at least four of the terms of this sequence are zero, and the remaining two terms are three arrows apart. Hence they are also zero, and the remaining action also has locally discrete K-theory. Q.E.D.

Unfortunately, the property of having locally discrete K-theory does not behave well in more general situations. We give two examples to show what can go wrong. Since these examples, and many later ones, involve the group $\mathbf{Z}/2\mathbf{Z}$, we first prove a lemma about actions of $\mathbf{Z}/2\mathbf{Z}$.

4.1.5 Lemma. Let α be an inner action of $\mathbf{Z}/2\mathbf{Z}$ on a C^*-algebra A. That is, $\alpha_g(a) = u_g a u_g^*$, where $g \longrightarrow u_g$ is a homomorphism from $\mathbf{Z}/2\mathbf{Z}$ to the unitary group of the multipler algebra of A. Then α has locally discrete K-theory if and only if $K_*(A)$ is a 2-torsion group.

Proof. Since α is exterior equivalent to the trivial action (see example 2.7.2), and since equivariant K-theory is invariant under exterior equivalence by theorem 2.8.3 (5), we may assume that α is the trivial action. Then $K_*^{\mathbf{Z}/2\mathbf{Z}}(A) \simeq K_*(A) \otimes R(\mathbf{Z}/2\mathbf{Z})$ by remark 2.8.5. Let τ be the nontrivial character of $\mathbf{Z}/2\mathbf{Z}$. Then $R(\mathbf{Z}/2\mathbf{Z}) = \mathbf{Z} + \mathbf{Z}\tau$ where $\tau^2 = 1$, that is, $R(\mathbf{Z}/2\mathbf{Z})$ is the set of all $m + n\tau$ for $m, n \in \mathbf{Z}$. The augmentation ideal $I(\mathbf{Z}/2\mathbf{Z})$ is generated by $1 - \tau$.

Assume that $K_*(A)$ is a 2-torsion group, and consider an element $\eta \otimes r$ of $K_*(A) \otimes R(\mathbf{Z}/2\mathbf{Z})$. Choose n such that $2^n \eta = 0$. Then $(1-\tau)^{n+1} = 2^n(1-\tau)$ by a calculation, using $\tau^2 = 1$. Therefore

$$(1-\tau)^{n+1}(\eta \otimes r) = 2^n \eta \otimes (1-\tau) r = 0 .$$

Since $K_*(A) \otimes R(\mathbf{Z}/2\mathbf{Z})$ consists of linear combinations of elements of this form, we have verified condition (2) of proposition 4.1.3. For the converse, let $K_*(A)$ have an element η such that $2^n \eta \neq 0$ for any n. Then the same calculation shows that $(1-\tau)^n (\eta \otimes 1) = 2^{n-1} \eta \otimes (1-\tau)$, which is never zero. Therefore α does not have locally discrete K-theory. Q.E.D.

4.1.6 Example. Let $G = \mathbf{Z}/2\mathbf{Z}$, let $A = C_0((0,1])$, and let α be the trivial action. Then $K_*(A) = 0$ since A is contractible. The previous lemma therefore implies that α has locally discrete K-theory, even though the corresponding action on $(0,1]$ is as far from being free as possible.

One might think that the problem with this example is that A is not unital. To dispel this idea, we present in the next example a unital type I C^*-algebra and an action of $\mathbf{Z}/2\mathbf{Z}$ on it which has locally discrete K-theory but whose restriction to an invariant ideal does not. Since ideals are the C^* equivalents of open subsets, this is like having a group act on a space in such a way that the action on the whole space is free but the action on an invariant subset is not.

4.1.7 Example. Let $A = C(S^1) \otimes M_2$, where S^1 is as usual the unit circle in $\mathbf{C}$. Let $D = C([0,1], M_2)$, and identify A with the set of elements $f \in D$ such that $f(0) = f(2\pi)$. Let $w \in D$ be the unitary given by

$$w(t) = \begin{bmatrix} 0 & \exp(\tfrac{1}{2}it) \\ \exp(-\tfrac{1}{2}it) & 0 \end{bmatrix}$$

for $t \in [0, 2\pi]$. Then w is not in A, but it is true that $waw^* \in A$ for all $a \in A$. This follows because conjugation by $w(0)$ has the same effect as conjugation by $w(2\pi) = -w(0)$, so that if $a(0) = a(2\pi)$, then the same is true of waw^*. Furthermore, $w^2 = 1$. Therefore there is an action $\alpha : \mathbf{Z}/2\mathbf{Z} \longrightarrow Aut(A)$ which sends the nontrivial element of $\mathbf{Z}/2\mathbf{Z}$ to the automorphism $a \longrightarrow waw^*$ of A, and $(\mathbf{Z}/2\mathbf{Z}, A, \alpha)$ is a $\mathbf{Z}/2\mathbf{Z}$-algebra. In order to compute its equivariant K-theory, we calculate the crossed product.

We also write α for the action of $\mathbf{Z}/2\mathbf{Z}$ on D defined by sending the nontrivial element of $\mathbf{Z}/2\mathbf{Z}$ to the automorphism $a \longrightarrow waw^*$ of D. On D, this action is inner, and therefore exterior equivalent to the trivial action ι. It follows that the crossed products of D by α and ι are isomorphic. In order to compute the image of $C^*(\mathbf{Z}/2\mathbf{Z}, A, \alpha)$ in $C^*(\mathbf{Z}/2\mathbf{Z}, D, \iota)$, we write down an isomorphism explicitly. Elements of $C^*(\mathbf{Z}/2\mathbf{Z}, D, \alpha)$ are functions from $\mathbf{Z}/2\mathbf{Z}$ into D, and we write them as functions from $[0,2\pi] \times \{0,1\}$ into M_2; similarly for $C^*(\mathbf{Z}/2\mathbf{Z}, D, \iota)$. With this notation, the operations in $C^*(\mathbf{Z}/2\mathbf{Z}, D, \alpha)$ become, for $t \in [0,2\pi]$ and $i \in \{0,1\}$,

$$(fg)(t,i) = f(t,0)g(t,i) + f(t,1)w(t)g(t,1-i)w(t)^*,$$

$$f^*(t,i) = w(t)^i f(t,i)^* (w(t)^i)^*,$$

and

$$\hat{\alpha}_\tau(f)(t,i) = (-1)^i f(t,i),$$

where τ is the nontrivial element of $\hat{G}$. The operations are the same in $C^*(\mathbf{Z}/2\mathbf{Z}, D, \iota)$ except that w is omitted. The map

$$\varphi : C^*(\mathbf{Z}/2\mathbf{Z}, D, \alpha) \longrightarrow C^*(\mathbf{Z}/2\mathbf{Z}, D, \iota)$$

defined by

$$\varphi(f)(t,i) = f(t,i)w(t)^i$$

is an isomorphism which intertwines the dual actions. Now define $\psi : C^*(\mathbf{Z}/2\mathbf{Z}, D, \iota) \longrightarrow D \oplus D$ by

$$\psi(f) = \left[\frac{f(\cdot,0) + f(\cdot,1)}{2}, \ \frac{f(\cdot,0) - f(\cdot,1)}{2} \right].$$

Again, one can check that ψ is a homomorphism — it is in fact the integrated form of the covariant representation sending d to $\begin{bmatrix} d & 0 \\ 0 & d \end{bmatrix}$ for $d \in D$ and the nontrivial element of G to $\begin{bmatrix} 1 & 0 \\ 0 & -1 \end{bmatrix}$. It is clearly bijective, and it is easily seen that $\psi \cdot \hat{\iota}_\tau = \beta \cdot \psi$, where $\beta(a,b) = (b,a)$ for $a, b \in D$.

We now trace what happens to $C^*(\mathbf{Z}/2\mathbf{Z}, A, \alpha)$ under these isomorphisms. As a subalgebra of $C^*(\mathbf{Z}/2\mathbf{Z}, D, \alpha)$, we have:

$$C^*(\mathbf{Z}/2\mathbf{Z}, A, \alpha) = \{f \in C^*(\mathbf{Z}/2\mathbf{Z}, D, \alpha) : f(2\pi, i) = f(0, i) \text{ for } i = 0,1\}.$$

Since $w(2\pi) = -w(0)$, the isomorphism φ sends this to

$$\{f \in C^*(\mathbf{Z}/2\mathbf{Z}, D, \iota) : f(2\pi, 0) = f(0,0) \text{ and } f(2\pi, 1) = -f(0,1)\}.$$

After applying ψ, we get

$$\{(a,b) \in D \oplus D : a(2\pi) = b(0) \text{ and } b(2\pi) = a(0)\}.$$

But this algebra is obviously just the set of M_2 valued continuous functions on a circle twice as large. Under under this identification, $\hat{\alpha}_\tau$ becomes the rotation half way around. Consequently, $\hat{\alpha}_\tau$ is homotopic to the identity, and so $(\hat{\alpha}_\tau)_* = id$. Therefore, α has locally discrete K-theory by proposition 4.1.3.

Now consider the ideal I in A consisting of all $f : [0, 2\pi] \longrightarrow M_2$ such that $f(0) = f(2\pi) = 0$. It is obviously α-invariant. Furthermore, w is an element of order 2 in its multiplier algebra, so that the restricted action is inner. It is well known that $K_1(I) \simeq \mathbf{Z}$. Therefore lemma 4.1.5 implies that the restriction $\alpha_{(\cdot)}|_I$ of α to I does not have locally discrete K-theory, as was promised.

4.2. K-Free and Totally K-Free Actions

We want a property which passes to invariant ideals and quotients by invariant ideals. We therefore make the following definition.

4.2.1 Definition. Let (G, A, α) be a G-algebra. We say the action α is K-free if for every G-invariant ideal I of A, the restricted action $\alpha_{(\cdot)}|_I : G \longrightarrow Aut(I)$ has locally discrete K-theory.

In the introduction (section 1.1), we said that an action α as above would be called K-free if the action it defines on A/I has locally discrete K-theory for every G-invariant ideal of A. Under both definitions, α itself must have locally discrete K-theory — take $I = A$ in definition 4.2.1 and $I = 0$ in the definition in the introduction. Now lemma 4.1.4 implies that for any I, the action on I has locally discrete K-theory if and only if the action on A/I does. So definition 4.2.1 agrees with the definition given in the introduction.

Since we have forced it to be true, we now obtain:

4.2.2 Proposition. Let (G,A,α) be a G-algebra, and let I be a G-invariant ideal in A. Then α is K-free if and only if the actions it induces on I and A/I are K-free.

Proof. First assume that α is K-free. Since a G-invariant ideal in I is a G-invariant ideal in A, it follows immediately that $\alpha_{(\cdot)}|_I$ is K-free. Now consider A/I, and let $\pi: A \longrightarrow A/I$ be the (equivariant) quotient map. Let $J \subset A/I$ be a G-invariant ideal. Then there is an equivariant exact sequence

$$0 \longrightarrow I \longrightarrow \pi^{-1}[J] \longrightarrow J \longrightarrow 0.$$

The actions on both I and $\pi^{-1}[J]$ have locally discrete K-theory because α is K-free. By lemma 4.1.4, the action on J also has locally discrete K-theory. Since J is arbitrary, the action on A/I is K-free.

Now assume that the actions on I and A/I are K-free, and let J be an arbitrary G-invariant ideal in A. There is an equivariant exact sequence

$$0 \longrightarrow I \cap J \longrightarrow J \longrightarrow (I+J)/I \longrightarrow 0.$$

The actions on the first and last terms have locally discrete K-theory, since they are G-invariant ideals in I and A/I respectively. So the action on the middle term also has locally discrete K-theory. Hence α is K-free. Q.E.D.

There is unfortunately another defect in our definition: as the next example shows, the restriction of a K-free action to a closed subgroup need not be K-free. This example is well-known — see for instance [74], remark 3.

4.2.3 Example. Let $G = \mathbf{Z}/2\mathbf{Z} \times \mathbf{Z}/2\mathbf{Z}$, and let $A = M_2$. Write the elements of G as e, h, k, and hk. Let $u, v \in M_2$ be the unitaries

$$u = \begin{bmatrix} 1 & 0 \\ 0 & -1 \end{bmatrix} \quad \text{and} \quad v = \begin{bmatrix} 0 & 1 \\ 1 & 0 \end{bmatrix}.$$

Define $\alpha: G \longrightarrow Aut(A)$ by $\alpha_e = id$, $\alpha_h = \mathrm{ad}(u)$, $\alpha_k = \mathrm{ad}(v)$, and $\alpha_{hk} = \mathrm{ad}(uv)$. (Recall that $\mathrm{ad}(w)(a) = waw^*$.) To check that α is a homomorphism, one needs to show that $\mathrm{ad}(uv) = \mathrm{ad}(vu)$, which follows from the equation $uv = -vu$.

We start by showing $C^*(C,A,\alpha)$ is simple, for which we use the Connes spectrum $\Gamma(\alpha)$ ([79], 8.8.2). The first step in computing it is to find the Arveson spectrum $Sp(\alpha)$ ([79], 8.1.6). Each of the elements $1, u, v$, and uv is easily seen to be in a different one of the eigenspaces

$$A_\tau = \{a \in A : \alpha_g(a) = \tau(g)a \text{ for } g \in G\}$$

for $\tau \in \widehat{G}$. Thus none of them is zero, and $Sp(\alpha) = \widehat{G}$. Since $1, u, v$, and uv span A as a vector space, the multiples of the identity constitute the entire eigenspace $A_1 = A^G$, and there are no nontrivial G-invariant projections. Since every G-invariant hereditary subalgebra of a matrix algebra must have the form pAp for a G-invariant projection p, there are no nontrivial G-invariant hereditary subalgebras. Therefore $\Gamma(\alpha) = Sp(\alpha) = \hat{G}$. By [79], 8.11.12, $C^*(G,A,\alpha)$ is simple.

Since A is finite dimensional and G is finite, $C^*(G,A,\alpha)$ is finite dimensional. Since it is simple, it must be isomorphic to a full matrix algebra M_n (in fact, $n = 4$). It follows that $(\hat{\alpha}_\tau)_* = id$ for all τ, since all automorphisms of M_n are trivial on K-theory. Therefore α has locally discrete K-theory, by proposition 4.1.3. Since A is simple, α is in fact K-free. Now let G_0 be any two element subgroup of G. Then the action of G_0 is inner. Since $K_*(A) = \mathbf{Z}$ is not a 2-torsion group, lemma 4.1.5 implies that $\alpha|_{G_0}$ does not even have locally discrete K-theory.

This example also shows that the isomorphism $K_0^G(K(H)\otimes B) \simeq K_0^G(B)$ need not hold if G acts on H via a projective unitary representation — take $H = \mathbf{C}^2$ with the projective unitary representation of G defined by u and v, and take $B = \mathbf{C}$, so that $K(H)\otimes B = A$. There can be no isomorphism because the trivial action of G on $\mathbf{C}$ does not have locally discrete K-theory. (We have $K_0^G(\mathbf{C}) = R(G)$.)

It is also worth pointing out that this example shows that the Connes spectrum as a measure of freeness of an action suffers the same disadvantage: even though $\Gamma(\alpha) = \hat{G}$, we have $\Gamma(\alpha|_{G_0}) = \{e\}$, rather than $\hat{G}_0$, for G_0 as above.

The following definition corrects the difficulty shown in the previous example.

4.2.4 Definition. Let (G,A,α) be a G-algebra. We say the action α is totally K-free is $\alpha|_H$ is K-free for every closed subgroup H of G.

4.2.5 Proposition. Let (G,A,α) be a G-algebra such that α is totally K-free. Then for any closed subgroup H of G, the restricted action $\alpha|_H$ is totally K-free.

Proof. Obvious. Q.E.D.

4.2.6 Proposition. Let (G,A,α) be a G-algebra, and let I be a G-invariant ideal in A. Then α is totally K-free if and only if the actions it induces on I and A/I are totally K-free.

Proof. Obvious from proposition 4.2.2. Q.E.D.

We will in fact use both K-freeness and total K-freeness. Clearly total K-freeness satisfies more of the properties that one would expect from a concept of freeness of group actions, but, as suggested by the last paragraph in example 4.2.3, K-freeness is a more appropriate concept to use when dealing with the Connes spectrum and related notions. We therefore state the following results for both properties; the proofs are usually essentially the same for both.

4.2.7 Proposition. Let $(G,C_0(X),\alpha)$ be a commutative G-algebra. Then the following are equivalent.

(1) The action on X defined by α is free.

(2) The action α is totally K-free.

(3) The action α is K-free.

Proof. (1)$\Longrightarrow$(2). If H is any closed subgroup of G, then the action of H on X is also free. The Localization Theorem 1.2.4 now implies that $K_*^H(C_0(U))_P = 0$ for any H-invariant open subset U of X and any prime ideal $P \subset R(H)$ whose support is not the trivial subgroup. Since P contains $I(H)$ if and only if its support is the trivial subgroup (see [97], proposition 3.7 and the example after it), we have proved that α is totally K-free.

(2)$\Longrightarrow$(3). Obvious.

(3)$\Longrightarrow$(1). From proposition 4.2.2, it follows that the action of G on $C(Y)$ is K-free for any compact G-invariant subset Y of X. Theorem 1.1.1 now implies that G acts freely on all such Y. Therefore G acts freely on X. Q.E.D.

4.2.8 Proposition. Let $\alpha,\beta : G \longrightarrow Aut(A)$ be two exterior equivalent (definition 2.7.1) actions of G on a C^*-algebra A. Then α is (totally) K-free if and only if β is.

Proof. By assumption, there is a map $g \longrightarrow u_g$ from G to the unitary group of the multiplier algebra $M(A)$ of A, satisfying a cocycle condition and a continuity property, such that $\beta_g(a) = u_g \alpha_g(a) u_g^*$ for all $a \in A$ and $g \in G$. Let H be any closed subgroup of G and let I be any $\alpha|_H$-invariant ideal of A. Then I is also an ideal in $M(A)$, and is therefore $\beta|_H$-invariant. By the proof of [79], 3.12.8, there is a canonical homomorphism $\varphi: M(A) \longrightarrow M(I)$, which is the identity on I. (The only place the hypothesis about I being an essential ideal is used there is to show that φ is injective. Of course, our φ is probably not.) Then it is easy to see that $g \longrightarrow \varphi(u_g)$, as a map from H to the unitary group of $M(I)$, is an exterior equivalence between the actions of H on I defined by α and β. Therefore, by theorem 2.8.3 (5), $K_*^H(I)$ does not depend on which action is chosen. So if β is (totally) K-free, then α is also. The reverse implication follows by symmetry (or directly from the above by noticing that any $\beta|_H$-invariant ideal is also $\alpha|_H$-invariant). Q.E.D.

Our next objective is to prove that K-freeness and total K-freeness are stable. That is, if (G,A,α) and $(G, K(V)\otimes A, \beta)$ are G-algebras, where β is the diagonal action on $K(V)\otimes A$ coming from a unitary representation of G on the Hilbert space V and the action α on A, then we want to show that α is (totally) K-free if and only if β is. Proving this requires that we identify the G-invariant ideals in $K(V)\otimes A$. It is intuitively clear that they are in one-to-one correspondence with the G-invariant ideals of A, but the proof is not obvious. We will obtain this fact as a corollary of the following more general result, which follows directly from results in the literature. Before stating it, we recall that the primitive ideal space $Prim(A)$ of C^*-algebra A, endowed with the hull-kernel topology, is a topological space whose open subsets are in one-to-one order-preserving correspondence with the ideals of A. (See [79], 4.1.2 and 4.1.3. Remember that according to our convention, ideals in C^*-algebras are closed.)

4.2.9 Proposition. Let A and B be C^*-algebras, and assume that A is nuclear ([62]). Then $Prim(A\otimes B)$ is canonically homeomorphic to $Prim(A)\times Prim(B)$ with the product topology. In particular, if U is the open subset of $Prim(A)$ corresponding to the ideal I of A, and V is the open subset of $Prim(B)$ corresponding to the ideal J of B, then this homeomorphism identifies $U \times V$ with the open subset of $Prim(A \otimes B)$ corresponding to the ideal $I \otimes J$ of $A \otimes B$.

Proof. We first point out that, since A is assumed to be nuclear, the tensor product C^*-algebra $A \otimes B$ is unambiguously defined. In [106], Tomiyama proves on page 225 that if A is a C^*-algebra such that, for every ideal I of A, the algebra A/I satisfies what he calls property T, then for any C^*-algebra B there is a canonical homeomorphism

$$Prim(A \otimes B) \simeq Prim(A) \times Prim(B),$$

as claimed in the conclusion of the proposition. Property T, defined on page 220 of [106], is what is now called nuclearity (see [62]), and it is now known ([19], corollary 4) that if A is nuclear, then so is A/I for every ideal I in A. Thus, Tomiyama's conclusion holds under the hypotheses of the proposition. Q.E.D.

4.2.10 Proposition. Let (G,A,α) be a G-algebra, let V be a Hilbert space with a unitary representation of G, and let $(G, K(V) \otimes A, \beta)$ be the tensor product G-algebra with the diagonal G-action. Then α is (totally) K-free if and only if β is.

Proof. The algebra $K(V)$ is type I, and therefore nuclear by [102], theorem 3. Since $K(V)$ is simple, its primitive ideal space consists of a single point. It follows from the previous proposition that, if H is any subgroup of G, then the H-invariant ideals of $K(V) \otimes A$ are exactly those of the form $K(V) \otimes J$ for H-invariant ideals J of A. By theorem 2.8.3 (4), we have $K_*^H(K(V) \otimes J) \simeq K_*^H(J)$ as $R(H)$-modules. The proposition is now obvious. Q.E.D.

4.2.11 Remark. Example 4.2.3 clearly shows that this proposition does not hold for K-free actions if the representation is merely required to be projective unitary. We will see later that, at least under certain circumstances, it does hold for totally K-free actions and projective unitary representations. (See corollary 5.2.7 and proposition 9.1.6.) We also point out that this proposition holds for arbitrary actions of finite cyclic groups on $K(V)$, because of the following (well known) lemma.

4.2.12. Lemma. Let G be a finite cyclic group, and let V be a Hilbert space. Then every action $\alpha: G \longrightarrow Aut(K(V))$ is inner.

Proof. It is well known that any automorphism of $K(V)$ is inner. (See for example [79], 8.7.4.) So let g be a generator of G. Then there is a unitary $u \in L(V)$ such that $\alpha_g(a) = uau^*$ for all $a \in K(V)$. Let n be the order of G. Since g^n is the identity, it follows that $\alpha_g^n = id_{K(V)}$. Therefore u^n commutes with all elements of $K(V)$, and is hence a multiple of the identity, that is, $u^n = \lambda \cdot 1$, with $|\lambda| = 1$. Let ζ be any complex number such that $\zeta^n = \lambda$, and define a unitary representation v of G on V by $v_{g^k} = \zeta^{-k} u^k$ for $0 \le k < n$. Since $\zeta^{-n} u^n = 1$, we see that v actually is a unitary representation of G, and clearly $\alpha_h(a) = v_h a v_h^*$ for $h \in G$. Q.E.D.

4.2.13 Proposition. Let (G,A,α) be a separable G-algebra. Let B be a G-invariant hereditary subalgebra of A. Then if α is (totally) K-free, so is its restriction to B. If B is full, then the converse is also true.

Proof. Let I be the closed ideal in A generated by B, which is G-invariant. Then B is a full hereditary subalgebra of I. The proof of the proposition in [80] shows that, with K being the algebra of compact operators on a separable Hilbert space and ι the trivial G-action on it, one has an isomorphism $K \otimes B \simeq K \otimes I$ carrying $\iota \otimes (\alpha_{(\cdot)}|_B)$ to an action exterior equivalent to $\iota \otimes \alpha_{(\cdot)}|_I$. (The hypothesis that G be abelian is not used except in the part of that proof dealing with the spectrum. The hypothesis that the large algebra be G-simple is used only to show that the hereditary subalgebra is full.) The result then follows from the previous two propositions, since the action on I must be (totally) K-free.

(An alternate proof proceeds by observing that the assignment $J \longrightarrow \overline{IJI}$ provides a one-to-one correspondence between H-invariant ideals J of B and H-invariant ideals of I, where H is any closed subgroup of G. The inverse is given by the assignment $J \longrightarrow J \cap B$. Since an ideal J in B is a full hereditary subalgebra of $\overline{IJI}$, it follows from proposition 2.9.1 that $K_*^H(J) \simeq K_*^H(\overline{IJI})$, and the result obviously follows.)

For B full in A, we have $I = A$, and the converse is obvious from either argument.

Q.E.D.

4.2.14 Lemma. Let $(G,A_\lambda,\alpha_\lambda)_{\lambda \in D}$ be a directed system of G-algebras. Suppose each α_λ has locally discrete K-theory. Then so does the action

$$\alpha = \varinjlim \alpha_\lambda : G \longrightarrow \mathit{Aut}(\varinjlim A_\lambda).$$

Proof. Let P be a prime ideal of $R(G)$ not containing $I(G)$. Then

$$K_*^G(\varinjlim A_\lambda)_P \simeq (\varinjlim K_*^G(A_\lambda))_P \simeq \varinjlim K_*^G(A_\lambda)_P.$$

Here the first step is theorem 2.8.3 (6), and the second follows from the fact that localization commutes with direct limits. (To see this, combine exercise 2.20 and proposition 3.5 of [5].) Since each α_λ has locally discrete K-theory, it follows that $K_*^G(\varinjlim A_\lambda)_P = 0$.

Q.E.D.

Unfortunately, it does not seem to be possible to control ideals in direct limits sufficiently well to be able to prove the analog of this lemma for K-freeness or total K-freeness. However, we do have the following result.

4.2.15 Proposition. Let (G,A,α) be a G-algebra, and suppose A has a composition series $(I_\lambda)_{\lambda \le \kappa}$ for some ordinal κ, consisting of G-invariant ideals. Suppose the induced action on each subquotient $I_{\lambda+1}/I_\lambda$ is (totally) K-free. Then α is (totally) K-free.

Proof. We first prove by transfinite induction that α has locally discrete K-theory. Clearly, the action on I_0 has locally discrete K-theory, so suppose μ is some ordinal, and the actions on I_λ have locally discrete K-theory for all $\lambda<\mu$. If μ is a limit ordinal, then $I_\mu = \varinjlim_{\lambda<\mu} I_\lambda$, and its action has locally discrete K-theory by the previous proposition. If μ is not a limit ordinal, then $\mu=\lambda+1$ for some λ, and consideration of the exact sequence

$$0 \longrightarrow I_\lambda \longrightarrow I_{\lambda+1} \longrightarrow I_{\lambda+1}/I_\lambda \longrightarrow 0$$

implies that the action on $I_\mu = I_{\lambda+1}$ has locally discrete K-theory by lemma 4.1.4. Therefore α has locally discrete K-theory.

Now let H be a closed subgroup of G, and let J an H-invariant ideal. Then $(I_\lambda \cap J)_{\lambda\le\kappa}$ is an H-invariant composition series for J, and each quotient

$$(I_{\lambda+1}\cap J)/(I_\lambda\cap J) \simeq ((I_{\lambda+1}\cap J)+I_\lambda)/I_\lambda$$

is an H-invariant ideal in the corresponding quotient $I_{\lambda+1}/I_\lambda$. So the result of the first paragraph applied to this composition series (with $H=G$ for the case of K-freeness) implies that the action of H on J has locally discrete K-theory. Thus, α is (totally) K-free. Q.E.D.

4.2.16 Corollary. Let (G,A,α) be a separable G-algebra. Suppose there are G-invariant hereditary subalgebras $(B_\lambda)_{\lambda\in D}$ which generate A as an ideal in A, and such that $\alpha_{(\cdot)}|_{B_\lambda}$ is (totally) K-free for each λ. Then α is (totally) K-free. If the B_λ are ideals in A, then the assumption that A be separable is unnecessary.

Proof. We may assume that D is a set of ordinal numbers $\{\lambda : \lambda<\kappa\}$. Let J_λ be the ideal generated by B_λ in A. Then, since B_λ is a full hereditary subalgebra of J_λ, proposition 4.2.13 implies that $\alpha_{(\cdot)}|_{J_\lambda}$ is (totally) K-free. For $\lambda\le\kappa$, let $I_\lambda = \sum_{\mu<\lambda} J_\mu$. Then $(I_\lambda)_{\lambda\le\kappa}$ is a G-invariant composition series for A, and

$$I_{\lambda+1}/I_\lambda = (I_\lambda + J_\lambda)/I_\lambda \simeq J_\lambda/(I_\lambda\cap J_\lambda),$$

which is a quotient of J_λ and hence has (totally) K-free action by proposition 4.2.2 or proposition 4.2.6. So α is (totally) K-free by the previous proposition, and the first part is proved. For the second part, notice that separability was used only in the application of proposition 4.2.13, which is unnecessary if the B_λ are already ideals in A. Q.E.D.

In the preceding propositions we have proved the analogs for (total) K-freeness of some obvious facts about free action of groups on spaces. Here we regard ideals, and even hereditary subalgebras, as the C^* equivalents of open subsets, and quotient algebras as the equivalents of closed subsets. The analogs of some other easy facts, however, are either much harder to prove, or are actually false.

We first consider the following fact about free actions of compact Lie groups on topological spaces. Let G act continuously on X; then for each $x \in X$ the stabilizer G_x is closed. If it is not trivial, then, because G is compact Lie, G_x must contain an element of some prime order. So if the restriction of the action to each prime order subgroup is free, then so is the entire action. The analogous result does not hold for either K-freeness or total K-freeness, as will be shown in several examples in chapter 9. One can nevertheless say something in this direction: we will show in the next chapter that if G is a finite p-group, that is, all elements have order equal to some power, depending on the element, of the prime number p, and if all cyclic subgroups of G act K-freely, then G acts totally K-freely.

Another fact we would like to generalize concerns actions on products. If a group G acts freely on a space X and arbitrarily on a space Y, then the diagonal action on $X \times Y$ is free. The analog for total K-freeness would be that if G acts totally K-freely on a C^*-algebra A and arbitrarily on a C^*-algebra B, then the diagonal action on $A \otimes B$ is totally K-free. We can prove this if G is a finite p-group and the actions satisfy certain technical conditions; the proof occupies most of chapter 6. The result fails disastrously if "totally K-free" is replaced by "K-free."

Finally, observe that if $f : X \longrightarrow Y$ is a G-equivariant function and G acts freely on Y, then G acts freely on X. The following example shows that, without extra hypotheses, there can be no analogous result for any reasonable notion of freeness on C^*-algebras.

4.2.17 Example. Let $A = \mathbf{C} \oplus \mathbf{C}$, let $B = M_2$, and let $G = \mathbf{Z}/2\mathbf{Z}$. Let g be the nontrivial element of G, and define actions α and β of G on A and B by $\alpha_g(a,b) = (b,a)$ for $a, b \in \mathbf{C}$, and $\beta_g(a) = uau^*$ for $a \in M_2$, where $u = \begin{pmatrix} 0 & 1 \\ 1 & 0 \end{pmatrix}$. Then the homomorphism $\varphi : A \longrightarrow B$ given by $\varphi(a,b) = \begin{pmatrix} a & 0 \\ 0 & b \end{pmatrix}$ is injective, equivariant, and unital. It is thus the C^* equivalent of an equivariant surjective map, going the other way, of compact spaces. The action α is totally K-free — in fact $A = C(X)$ for a free G-space X. However, β is actually inner, which is the opposite extreme. In particular, lemma 4.1.5 implies that β doesn't even have locally discrete K-theory.

The extra hypothesis needed in order to prove such results is that the image of the map from A to B be in the center of B. We will prove several theorems related to this idea, namely theorems 4.3.8, 4.5.8, and 4.6.1. At the cost of some extra technical conditions (which may not be necessary), we can prove a result for totally K-free actions on A under the weaker hypothesis that B is generated as a C^*-algebra by the image of A and the relative commutant in B of the image of A. Note that the image of A is no longer forced to be abelian. The result in question, theorem 6.6.7, uses the results on tensor products in chapter 6.

4.3. Twisted Products and Free Action on the Primitive Ideal Space

In this section, we will relate the K-freeness of an action of a closed subgroup H of G on an algebra A to K-freeness of the twisted product action of G on $C(G\times_H A)$. Using a closely related result, we will then prove for finite groups a substantial generalization of one direction of proposition 4.2.7, and thus show the existence of a large number of totally K-free actions. We start with several preliminary results.

4.3.1 Lemma. Let H be a closed subgroup of G, and let I be the ideal in $R(H)$ generated by the image of $I(G)$ under the usual restriction map from $R(G)$ to $R(H)$. Then there is n such that $I(H)^n \subset I \subset I(H)$.

Proof. Obviously $I \subset I(H)$. By [97], proposition 3.8, the prime ideals in $R(H)$ containing I are the same as the ones containing $I(H)$. By [5], proposition 1.14, the radical of I is therefore equal to $I(H)$. Since $R(H)$ is Noetherian by [97], corollary 3.3, it follows that I contains a power of its radical by [5], proposition 7.14, and the lemma is proved. Q.E.D.

4.3.2 Proposition. Let (G,A,α) be a G-algebra, and let X be a locally compact space on which G acts trivially. Then the diagonal action β of G on $C_0(X)\otimes A$ is (totally) K-free if and only if α is (totally) K-free.

Proof. Since A is the quotient of $C_0(X)\otimes A$ by the G-invariant ideal $C_0(X-\{x\})\otimes A$ for any $x \in X$, the (total) K-freeness of β implies that of α by propositions 4.2.2 and 4.2.6.

The first step in proving the other direction is to prove the following claim: if α has locally discrete K-theory, then so does β. One way to do this is to compute the K-theory of

$$C^*(G, C_0(X)\otimes A, \beta) \simeq C_0(X)\otimes C^*(G,A,\alpha)$$

via the Künneth formula, theorem 4.1 of [95]. However, it seems easier to imitate the proof of proposition 2.11 in [95] to produce a direct proof. Accordingly, we observe that the claim is true if $X = \mathbf{R}^n$, since $K_*^G(C_0(\mathbf{R}^n)\otimes A) \simeq K_*^G(A)$ (with a degree shift if n is odd). Lemma 4.1.4 now shows that the claim holds for $X = S^n$, the sphere of dimension n. Repeated use of lemma 4.1.4 and induction on the number of cells now shows that the action on $C(X)\otimes A$ has locally discrete K-theory when X is a finite cell complex. If X is an arbitrary compact space, then we use theorem X.10.1 of [33] to write X as an inverse limit of finite complexes X_λ, whence $C(X) = \varinjlim C(X_\lambda)$. Since $C(X)\otimes A = \varinjlim C(X_\lambda)\otimes A$, we can apply lemma 4.2.14 to conclude that the claim holds for this X also. If now X is an arbitrary locally compact space, then an application of lemma 4.1.4 to the sequence

$$0 \longrightarrow C_0(X)\otimes A \longrightarrow C(X^+)\otimes A \longrightarrow A \longrightarrow 0$$

completes the proof of the claim.

Now let I be a G-invariant ideal in $C_0(X)\otimes A$, and let U be the corresponding G-invariant open subset of $Prim\,(C_0(X)\otimes A)$. Proposition 4.2.9 enables us to identify $Prim\,(C_0(X)\otimes A)$ with $X\times Prim\,(A)$, with the product topology. Therefore U is the union of rectangular open sets $V_\lambda\times W_\lambda$, where each V_λ is an open subset of X and each W_λ is an open subset of $Prim\,(A)$. If we replace W_λ with GW_λ, then we still have $V_\lambda\times W_\lambda\subset U$, since U is G-invariant. (We are using the fact that $Prim\,(A)$ is a G-space for the action defined by $gJ=\alpha_g\,[J]$ for a primitive ideal J of A.) Now let J_λ be the G-invariant ideal in A corresponding to W_λ. Then $C_0(V_\lambda)\otimes J_\lambda\subset I$, and the ideals $C_0(V_\lambda)\otimes J_\lambda$ generate I as a subalgebra.

We now assume that α is K-free. In view of the claim proved in the second paragraph, we then know that the action of G on any ideal of the form $C_0(V)\otimes J$ has locally discrete K-theory. We will further assume that the index set of the previous paragraph is $\{\lambda:\lambda<\kappa\}$ for some ordinal κ. We will now prove by induction on μ that the action of G on the ideal I_μ generated by all $C_0(V_\lambda)\otimes J_\lambda$ for $\lambda<\mu$ has locally discrete K-theory, where the V_λ are arbitrary open subsets of X and the J_λ are arbitrary G-invariant ideals in A. The case $\mu=1$ has already been proved. If μ is a limit ordinal, then $I_\mu=\varinjlim_{\lambda<\mu} I_\lambda$, and the action of G on I_λ has locally discrete K-theory by lemma 4.2.14. If μ is not a limit ordinal, then $\mu=\eta+1$ for some ordinal η, and we set

$$L=\overline{\sum_{\lambda<\eta} C_0(V_\lambda\cap V_\eta)\otimes(J_\lambda\cap J_\eta)}\,.$$

Note that the action of G on L has locally discrete K-theory by the induction hypothesis. We furthermore have

$$L=I_\eta\cap(C_0(V_\eta)\otimes J_\eta)\quad\text{and}\quad I_\mu=I_\eta+(C_0(V_\eta)\otimes J_\eta),$$

whence

$$I_\eta/L\simeq I_\mu/(C_0(V_\eta)\otimes J_\eta)\,.$$

We apply lemma 4.1.4 to the exact sequence

$$0\longrightarrow L\longrightarrow I_\eta\longrightarrow I_\eta/L\longrightarrow 0$$

to conclude that the action of G on I_η/L has locally discrete K-theory. Furthermore, the action of G on $C_0(V_\eta)\otimes J_\eta$ has locally discrete K-theory, so we can apply lemma 4.1.4 to the exact sequence

$$0\longrightarrow C_0(V_\eta)\otimes J_\eta\longrightarrow I_\mu\longrightarrow I_\eta/L\longrightarrow 0$$

to conclude that the action of G on I_μ has locally discrete K-theory. This completes the induction, and proves that the action of G on I has locally discrete K-theory. Since I was an arbitrary G-invariant ideal in $C_0(X)\otimes A$, we have proved that β is K-free.

Now assume that α is totally K-free, and let H be a closed subgroup of G. Then $\alpha|_H$ is K-free, and the part of the proposition we have already proved implies that $\beta|_H$ is K-free. So β is totally K-free. Q.E.D.

In chapter 6, where K-freeness of actions on tensor products is considered, we will improve the "if" part of this proposition, for a restricted class of groups, and for totally K-free actions, by allowing in place of $C_0(X)$ many noncommutative C^*-algebras with nontrivial actions of G. The question of what happen when G acts freely on X is taken up in sections 4.5 and 4.6. Another possible generalization is the following conjecture.

4.3.3 Conjecture. Let X be a locally compact space, and let A be a bundle of C^*-algebras over X. (See [49] for the definition.) Let G act continuously on A via bundle automorphisms over the trivial action on the base, that is, each $g \in G$ maps the fiber A_x over $x \in X$ to itself via an automorphism α_x depending continuously on x in the appropriate sense. Let α be the induced action of G on the C^*-algebra $\Gamma_0(A)$ of continuous sections of A which vanish at infinity on X. We conjecture that if each α_x is (totally) K-free, then α is (totally) K-free.

The converse of this conjecture, that if α is (totally) K-free then each α_x is also, is obvious from the relation

$$A_x \simeq \Gamma_0(A)/\{f \in \Gamma_0(A) : f(x) = 0\}.$$

In the proof of the next theorem we will use an argument which, in effect, shows that the conjecture is true if there is an increasing sequence U_n of open subsets of X such that for each n, $A|_{U_{n+1} - U_n}$ is locally trivial in the appropriate equivariant sense. In general, however, we do not know how to approach this conjecture.

We are now ready to consider K-freeness and total K-freeness of twisted products.

4.3.4 Theorem. Let H be a closed subgroup of G, let (H,A,α) be an H-algebra, and let $(G, C(G\times_H A), \beta)$ be the corresponding twisted product G-algebra, as in definition 2.9.2. Then β has locally discrete K-theory if and only if α does, and β is (totally) K-free if and only if α is.

Proof. Let $B = C(G\times_H A)$. By proposition 2.9.4, we have $K_*^H(A) \simeq K_*^G(B)$ as $R(G)$-modules, where $R(G)$ acts on $K_*^H(A)$ via the restriction map from $R(G)$ to $R(H)$. Let I be the ideal in $R(H)$ generated by the image of $I(G)$ under this restriction, and use lemma 4.3.1 to choose n such that $I(H)^n \subset I \subset I(H)$. If α has locally discrete K-theory, then every element of $K_*^H(A)$ is annihilated by some power of $I(H)$, and the same power of $I(G)$ annihilates the corresponding element in $K_*^G(B)$. Thus, β has locally discrete K-theory. For the opposite direction, let β have locally discrete K-theory, let $\eta \in K_*^G(B)$, and let $I(G)^m \cdot \eta = 0$. If λ is the corresponding element of $K_*^H(A)$, then $I^m \cdot \lambda = 0$, whence $I(H)^{mn} \cdot \lambda = 0$. Thus, α has locally discrete K-theory.

Next, we prove that α is K-free if and only if β is K-free. If I is an H-invariant ideal of A, then $C(G\times_H I)$ is a G-invariant ideal of B. Conversely, given a G-invariant ideal J of B, we find that $J = C(G\times_H I)$, where, with e being the identity of G, we define $I = \{f(e) : f \in J\}$. (We omit the easy calculation needed here.) By the first part of the lemma, $\beta_{(\cdot)}|_{C(G\times_H I)}$ has locally discrete K-theory if and only if $\alpha_{(\cdot)}|_I$ has locally discrete K-theory. Therefore β is K-free if and only if α is.

Finally, we consider total K-freeness. If β is totally K-free, then so is $\beta|_H$. Since there is an H-equivariant isomorphism

$$A \simeq C(G\times_H A)/ \{f \in C(G\times_H A) : f|_H = 0\},$$

we conclude that α is totally K-free.

The other direction is much harder. Assume that α is totally K-free, and let G_0 be a closed subgroup of G. Let $G_0\times H$ act on G via the formula $(g_0, h)g = g_0 g h^{-1}$ for $g \in G$, $g_0 \in G_0$, and $h \in H$. Let X be the orbit space $G/(G_0\times H)$, which is the space of $G_0 - H$ double cosets in G, and let $\pi : G \longrightarrow X$ be the quotient map. To each open subset U of X, there corresponds a G_0-invariant ideal $I(U)$ of $C(G\times_H A)$ consisting of all functions in $C(G\times_H A)$ which vanish outside the open subset $\pi^{-1}[U]$ of G. We are going to construct a transfinite increasing sequence $\{U_\lambda : \lambda \le \kappa\}$ for some ordinal κ such that the ideals $\{I(U_\lambda) : \lambda \le \kappa\}$ form a composition series for which the action of G_0 on the subquotients $I(U_{\lambda+1})/ I(U_\lambda)$ is K-free. It will then follow from proposition 4.2.15 that the action of G_0 on $C(G\times_H A)$ is K-free, and, since G_0 is an arbitrary closed subgroup of G, we conclude that β is totally K-free.

We start by choosing $U_0 = \emptyset$. If λ is a limit ordinal, and U_μ has been defined for $\mu < \lambda$, we simply choose $U_\lambda = \bigcup_{\mu<\lambda} U_\mu$. It follows that $I(U_\lambda) = \overline{\bigcup_{\mu<\lambda} I(U_\mu)}$. It remains to show how to find $U_{\lambda+1}$ given U_λ. That is, given a closed subset $Y = X - U_\lambda$, we want to find an open subset U of Y such that the action of G_0 on $D = I((X-Y)\cup U)/ I(X-Y)$ is K-free. Observe that D is isomorphic to the set of continuous functions $f : \pi^{-1}[U] \longrightarrow A$ which vanish at infinity on $\pi^{-1}[U]$ and satisfy $\alpha_h(f(gh)) = f(g)$ for all $g \in \pi^{-1}[U]$ and $h \in H$. The G_0-action is given by $\delta_{g_0}(f)(g) = f(g_0^{-1}g)$ for $g_0 \in G_0$ and $g \in \pi^{-1}[U]$. We will now proceed to construct such a set U.

By combining proposition IV.1.2 and corollary VI.2.4 of [14], we see that the action of $G_0\times H$ on $\pi^{-1}[Y]$ has only finitely many orbit types, that is, the stabilizer subgroups $(G_0\times H)_x$ for $x \in \pi^{-1}[Y]$ fall into only finitely many conjugacy classes. In particular, there is $x \in \pi^{-1}[Y]$ such that the stabilizer $K = (G_0\times H)_x$ does not contain any other stabilizer subgroups for the action of $G_0\times H$ on $\pi^{-1}[Y]$. (The discussion following corollary I.4.4 of [14] implies that the conjugacy classes of a compact group are partially ordered by the relation that an element of one class be contained in some element of the other. We are then merely choosing x such that the conjugacy class of $(G_0\times H)_x$ is minimal.) By theorem II.5.4 of [14], there is a subset S of $\pi^{-1}[Y]$ containing x and such that the map $(m, y) \longrightarrow my$ from $G_0\times H\times S$ to the open subset $V = (G_0\times H)S$ of $\pi^{-1}[Y]$ defines an equivariant homeomorphism from

$(G_0 \times H) \times_K S$ to V. Now the stabilizers of the points of S are all contained in K, and by the choice of x they must equal K. The set S is therefore K-invariant, and V is equivariantly homeomorphic to $((G_0 \times H)/K) \times S$ via $(mK, y) \longrightarrow my$. Set $U = \pi[V]$, which is of course also equal to $\pi[S]$.

Let $H_0 = x^{-1}Hx \cap G_0$, which is a subgroup of G_0. It is easily seen that $K = \{(h_0, xh_0x^{-1}) : h_0 \in H_0\}$. We are going to show that $D \simeq C(G_0 \times_{H_0} (C_0(S) \otimes A))$, where H_0 acts on $C_0(S) \otimes A$ via the action $\gamma_{h_0}(f \otimes a) = f \otimes \alpha_{xh_0x^{-1}}(a)$. Set $C = C_0(S) \otimes A$, and identify $C(G_0 \times_{H_0} C)$ with the set of continuous functions $f : G_0 \times S \longrightarrow A$ which vanish at infinity and satisfy $\alpha_{xhx^{-1}}(f(gh, s)) = f(g, s)$ for $g \in G_0$, $h \in H_0$, and $s \in S$. Define $\varphi : D \longrightarrow C(G_0 \times_{H_0} C)$ by $\varphi(f)(g_0, s) = f(g_0 s)$ for $g_0 \in G_0$ and $s \in S$. To see that $\varphi(f)$ really is in $C(G_0 \times_{H_0} C)$, let $h \in H_0$ and compute:

$$\alpha_{xhx^{-1}}(\varphi(f)(gh, s)) = \alpha_{xhx^{-1}}(f(ghs)) = \alpha_{xhx^{-1}}(f(gsxhx^{-1})) = \varphi(f)(g, s)$$

for $g \in G_0$ and $s \in S$. The middle step follows from the fact that $(h, xhx^{-1}) \in K$, whence $hs(xhx^{-1})^{-1} = s$. Also the compactness of G_0 insures that $\varphi(f)$ does in fact vanish at infinity on $G_0 \times S$. It is clear that φ is equivariant.

The inverse of φ is the map ψ given by $\psi(f)(g_0 sh) = \alpha_h^{-1}(f(g_0, s))$ for $g_0 \in G_0$, $s \in S$, and $h \in H$. We first prove that ψ is well defined. If $g_1, g_2 \in G_0$, $s_1, s_2 \in S$, and $h_1, h_2 \in H$ satisfy $g_1 s_1 h_1 = g_2 s_2 h_2$, then we have $(g_2^{-1} g_1, h_2 h_1^{-1}) s_1 = s_2$. Since S intersects each orbit of $G_0 \times H$ exactly once, we obtain $s_1 = s_2$, whence $(g_2^{-1} g_1, h_2 h_1^{-1}) \in K$. Therefore $g_2^{-1} g_1 = x^{-1} h_2 h_1^{-1} x$, and it is now easily seen that

$$\alpha_{h_1}^{-1}(f(g_1, s_1)) = \alpha_{h_2}^{-1}(f(g_2, s_2)).$$

Another short calculation now shows that $\psi(f)$ is in fact in D for $f \in C(G_0 \times_{H_0} C)$. That ψ is the inverse of φ is obvious.

We have thus shown that $D \simeq C(G_0 \times_{H_0} (C_0(S) \otimes A))$ as a G_0-algebra. The action of H_0 on A is K-free because the map $h_0 \longrightarrow xh_0x^{-1}$ identifies this action with the action of the subgroup $H \cap xG_0x^{-1}$ of H. Proposition 4.3.2 now shows that the action of H_0 on $C_0(S) \otimes A$ is K-free, and the part of this theorem concerning K-freeness, which we have already proved, now implies that the action of G_0 on D is K-free. This completes the proof. Q.E.D.

4.3.5 Corollary. Let A be a C^*-algebra, and let G act on $C(G) \otimes A$ via the diagonal action $\alpha = \tau \otimes \iota$, where τ is the action on $C(G)$ given by translation on G and ι is the trivial action. Then α is totally K-free.

Proof. We have $C(G) \otimes A \simeq C(G \times_H A)$ where H is the one element subgroup. The action of H on A is trivially totally K-free, so α is totally K-free by the theorem. Q.E.D.

We now turn to the second object of this section. We require the following lemma, which is closely related to theorem 4.3.4.

4.3.6 Lemma. Let H be a closed subgroup of G, and let (H,A,α) be an H-algebra. Then $K_*^G(C(G\times_H A))_P = 0$ for any prime ideal P in $R(G)$ whose support is not contained in a conjugate of H.

Proof. By proposition 2.9.4, we have $K_*^G(C(G\times_H A)) \simeq K_*^H(A)$ as $R(G)$-modules, the $R(G)$-module structure on $K_*^H(A)$ being obtained via the restriction map from $R(G)$ to $R(H)$. By [97], proposition 3.7, we have $R(H)_P = 0$. Therefore, $K_*^H(A)_P = 0$, as desired.

Q.E.D.

If we knew that $R(G) \longrightarrow R(H)$ was surjective, which is true if G is abelian but fails in general, then we would be able to use this lemma in place of lemma 4.3.1 at the beginning of the proof of theorem 4.3.4. (The difficulty occurs when trying to deal with the prime ideals whose supports are contained in a conjugate of H.) We also point out that K-freeness in corollary 4.3.5 follows from this lemma by taking $H = \{e\}$.

In preparation for the next theorem, we recall some elementary properties of the topology on the primitive ideal space $Prim(A)$ of a C^*-algebra A. It is locally compact but not necessarily Hausdorff ([29], 3.3.8). Also, it is a T_0 space ([79], 4.1.4), that is, if x and y are distinct points of $Prim(A)$, then there is an open subset containing x but not y or there is an open subset containing y but not x.

We can now prove the following analog of part of the Localization Theorem 1.2.4.

4.3.7 Proposition. Let G be a finite group, let (G,A,α) be a separable G-algebra, and let P be a prime ideal in $R(G)$ whose support acts on $Prim(A)$ with no fixed points. Then, for any nuclear G-algebra (G,B,β), we have $K_*^G(A\otimes B)_P = 0$.

Proof. We begin by showing that the hypotheses imply the existence of a nonzero G-invariant ideal I in A such that $K_*^G(I\otimes B)_P = 0$. By [28], proposition 1, there is a separated point $x \in Prim(A)$, that is, a point x such that if $y \in Prim(A)$ and $y \notin \overline{\{x\}}$, then x and y have disjoint neighborhoods. (In [28], Dixmier actually refers to $\widehat{A}$, the space of equivalence classes of irreducible representations of A, but the map assigning to a representation its kernel defines a one-to-one correspondence between the open subsets of $\widehat{A}$ and those of $Prim(A)$, so that the result carries over to $Prim(A)$.)

Let x be a separated point, and let $S = \{g \in G : gx \in \overline{\{x\}}\}$. We claim S is a subsemigroup of G. To see this, let $g, h \in S$. Then $hx \in \overline{\{x\}}$, whence $ghx \in \overline{\{gx\}}$ (since G acts by homeomorphisms). Since $gx \in \overline{\{x\}}$ it follows $\overline{\{gx\}} \subset \overline{\{x\}}$, so that $ghx \in \overline{\{x\}}$. Thus, $gh \in S$ as desired, and S is a semigroup. Since G is a finite group, S is actually a subgroup. If $g \in S$, then $g^{-1} \in S$, and $x = g^{-1}(gx) \in \overline{\{gx\}}$. Since $Prim(A)$ is a T_0 space and $gx \in \overline{\{x\}}$, it follows that $gx = x$. Thus, we have shown that $S \subset G_x$. Since the opposite inclusion is trivial, we have in fact $S = G_x$. Because x is separated, we conclude that for every $g \notin G_x$ there are disjoint open sets U_g containing x and V_g containing gx.

Set

$$U = \bigcap_{h \in G_x} h \cdot \left[\bigcap_{g \notin G_x} (U_g \cap g^{-1} V_g) \right].$$

Since G is finite, U is an open neighborhood of x, and furthermore GU is the disjoint union of the translates of U by the elements of a system R of left coset representatives of G_x in G. (The translates of $\bigcap_{g \notin G_x} (U_g \cap g^{-1} V_g)$ by elements of R are already pairwise disjoint, and U is a G_x-invariant subset of this intersection.) Let J be the ideal in A corresponding to U as an open subset of $Prim(A)$, and let I be the ideal corresponding to GU. Then I is a nonzero G-invariant ideal such that $I = \bigoplus_{g \in R} \alpha_g [J]$. Therefore

$$I \otimes B = \bigoplus_{g \in R} (\alpha_g \otimes \beta_g) [J \otimes B].$$

By proposition 2.9.6, we therefore have $I \otimes B \simeq C(G \times_{G_x} (J \otimes B))$. Since the support of P acts with no fixed points on $Prim(A)$, while every conjugate $hG_x h^{-1}$ of G_x does have a fixed point in $Prim(A)$, namely hx, we see that the support of P cannot be contained in any conjugate of G_x. Therefore $K_*^G(I \otimes B)_P = 0$ by the previous lemma.

We now construct by transfinite induction a G-invariant composition series $(I_\lambda)_{\lambda \le \kappa}$ for A such that $K_*^G(I_\lambda \otimes B)_P = 0$ for all λ. Let I_1 be the ideal constructed above. If I_λ has been constructed for all $\lambda < \mu$ for some limit ordinal μ, then we set $I_\mu = \overline{\bigcup_{\lambda < \mu} I_\lambda}$. Since B is nuclear, we then have $I_\mu \otimes B = \overline{\bigcup_{\lambda < \mu} I_\lambda \otimes B}$, whence

$$K_*^G(I_\mu \otimes B)_P = \varinjlim_{\lambda < \mu} K_*^G(I_\lambda \otimes B)_P = 0.$$

Also, if we are given I_λ, then the G-algebra A / I_λ satisfies the hypotheses of the theorem, so that there is a nonzero G-invariant ideal J of A / I_λ such that $K_*^G(J \otimes B)_P = 0$. Let π be the quotient map from A to A / I_λ and set $I_{\lambda+1} = \pi^{-1}[J]$. Then we have an exact sequence

$$0 \longrightarrow I_\lambda \otimes B \longrightarrow I_{\lambda+1} \otimes B \longrightarrow J \otimes B \longrightarrow 0.$$

Consequently there is an exact sequence

$$\begin{array}{ccccc} K_0^G(I_\lambda \otimes B)_P & \longrightarrow & K_0^G(I_{\lambda+1} \otimes B)_P & \longrightarrow & K_0^G(J \otimes B)_P \\ \uparrow & & & & \downarrow \\ K_1^G(J \otimes B)_P & \longleftarrow & K_1^G(I_{\lambda+1} \otimes B)_P & \longleftarrow & K_1^G(I_\lambda \otimes B)_P \end{array}.$$

The four terms in the corners are all zero, so that the other two are as well. Thus $K_*^G(I_{\lambda+1} \otimes B)_P = 0$, as required.

Since $I_{\lambda+1}$ is strictly larger than I_λ for every λ, there must be some ordinal κ such that $I_\kappa = A$. We then have $K_*^G(A \otimes B)_P = 0$. Q.E.D.

As a corollary, we obtain the following theorem.

4.3.8 Theorem. Let G be a finite group, and let (G,A,α) be a separable G-algebra such that the action of G on $Prim(A)$ is free. Then α is totally K-free.

Proof. Let H be any subgroup of G, and let I be any H-invariant ideal in A. Then $Prim(I)$ is a subset of $Prim(A)$, and so H acts freely on $Prim(I)$. Let $P \subset R(H)$ be a prime ideal which does not contain the augmentation ideal. Then the support of P is not the one element subgroup, and therefore acts on $Prim(I)$ with no fixed points. By the proposition, we therefore have $K_*^H(I)_P = 0$. Thus α is totally K-free. Q.E.D.

The Localization Theorem for spaces asserts that if G acts on X, if $P \subset R(G)$ is a prime ideal, and if Y is the subspace of X consisting of those points in X whose stabilizers contain some conjugate of the support of P, then the map $K_G^*(X)_P \longrightarrow K_G^*(Y)_P$ is an isomorphism. This make sense because Y is a closed subset of X. Unfortunately, the corresponding statement for C^*-algebras does not make sense, because if X is not Hausdorff then Y need not be closed in X.

It would also be nice to remove the hypotheses that G be finite and that A be separable. The separability assumption was used to produce the separated point in A, without which our proof breaks down. Finiteness of G was also used in the construction of an open set of the form $G \times_H U$; the general case would require an analog of the results on slices in section II.5 of [14] for actions of compact groups on T_0 spaces, but nothing of the sort seems to be known. It is not even apparent what to do in the case of a free action of an infinite compact Lie group on $Prim(A)$, so we ask the following question.

4.3.9 Question. Does theorem 4.3.8 hold for nonseparable G-algebras? Does it hold for actions of general compact Lie groups?

If the primitive ideal space is assumed to be Hausdorff, then all three of the difficulties discussed above can be avoided. We will, however, adopt a different approach to the problem, which has the (not very serious) disadvantage of applying only to separable algebras, but which has the advantage of giving information about KK-theory as well. The next section therefore develops the basic properties of KK-freeness.

4.4. KK-Free Actions

4.4.1 Definition. Let (G,A,α) be a separable G-algebra. We say that the action α has discrete KK-theory if $KK_G^0(A,A)_P = 0$ for all prime ideals $P \subset R(G)$ such that $I(G) \not\subset P$. We say that α is KK-free if $\alpha_{(\cdot)}|_I$ has discrete KK-theory for all G-invariant ideals I of A.

The justification for saying "discrete KK-theory" rather than "locally discrete KK-theory" will appear in proposition 4.4.4. We do not introduce the concept of total KK-freeness because, as will be seen in the next chapter, KK-freeness implies total KK-freeness, at least for actions of finite groups on separable nuclear C^*-algebras.

The following algebraic lemma will be needed to exploit the existence of the product in KK-theory.

4.4.2 Lemma. Let R be a commutative ring with unit, and let $S \subset R$ be a multiplicative system. If M is a (not necessarily commutative) unital algebra over R, then $S^{-1}M$ has a natural $S^{-1}R$-algebra structure such that the map $M \longrightarrow S^{-1}M$ is an R-algebra homomorphism. If furthermore N is a module over M, then $S^{-1}N$ is a module over $S^{-1}M$.

Proof. The only part of the first assertion which is not standard (see for example [5], chapter 3) concerns the properties of multiplication in $S^{-1}M$. We will prove that the obvious multiplication in $S^{-1}M$, namely

$$(a_1/s_1)\cdot(a_2/s_2) = (a_1a_2)/(s_1s_2)$$

for $a_1, a_2 \in M$ and $s_1, s_2 \in S$, is well defined. The remaining statements, except for the one about the module N, then become obvious.

Recall that $a/s = b/t$ in $S^{-1}M$ if and only if there is $r \in S$ and that $r(ta - sb) = 0$. Accordingly, let $a_i/s_i = b_i/t_i$ for $i = 1, 2$. Then there is $r_i \in S$ such that $r_i(t_i a_i - s_i b_i) = 0$, and we have

$$r_1r_2(t_1t_2a_1a_2 - s_1s_2b_1b_2) = r_2t_2[r_1(t_1a_1 - s_1b_1)]a_2 + r_1s_1b_1[r_2(t_2a_2 - s_2b_2)],$$

where we have used the fact that elements of R commute with elements of M. Each expression in square brackets in zero, and $r_1r_2 \in S$, so that

$$(a_1a_2)/(s_1s_2) = (b_1b_2)/(t_1t_2),$$

as was to be shown.

The last statement, about modules over M, follows by a similar calculation.

Q.E.D.

4.4.3 Corollary. Let R, S, M, and N be as in the lemma. If $S^{-1}M = 0$, then $S^{-1}N = 0$.

Proof. By assumption $1 = 0$ in $S^{-1}M$. Therefore, for all $x \in S^{-1}N$, we have $x = 1 \cdot x = 0 \cdot x = 0$.

Q.E.D.

The next proposition is the analog of proposition 4.1.3.

4.4.4 Proposition. Let (G, A, α) be a separable G-algebra. Then the following conditions are equivalent.

(1) The action α has discrete KK-theory.

(2) For all $\eta \in KK_G^0(A,A)$ there is n such that $I(G)^n \cdot \eta = 0$.

(3) There is n such that $I(G)^n \cdot KK_G^0(A,A) = 0$.

(4) There is n such that $I(G)^n \cdot 1_A = 0$ in $KK_G^0(A,A)$.

If G is abelian, conditions (1) through (4) are equivalent to:

(5) For all $\tau \in \hat{G}$, there is n such that $(1-\tau)^n \cdot 1_A = 0$ in $KK_G^0(A,A)$.

Proof. The equivalence of (1), (2), and (3) follows from lemma 1.2.2, since $KK_G^0(A,A)$ is a unital algebra over $R(G)$. The implication (2)$\Longrightarrow$(4) is trivial, and (4)$\Longrightarrow$(3) because if $I(G)^n$ annihilates the identity of $KK_G^0(A,A)$, then $I(G)^n$ annihilates all of $KK_G^0(A,A)$.

Now let G be abelian. Since the elements $1-\tau$ are all in $I(G)$, the implication (4)$\Longrightarrow$(5) is obvious. The proof of the converse is exactly the same as the proof of (3)$\Longrightarrow$(2) in proposition 4.1.3, except that we are unable to identify multiplication by τ with $(\hat{\alpha}_\tau)_*$. Q.E.D.

4.4.5 Proposition. Let (G,A,α) be a separable G-algebra. If α has discrete KK-theory, then for any other separable G-algebra (G,B,β), we have

$$KK_G^i(A,B)_P = 0 \quad \text{and} \quad KK_G^i(B,A)_P = 0,$$

for $i = 0,1$ and for any prime ideal $P \subset R(G)$ not containing $I(G)$.

Proof. Since $KK_G^i(A,B)$ and $KK_G^i(B,A)$ are modules over $KK_G^0(A,A)$ by corollary 3.6.4, we need only apply corollary 4.4.3 with $S = R - P$. Q.E.D.

Of course, we actually get that $I(G)^n$ annihilates these groups for some n. For general KK-groups, that is of little interest, but we state the result for $B = \mathbf{C}$:

4.4.6 Proposition. Let (G,A,α) be a separable G-algebra. If α has discrete KK-theory, then α has discrete K-theory. If α is KK-free, then α is K-free.

Proof. If α has discrete KK-theory then for some n, the ideal $I(G)^n$ annihilates $KK_G^0(A,A)$. Now $K_*^G(A) \simeq KK_G^*(\mathbf{C},A)$ by theorem 3.6.12, and $KK_G^*(\mathbf{C},A)$ is a module over $KK_G^0(A,A)$. It follows that $I(G)^n$ annihilates $K_G^*(A)$, so that α has discrete K-theory.

The second statement follows by applying the first to G-invariant ideals. Q.E.D.

We can now prove the analogs of some of the properties of K-freeness proved earlier. However, we must now assume that everything is not only separable but also nuclear.

4.4.7 Lemma. (Compare 4.1.4.) Let

$$0 \longrightarrow I \longrightarrow A \longrightarrow B \longrightarrow 0$$

be an equivariant exact sequence of separable nuclear G-algebras. If the actions on two of the three terms have discrete KK-theory, then so does the action on the third.

Proof. Let Q be the algebra whose action is not known to have discrete KK-theory, so Q is one of I, A, or B. Then there is an exact sequence

$$\begin{array}{ccccc} KK_G^0(Q,I) & \longrightarrow & KK_G^0(Q,A) & \longrightarrow & KK_G^0(Q,B) \\ \uparrow & & & & \downarrow \\ KK_G^1(Q,B) & \longleftarrow & KK_G^1(Q,A) & \longleftarrow & KK_G^1(Q,I) \end{array}$$

We localize it at a prime $P \subset R(G)$ which does not contain $I(G)$; it remains exact. Furthermore, the four terms in which the second argument is not Q become 0 when localized. Therefore so do the terms in which the second argument is Q. In particular, $KK_G^0(Q,Q)_P = 0$, as was to be shown. Q.E.D.

4.4.8 Proposition. (Compare 4.2.2.) Let (G,A,α) be a separable nuclear G-algebra, and let I be a G-invariant ideal in A. Then α is KK-free if and only if the actions it induces on I and A/I are KK-free.

Proof. Exactly the same as the proof of proposition 4.2.2, except that we use lemma 4.4.7 instead of lemma 4.1.4. (Note that ideals and quotients of a nuclear C^*-algebra are nuclear.) Q.E.D.

Let (G,A,α) be a separable G-algebra, let V be a separable Hilbert space with a representation of G, and let $(G, K(V)\otimes A, \beta)$ be the tensor product G-algebra with the diagonal G-action. Then the method of proposition 4.2.10 applies to show that β is KK-free if and only if α is. However, we can do better: the result is still true even if the action of G on $K(V)$ only comes from a projective unitary representation of G. We use the following lemma, whose proof was suggested to us by Jonathan Rosenberg.

4.4.9 Lemma. Let V be a separable Hilbert space, and let β be any continuous action of G on $K(V)$. Then for any separable G-algebra (G,A,α), the map

$$\tau_{K(V)} : KK_G^0(A,A) \longrightarrow KK_G^0(K(V)\otimes A, K(V)\otimes A)$$

is an isomorphism of $R(G)$-algebras.

Proof. If the action β is inner, that is, if β comes from a unitary representation of G on V, then theorem 3.5.10 implies that there are $\lambda \in KK_G^0(\mathbf{C}, K(V))$ and $\mu \in KK_G^0(K(V), \mathbf{C})$ such that $\lambda\mu = 1_{\mathbf{C}}$ and $\mu \otimes \lambda = 1_{K(V)}$. The map $\tau_{K(V)}$ then satisfies $\tau_{K(V)}(\eta) = \mu \otimes \lambda \otimes \eta$, and is therefore bijective because the multiplications by both μ and λ are isomorphisms. Also $\tau_{K(V)}$ is an $R(G)$-algebra homomorphism by proposition 3.5.2.

Now let β be an arbitrary action of G on $K(V)$. Since all automorphisms of $K(V)$ are inner, there is a function $u : G \longrightarrow L(V)$ such that $\beta_g(x) = u_g x u_g^*$ for $g \in G$. Furthermore, the unitaries u_g are unique up to multiplication by scalars of absolute value 1, so that there is a function (actually a cocycle) $\lambda : G \times G \longrightarrow \mathbf{C}$ such that $|\lambda(g,h)| = 1$ and $u_g u_h = \lambda(g,h) u_{gh}$ for all $g, h \in G$. Let $\widetilde{V}$ denote the Hilbert space V with the complex conjugate operations, and let $\widetilde{u}$ be the same function as u, but regarded as a map from G to $L(\widetilde{V})$. (Note that $L(\widetilde{V}) = L(V)$, but again with the complex conjugate operations.) Then it is easily seen that $\widetilde{u}_g \widetilde{u}_h = \overline{\lambda(g,h)} \widetilde{u}_{gh}$ in $L(\widetilde{V})$. Let $v_g = \widetilde{u}_g \otimes u_g \in L(\widetilde{V} \otimes V)$. Then $v_g v_h = v_{gh}$ for $g, h \in G$, since $|\lambda(g,h)| = 1$. Thus, v is a unitary representation of G on $\widetilde{V} \otimes V$. Writing $K = K(V)$ and $\widetilde{K} = K(\widetilde{V})$, we obtain an isomorphism

$$\tau_{\widetilde{K}\otimes K} : KK_G^0(A,A) \longrightarrow KK_G^0(\widetilde{K} \otimes K \otimes A, \widetilde{K} \otimes K \otimes A).$$

by the first paragraph of the proof. Since $\tau_{\widetilde{K}\otimes K} = \tau_{\widetilde{K}} \cdot \tau_K$, we have shown that $\tau_{\widetilde{K}}$ is surjective. A similar argument shows that the map

$$\tau_K \cdot \tau_{\widetilde{K}} : KK_G^0(K \otimes A, K \otimes A) \longrightarrow KK_G^0(K \otimes \widetilde{K} \otimes K \otimes A, K \otimes \widetilde{K} \otimes K \otimes A)$$

is an isomorphism, whence $\tau_{\widetilde{K}}$ is injective and therefore an isomorphism. Since $\tau_{\widetilde{K}} \cdot \tau_K$ is also an isomorphism, so is τ_K. Q.E.D.

4.4.10 Proposition. Let (G,A,α) be a separable G-algebra, let V be a separable Hilbert space, and let β be any continuous action of G on $K(V)$. Then the diagonal action $\beta \otimes \alpha$ on $K(V) \otimes A$ is KK-free if and only if α is.

Proof. As in the proof of proposition 4.2.10, the G-invariant ideals in $K(V) \otimes A$ are exactly the ideals $K(V) \otimes J$ for G-invariant ideals of J of A. By the previous lemma, we have

$$KK_G^0(K(V) \otimes J, K(V) \otimes J) \simeq KK_G^0(J,J)$$

for all such J. The result is now obvious. Q.E.D.

The analogs of propositions 4.2.8 and 4.2.13 are presumably also true, although the methods for the first of these would be different. Unfortunately, the analogs of several of our other properties of K-freeness are false, as is the converse of proposition 4.4.6.

4.4.11 Example. Let $G = \mathbf{Z}/2\mathbf{Z} \times \mathbf{Z}/2\mathbf{Z}$, let $A = M_2$, and let α be the action in example 4.2.3. It was shown there that α is K-free. But α is not KK-free, since $KK_G^0(A,A) \simeq KK_G^0(\mathbf{C},\mathbf{C}) \simeq R(G)$ by lemma 4.4.9. Thus, K-freeness does not imply KK-freeness.

The next example shows that the analogs of lemma 4.2.14 and proposition 4.2.15 are false. The proof used for lemma 4.2.14 does not work here, since KK_G^* is not continuous in its second variable, that is, one does not in general have an isomorphism

$$KK_{\mathcal{G}}^{*}(B,\varinjlim A_{\lambda}) \simeq \varinjlim KK^{*}(B,A_{\lambda}).$$

It is continuous in its first variable, in the sense that $KK_{\mathcal{G}}^{*}(\varinjlim A_{\lambda}, B)$ is given by a Milnor $\varprojlim^{1}$-sequence. (See section 7 of [96], where this possibility is considered for general cohomology theories, and see corollary 9.7.6.) However, this doesn't help since localization doesn't commute with inverse limits. As we now show, these difficulties cannot be circumvented.

4.4.12 Example. Let $G = S^1$ be the circle group in $\mathbf{C}$. Let X be the disjoint union $\coprod_{n=0}^{\infty} S^{2n+1}$ of the unit spheres $S^{2n+1} = \{z \in \mathbf{C}^{n+1} : \| z \|_2 = 1\}$. Let G act on S^{2n+1} by letting gz be the ordinary scalar product of $g \in \mathbf{C}$ and $z \in \mathbf{C}^{n+1}$. Then X, with this action on each of its components, is a locally compact G-space. Since the action is free, the induced action α on $C_0(X)$ is totally K-free by proposition 4.2.7.

In order to do more, we are actually going to need to compute $K_{\mathcal{G}}^{*}(S^{2n+1})$. Since the action is free, we have

$$K_{\mathcal{G}}^{*}(S^{2n+1}) \simeq K^{*}(S^{2n+1}/G) = K^{*}(\mathbf{C}P^{n}),$$

where $\mathbf{C}P^n$ is the complex n-dimensional projective space. We deal with K^1 first. The space $\mathbf{C}P^n$ is a cell complex with no odd dimensional cells (see the example on p. 47 of [107] and the exercise after it), and therefore by [2], proposition 2.5.2, we have $K^1(\mathbf{C}P^n) = 0$.

To compute $K^0(\mathbf{C}P^n)$, we use [2], corollary 2.5.4, which asserts that as a ring it is isomorphic to $\mathbf{Z}[t]/<(t-1)^{n+1}>$, where $< >$ means "ideal generated by", and where the isomorphism is defined by $t \longrightarrow [H]$ for a certain line bundle H on $\mathbf{C}P^n$. (Note that $P(\mathbf{C}^n)$ in [2] is our $\mathbf{C}P^{n-1}$). In order to identify the $R(G)$-module structure, we will determine what $R(G)$ does to $[H]$, for which we need to analyze H. It is defined in [2] as the bundle $Hom\,(E,\mathbf{C})$ dual to the bundle E obtained by assigning to every point in $\mathbf{C}P^n$ the line in $\mathbf{C}^{n+1}$ which it represents. (Throughout this argument, $Hom\,(P,Q)$ stands for the bundle whose fibers are given by $Hom\,(P,Q)_x = Hom\,(P_x, Q_x)$.) One checks easily that $E = E_0/G$, where E_0 is the line bundle over S^{2n+1} given by

$$E_0 = \{(z,v) \in S^{2n+1} \times \mathbf{C}^{n+1} : v = \lambda x \text{ for some } \lambda \in \mathbf{C}\}$$

and the action of G on E_0 is $g\cdot(z,v) = (gz,v)$. Let $H_0 = Hom\,(E_0,\mathbf{C})$. Then $H = H_0/G$. Furthermore, $H_0 \otimes E_0$ is isomorphic to the one-dimensional trivial bundle $Hom\,(E_0, E_0)$, so that $[H_0][E_0] = 1$.

We are now in a position to compute the $R(G)$-module structure. We have $R(G) \simeq \mathbf{Z}[\widehat{G}] = \mathbf{Z}[\tau,\tau^{-1}]$ where τ is the character of S^1 defined by $\tau(g) = g$. Let F_0 be the trivial one-dimensional G-vector bundle on S^{2n+1}, that is, $F_0 = \mathbf{C} \times S^{2n+1}$. Let V be the one-dimensional representation space of G with representation τ^{-1}, so that $[V]$

corresponds to τ^{-1} in the isomorphism above. Define $\varphi: V \otimes F_0 \longrightarrow E_0$ by $\varphi(\lambda \otimes (z, \mu)) = (z, \lambda \mu z)$ for $\lambda \in V = \mathbf{C}$, $\mu \in \mathbf{C}$, and $z \in S^{2n+1}$. The calculation

$$\varphi(g \cdot (\lambda \otimes (z, \mu)) = \varphi(g^{-1}\lambda \otimes (gz, \mu)) = (gz, \lambda \mu z) = g \cdot (z, \lambda \mu z)$$

shows that φ is equivariant, and it is obviously a bundle isomorphism. So $\tau^{-1} \cdot 1 = [E_0]$, whence $\tau \cdot 1 = [H_0]$ in $K_G^0(S^{2n+1})$, and the $R(G)$-module structure on $\mathbf{Z}[t]/<(t-1)^{n+1}>$ is determined by $\tau \cdot \eta(t) = t\eta(t)$ for $\eta(t) \in \mathbf{Z}[t]/<(t-1)^{n+1}>$.

We now claim that

$$\mathbf{Z}[t, t^{-1}]/<(t-1)^{n+1}> = \mathbf{Z}[t]/<(t-1)^{n+1}>.$$

We write

$$t^{-1} = t^{-1}(1-(1-t)^{n+1}) + t^{-1}(1-t)^{n+1},$$

where the first term on the right is a polynomial in t and the second term is zero modulo $<(t-1)^{n+1}>$. Thus $t^{-1} \in \mathbf{Z}[t]/<(t-1)^{n+1}>$, and the claim follows. It is now easy to see that $K_G^0(S^{2n+1}) \simeq R(G)/I(G)^{n+1}$ as a ring and as an $R(G)$-module, since $I(G) = <1-\tau>$.

It now follows that $K_G^0(X) \simeq \bigoplus_{n=1}^{\infty} R(G)/I(G)^n$. Since $I(G)^n \neq 0$ for all n, we conclude that the action α, which certainly has locally discrete K-theory, does not have discrete K-theory. Hence α cannot be KK-free, by proposition 4.4.6. Also, $C_0(X)$ has a composition series in which the factors are $C(S^{2n+1})$, and, as will be shown in the next section, the actions of G on these *are* KK-free. We have thus found: (1) a free, and hence totally K-free, action on a locally compact space which is not KK-free, and (2) a G-algebra having a composition series such that the actions on all the factors are KK-free but the action on the whole algebra is not.

4.5. $C(X)$-Algebras and KK-Freeness

The version of bundles or sheaves of C^*-algebras most convenient for our purposes is given in the following definition. (For other versions, see [29], chapter 10, and [49]. Our definition is most closely related to the $C(X)$-modules in section 4 of [49].)

4.5.1 Definition. Let X be a locally compact Hausdorff space, and let A be a C^*-algebra. We say A is a $C_0(X)$-algebra if there is a pairing $(f, a) \longrightarrow fa$ from $C_0(X) \times A$ to A such that:

(1) A is algebraically a $C_0(X)$-algebra (except for conditions on units, because neither A nor $C_0(X)$ need have one).

(2) Unit conditions: if $\{f_\lambda\}$ is an approximate identity in $C_0(X)$ and $a \in A$, then $f_\lambda a \longrightarrow a$ in norm.

(3) Adjoint property: $(fa)^* = \bar{f} a^*$ for $f \in C_0(X)$ and $a \in A$.

(4) Continuity: $\| fa \| \le \| f \| \, \| a \|$ for $f \in C_0(X)$ and $a \in A$.

If X is a G-space, with β being the corresponding action on $C_0(X)$, and (G,A,α) is a G-algebra, we say that (G,A,α) is a $(G, C_0(X),\beta)$-algebra if in addition the following property holds:

(5) $\alpha_g(fa) = \beta_g(f)\alpha_g(a)$ for $f \in C_0(X)$, $a \in A$, and $g \in G$.

4.5.2 Example. If A is a C^*-algebra, then A is always a Z-algebra, where Z is the center of the multiplier algebra $M(A)$. Similarly, if (G,A,α) is a G-algebra, then (G,A,α) is a (G,Z_0,β)-algebra, where Z_0 is the center of $M_G(A)$, the set of elements in $M(A)$ on which the extension of α to $M(A)$ is continuous.

4.5.3 Remark. Let A be a $C_0(X)$-algebra. Then the multiplications by functions $f \in C_0(X)$ clearly define elements of $M(A)$ which commute with the elements of A and are therefore in the center Z of $M(A)$. Thus, every $C_0(X)$-algebra is given by a homomorphism from $C_0(X)$ to Z. Since multipliers are automatically continuous and homomorphisms are norm reducing, we see that condition (4) (continuity) actually follows from conditions (1) and (3). Furthermore, if X is compact, and we represent $M(A)$ on a Hilbert space, we can derive the inequality

$$\| fa + (1-f)b \| \le \max(\| a \|, \| b \|),$$

for $f \in C(X)$ with $0 \le f \le 1$; the details are left to the reader. It follows that A is a locally convex $C(X)$-module in the sense of [49], definition 6.1.

It is, however, not true that every homomorphism φ from $C_0(X)$ to the center of $M(A)$ defines a $C_0(X)$-algebra structure on A, since condition (2) need not be satisfied. A necessary and sufficient condition for φ to define a $C_0(X)$-algebra structure is that the identity in $M(A)$ be in the closure of the image of $C_0(X)$ for the strict topology on $M(A)$. Again, we omit the details.

There are of course analogs of these facts for $(G,C_0(X))$-algebras, where one must use $M_G(A)$ instead of $M(A)$. However, it is not very useful to consider A as a (G,Z_0,β)-algebra as in the previous example (where Z_0 is the center of $M_G(A)$), since Z_0 is very unlikely to be non-trivial but separable.

4.5.4 Lemma. Let I be an ideal (in the C^*-algebra sense) in a $C_0(X)$-algebra A. Then I is also a $C_0(X)$-algebra.

Proof. The only nonobvious property is that $fa \in I$ for $f \in C_0(X)$ and $a \in I$. To prove this, let $\{e_\lambda\}$ be an approximate identity in A. Then $fe_\lambda \in A$ and $a \in I$ imply $fe_\lambda a \in I$. Also $fe_\lambda a = e_\lambda \cdot fa$ converges to fa in norm, so also $fa \in I$. Q.E.D.

The main lemma on $C_0(X)$-algebras is the following (we do not worry about the case in which X is not compact since it is not needed):

4.5.5 Lemma. Let X be compact, and let A be a $C(X)$-algebra. Then there is a homomorphism $m : C(X)\otimes A \longrightarrow A$ such that $m(f\otimes a) = fa$. If X is a G-space, and (G,A,α) is a $(G,C(X),\beta)$-algebra, then m is equivariant.

Proof. We first consider the case in which A is unital. Then there is a homomorphism $\varphi: C(X) \longrightarrow A$ defined by $\varphi(f) = f\cdot 1$. Since the image of φ is in the center of A, the images of $\varphi: C(X) \longrightarrow A$ and $id_A : A \longrightarrow A$ commute. By [104], proposition IV.4.7, there is a homomorphism $m : C(X)\otimes_{\max} A \longrightarrow A$ such that $m(f\otimes a) = fa$, where $C(X)\otimes_{\max} A$ is the completion of the algebraic tensor product in the largest C^* cross norm. But all C^* cross norms on the algebraic tensor product agree because $C(X)$ is commutative. (See, for example, [104], lemma IV.4.18, or [41], theorems 3 and 4.)

Now consider the non-unital case. As in remark 4.5.3, there is a homomorphism $\varphi: C(X) \longrightarrow M(A)$, and clearly $\varphi(1) = 1$. It is now trivial to check that $M(A)$ becomes a $C(X)$-algebra, with $f\cdot a = \varphi(f)a$ for $f \in C(X)$ and $a \in M(A)$. By the first part of the proof there is a function $m_0: C(X)\otimes M(A) \longrightarrow M(A)$ such that $m_0(f\otimes a) = fa$. Since $fa \in A$ for $f \in C(X)$ and $a \in A$, the image under m_0 of $C(X)\otimes A$ is in A, and we can take $m = m_0 |_{C(X)\otimes A}$.

Now suppose (G,A,α) is a $(G,C(X),\beta)$-algebra. It is trivial to check that m is equivariant on the elementary tensors $f\otimes a$ for $f \in C(X)$ and $a \in A$. Since their linear span is dense in $C(X)\otimes A$, it follows that m is equivariant. Q.E.D.

4.5.6 Lemma. Let (G,A,α) be a separable $(G,C(X),\beta)$-algebra, where X is compact and second countable. Then $KK_G^0(A,A)$ is a $K_G^0(X)$-module, in such a way that the $R(G)$-algebra structures are compatible.

Proof. We identify $K_G^0(X)$ with $KK_G^0(\mathbf{C},C(X))$ as in theorem 3.6.12. To avoid confusion among the various products defined on KK-groups, we will write the product of $\eta\in KK_G^0(A_1,B_1)$ and $\lambda\in KK_G^0(A_2,B_2)$ as $\eta\otimes\lambda\in KK_G^0(A_1\otimes A_2,B_1\otimes B_2)$, and the product of $\eta\in KK_G^0(A,B)$ and $\lambda\in KK_G^0(B,C)$ as $\eta\lambda\in KK_G^0(A,C)$. With this notation, we recall from proposition 3.6.15 that the product in $KK_G^0(\mathbf{C},C(X))$ obtained by regarding it as $K_G^0(X)$ is defined as follows. Let $\Delta: C(X)\otimes C(X) \longrightarrow C(X)$ be the diagonal map $\Delta(f\otimes g) = fg$, which is a homomorphism (either by the preceding lemma, or because it comes from the map $x \longrightarrow (x,x)$ on spaces). Then $rs = \Delta_*(r\otimes s)$ for $r,s \in KK_G^0(\mathbf{C},C(X))$, where $r\otimes s$ is of course an element of $KK_G^0(\mathbf{C},C(X)\otimes C(X))$. (Note that $C(X)$ is separable because X is second countable.)

To make $KK_G^0(A,A)$ a module over $KK_G^0(\mathbf{C},C(X))$, we need a pairing

$$n : KK_G^0(\mathbf{C},C(X))\times KK_G^0(A,A) \longrightarrow KK_G^0(A,A).$$

It is defined by $n(r,\eta) = m_*(r\otimes\eta)$, where $m : C(X)\otimes A \longrightarrow A$ is the homomorphism of the previous lemma. That is, n is the composite

$$KK^0_G(\mathbf{C},C(X))\times KK^0_G(A,A)\longrightarrow KK^0_G(A,C(X)\otimes A)\xrightarrow{m_*} KK^0_G(A,A).$$

There are now two nontrivial things to be proved: $n(r,n(s,\eta))=n(rs,\eta)$ for $r,s\in KK^0_G(\mathbf{C},C(X))$ and $\eta\in KK^0_G(A,A)$ (associativity), and $n(d\cdot 1,\eta)=d\cdot\eta$ for $d\in R(G)$ and $\eta\in KK^0_G(A,A)$, where 1 is the identity in $KK^0_G(\mathbf{C},C(X))$ (agreement of module structures).

For associativity, notice that the equality $(f_1 f_2)a=f_1(f_2 a)$, for $f_1,f_2\in C(X)$ and $a\in A$, implies that $m\circ(\Delta\otimes id_A)=m\circ(id_{C(X)}\otimes m)$. Consequently, for $r,s\in KK^0_G(\mathbf{C},C(X))$ and $\eta\in KK^0_G(A,A)$, we have

$$n(rs,\eta)=m_*(\Delta_*(r\otimes s)\otimes\eta)=m_*(r\otimes m_*(s\otimes\eta))=n(r,n(s,\eta)),$$

as desired.

To prove that the two $R(G)$-module structures on $KK^0_G(A,A)$ agree, we recall that the pairings defined by the Kasparov product are all $R(G)$-bilinear by corollary 3.6.3, and that m_* is an $R(G)$-module homomorphism by theorem 3.2.12. The equation $n(d\cdot 1,\eta)=d\cdot\eta$ will now follow if we can prove that $m_*(1\otimes\eta)=\eta$. To prove this last equation, represent η by an equivariant quasihomomorphism

$$\Phi=(\varphi,\psi):A\longrightarrow E\vartriangleright J\xrightarrow{\mu} K\otimes A,$$

where K is the algebra of compact operators on a separable Hilbert space with the unitary representation of G given in 3.1.5. The identity of $KK^0_G(\mathbf{C},C(X))$ is represented by the homomorphism $i:\mathbf{C}\longrightarrow C(X)$ defined by $i(1)=1$. Therefore $1\otimes\eta=[i]\otimes\eta$. The definition of the general form of the product, definition 3.5.3, together with definition 3.5.1 and proposition 3.4.6, now shows that

$$m_*(1\otimes\eta)=m_*(\tau_A([i])\,\tau_{C(X)}(\eta))=m_*((i\otimes id_A)^*([id_{C(X)}\otimes\Phi])). \qquad (*)$$

Here, $id_{C(X)}\otimes\Phi$ is the quasihomomorphism

$$(id_{C(X)}\otimes\varphi,id_{C(X)}\otimes\psi):C(X)\otimes A\longrightarrow C(X)\otimes E\vartriangleright C(X)\otimes K\xrightarrow{\mu_0} K\otimes C(X)\otimes A,$$

where μ_0 is the composite of $id_{C(X)}\otimes\mu$ and the canonical isomorphism $C(X)\otimes K\otimes A\simeq K\otimes C(X)\otimes A$. The last expression in (*) is perhaps clearer if it is rewritten as $[i\otimes id_A][id_{C(X)}\otimes\Phi][m]$. We compute it according to definition 3.2.10, and find that it is represented by the following prequasihomomorphism, in which $\sigma=(id_{C(X)}\otimes\varphi)\circ(i\otimes id_A)$ and $\tau=(id_{C(X)}\otimes\psi)\circ(i\otimes id_A)$:

$$\Sigma=(\sigma,\tau):A\longrightarrow C(X)\otimes E\vartriangleright C(X)\otimes J\xrightarrow{\mu_0} K\otimes C(X)\otimes A\xrightarrow{id_K\otimes m} K\otimes A.$$

It is now easily seen that the quasihomomorphism defined by Σ is exactly Φ. Indeed, let $\overline{m}:C(X)\otimes K\otimes A\longrightarrow K\otimes A$ be the homomorphism defined by $\overline{m}(f\otimes k\otimes a)=k\otimes fa$ for $f\in C(X)$, $k\in K$, and $a\in A$. Then $(id_K\otimes m)\circ\mu_0=\overline{m}\circ(id_{C(X)}\otimes\mu)$ and we have, for $a\in A$, using the notation from definition 3.2.2,

$$D_\Sigma(a) = (id_K \otimes m)\cdot\mu_0(\sigma(a) - \tau(a))$$
$$= \overline{m}\cdot(id_{C(X)}\otimes\mu)(1\otimes\varphi(a) - 1\otimes\psi(a)) = \overline{m}(1\otimes D_\Phi(a)) = D_\Phi(a),$$

and similarly $Q_\Sigma = Q_\Phi$. Now use proposition 3.2.4. So $[\Sigma] = [\Phi]$, and we have shown that $m_*(1\otimes\eta) = \eta$. This completes the proof that the $R(G)$-module structures agree.

Q.E.D.

4.5.7 Remark. Notice that we do not claim that $KK_G^0(A,A)$ is a $K_G^0(X)$-algebra in the previous lemma. In fact, the elements $m_*(r\otimes 1)$ are in general not in the center of $KK_G^0(A,A)$. (The simplest example is $A = C(X) = \mathbf{C}^2$ and G trivial, for which $KK^0(A,A)$ is the ring $M_2(\mathbf{Z})$ of 2×2 matrices with integer entries. In this case, the image of $K^0(X)$ under $r \longrightarrow m_*(r\otimes 1)$ looks like the diagonal matrices.) It is, however, true that the map $\varphi: K_G^0(X) \longrightarrow KK_G^0(A,A)$ is a unital $R(G)$-algebra homomorphism. This fact implies the result of the previous lemma, and we sketch the (somewhat less conceptual) proof.

First, let A_i and B_i be G-algebras for $i = 1, 2$, and let $\eta \in KK_G^0(A_1, A_2)$ and $\lambda \in KK_G^0(B_1, B_2)$. Using proposition 3.5.5, it is easy to show that

$$(1_{A_1}\otimes\lambda)(\eta\otimes 1_{B_2}) = (\eta\otimes 1_{B_1})(1_{A_2}\otimes\lambda). \qquad (*)$$

Now let $r, s \in KK_G^0(\mathbf{C}, C(X))$. Then $rs = sr$, where the product is the operation in $K_G^0(X)$. We then obtain

$$\varphi(rs) = m_*(sr\otimes 1_A) = (((s\otimes r)[\Delta])\otimes 1_A)[m]$$
$$= (s\otimes r\otimes 1_A)([\Delta]\otimes 1_A)[m] = (s\otimes r\otimes 1_A)(1_{C(X)}\otimes[m])[m]$$
$$= (1_{\mathbf{C}}\otimes(r\otimes 1_A)[m])(s\otimes 1_A)[m] = \varphi(r)\varphi(s).$$

In this computation, the fourth step follows from the equation

$$m\cdot(id_{C(X)}\otimes m) = m\cdot(\Delta\otimes id_A),$$

and the fifth step follows from (*). The remaining verifications are easy and are omitted.

We are now ready to prove the main theorem of this section.

4.5.8 Theorem. Let (G, A, α) be a separable $(G, C(X), \beta)$-algebra, where X is compact and second countable. If β induces a free action on X, then α is KK-free.

Proof. Let I be a G-invariant ideal in A. Then by lemma 4.5.4, I is also a $(G, C(X), \beta)$-algebra. Therefore, by the previous lemma, $KK_G^0(I,I)$ is a module over $K_G^0(X)$. If $P\subset R(G)$ is a prime ideal not containing $I(G)$, then $K_G^0(X)_P = 0$ by theorem 1.1.1, so $KK_G^0(I,I)_P = 0$ by corollary 4.4.3.

Q.E.D.

4.5.9 Corollary. Let X be a compact second countable G-space. If G acts freely on X, then the induced action on $C(X)$ is KK-free.

Proof. Take $A = C(X)$ in the theorem. Q.E.D.

4.5.10 Corollary. Let (G,A,α) be a separable unital G-algebra, and suppose that G acts freely on the maximal ideal space of the center of A. Then α is KK-free.

Proof. (G,A,α) is clearly a $(G,Z,\alpha_{(\cdot)}|_Z)$-algebra, where Z is the center of A. Since A is unital, its maximal ideal space X is compact, and since A is separable, X is second countable. So we can apply the theorem. Q.E.D.

4.6. $C_0(X)$-Algebras and Total K-Freeness

As example 4.4.12 showed, the hypothesis that X be compact in theorem 4.5.8 cannot be dropped. However, we still obtain total K-freeness.

4.6.1 Theorem. Let (G,A,α) be a separable $(G,C_0(X),\beta)$-algebra, with X second countable. If G acts freely on X then α is totally K-free.

Proof. The hypotheses hold for the restrictions of the actions to any closed subgroup H, and by lemma 4.5.4 they still hold if A is then replaced by any H-invariant ideal. Therefore it is sufficient to prove that α has locally discrete K-theory. To do this, we will express A as a direct limit of $(G,C(Y))$-algebras for *compact* free G-spaces Y.

Since X is second countable and locally compact, it is σ-compact, and in fact there is an increasing sequence of open sets U_n with compact closures such that $\overline{U}_n \subset U_{n+1}$ and $X = \bigcup_{n=1}^{\infty} U_n$. Since G is compact, the sets U_n may be taken to be G-invariant. Regard $C_0(U_n)$ as an ideal of $C_0(X)$ in the obvious way. Then we have $C_0(X) = \overline{\bigcup_{n=1}^{\infty} C_0(U_n)}$. Let

$$A_n = \overline{C_0(U_n)A} = \overline{span}\{fa : f \in C_0(U_n),\, a \in A\}$$

($\overline{span}$ means closed linear span). Then A_n is a C^*-subalgebra of A, and is obviously G-invariant.

We first show that (G,A_n) is a $(G,C(\overline{U}_n))$-algebra, and for this it is necessary to define the product fa, for $f \in C(\overline{U}_n)$ and $a \in A_n$. Since X^+ is compact, second countable, and Hausdorff, there is by the Tietze extension theorem ([57], page 242, problem O) a continuous function $f_0 \in C(X^+)$ such that $f_0|_{\overline{U}_n} = f$. Multiplying f_0 by a continuous function which is 1 on $\overline{U}_n$ and 0 outside U_{n+1}, we may assume that $f_0 \in C_0(X)$. We then define $fa = f_0 a$, where the second expression is the given product of $f_0 \in C_0(X)$ and $a \in A$. To show this is well defined, let f_1 be another element of $C_0(X)$ with $f_1|_{\overline{U}_n} = f$, and take a to be of the form $a = hb$ for $h \in C_0(U_n)$ and $b \in A$.

(This restriction is permissible, since such elements span a dense subalgebra of A_n). Then $(f_0 - f_1) \mid_{\overline{U}_n} = 0$, whence $f_0 ha - f_a ha = ((f_0 - f_1)h)a = 0$. The axioms for a $(G, C(\overline{U}_n))$-algebra are now obviously satisfied. (The axiom involving approximate identities here takes the form $1 \cdot a = a$, and follows from the fact that if $h \in C_0(U_n)$ and $f \in C_0(X)$ with $f \mid_{\overline{U}_n} = 1$, then $fh = h$.) Since $\overline{U}_n$ is a second countable compact free G-space, theorem 4.5.8 implies that $\alpha_{(\cdot)} \mid_{A_n}$ is KK-free, and hence has (locally) discrete K-theory by proposition 4.4.5.

We now claim that $A = \overline{\bigcup_{n=1}^{\infty} A_n}$. Let f_n be a continuous function on X such that $f_n = 1$ on $\overline{U}_{n+1}$, $f_n = 0$ outside U_n, and the range of f_n is in $[0,1]$. Then $\{f_n\}_{n=1}^{\infty}$ is an approximate identity for $C_0(X)$, and for $a \in A$ one has $f_n a \in A_n$ and $f_n a \longrightarrow a$ as $n \longrightarrow \infty$. Therefore $a \in \overline{\bigcup_{n=1}^{\infty} A_n}$, and the claim is proved. It follows that $A = \varinjlim A_n$. Since the restrictions of α to the A_n all have locally discrete K-theory, lemma 4.2.14 implies α has locally discrete K-theory. Q.E.D.

4.6.2 Remark. In this situation, there is also a localization theorem. Let (G,A,α) be a separable $(G,C_0(X),\beta)$-algebra, with X second countable. Let $P \subset R(G)$ be a prime ideal with support S, and let U be the set of all points $x \in X$ whose stabilizer groups for the action of G on X contain no conjugate of S. Then U is a G-invariant open subset of X. Let $I = \overline{C_0(U)A}$, which is a G-invariant ideal in A. Then the map $A \longrightarrow A/I$ induces an isomorphism $K_*^G(A)_P \simeq K_*^G(A/I)_P$. To prove this, it is enough to show that $K_*^G(I)_P = 0$. Now let V be a G-invariant open set whose closure is compact and contained in U. Then $K_G^0(\overline{V})_P = 0$ by the Localization Theorem for spaces, theorem 1.2.4. It follows that $KK_G^0(\overline{C_0(V)I}, \overline{C_0(V)I})_P = 0$, and thence that $K_*^G(\overline{C_0(V)I})_P = 0$. An argument as in the proof of the previous theorem shows that $K_*^G(I)_P = 0$.

4.6.3 Remark. As mentioned at the end of section 4.3, separability and countability are not really needed for 4.6.1 and 4.6.2. Indeed, the following result enables one to use the techniques of section 4.3 and the slice theorem ([14], theorem II.5.4), neither of which requires these hypotheses, and neither of which requires KK-theory. The result is: if (G,A,α) is a $(G,C_0(X),\beta)$-algebra, and if U is an open subset of X which is equivariantly homeomorphic to $G \times_H S$ for some closed subgroup H of G and some closed subset S of U, then $\overline{C_0(U)A}$ is equivariantly isomorphic to $C(G \times_H B)$, where $B = \overline{C_0(U)A} / \overline{C_0(U-S)A}$.

Theorem 4.6.1 has the following corollaries.

4.6.4 Corollary. Let (G,A,α) be a separable G-algebra. If G acts freely on the maximal ideal space of the center of A, then α is totally K-free.

Proof. Just like corollary 4.5.10, but using theorem 4.6.1 instead of theorem 4.5.8. Q.E.D.

4.6.5 Corollary. Let (G,A,α) and $(G,C_0(X),\beta)$ be separable G-algebras. If β is free, then the action $\beta\otimes\alpha$ of G on $C_0(X)\otimes A$ is totally K-free.

Proof. $(G, C_0(X)\otimes A, \beta\otimes\alpha)$ is a $(G,C_0(X),\beta)$-algebra. Q.E.D.

4.6.6 Corollary. Let A be the algebra of global sections vanishing at infinity of a continuous field of separable C^*-algebras on a second countable locally compact Hausdorff G-space X. (See [29], definitions 10.3.1 and 10.4.1.) Suppose that G acts on the continuous field in such a way that the fiber over x is sent to the fiber over gx. Then if X is a free G-space, the induced action on A is totally K-free.

Proof. Again, (G,A) is a $(G,C_0(X))$-algebra. Q.E.D.

We conclude this chapter with a discussion of what might be called the "free compactification" of a G-algebra. When considering K-theory, it is often necessary to work with unital algebras, and to derive results for non-unital algebras by considering the unitization. (This was done frequently in chapter 2.) Unfortunately, if (G,A,α) is a G-algebra, then the unitization is not K-free even if α is, since $A^+/A \simeq \mathbf{C}$ with the trivial G-action. The next proposition shows that if one is willing to replace A by $K(L^2(G))\otimes A$, where $L^2(G)$ carries the left regular representation of G, and if G is abelian, then one can find a unital C^*-algebra containing A on which the action is K-free.

4.6.7 Proposition. Let (G,A,α) be a G-algebra, with G abelian. Then there is a naturally split equivariant exact sequence

$$0 \longrightarrow K(L^2(G))\otimes A \longrightarrow B \longrightarrow C(G) \longrightarrow 0,$$

where $B = C^*(\hat{G}, C^*(G,A,\alpha)^+, \hat{\alpha}^+)$ is a unital G-algebra. The action on $C(G)$ comes from the action of G on itself by translation. The action on B is (totally) K-free if and only if α is. If A is separable, then the action on B is KK-free if and only if α is.

Proof. Let $\hat{\alpha}^+ : \hat{G} \longrightarrow Aut(C^*(G,A,\alpha)^+)$ be the extension to the unitization of the dual action of $\hat{G}$ on $C^*(G,A,\alpha)$. Then there is a natural $\hat{G}$-equivariantly split exact sequence

$$0 \longrightarrow C^*(G,A,\alpha) \longrightarrow C^*(G,A,\alpha)^+ \longrightarrow \mathbf{C} \longrightarrow 0.$$

Taking crossed products with $\hat{G}$, we obtain a G-equivariantly split exact sequence

$$0 \longrightarrow C^*(\hat{G}, C^*(G,A,\alpha),\hat{\alpha}) \longrightarrow B \longrightarrow C^*(\hat{G}) \longrightarrow 0,$$

where $B = C^*(\hat{G}, C^*(G,A,\alpha)^+, \hat{\alpha}^+)$. The algebra B is unital because $C^*(G,A,\alpha)^+$ is unital and $\hat{G}$ is discrete. It is well known that $C^*(\hat{G}) \simeq C(G)$, with the action described in the statement of the proposition. Takai duality ([79], 7.9.3) implies that the first term is equivariantly isomorphic to $K(L^2(G))\otimes A$ with the action specified in the discussion before the proposition. The action on $C(G)$ is KK-free by corollary 4.5.9 and totally K-free by corollary 4.6.5. By stability (propositions 4.2.10 and 4.4.10), the action on the first term is (totally) K-free or KK-free if and only if α is. Therefore the action on B is (totally) K-free or KK-free if and only if α is, by propositions 4.2.2 and 4.4.8. Q.E.D.

One would hope that one could find an exact sequence

$$0 \longrightarrow A \longrightarrow B \longrightarrow C(G) \longrightarrow 0,$$

or even a sequence

$$0 \longrightarrow A \longrightarrow B \longrightarrow C(Z) \longrightarrow 0 \qquad (*)$$

for some compact free G-space Z, satisfying the properties of the proposition. In general, that is not possible: we present an example of a locally compact free G-space X such that, with $A = C_0(X)$, no sequence $(*)$ can exist. In particular, there is no injective continuous equivariant map from X to any compact free G-space.

4.6.8 Example. Let $G = S^1$, let $X = \coprod_{n=0}^{\infty} S^{2n+1}$, and let the action of G on X be as in example 4.4.12. Suppose there were an equivariant exact sequence

$$0 \longrightarrow C_0(X) \longrightarrow B \longrightarrow C(Z) \longrightarrow 0, \qquad (**)$$

with B unital and Z a free G-space. We will derive a contradiction.

The first step is to prove that B is commutative. This is surely known, but we have never seen it in print, so we prove it here. We need a lemma:

4.6.9 Lemma. Let a and b be selfadjoint elements of a C^*-algebra A. If a commutes with $ab - ba$, then a commutes with b.

Proof. By the Kleinecke-Shirokov theorem ([42], problem 184), the hypotheses imply that $ab - ba$ is quasinilpotent. (I am indebted to Donald Sarason for pointing this theorem out to me.) Also $(ab - ba)^* = -(ab - ba)$, so $ab - ba$ is normal. A normal quasinilpotent operator is necessarily zero. So $ab = ba$. Q.E.D.

4.6.10 Proposition. Let

$$0 \longrightarrow I \longrightarrow B \xrightarrow{\pi} A \longrightarrow 0$$

be an exact sequence of C^*-algebras, with I and A commutative. Then B is commutative.

Proof. Identify I with its image in B. If $a \in B$ and $x \in I$, then $ax - xa \in I$, so x commutes with $ax - xa$. If a and x are selfadjoint, then $ax = xa$ by the lemma. Since every element of a C^*-algebra has the form $c + id$ for selfadjoint elements c and d, it follows that an arbitrary element of B commutes with an arbitrary element of I.

Now let $a, b \in B$. Then $\pi(ab - ba) = 0$ since the image of π is commutative. Therefore $ab - ba \in I$, so a commutes with $ab - ba$ by the above. Another application of the lemma shows that if a and b are selfadjoint then they commute. Just as before, we conclude that arbitrary elements of B commute. Q.E.D.

We now return to our exact sequence (**). By the proposition, B is commutative. Since B is unital, it is isomorphic to $C(Y)$ for some compact G-space Y. We can then identify X with an open subset of Y, and Z with its complement $Y - X$. Therefore there is an exact sequence

$$\begin{array}{ccccc} K_G^0(X) & \longrightarrow & K_G^0(Y) & \longrightarrow & K_G^0(Z) \\ \uparrow & & & & \downarrow \\ K_G^1(Z) & \longleftarrow & K_G^1(Y) & \longleftarrow & K_G^1(X) . \end{array}$$

The actions of G on X and Z are free, so the action of G on Y is also free. Since Y and Z are compact, theorem 1.1.1 implies that there are integers m and n such that $I(G)^m K_G^*(Y) = 0$ and $I(G)^n K_G^*(Z) = 0$. It follows from the exact sequence above that $I(G)^{m+n} K_G^*(X) = 0$. But this contradicts example 4.4.12. Therefore no equivariant exact sequence

$$0 \longrightarrow C_0(X) \longrightarrow B \longrightarrow C(Z) \longrightarrow 0 ,$$

with B unital and Z a free G-space, can exist.

Chapter 5

Subgroups

Let G be a compact Lie group and let X be a Hausdorff space. Then there are two easy relations between freeness of an action λ of G on X and freeness of the actions $\lambda|_H$ for H a subgroup of G:

(1) If λ is free, so is $\lambda|_H$ for every closed subgroup H.

(2) If $\lambda|_H$ is free for every finite subgroup H of prime order, then λ is free.

In this chapter, we attempt to generalize these properties to the various forms of K-theoretic freeness defined in the previous chapter. As we have seen, the analog of (1) fails for K-freeness, and it holds essentially by definition for total K-freeness. We are able to show that it holds for KK-freeness, provided we consider only G-algebras (G,A,α) with G finite and A separable and nuclear. The analog of (2) also fails, for all three notions; the examples are more complicated than those presented so far and appear in chapter 9. Nevertheless, something can be salvaged: if G is a finite group and (G,A,α) is a G-algebra such that $\alpha|_H$ is K-free for all cyclic subgroups, then α is "almost" totally K-free, and is in fact totally K-free if G is a p-group.

In the first section of this chapter we define the induction and restriction, or "change of group" homomorphisms in K-theory and KK-theory, and prove their basic properties. The second section is devoted to the proof of the weaker analog of fact (2) mentioned above. In the third section we prove the analog of (1) for KK-freeness; the main technical difficulties are concentrated in the "Ideal Decomposition Lemma," which will also be needed in the next chapter.

5.1. Restriction and Induction in K-Theory and KK-Theory

We first consider restriction and induction in K-theory. The following material is the analog of restriction and induction of group representations. In order to stay within the category of finitely generated projective modules without resorting to elliptic operators (as used, for example, in section 2 of [97]), we assume that all groups are finite. (Also see [52].) The results we generalize can all be found in [99].

5.1.1 Definition. Let G be a finite group, and let H be a subgroup of G. Let (G,A,α) be a unital G-algebra. If (G,E,λ) is a finitely generated projective (G,A,α)-module, its restriction $Res_H^G(G,E,\lambda)$ to H is defined to be $(H,E,\lambda|_H)$. If (H,F,μ) is a finitely generated projective $(H,A,\alpha|_H)$-module, its induction $Ind_H^G(H,F,\mu)$ to G is the

(G,A,α)-module (G,E,λ), where

$$E = \{x : G \longrightarrow F :\ x(g) = \mu_h(x(gh)) \text{ for } h \in H \text{ and } g \in G\}$$

and the actions of A and G on E are given by

$$(xa)(g) = x(g)\alpha_g^{-1}(a) \quad \text{and} \quad \lambda_g(x)(g_0) = x(g^{-1}g_0)$$

for $g, g_0 \in G$, $a \in A$, and $x \in E$. (Compare with the definition of $C(G \times_H A)$, definition 2.9.2.)

5.1.2 Remark. Let $A = \mathbf{C}$, with the trivial action of G. Then a finitely generated projective (G,A)-module is just a finite dimensional representation space of G. It is then easily seen that Res_H^G and Ind_H^G are just restriction and induction of (finite dimensional) representations of G and H. In particular, Res_H^G defines a restriction map $Res_H^G : R(G) \longrightarrow R(H)$, which we have already used several times. It is obviously a ring homomorphism.

5.1.3 Proposition. Let G be a finite group, and let H a subgroup. We regard all $R(H)$-modules as $R(G)$-modules via the ring homomorphism $Res_H^G : R(G) \longrightarrow R(H)$ defined by restriction of representations. Then the functors Ind_H^G and Res_H^G of the previous definition give rise to, for any G-algebra (G,A,α), natural $R(G)$-module homomorphisms

$$Res_H^G : K_*^G(A) \longrightarrow K_*^H(A) \quad \text{and} \quad Ind_H^G : K_*^H(A) \longrightarrow K_*^G(A).$$

These homomorphism satisfy the following properties:

(1) Transitivity: if K is a subgroup of H, then $Res_K^H \circ Res_H^G = Res_K^G$ and $Ind_H^G \circ Ind_K^H = Ind_K^G$. (The second formula is also called "induction in stages".)

(2) $Ind_H^G(\chi Res_H^G(\eta)) = Ind_H^G(\chi)\eta$ for $\chi \in R(H)$ and $\eta \in K_*^G(A)$. In particular, $Ind_H^G \circ Res_H^G$ is multiplication by the class in $R(G)$ of the induction to G of the trivial one-dimensional representation of H.

Proof. We do everything for K_0^G and unital algebras. Since the constructions are natural, as will be shown, the result holds for non-unital algebras by considering the unitization, and for K_1^G by considering suspensions. To simplify the notation, we write Ind for Ind_H^G and Res for Res_H^G, when no confusion can arise.

We first consider restriction. If (G,E,λ) is a finitely generated projective (G,A,α)-module, then it is obvious that $(H,E,\lambda|_H)$ is a finitely generated projective $(H,A,\alpha|_H)$-module, that this construction is functorial in A, and that it is a group homomorphism on K-theory. Clearly $Res(V \otimes E) \simeq Res(V) \otimes Res(E)$ for any finite dimensional representation space V of G and any finitely generated projective (G,A,α)-module E. Thus $Res : K_0^G(A) \longrightarrow K_0^H(A)$ is an $R(G)$-module homomorphism, since for $\eta \in K_0^H(A)$, the product $[V]\eta$ is defined to be $[Res(V)]\eta$. Finally, transitivity of restriction of obvious.

Now we consider induced modules. Let (H,F,μ) be a finitely generated projective $(H,A,\alpha|_H)$-module. The first step is to check that $Ind\,(H,F,\mu)=(G,E,\lambda)$ really is a finitely generated projective (G,A,α)-module. If R is a set of left coset representatives of H in G, let $F^R=\prod_{r\in R}F$ be the set of all functions from R to F. Then the function $\varphi: E \longrightarrow F^R$ defined by $\varphi(x)(r)=\mu_r(x(r))$ is easily seen to be an isomorphism of A-modules (ignoring the groups), so E is finitely generated projective because R is finite. If $a\in A$ and $x\in E$, then for $h\in H$ and $g\in G$ we have

$$\mu_h((xa)(gh))=\mu_h(x(gh)\alpha_{h^{-1}g^{-1}}(a))=(xa)(g),$$

so $xa\in E$. A similar calculation shows $\lambda_g(x)\in E$ for $x\in E$ and $g\in G$. Finally, the calculation

$$\lambda_g(xa)(g_0)=x(g^{-1}g_0)\alpha_{g_0^{-1}g}(a)=(\lambda_g(x)\alpha_g(a))(g_0)$$

shows that (G,E,λ) is a (G,A,α)-module. It is now obvious that Ind defines a natural group homomorphism from $K_0^H(A)$ to $K_0^G(A)$.

The next step is to check that Ind is an $R(G)$-module homomorphism. This amounts to proving that if V is a finite dimensional representation space of G, with representation π, and if (H,F,μ) is a finitely generated projective $(H,A,\alpha|_H)$-module, then

$$Ind\,(Res\,(V)\otimes F)\simeq V\otimes Ind\,(F).$$

Define

$$\varphi: V\otimes Ind\,(F)\longrightarrow Ind\,(Res\,(V)\otimes F)$$

by $\varphi(v\otimes x)(g)=\pi_{g^{-1}}(v)\otimes x(g)$. There are four things to check. First, $\varphi(v\otimes x)$ really is in $Ind\,(Res\,(V)\otimes F)$, because:

$$(\pi\otimes\mu)_h(\varphi(v\otimes x)(gh))=\pi_g^{-1}(v)\otimes\mu_h(x(gh))=\varphi(v\otimes x)(g)$$

since $\mu_h(x(gh))=x(g)$, for $x\in Ind\,(F)$, $h\in H$, $g\in G$, and $v\in V$. Secondly, the following calculation shows that φ is an A-module homomorphism:

$$\varphi((v\otimes x)a)(g)=\pi_g^{-1}(v)\otimes(x(g)\alpha_g^{-1}(a))=(\varphi(v\otimes x)a)(g).$$

Next, we check that φ is equivariant: let $\tilde{\lambda}$ be the G-action on $Ind\,(Res\,(V)\otimes F)$. Then for $g, g_0\in G$,

$$\tilde{\lambda}_g(\varphi(v\otimes x))(g_0)=\pi_{g_0^{-1}g}(v)\otimes x(g^{-1}g_0)=\varphi((\pi_g\otimes\lambda_g)(v\otimes x))(g_0).$$

Finally, we must show that φ is bijective. If R is a set of left coset representatives of H in G, then the restriction of functions defined on G to R defines bijections from $Ind\,(F)$ to F^R and from $Ind\,(Res\,(V)\otimes F)$ to $(V\otimes F)^R$, as complex vector spaces. If we transport φ to a map ψ from $V\otimes F^R$ to $(V\otimes F)^R$, it becomes $\psi(v\otimes x)(r)=\pi_r(v)\otimes x(r)$ for $r\in R$, which is obviously bijective. Consequently φ is an isomorphism, and we have shown that Ind is an $R(G)$-module homomorphism from $K_0^H(A)$ to $K_0^G(A)$. Naturality is again easy, and we omit the proof.

Next we must show that induction is transitive, that is, if H and K are subgroups of G, with $K \subset H$, and (K,F,μ) is a finitely generated projective $(K,A,\alpha|_K)$-module, then $Ind_H^G(Ind_K^H(F)) \simeq Ind_K^G(F)$. Let (G,E,λ) be the doubly induced module $Ind_H^G(Ind_K^H(F))$, which is easily seen to consist of all functions $x : H \times G \longrightarrow F$ such that

$$x(h,g) = \mu_k(x(hk,g)) \quad \text{and} \quad x(h,g) = x(l^{-1}h, gl)$$

for $k \in K$, $h, l \in H$, and $g \in G$. The G-action and A-module structure are given by

$$\lambda_g(x)(h,l) = x(h, g^{-1}l) \quad \text{and} \quad (xa)(h,g) = x(h,g)\,\alpha_{g^{-1}h^{-1}}(a)$$

for $h \in H$, $g, l \in G$, and $a \in A$. Define $\varphi : E \longrightarrow Ind_K^G(F)$ by $\varphi(x)(g) = x(e,g)$ for $g \in G$ and $x \in E$. It clear that φ is a homomorphism of (G,A,α)-modules, and its image is actually in $Ind_K^G(G)$, because we have, for $k \in K$ and $g \in G$,

$$\mu_k(\varphi(x)(gk)) = \mu_k(x(e,gk)) = \mu_k(x(k,g)) = x(e,g) = \varphi(x)(g).$$

Furthermore, φ is bijective because it has an inverse ψ defined by $\psi(x)(h,g) = x(gh)$ for $h \in H$, $g \in G$, and $x \in Ind_K^G(F)$; we omit the calculation that $\psi = \varphi^{-1}$.

It now remains only to prove the assertion (2). If V is a finite dimensional representation space of H, and (G,E,λ) is a finitely generated projective (G,A,α)-module, we have to show that $Ind\,(V \otimes Res E) \simeq Ind(V) \otimes E$. This is the same as the proof that Ind is an $R(G)$-module homomorphism, except that the vector space and the A-module have been interchanged. The isomorphism is given by

$$\varphi(w \otimes m)(g) = w(g) \otimes \lambda_g^{-1}(m)$$

for $w \in Ind(V)$ and $m \in E$, and we omit the proof that it really is an isomorphism, except that we show that it is an A-module homomorphism (since here the proof is a little different):

$$\begin{aligned}\varphi(w \otimes ma)(g) &= w(g) \otimes \lambda_g^{-1}(m)\,\alpha_g^{-1}(a) \\ &= (\varphi(w \otimes m)(g))\,\alpha_g^{-1}(a) = (\varphi(w \otimes m)a)(g).\end{aligned}$$

The second part of (2) follows immediately from the first, as seen by taking χ to be the class in $R(H)$ of the trivial one-dimensional representation of H. Q.E.D.

5.1.4 Warning. If G is a finite *abelian* group, and E is a complex representation space of a subgroup H of G, then there is an isomorphism $Res_H^G(Ind_H^G(E)) \simeq E^n$ as representation spaces of H, where n is the index of H in G. This need no longer be true if E is instead a (G,A,α)-module for a nontrivial action α of G on a C^*-algebra A. The correct formula (which we do not actually need) is analogous to what happens for a normal subgroups of a nonabelian group. It is

$$Res_H^G(Ind_H^G(E)) \simeq \bigoplus_{r \in R} (\alpha_r)_*(E),$$

where R is a set of coset representatives of H in G. (Observe that α_r is an H-equivariant automorphism of A because G is abelian).

We now turn to restriction and induction in KK-theory.

5.1.5 Definition. Let G be a compact group, and let H be a closed subgroup of G. Let (G,A,α) and (G,B,β) be separable G-algebras. Then the restriction map $Res_H^G\colon KK_G^0(A,B) \longrightarrow KK_H^0(A,B)$ is the map obtained by regarding a G-equivariant prequasihomomorphism from A to B as an H-equivariant prequasihomomorphism.

5.1.6 Proposition. The maps Res_H^G are natural well defined $R(G)$-module homomorphisms. (As usual, $KK_H^0(A,B)$ is regarded as an $R(G)$-module via the restriction $R(G) \longrightarrow R(H)$.) They furthermore satisfy the following properties:

(1) Transitivity: If K is a subgroup of H, then $Res_K^H \cdot Res_H^G = Res_K^G$.

(2) If A, B, and C are separable G-algebras, and $\eta \in KK_G^0(A,B)$ and $\lambda \in KK_G^0(B,C)$, then $Res_H^G(\eta\lambda) = Res_H^G(\eta)\, Res_H^G(\lambda)$. In particular, $Res_H^G\colon KK_G^0(A,A) \longrightarrow KK_H^0(A,A)$ is a unital $R(G)$-algebra homomorphism.

Proof. This is essentially obvious, since all G-equivariant maps occurring in the definition of KK-theory and the construction of the product can be regarded as H-equivariant maps. The fact that Res_H^G is an $R(G)$-module homomorphism follows just as it does for K-theory. That $Res_H^G\colon KK_G^0(A,A) \longrightarrow KK_H^0(A,A)$ is unital is immediate.

Q.E.D.

Of course, these results are also applicable to the general form of the product and to KK^1.

The situation for induction is a bit more complicated, and it seems best to proceed in several steps. Recall that if G is a compact group, H is a closed subgroup, and A is an H-algebra, then we defined the twisted product G-algebra $C(G\times_H A)$ in 2.9.2. It is the C^*-analog of the space of the representation of G induced from the action of H on A. We will need:

5.1.7 Lemma. (See [56], section 5, lemma 1.) Let G be a compact group, let H be a closed subgroup of G, let (G,A,α) be a G-algebra, and let (H,B,β) be an H-algebra. Then the map

$$\omega\colon A \otimes C(G\times_H B) \longrightarrow C(G\times_H (A\otimes B))$$

given by $\omega(a\otimes f)(g) = \alpha_g^{-1}(a)\otimes f(g)$ for $a \in A$, $f \in C(G\times_H B)$, and $g \in G$, is an isomorphism of G-algebras.

Proof. The proof is similar to some of the computations involving inducing restricted representations in the proof of proposition 5.1.3, and is omitted. Q.E.D.

5.1.8 Proposition. (See [56], section 5, theorem 1.) Let G be a second countable compact group, and let H be a closed subgroup of G. Then for separable H-algebras A and B there is a natural induction homomorphism

$$I_H^G\colon KK_H^0(A,B) \longrightarrow KK_G^0(C(G\times_H A),\, C(G\times_H B)),$$

satisfying the following properties:

(1) I_H^G is an $R(G)$-module homomorphism, when $KK_H^0(A,B)$ is regarded as an $R(G)$-module in the usual way.

(2) If A, B, and C are separable H-algebras, and $\eta \in KK_H^0(A,B)$, $\lambda \in KK_H^0(B,C)$, then $I_H^G(\eta\lambda) = I_H^G(\eta)\, I_H^G(\lambda)$.

(3) If A and B are separable G-algebras, and D is a separable H-algebra, then $I_H^G \cdot \tau_D \cdot Res_H^G = \tau_{C(G\times_H D)}$ as maps from $KK_G^0(A,B)$ to $KK_G^0(C(G\times_H(A\otimes D)), C(G\times_H(B\otimes D)))$.

(4) Induction in stages: If K is a closed subgroup of H, and if A and B are separable K-algebras, then $I_H^G \cdot I_K^H = I_K^G$ as maps from $KK_K^0(A,B)$ to $KK_G^0(C(G\times_K A), C(G\times_K B))$.

(5) If A is a separable H-algebra then $I_H^G(1_A) = 1_{C(G\times_H A)}$.

Proof. Recall that $KK_G^0(A,B)$ is defined in terms of prequasihomomorphisms from A to $K\otimes B$, where K is the algebra of compact operators on the Hilbert space $V = l^2 \otimes L^2(G)$, and where G acts on V by the regular representation λ in the second factor, and on $K = K(V)$ by conjugation. (See 3.1.5 and 3.1.6.) Restricting to the subgroup H, we find that $L^2(G) \simeq L^2(G/H) \otimes L^2(H)$, where H acts trivially on the first factor and via the regular representation on the second factor. Therefore, as a representation space of H, we have

$$V \simeq l^2 \otimes L^2(G/H) \otimes L^2(H) \simeq l^2 \otimes L^2(H).$$

Thus, the restriction to H of the universal representation of G on V is the universal representation of H. Consequently, whenever we consider an H-equivariant prequasihomomorphism

$$\Phi = (\varphi, \psi) : A \longrightarrow E \rhd J \longrightarrow K \otimes B, \tag{*}$$

we may assume that the action of H on K is the restriction of an action of G on K such that K is isomorphic to the algebra of compact operators on the universal representation space of G.

Given the prequasihomomorphism (*), we define $I_H^G([\Phi])$ to be the equivalence class of the following prequasihomomorphism, obtained by applying the functor $C(G\times_H \cdot)$ everywhere:

$$(\tilde{\varphi}, \tilde{\psi}) : C(G\times_H A) \longrightarrow C(G\times_H E) \rhd C(G\times_H J) \longrightarrow C(G\times_H(K\otimes B)) \simeq K\otimes C(G\times_H B).$$

The last isomorphism follows from lemma 5.1.7, since, by the previous paragraph, K may be considered to be a G-algebra.

We show that I_H^G is well defined. Since the functor $C(G\times_H \cdot)$ sends H-homotopies to G-homotopies, and also behaves properly with respect to subalgebras and ideals generated by H-invariant subsets, we need only show that $I_H^G([\Phi])$ doesn't depend on the choice of G-action on K. Accordingly, let $K = K(V)$, where $V = l^2 \otimes L^2(H)$, let λ be the action of H on V, and let σ and π be two representations of G on V, each equivalent to the universal representation of G and each restricting to λ on H.

Then there is a unitary $u \in L(V)$ such that $u\sigma_g u^* = \pi_g$ for all $g \in G$. In particular, $u\lambda_h u^* = \lambda_h$ for all $h \in H$. Thus, u is H-invariant, and as in the proof of lemma 3.1.7 (1), there is a continuous path $t \longrightarrow u_t$ of H-invariant unitaries with $u_0 = 1$ and $u_1 = u$. Let $\vartheta_t(a) = u_t^* a u_t$ for $a \in K$. Then an application of the functor $C(G\times_H \cdot)$ to the homotopy of prequasihomomorphisms

$$\Phi_t = (\varphi, \psi): A \longrightarrow E \triangleright J \longrightarrow K\otimes B \xrightarrow{\vartheta_t \otimes id_B} K\otimes B$$

yields a homotopy between the two possible choices of the prequasihomomorphism representing $I_H^G([\Phi])$, where the action of G on the first K comes from the representation $g \longrightarrow u_t \sigma_g u_g^*$ and the action on the second K comes from σ. Thus I_H^G is well defined.

It is now obvious that I_H^G is a homomorphism of abelian groups. Also, property (2) is obvious from the functoriality of the twisted product, property (5) is clear, and property (4) follows from the canonical isomorphism

$$C(G\times_H C(H\times_K A)) \simeq C(G\times_K A)$$

for K-algebras A. (See [56], section 5, lemma 1. We do not prove this last fact, since its proof is very similar to the induction in stages calculation we have already done in the proof of proposition 5.1.3.)

To prove (1), let

$$\Phi = (\varphi, \psi): A \longrightarrow E \triangleright J \longrightarrow K\otimes B$$

be an H-equivariant prequasihomomorphism, and let V be a finite dimensional representation space of G. Then $[V]\cdot[\Phi]$ is represented by the prequasihomomorphism

$$(\varphi, \psi): A \longrightarrow E \triangleright J \longrightarrow K\otimes B \xrightarrow{\mu} L(V)\otimes K\otimes B \simeq K\otimes B,$$

where $\mu(a) = 1_V \otimes a$. Applying $C(G\times_H \cdot)$ and using lemma 5.1.7, we obtain

$$(\tilde{\varphi}, \tilde{\psi}): C(G\times_H A) \longrightarrow C(G\times_H E) \triangleright C(G\times_H J) \longrightarrow K\otimes C(G\times_H B)$$
$$\longrightarrow L(V)\otimes K\otimes C(G\times_H B) \simeq K\otimes C(G\times_H B),$$

which also represents the class $[V]\cdot I_H^G([\Phi])$. This proves (1).

To see (3), let A and B be G-algebras, and let

$$\Phi = (\varphi, \psi): A \longrightarrow E \triangleright J \longrightarrow K\otimes B$$

be a G-equivariant prequasihomomorphism. Then $I_H^G \circ \tau_D \circ Res_H^G([\Phi])$ is represented by the prequasihomomorphism

$$((\varphi\otimes id_D)^\sim, (\psi\otimes id_D)^\sim): C(G\times_H (A\otimes D)) \longrightarrow C(G\times_H (E\otimes D))$$
$$\triangleright\ C(G\times_H (J\otimes D)) \longrightarrow K\otimes C(G\times_H (B\otimes D)).$$

Since A, E, J, and B are all G-algebras, and the original maps are all G-equivariant, lemma 5.1.7 shows that this is the same as the prequasihomomorphism

$$(\varphi,\psi)\otimes id_{C(G\times_H D)}: A\otimes C(G\times_H D) \longrightarrow E\otimes C(G\times_H D)$$
$$\triangleright J\otimes C(G\times_H D) \longrightarrow K\otimes B\otimes C(G\times_H D),$$

whose class is just $\tau_{C(G\times_H D)}([\Phi])$. Q.E.D.

We are now almost in a position to define an induction homomorphism similar to the induction we have already defined in K-theory. We still, however, need to define certain homomorphisms.

5.1.9 Definition. Let G be a compact group, and let H be a closed subgroup, which for convenience will be assumed to have finite index in G. Let $V = l^2(G/H)$ carry the obvious representation of G, namely $u_g\,\xi(g_0H) = \xi(g^{-1}g_0H)$ for $g, g_0 \in G$. For a G-algebra A, we define three natural transformations. For the first two, we identify $C(G\times_H A)$ with $C(G/H)\otimes A$ as in lemma 5.1.7 (with $B = \mathbf{C}$). Define $i_A : A \longrightarrow C(G\times_H A)$ by $i_A(a) = 1\otimes a$ and $\pi_A : C(G\times_H A) \longrightarrow L(V)\otimes A$ by $\pi_A(f\otimes a) = m(f)\otimes a$, where, for $f \in C(G/H)$, we let $m(f) \in L(V)$ be the operator of multiplication by f. Finally, we let $\xi \in V$ be the sum of all the standard basis vectors ξ_{gH} of V, and let $p \in L(V)$ be the projection onto $\mathbf{C}\xi$. Then $d_A : A \longrightarrow L(V)\otimes A$ is defined by $d_A(a) = p\otimes a$. If the groups G and H need to be specified in the notation, we will write $i_{G,H,A}$, $\pi_{G,H,A}$, and $d_{G,H,A}$ for i_A, π_A, and d_A.

5.1.10 Definition. Let G be a finite group, let H be a subgroup of G, and let (G,A,α) and (G,B,β) be separable G-algebras. Recall from definition 3.2.13 that if φ is an equivariant homomorphism, then $[\varphi]$ denotes the class in KK-theory defined by φ. Define

$$Ind_H^G: KK_H^0(A,B) \longrightarrow KK_G^0(A,B)$$

by

$$Ind_H^G(\lambda) = [i_A]\, I_H^G(\lambda)\, [\pi_B]\, [d_B]^{-1}.$$

(The proof of theorem 3.5.10, stability for KK-theory, shows that $[d_B]$ is in fact invertible.)

To prove that this map has the correct properties, we need a lemma concerning the properties of the maps introduced in definition 5.1.9.

5.1.11 Lemma. Let G be a compact group, and let H be a closed subgroup of finite index in G. Then the maps of definition 5.1.9 satisfy:

(1) Transitivity: If A is a G-algebra and K is a closed subgroup of H of finite index, then

(1a) $I_H^G(i_{H,K,A})\cdot i_{G,H,A} = i_{G,K,A}$.

(1b) If A is separable, then

$$I_H^G([\pi_{H,K,A}][d_{H,K,A}]^{-1})[\pi_{G,H,A}][d_{G,H,A}]^{-1} = [\pi_{G,K,A}][d_{G,K,A}]^{-1}$$

in $KK_G^0(C(G\times_K A), A)$.

(2) Naturality: If A and B are separable G-algebras, and $\lambda \in KK_G^0(A, B)$, then, with $V = l^2(G/H)$, we have

(2a) $\lambda \cdot [i_B] = [i_A] \cdot \tau_{C(G/H)}(\lambda)$.

(2b) $\tau_{C(G/H)}(\lambda) \cdot [\pi_B] = [\pi_A]\tau_{L(V)}(\lambda)$.

(2c) $\lambda \cdot [d_B] = [d_A] \cdot \tau_{L(V)}(\lambda)$.

Proof. The statements in (2) follow quickly from the naturality of the maps i_A, π_A, and d_A for homomorphisms between C^*-algebras. We therefore prove only (1). In (1), note that lemma 5.1.7, together with some easy calculations, implies that the G-algebra A can be factored out of everything, and it is therefore sufficient to consider the case $A = \mathbf{C}$. We will suppress A in the notation when $A = \mathbf{C}$.

(1a) The map $i_{G,H}$ is the inclusion of $\mathbf{C}$ in $C(G/H)$ via $i_{G,H}(1) = 1$. A similar statement holds for $i_{H,K}$, so that $I_H^G(i_{H,K})$ sends $1 \in C(G/H)$ to $1 \in C(G\times_H C(H/K))$, which is the same as $C(G/K)$. So $I_H^G(i_{H,K})\cdot i_{G,H}$ sends $1 \in \mathbf{C}$ to $1 \in C(G/K)$, and is therefore equal to $i_{G,K}$.

(1b) We start by constructing a commutative diagram of G-algebras:

$$\begin{array}{ccc} C(G/H) & \xrightarrow{\tilde{d}_{H,K}} & C(G\times_H L(l^2(H/K)) \\ \pi_{G,H} \downarrow & & \downarrow \varphi \\ L(l^2(G/H)) & \xrightarrow[\psi]{} & L(l^2(G/K)) \end{array} \qquad (*)$$

Here the G-actions are in all cases the obvious ones. The map $\tilde{d}_{H,K}$ is just $I_H^G(d_{H,K})$. The other two maps require more effort to define. Let u be the obvious representation of H on $l^2(H/K)$, and let $\tilde{u}$ be the induced representation of G on the space

$$W = \{\xi : G \longrightarrow l^2(H/K) : u_h\, \xi(gh) = \xi(g) \text{ for } g \in G \text{ and } h \in H\}.$$

Observe that $W \simeq l^2(G/K)$ via the unitary w given by $(w\xi)(gK) = \xi(g)(K)$ for $\xi \in W$ and $g \in G$, and that w intertwines the representation $\tilde{u}$ with the usual representation of G on $l^2(G/K)$. Define

$$\sigma : C(G\times_H L(l^2(H,K))) \longrightarrow L(W)$$

by $(\sigma(f)\xi)(g) = f(g)(\xi(g))$ for $f \in C(G\times_H L(l^2(H/K)))$, $\xi \in W$, and $g \in G$. Then σ is easily seen to be equivariant, and we define φ by $\varphi(f) = w\,\sigma(f)\,w^*$. Next, define an isometry x from $l^2(G/H)$ to $l^2(G/K)$ by

$$x(\xi_{gH}) = \frac{1}{\sqrt{n}} \sum_{hK\subset gH} \xi_{hK},$$

where n is the index of K in H, and ξ_{gH} and ξ_{hK} are the standard basis vectors in $l^2(G/H)$ and $l^2(G/K)$. Then set $\psi(a) = xax^*$ for $a \in L(l^2(G/H))$.

We now show that the diagram commutes. If $f \in C(G/H)$, then clearly $\psi \cdot \pi_{G,H}(f)$ is an operator on $l^2(G/K)$ sending $x(\xi_{gH})$ to $f(gH)x(\xi_{gH})$ and vanishing on the orthogonal complement of the range of x. Now $w^*x(\xi_{gH})$ is the element of W given by

$$(w^*x(\xi_{gH}))(g_0)(hK) = \begin{cases} \frac{1}{\sqrt{n}} & g_0H \subset gH \\ 0 & \text{otherwise}. \end{cases}$$

A calculation now shows that $\sigma \cdot \tilde{d}_{H,K}(f)$ multiplies this vector by $f(gH)$ and vanishes on the orthogonal complement of all of the $w^*x(\xi_{gH})$. This shows that the diagram (*) commutes.

We also observe that $\psi \cdot d_{G,H} = d_{G,K}$ and that $\varphi \cdot I_H^G(\pi_{H,K}) = \pi_{G,K}$. (The first is immediate, while the second requires a short computation.) We therefore obtain

$$\begin{aligned} I_H^G([\pi_{H,K}][d_{H,K}]^{-1})[\pi_{G,H}][d_{G,H}]^{-1} &= I_H^G([\pi_{H,K}])[\tilde{d}_{H,K}]^{-1}[\pi_{G,H}][d_{G,H}]^{-1} \\ &= I_H^G([\pi_{H,K}])[\varphi][\psi]^{-1}[d_{G,H}]^{-1} = [\pi_{G,K}][d_{G,K}]^{-1}. \end{aligned}$$

Here the first step follows from proposition 5.1.8 (2), while the second step is the commutativity of (*). Also notice that $[\psi] = [d_{G,K}][d_{G,H}]^{-1}$ and therefore is in fact invertible.

Q.E.D.

5.1.12 Proposition. Let G be a finite group, and let H be a subgroup of G. Then the map Ind_H^G of definition 5.1.10 is a natural $R(G)$-module homomorphism satisfying the following properties:

(1) Induction in stages: If K is a subgroup of H, then $Ind_H^G \cdot Ind_K^H = Ind_K^G$.

(2) If A, B, and C are separable G-algebras, then $Ind_H^G(\eta \cdot Res_H^G(\lambda)) = Ind_H^G(\eta) \cdot \lambda$ for $\eta \in KK_H^0(A, B)$ and $\lambda \in KK_G^0(B, C)$, and $Ind_H^G(Res_H^G(\eta) \cdot \lambda) = \eta \cdot Ind_H^G(\lambda)$ for $\eta \in KK_G^0(A, B)$ and $\lambda \in KK_H^0(B, C)$.

Proof. That Ind_H^G is an $R(G)$-module homomorphism follows from the $R(G)$-bilinearity of the product and the fact that I_H^G is an $R(G)$-module homomorphism. Naturality in this case is a special case of (2).

(1) We have, for $\lambda \in KK_K^0(A, B)$,

$$\begin{aligned} &Ind_H^G(Ind_K^H(\lambda)) \\ &= [i_{G,H,A}]\, I_H^G([i_{H,K,A}]\, I_K^H(\lambda)[\pi_{H,K,A}][d_{H,K,A}]^{-1})[\pi_{G,H,A}][d_{G,H,A}]^{-1} \\ &= [i_{G,K,A}]\, I_H^G(I_K^H(\lambda))[\pi_{G,K,A}][d_{G,K,A}]^{-1} = Ind_K^G(\lambda), \end{aligned}$$

where the second step follows from proposition 5.1.8 (2) and lemma 5.1.11 (1), and the last step follows from proposition 5.1.8 (4).

(2) First, let $\eta \in KK_H^0(A,B)$ and let $\lambda \in KK_G^0(B,C)$. Then

$$Ind_H^G(\eta \cdot Res_H^G(\lambda)) = [i_A] I_H^G(\eta) I_H^G(Res_H^G(\lambda)) [\pi_C][d_C]^{-1}$$
$$= [i_A] I_H^G(\eta) \tau_{C(G/H)}(\lambda)[\pi_C][d_C]^{-1} = [i_A] I_H^G(\eta)[\pi_B][d_B]^{-1} \cdot \lambda = Ind_H^G(\eta) \cdot \lambda .$$

Here, the second step is a consequence of proposition 5.1.8 (3), and the third step follows from lemma 5.1.11 (2b) and (2c). The calculation for the other part of (2) is similar, using part (2a) of lemma 5.1.11 instead of parts (2b) and (2c). Q.E.D.

Similar results, of course, hold for the general form of the product, from $KK_G^*(A_1, B_1 \otimes D) \times KK_G^*(D \otimes A_2, B_2)$ to $KK_G^*(A_1 \otimes A_2, B_1 \otimes B_2)$.

5.2. Actions Whose Restrictions to Cyclic Subgroups are K-Free

Throughout this section, G is a finite group, (G,A,α) is a G-algebra, and C is the set of cyclic subgroups of G. We start out with:

5.2.1 Lemma. Let G have order n. Define

$$r : K_*^G(A) \longrightarrow \bigoplus_{S \in C} K_*^S(A)$$

by $r(\eta)_S = Res_S^G(\eta)$. Then there is an $R(G)$-module homomorphism

$$i : \bigoplus_{S \in C} K_*^G(A) \longrightarrow K_*^G(A)$$

such that $i \cdot r$ is multiplication by n on $K_*^G(A)$.

Proof. By Artin's theorem (propositions 27 and 28 in [99]), there are elements $\chi_S \in R(S)$ for $S \in C$ such that, in $R(G)$,

$$\sum_{S \in C} Ind_S^G(\chi_S) = n . \qquad (*)$$

(Here of course n is considered to be the element of $R(G)$ corresponding to the n dimensional trivial representation of G). Define i by

$$i((\eta_S)_{S \in C}) = \sum_{S \in C} Ind_S^G(\chi_S \eta_S) .$$

Then i is an $R(G)$-module homomorphism by proposition 5.1.3. Furthermore, if $\eta \in K_*^G(A)$, we have

$$i \cdot r(\eta) = \sum_{S \in C} Ind_S^G(\chi_S Res_S^G(\eta)) = \sum_{S \in C} Ind_S^G(\chi_S)\eta = n\eta$$

by proposition 5.1.3 (2) and by (*). Q.E.D.

5.2.2 Proposition. Suppose $K_*^G(A)$ is torsion-free as an abelian group. If for each $S \in C$, the restriction $\alpha|_S$ has locally discrete K-theory, then α has locally discrete K-theory.

Proof. We first show that if S is a cyclic subgroup of G and $\eta \in K_*^S(A)$, then some power of $I(G)$ annihilates η. Let I be the ideal in $R(S)$ generated by $Res_S^G[I(G)]$. Then it is enough to show that some power of I annihilates η. But $I \subset I(H)$ by lemma 4.3.1, and $I(H)^m$ annihilates η for some m, so that I^m annihilates η. It now follows from lemma 1.2.2 that $K_*^S(A)_P = 0$ for all primes P in $R(G)$ not containing $I(G)$.

Let i and r be the homomorphisms of the previous lemma. Then $i \cdot r(\eta) = n\eta$ for $\eta \in K_*^G(A)$, where n is the order of G. Since $K_*^G(A)$ is torsion-free, $i \cdot r$ is injective. Since localization is exact, $(i \cdot r)_P : K_*^G(A)_P \longrightarrow K_*^G(A)_P$ is also injective for any prime P. If P does not contain $I(G)$, the first paragraph of the proof shows that

$$\bigoplus_{S \in C} K_*^S(A)_P = 0 .$$

Therefore $(i \cdot r)_P = i_P \cdot r_P = 0$. Since $(i \cdot r)_P$ is injective, we conclude $K_*^G(A)_P = 0$, as required. Q.E.D.

We will see in chapter 9 that this proposition is applicable to "good" actions on AF algebras. In general, however, it is not very useful. There are two alternative approaches available, the first of which is a modification of our definitions.

5.2.3 Definition. We say that α is K-free up to torsion if for each prime ideal P in $R(G)$ not containing $I(G)$, and each G-invariant ideal I in A, the $R(G)_P$-module $K_*^G(I)_P$ is a torsion group as an abelian group, that is, every element has finite order. We say that α is totally K-free up to torsion if $\alpha|_H$ is K-free up to torsion for every subgroup H.

In order to use the definition, we need an analog of lemma 1.2.2.

5.2.4 Lemma. Let R be a Noetherian ring, let I be a prime ideal in R, and let M be an R-module. Then the following are equivalent:

(1) M_P is a torsion group (in the sense used in the previous definition) for every prime ideal $P \subset R$ such that $P \not\supset I$.

(2) For every $\eta \in M$ there is $n \in \mathbf{Z}$ such that $I^n \eta$ is a torsion group.

Proof. We proceed as in the proof of lemma 1.2.2.

(1)$\Longrightarrow$(2). Let $\eta \in M$ and let

$$J = \{r \in I : mr\eta = 0 \text{ for some } m \in \mathbf{Z},\ m \neq 0\} .$$

Again, $rad(J)$ is the intersection of all prime ideals containing J, so we show that if a prime P does not contain I then P does not contain J. Let P be such a prime. Then by hypothesis there is $m \in \mathbf{Z}$, $m \neq 0$, such that the image of $m\eta$ is zero in M_P. Therefore there is $r \notin P$ such that $mr\eta = 0$, and replacing r by rs, where $s \in I$ but $s \notin P$, we may assume $r \in I$. Therefore $r \in J$, and P does not contain J. So $rad(J) = I$, and thus by [5], proposition 7.14, $I^n \subset J$ for some n. Therefore $I^n\eta$ is a torsion group.

(2)$\Rightarrow$(1). Let $P \subset R$ be a prime ideal with $P \not\supset I$. Let $r \in I$, $r \notin P$. Then for $\eta \in M$ there are $m, n \in \mathbf{Z}$, $m \neq 0$, such that $mr^n\eta = 0$. Since $r^n \notin P$, it follows that η is a torsion element in M_P. Q.E.D.

Accordingly, we obtain:

5.2.5 Proposition. If each restriction $\alpha|_S$ for $S \in C$ is K-free up to torsion, then α is totally K-free up to torsion.

Proof. The proof is essentially the same as the proof of proposition 5.2.2. We first prove that $K_*^G(A)_P$ is a torsion group for each prime ideal P not containing $I(G)$. To do this, we use the method of the proof of 5.2.2 to conclude that for each S and each $\eta \in K_*^S(A)$, there is n such that $I(G)^n\eta$ is a torsion group (using lemma 5.2.4 instead of lemma 1.2.2), and therefore that $K_*^S(A)_P$ is a torsion group for all cyclic subgroups S (again using lemma 5.2.4). Consequently the image of the localized map $(i{\cdot}r)_P$ in $K_*^G(A)_P$ is a torsion group. Since it is just $n \cdot K_*^G(A)_P$, we conclude also that $K_*^G(A)_P$ is a torsion group, as claimed.

It is clear that the hypotheses of the proposition hold for any subgroup H of G and any H-invariant ideal in A. (A cyclic subgroup of H is a cyclic subgroup of G). Therefore α is totally K-free up to torsion, as desired. Q.E.D.

The methods of [94] can presumably be used to extend results of this sort to arbitrary compact Lie groups, at the cost of introducing elliptic operators and possibly using KK-theory. We will not pursue this here, and we will also not develop the general properties of K-freeness up to torsion, although it is clear that many of the results of chapter 4 are also valid for it. Instead, we consider a different way to improve proposition 5.2.2, which involves restricting the class of groups we consider. Recall that a p-group is a group in which every element has order equal to some power, depending on the element, of the prime number p.

5.2.6 Theorem. Assume that G is a p-group. If $\alpha|_S$ is K-free for every cyclic subgroup S of G, then α is totally K-free.

Proof. Let $P \subset R(G)$ be a prime ideal not containing $I(G)$. Exactly as in the proof of proposition 5.2.2, we conclude that multiplication by the order of G is the zero map on $K_*^G(A)_P$.

We are now going to prove that p is invertible in $R(G)_P$, that is, $p \notin P$. Since the order of G has the form p^n, it will follow that $K_*^G(A)_P = 0$. Let S be the support of P, which is a cyclic subgroup of G such that $P = (Res_S^G)^{-1}$ [Q] for some prime ideal $Q \subset R(S)$. (See [97], proposition 3.7.) Since S is abelian, we can consider the set

$\Gamma = \{\tau \in \widehat{S}: 1-\tau \in Q\}$. If $\sigma, \tau \in \widehat{S}$ and $1-\sigma$, $1-\tau \in Q$, then $1-\sigma\tau = 1-\sigma+\sigma(1-\tau) \in Q$. Therefore Γ is a subgroup of $\widehat{S}$. It is not all of $\widehat{S}$, since Q does not contain $I(S)$. ($Q \supset I(S)$ would imply $P \supset I(G)$.) Since $\widehat{S}$ is a cyclic group of prime power order, there is $\tau \in \widehat{S}$ such that $\tau \notin \Gamma$ but $\tau^p \in \Gamma$.

There are now two cases. If $p \neq 2$, then $(1-\tau)^p \notin Q$, since Q is prime and $1-\tau \notin Q$. Since $1-\tau^p \in Q$, we obtain $(1-\tau)^p - (1-\tau^p) \notin Q$. Since the binomial coefficients $\binom{p}{k}$ for $1 \le k \le p-1$ are all divisible by p, it follows that $(1-\tau)^p - (1-\tau^p) = p\eta$ for some $\eta \in R(S)$. Since $p\eta \notin Q$, we must have $p \notin Q$.

If $p = 2$, we notice that $(1-\tau)(1+\tau) = 1-\tau^2 \in Q$. Since Q is prime and $1-\tau \notin Q$, it follows $1+\tau \in Q$. Therefore $4\tau = (1+\tau)^2 - (1-\tau)^2 \notin Q$ (as $(1-\tau)^2 \notin Q$), and it follows that $2 \notin Q$.

Thus in either case $p \notin Q$. Therefore $p \notin P = (Res_S^G)^{-1}[Q]$, since $Res_S^G(p) = p$. It follows that p is invertible in $R(G)_P$, so, as mentioned at the beginning of the proof, it follows that $K_*^G(A)_P = 0$. We have so far shown that α has locally discrete K-theory. But the hypotheses of the theorem clearly hold for any subgroup H of G and any H-invariant ideal I of A. Therefore α is totally K-free, as desired. Q.E.D.

5.2.7 Corollary. (Compare proposition 4.2.10 and remark 4.2.11.) Let (G,A,α) be a G-algebra, where G is a p-group. Let K be the algebra of compact operators on some Hilbert space, and let $\beta: G \longrightarrow Aut(K)$ be any action. Then the diagonal action $\beta\otimes\alpha$ on $K\otimes A$ is totally K-free if and only if α is.

Proof. It follows immediately from the theorem that an action of G is totally K-free if and only if its restrictions to cyclic subgroups are also totally K-free. It is therefore enough to prove the corollary when G is cyclic. By lemma 4.2.12, the action of a cyclic group on K is inner, so the corollary follows from proposition 4.2.10. Q.E.D.

This corollary certainly ought to hold for arbitrary compact Lie groups G, not just finite p-groups, especially in view of proposition 4.4.10. However, we have been unable to prove it in this generality.

5.2.8 Remark. If, in theorem 5.2.6, we are willing to use elementary subgroups instead of cyclic subgroups, we can do better. Recall ([99], page 74) that a group is called elementary if it is the product of a p-group and a cyclic group of order relatively prime to p. Instead of using Artin's theorem, we use Brauer's theorem ([99], theorem 19), which says that if E is the set of elementary subgroups of G, then there are elements $\chi_S \in R(S)$ for $S \in E$ such that

$$\sum_{S \in E} Ind_S^G(\chi) = 1 \quad \text{in } R(G).$$

Accordingly, the restriction map

$$r : K_*^G(A) \longrightarrow \bigoplus_{S \in E} K_*^S(A)$$

actually has a right inverse (by the method of proof of lemma 5.2.1), and the method of proof of the other propositions in this section then shows that if $\alpha|_S$ is K-free for every elementary subgroup S of G, then α is totally K-free. Thus, the only obstruction to being able to prove proposition 5.2.5 for K-freeness rather than K-freeness up to torsion is the gap between p-groups and elementary groups. We have neither theorems nor counterexamples; all we can say is that the method of the previous theorem does not work even for the simplest case, namely $(\mathbf{Z}/2\mathbf{Z})^2 \times \mathbf{Z}/3\mathbf{Z}$.

We now turn to the KK-theory analogs of these results. From now on, A will be assumed to be separable. Since the proofs of the following results are similar to the case of K-theory, we will be rather sketchy. The first step is the analog of lemma 5.2.1.

5.2.9 Lemma. Let G have order n. Define

$$r : KK_G^0(A,A) \longrightarrow \bigoplus_{S \in C} KK_S^0(A,A)$$

by $r(\eta)_S = Res_S^G(\eta)$. Then there is an $R(G)$-module homomorphism

$$i : \bigoplus_{S \in C} KK_S^0(A,A) \longrightarrow KK_G^0(A,A)$$

such that $i \circ r$ is multiplication by n on $KK_G^0(A,A)$.

Proof. The proof is the same as that of lemma 5.2.1, using the same elements $\chi_S \in R(S)$, except that the equation $Ind_S^G(\chi_S\, Res_S^G(\eta)) = Ind_S^G(\chi_S)\eta$ is justified by the remark following proposition 5.1.12, rather than proposition 5.1.2. Q.E.D.

The analog of proposition 5.2.2 will also be used on AF algebras. It is:

5.2.10 Proposition. Suppose $KK_G^0(A,A)$ is torsion-free as an abelian group. If for each $S \in C$, the restriction $\alpha|_S$ has discrete KK-theory, then α has discrete KK-theory.

Proof. The proof is essentially the same as the proof of proposition 5.2.2, and is omitted. Q.E.D.

We similarly have an analog of proposition 5.2.5, and the following analog of theorem 5.2.6, whose proof we also omit.

5.2.11 Theorem. Assume that G is a p-group. If $\alpha|_S$ is KK-free for every cyclic subgroup S of G, then α is KK-free.

5.3. The Ideal Decomposition Lemma and the KK-Subgroup Theorem

We start off by showing that if an action has discrete KK-theory then so does its restriction to any closed subgroup. While extremely easy, this step fails for K-theory, because $K_0^G(A)$ is not in general a ring. (See example 4.2.3.)

5.3.1 Lemma. Let (G,A,α) be a G-algebra, with G compact Lie and A separable. Let H be a closed subgroup of G. If α has discrete KK-theory, then so does $\alpha|_H$.

Proof. By proposition 4.4.4 there is n such that $I(G)^n \cdot 1 = 0$ in $KK_G^0(A,A)$. Let I be the ideal in $R(H)$ generated by the restriction of $I(G)$. Then by restricting this equation to H we obtain $I^n \cdot 1 = 0$ in $KK_H^0(A,A)$. By lemma 4.3.1 there is m such that $I(H)^m \subset I$. Therefore $I(H)^{mn} \cdot 1 = 0$ in $KK_H^0(A,A)$, so $\alpha|_H$ has discrete KK-theory.

Q.E.D.

Using this lemma, we can obviously prove that if α is KK-free, then $KK_H^0(I,I)_P = 0$ for every prime P of $R(H)$ not containing $I(H)$ and every G-invariant ideal I of A. To show that $\alpha|_H$ is KK-free, we must, however, obtain this result for all H-invariant ideals I in A. The main technical lemma in this section is a tool for getting $KK_H^0(I,I)_P = 0$ for all H-invariant ideals when it is known only for G-invariant ideals, provided that G is finite.

Before stating and proving it, we illustrate what happens in an especially simple case, namely when the index of H in G is 2. Let I be an H-invariant ideal in A, and let g be any element of G not in H. The ideals $I \cap \alpha_g[I]$ and $I + \alpha_g[I]$ are both G-invariant (H is normal since it has index 2), and therefore $(\alpha|_H)|_{I\cap\alpha_g[I]}$ and $(\alpha|_H)|_{I+\alpha_g[I]}$ have discrete KK-theory. Furthermore,

$$(I + \alpha_g[I])/(I \cap \alpha_g[I]) = \left[I/(I\cap\alpha_g[I])\right] \oplus \left[\alpha_g[I]/(I\cap\alpha_g[I])\right].$$

Since the action of H on the direct sum on the right hand side of the equation has discrete KK-theory, it follows that the action on each of the summands also has discrete KK-theory. (Note that $KK_G^0(R\oplus S, R\oplus S)$ is a direct sum of 4 terms, one of which is $KK_G^0(R,R)$.) It then follows that $(\alpha|_H)|_I$ has discrete KK-theory, by lemma 4.4.7. The Ideal Decomposition Lemma is the technical tool needed to carry out this argument in the general case. It will also play a crucial role in the next chapter. We first put on record the following important fact, which we will frequently use without comment in the rest of this section:

5.3.2 Lemma. Let I, J, and L be ideals in a C^*-algebra A. Then

$$(I+J)\cap L = (I\cap L)+(J\cap L) \quad \text{and} \quad (I\cap J)+L = (I+L)\cap(J+L).$$

Proof. We are assuming as usual that all ideals are closed. The ideals in A are in one-to-one order preserving correspondence with the open subsets of $Prim(A)$, in such a way that intersections of ideals correspond to intersections of sets, and sums of ideals correspond to unions of sets. Since the analogous properties hold for sets, the lemma is proved. Q.E.D.

Of course, in arbitrary rings these two distributive laws generally do not hold.

5.3.3 Ideal Decomposition Lemma. Let (G,A,α) be a G-algebra, with G finite. Let I be an arbitrary ideal in A (recall that we assume all ideals are closed). For each subset S of G containing the identity, define ideals I_S and $I_{\bar{S}}$ in A by the formulas

$$I_S = \sum_{g \in G} \alpha_g \left[\bigcap_{h \in S} \alpha_h [I] \right] \quad \text{and} \quad I_{\bar{S}} = \sum_{g \notin S} I_{S \cup \{g\}} .$$

Then the following properties hold:

(1) I_S and $I_{\bar{S}}$ are G-invariant.

(2) If $S \subsetneq T$ then $I_T \subset I_{\bar{S}} \subset I_S$.

(3) $I_G \subset I \subset I_{\{e\}}$.

(4) Let S and T be disjoint subsets of G, with $e \in S$, and let $k \notin S \cup T$. Let $L = I_{S \cup \{k\}} \cap I$. Then

(4a) $I_{S \cup \{k\}} \cap \sum_{g \in T} I_{S \cup \{g\}} = \sum_{g \in T} L_{S \cup \{g\}}$, and

(4b) $(I_{S \cup \{k\}} \cap I) \cap \left(\sum_{g \in T} I_{S \cup \{g\}} \cap I \right) = \sum_{g \in T} L_{S \cup \{g\}} \cap L$.

(5) For each S there is a subgroup H_S of G and an H_S-invariant ideal $M \subset I_S / I_{\bar{S}}$ such that $I_S / I_{\bar{S}} \simeq C(G \times_{H_S} M)$. In particular, if R is a set of coset representatives of H_S in G, then

$$I_S / I_{\bar{S}} \simeq \bigoplus_{g \in R} \alpha_g [M],$$

and under this identification,

$$((I_S \cap I) + I_{\bar{S}}) / I_{\bar{S}} = \bigoplus_{g \in R, g^{-1} \in S} \alpha_g [M] .$$

One has $H_S = G$ if and only if $S = G$.

(6) If J is any G-invariant ideal of A, the $(I + J)_S = I_S + J$ and $(I \cap J)_S = I_S \cap J$. Similarly $(I + J)_{\bar{S}} = I_{\bar{S}} + J$ and $(I \cap J)_{\bar{S}} = I_{\bar{S}} \cap J$.

The ideals I_S of this lemma are the most obvious G-invariant ideals related to I after $I_G = \bigcap_{g \in G} \alpha_g [I]$ and $I_{\{e\}} = \sum_{g \in G} \alpha_g [I]$. (Notice that the summation over the images under the α_g is superfluous in the definition of I_G, since the intersection is clearly G-invariant.) One might hope that for $g \notin S$ the image of $I \cap I_S$ in $I_S / I_{S \cup \{g\}}$ would be easy to deal with. Unfortunately, that is not the case — one must use the sum of all $I_{S \cup \{g\}}$, which we have called $I_{\bar{S}}$, in order to obtain a reasonable result. What we do get then is very satisfying: condition (5) essentially says that the action of G on $I_S / I_{\bar{S}}$ has a system of imprimitivity based on G/H such that the image of $I \cap I_S$ is the sum of certain of the subspaces in the associated decomposition. We are then able to use the approach in the discussion following lemma 5.3.1. Unfortunately one must now also get a handle on the ideals $I_{\bar{S}}$, which is what condition (4) is used for. (Actually, we only use (4b) here; (4a) will be needed for applications to tensor products.)

Before proving this lemma, we will illustrate how it is used. We let G be a finite group, H a subgroup, and we let h_* be a homology theory (in the sense of [96], appropriately modified) on either all H-algebras or separable nuclear H-algebras. (In applications, h_* will always be either $K_*^H(\cdot)_P$ or $KK_H^*(A,\cdot)_P$ for some fixed H-algebra A, where P is some prime ideal is $R(H)$.) We need a preliminary lemma, which has nothing to do with G or finiteness of the group.

5.3.4 Lemma. Let (H,A,α) be a (separable nuclear) H-algebra, and let h_* be a homology theory defined on (separable nuclear) H-algebras. Let I and J be H-invariant ideals in A. If $h_*(I)$, $h_*(J)$ and $h_*(I\cap J)$ are all zero, then $h_*(I+J)=0$.

Proof. Consideration of the long exact sequence for h_* corresponding to the equivariant exact sequence

$$0 \longrightarrow I\cap J \longrightarrow I \longrightarrow I/(I\cap J) \longrightarrow 0$$

implies that $h_*(I/I\cap J))=0$. Since there is a canonical isomorphism $I/(I\cap J)\simeq(I+J)/J$, we obtain from the long exact sequence for h_* corresponding to

$$0 \longrightarrow J \longrightarrow I+J \longrightarrow (I+J)/J \longrightarrow 0$$

the fact that $h_*(I+J)=0$. Q.E.D.

5.3.5 Lemma. Let G be a finite group, let H be a subgroup of G, and let (G,A,α) be a (separable nuclear) G-algebra. Let h_* be a homology theory defined on (separable nuclear) H-algebras, and suppose that $h_*(I)=0$ for all G-invariant ideals I in A. Then $h_*(I)=0$ for all H-invariant ideals I in A.

Proof. We first observe that if A is nuclear, then so are all its ideals, so that $h_*(I)$ makes sense in the case that h_* is defined on separable nuclear H-algebras. Using the notation of the Ideal Decomposition Lemma, we will prove that $h_*(I_S\cap I)=0$ for all $S\subset G$ and all H-invariant ideals I in A, by downward induction on the size of S.

The start of the induction is easy: $I_G\cap I=I_G$, and I_G is G-invariant, so $h_*(I_G\cap I)=0$. We therefore assume we have shown $h_*(I_T\cap I)=0$ for all H-invariant ideals I of A and all $T\underset{\neq}{\supset} S$. We must now use an inner induction argument: we prove by induction on the number of elements of a set V disjoint from S that

$$h_*\Big(\sum_{g\in V} I_{S\cup\{g\}}\cap I\Big)=0.$$

We start the inner induction: if V has only one element, the result follows from the outer induction hypothesis. So assume the result has been proved for a set V, and let $k \notin S \cup V$. Define ideals J and L by

$$J = \sum_{g \in V} I_{S \cup \{g\}} \cap I \quad \text{and} \quad L = I_{S \cup \{k\}} \cap I .$$

Then we have $h_*(J) = 0$ and $h_*(L) = 0$ by the inner induction hypothesis. Furthermore,

$$L \cap J = \sum_{g \in V} L_{S \cup \{g\}} \cap L$$

by part (4b) of the Ideal Decomposition Lemma, and the inner induction hypothesis, applied to the H-invariant ideal L, now implies that $h_*(L \cap J) = 0$. Therefore $h_*(L + J) = 0$ by the previous lemma. Since

$$L + J = \sum_{g \in V \cup \{k\}} I_{S \cup \{g\}},$$

the inner induction is complete.

We can now return to the outer induction. We have

$$(I_S \cap I) / (I_{\bar{S}} \cap I) \simeq ((I_S \cap I) + I_{\bar{S}}) / I_{\bar{S}} .$$

By (5) of the Ideal Decomposition Lemma, this algebra is a direct summand in $I_S / I_{\bar{S}}$. Since both I_S and $I_{\bar{S}}$ are G-invariant, $h_*(I_S / I_{\bar{S}}) = 0$. Therefore also

$$h_*((I_S \cap I) / (I_{\bar{S}} \cap I)) = 0 .$$

Now

$$I_{\bar{S}} \cap I = \sum_{g \notin S} I_{S \cup \{g\}} \cap I ,$$

so the result of the inner induction argument shows that $h_*(I_{\bar{S}} \cap I) = 0$. It follows that $h_*(I_S \cap I) = 0$, completing the outer induction. Therefore $h_*(I) = h_*(I_{\{e\}} \cap I) = 0$, as desired. Q.E.D.

As corollaries to this lemma we obtain the following two results:

5.3.6 KK-Subgroup Theorem. Let (G,A,α) be a separable nuclear G-algebra, with G finite. If α is KK-free, then so is $\alpha\,|_H$ for any subgroup H.

Proof. Let P be a prime ideal in $R(H)$ not containing $I(H)$, and let I be an H-invariant ideal in A. By lemma 5.3.1, $KK_H^0(J,J)_P = 0$ for all G-invariant ideals J in A. Since $KK_H^*(I,J)$ is a module over $KK_H^0(J,J)$, it follows that $KK_H^*(I,J)_P = 0$ for all G-invariant ideals J. By the previous lemma, this also holds for all H-invariant ideals; in particular $KK_H^0(I,I)_P = 0$ as desired. Q.E.D.

5.3.7 Proposition. Let (G,A,α) be a G-algebra with G finite. Suppose that for all subgroups H of G, and all G-invariant ideals I of A, $(\alpha\,|_H)\,|_I$ has locally discrete K-theory. Then α is totally K-free.

In other words, one does not have to look at any ideals which are invariant under subgroups of G but not under G itself when trying to show that an action of a finite group is totally K-free.

Proof of proposition 5.3.7. By assumption, if H is any subgroup of G then $K_*^H(I)_P = 0$ for all primes $P \subset R(H)$ not containing $I(H)$ and all G-invariant ideals I. By lemma 5.3.5, this also holds for H-invariant ideals I. Q.E.D.

We now return to the proof of the Ideal Decomposition Lemma.

Proof of lemma 5.3.3. We recall the formulas

$$I_S = \sum_{g \in G} \alpha_g \left[\bigcap_{h \in S} \alpha_h [I]\right] \quad \text{and} \quad I_{\bar{S}} = \sum_{g \notin S} I_{S \cup \{g\}} .$$

It is obvious that I_S is G-invariant for all S, so that $I_{\bar{S}}$ is also, proving (1). To prove (2), observe that it is obvious that $I_T \subset I_S$ for $S \subset T$, so that also $I_{\bar{S}} \subset I_S$. Furthermore, if $S \subsetneq T$, then $S \cup \{g\} \subset T$ for some $g \notin S$, proving that $I_T \subset I_{\bar{S}}$. Since $I_{\{e\}} = \sum_{g \in G} \alpha_g [I]$ and $I_G = \bigcap_{g \in G} \alpha_g [I]$, part (3), namely $I_G \subset I \subset I_{\{e\}}$, is also obvious. To prove (6), let J be any G-invariant ideal. Then

$$(I \cap J)_S = \sum_{g \in G} \alpha_g \left[\bigcap_{h \in S} \alpha_h [I \cap J]\right] = \sum_{g \in G} \alpha_g \left[\bigcap_{h \in S} \alpha_h [I]\right] \cap J = I_S \cap J .$$

Here, the second step follows from lemma 5.3.2. The formulas for $(I+J)_S$, $(I \cap J)_{\bar{S}}$, and $(I+J)_{\bar{S}}$ are proved similarly.

We now turn to the harder parts of the lemma, starting with (4). Let S, T, and k be as in (4), and let $L = I_{S \cup \{k\}} \cap I$. Then $L_{S \cup \{g\}} = I_{S \cup \{k\}} \cap I_{S \cup \{g\}}$ by (1) and (6), whence

$$L_{S \cup \{g\}} \cap L = (I_{S \cup \{k\}} \cap I) \cap (I_{S \cup \{g\}} \cap I) .$$

Consequently we have by distributivity

$$\sum_{g \in T} L_{S \cup \{g\}} = I_{S \cup \{k\}} \cap \sum_{g \in T} I_{S \cup \{g\}} ,$$

and

$$\sum_{g \in T} L_{S \cup \{g\}} \cap L = (I_{S \cup \{k\}} \cap I) \cap \left[\sum_{g \in T} I_{S \cup \{g\}} \cap I\right] ,$$

which is the two parts of (4).

What remains to be proved is (5). Let $H_S = \{g \in G : gS = S\}$, which is obviously a subgroup of G. Let R be a set of left coset representatives of H_S in G, so that G is the disjoint union $\bigcup_{g \in R} gH_S$. The set S^{-1} of inverses of elements of S satisfies $S^{-1}h = S^{-1}$ for all $h \in H_S$, and is therefore the union of left cosets of H_S in G,

$$S^{-1} = \bigcup_{g \in R \cap S^{-1}} gH_S .$$

In fact, H_S is the largest subgroup of G such that this is true. Obviously $H_S = G$ if and only if $S = G$ or $S = \emptyset$; the latter is ruled out by the requirement $e \in S$.

Let $B = I_S / I_{\bar{S}}$, and let $J = ((I_S \cap I) + I_{\bar{S}}) / I_{\bar{S}}$, which is isomorphic to $(I_S \cap I) / (I_{\bar{S}} \cap I)$. Then J is an ideal in B, and it is clear that $J_S = B$ while $J_{S \cup \{k\}} = 0$ for $k \notin S$. (The second equation follows from the inclusion $I_{S\cup\{k\}} \subset I_{\bar{S}}$.) Write α_g also for the induced action of G on B, and let $M = \bigcap_{g \in S} \alpha_g[J]$. The equation $J_S = B$ then becomes

$$\sum_{g \in G} \alpha_g[M] = B. \tag{*}$$

Now let $g \in G$. We have

$$M \cap \alpha_g[M] = \bigcap_{h \in S \cup gS} \alpha_h[J] \subset J_{S \cup gS}.$$

If $g \notin H_S$ then $S \cup gS$ is strictly larger than S, and hence $J_{S\cup gS} = 0$. Otherwise $S \cup gS = S$ and it follows that $\alpha_g[M] = M$. It now follows that if $g, h \in G$ are in the same left coset of H_S, then $\alpha_g[M] = \alpha_h[M]$ (since $g^{-1}h \in H_S$), while if they are in different left cosets, then $\alpha_g[M] \cap \alpha_h[M] = 0$ (apply $\alpha_{g^{-1}}$ to both sides). Recalling that R is a set of left coset representatives of H_S, we find that (*) becomes $\bigoplus_{g \in R} \alpha_g[M] = B$. Therefore $B \simeq C(G \times_{H_S} M)$ by proposition 2.9.6.

It remains to identify J as

$$J = \bigoplus_{g \in R \cap S^{-1}} \alpha_g[M].$$

In view of the direct sum decomposition of B and the fact that S^{-1} is a union of left cosets of H_S, it is sufficient to show that $J = \sum_{g \in S^{-1}} \alpha_g[M]$. So let $L = \sum_{g \in S^{-1}} \alpha_g[M]$. Then we have:

$$\begin{aligned} J &= \left[J \cap \sum_{g \in S} \alpha_{g^{-1}}[M]\right] + \left[J \cap \sum_{g \notin S} \alpha_{g^{-1}}[M]\right] \\ &= J \cap L + \sum_{g \notin S} \alpha_{g^{-1}}[\alpha_g[J] \cap M] \\ &= J \cap L + \sum_{g \notin S} \alpha_{g^{-1}}\left[\bigcap_{h \in S \cup \{g\}} \alpha_h[J]\right] \subset J \cap L + \sum_{g \notin S} J_{S \cup \{g\}} = J \cap L, \end{aligned}$$

since $J_{S\cup\{g\}} = 0$ for $g \notin S$. Therefore $J \subset L$. For the reverse inclusion, let $g \in S$. Then

$$\alpha_{g^{-1}}[M] = \alpha_{g^{-1}}\left[\bigcap_{h \in S} \alpha_h[J]\right] \subset J,$$

whence $L \subset J$. We thus have $L = J$, completing the proof of (5), and of the lemma.

Q.E.D.

5.3.8 Question. Can the methods of this section be generalized to arbitrary compact Lie groups?

The major difficulty clearly lies with the Ideal Decomposition Lemma. It can be easily modified to handle closed (and therefore also open) subgroups H of finite index: one finds the largest normal subgroup N of G contained in H, which is again closed of finite index, and considers subsets of G/N rather than of G. However, the general case seems to be difficult.

Chapter 6

Tensor Products

Let G be a compact Lie group, let X be a compact free G-space, and let Y be an arbitrary compact G-space. Then $X\times Y$ with the diagonal action is a free G-space, and this can be proved using theorem 1.1.1 as follows. It is easy to construct a unital $R(G)$-algebra homomorphism from $K_G^0(X)$ to $K_G^0(X\times Y)$. (One assigns to the class $[E]$ of a G-vector bundle E over X the class of the external tensor product bundle $E\otimes(\mathbf{C}\times Y)$ over $X\times Y$, where $\mathbf{C}\times Y$ is the trivial one-dimensional bundle on Y.) Therefore if $I(G)^n$ annihilates $K_G^0(X)$, it also annihilates the identity in $K_G^0(X\times Y)$ and therefore annihilates all of $K_G^*(X\times Y)$. (This result of course does not depend on the topological properties of G, X, and Y, although the proof just sketched does.) Toward the end of section 4.2, the question of whether analogous results hold for K-free and totally K-free actions on noncommutative G-algebras was raised. This chapter is devoted to some partial answers to that question.

To begin with, if A is an arbitrary G-algebra, then $K_0^G(A)$ is unlikely to be a ring. We therefore need a way of computing $K_*^G(A\otimes B)$ in terms of $K_*^G(A)$ and $K_*^G(B)$ for G-algebras A and B, that is, an equivariant Künneth theorem along the lines of the main result of [95]. Even for G-spaces, the results available (see [48]) are far too weak for our purposes. In [48], the group G is assumed to be connected and have torsion-free fundamental group $\pi_1(G)$, and all spaces are assumed to be locally contractible and to have finite covering dimension. For the conditions on the spaces, no analog for noncommutative C^*-algebras is known, and later steps in our proof force us to assume G is finite. The way around this difficulty is to prove a Künneth formula for certain localizations of equivariant K-theory, specifically $K_*^G(\cdot)_P$, where G is a cyclic group of finite order and P is a prime ideal in $R(G)$ with support G. We must also assume that certain technical conditions analogous to those in [95] are satisfied. (A preprint [93] by Rosenberg and Schochet, generalizing the results of [48] to noncommutative C^*-algebras, has recently appeared. They do not need any analog of local contractibility or finite covering dimension, but they retain the assumptions on the group. Thus, there is no overlap between this chapter and their preprint.)

One of the crucial factors is the homological dimension of the ring of coefficients. (See the discussion preceding lemma 6.4.2). For the Künneth formula of [95], this ring is $\mathbf{Z}$, whose homological dimension is 1. For equivariant K-theory, the relevant ring is $R(G)$, whose homological dimension is usually infinite for nontrivial finite groups. For equivariant K-theory localized at the prime P, the relevant ring is $R(G)_P$, and its homological dimension is again 1 if G is cyclic and the support of P is G. Since the definition of K-freeness requires that one consider all prime ideals whose support is not the one element group, one might think that we would only get a theorem for cyclic groups of prime order. In fact, certain tricks and theorem 5.2.6 enable us to get a theorem for actions of arbitrary finite p-groups, provided we assume total K-freeness of one of the actions. (It should be pointed out that the usefulness of localizing equivariant K-theory in such a way that the coefficient ring has homological dimension 1 has already been noticed by Iberkleid and Petrie [50] in the case of actions of the circle group S^1 on spaces. They obtain a universal coefficient theorem, in the form of a short exact sequence, involving the localization of S^1-equivariant K-theory at an arbitrary prime ideal, or set of prime ideals, not containing the augmentation ideal $I(S^1)$.)

The Künneth formula enables us to prove that if A and B are G-algebras, satisfying the appropriate technical conditions, and G and P are as above, then $K_*^G(A)_P = 0$ implies that $K_*(A \otimes B)_P = 0$. In particular, if the action of G on A is K-free, then $K_*^G(I \otimes J)_P = 0$ for all G-invariant ideals I of A and J of B. Unfortunately, we must show that $K_*^G(L)_P = 0$ for all G-invariant ideals L of $A \otimes B$, not just those of the form $I \otimes J$. To do this, we need the Ideal Decomposition Lemma 5.3.3.

If we are willing to assume that one of the actions is KK-free instead of totally K-free, then we can bypass the equivariant Künneth formula, using an argument essentially identical to the one sketched in the first paragraph. We still have to get information about all G-invariant ideals in $A \otimes B$ from information about only those of the form $I \otimes J$. The lemma used to do so does not work for KK-theory, so we wind up concluding that, if G is a finite group acting KK-freely on A and arbitrarily on B, where A and B are separable and A is nuclear, then the diagonal action on $A \otimes B$ is totally K-free.

This chapter is organized as follows. The first four sections are devoted to the proof of the Künneth formula, roughly following the procedure used by Schochet in [95]. In the first section, we construct the pairing from $K_*^G(A) \otimes_{R(G)} K_*^G(B)$ to $K_*^G(A \otimes B)$, and in the second section we consider certain special cases of the Künneth formula, in which $K_*^G(B)_P$ is a projective $R(G)_P$-module. In section 3 we prove the analogs for equivariant K-theory of the Connes isomorphism theorem ([20]) and the Pimsner-Voiculescu exact sequences ([82], [83]). These results suggest that one might hope that if (G, A, α) is a G-algebra such that α is K-free, and, for example, $\beta : \mathbf{Z} \longrightarrow Aut(A)$ is an action commuting with α, then the action of G on $C^*(\mathbf{Z}, A, \beta)$ might also be K-free. We include an example to show that this is in general not the case. In section 4 we prove our Künneth formula, subject to the analogs of the conditions in [95]. The next section contains the description of the category for which the theorem on total K-freeness of diagonal actions on tensor products works. This category is somewhat smaller than the one for which the Künneth theorem works, since it must be closed under passage to invariant ideals. Finally, the last section applies the Künneth formula and the Ideal Decomposition Lemma to prove the total K-freeness of diagonal actions, assuming total K-freeness of one of the actions and the appropriate technical conditions. The result obtained by starting with KK-freeness is also proved here.

Throughout this chapter, all tensor products of G-algebras are minimal tensor products ([104], definition IV.4.8), unless otherwise specified, and carry the diagonal G-action. We also adopt the convention that a cyclic subgroup of a given group is required to have more than one element.

6.1. A Pairing in Equivariant K-Theory

6.1.1 Convention. If A is a G-algebra, then we consider $K_*^G(A)$ to be $\mathbf{Z}/2\mathbf{Z}$-graded by the direct sum decomposition $K_*^G(A)=K_0^G(A)\oplus K_1^G(A)$, with $K_i^G(A)$ being the elements of degree i. The tensor product $K_*^G(A)\otimes K_*^G(B)$ is graded by taking the degree of an element of $K_i^G(A)\otimes K_j^G(B)$ to be $i+j$. A similar convention applies to $K_*^G(A)\otimes_{R(G)} K_*^G(B)$. A homomorphism of degree 0 is one which preserves the grading, and a homomorphism of degree 1 is one which reverses it. All homomorphisms are assumed to be of degree 0 unless otherwise specified.

As examples, notice that if $\varphi: A \longrightarrow B$ is an equivariant homomorphism of G-algebras, then φ_* has degree 0, while the connecting homomorphism on K-theory associated to an equivariant short exact sequence has degree 1.

Before constructing our pairing, we need one preliminary lemma.

6.1.2 Lemma. Let $\varphi: C_0(\mathbf{R})\otimes C_0(\mathbf{R}) \longrightarrow C_0(\mathbf{R})\otimes C_0(\mathbf{R})$ be the flip, that is, $\varphi(f_1\otimes f_2)=f_2\otimes f_1$. Let the compact group G act trivially on everything. Then φ_* is multiplication by -1 on $K_*^G(C_0(\mathbf{R})\otimes C_0(\mathbf{R}))$.

Proof. In ordinary K-theory this is a standard fact. (See for example [46], proposition VII.8). Since the action of G is trivial, the equivariant K-theory is obtained by tensoring everything over $\mathbf{Z}$ with $R(G)$. (See remark 2.8.5.) So φ_* is multiplication by -1 in this case also. Q.E.D.

6.1.3 Proposition. (Compare [46], section 7.) Let G be a compact group. Then there is a pairing (of degree 0)

$$\omega: K_*^G(A)\otimes_{R(G)} K_*^G(B) \longrightarrow K_*^G(A\otimes B)$$

for G-algebras A and B, satisfying the following conditions:

(1) ω is an $R(G)$-module homomorphism.

(2) ω is natural with respect to pairs of equivariant maps, suspensions, and connecting homomorphisms.

(3) ω is associative, in the sense that the two maps from $K_*^G(A)\otimes_{R(G)} K_*^G(B)\otimes_{R(G)} K_*^G(C)$ to $K_*^G(A\otimes B\otimes C)$ determined by ω are the same.

(4) If A and B are unital, if V and W are finite dimensional representation spaces of G, and if $p\in L(V)\otimes A$ and $q\in L(W)\otimes B$ are G-invariant projections, then

$\omega([p]\otimes[q]) = [p\otimes q]$, where $p\otimes q$ is regarded as an element of $L(V\otimes W)\otimes A\otimes B$.

Proof. We construct ω in stages. First, suppose A and B are unital G-algebras. Then the formula $([p],[q]) \longrightarrow [p\otimes q]$, for p and q as in (4), clearly defines a $\mathbf{Z}$-bilinear map from $K_0^G(A)\times K_0^G(B)$ to $K_0^G(A\otimes B)$. It is natural with respect to pairs of maps, that is, the following diagram commutes:

$$\begin{array}{ccc} K_0^G(A_1)\times K_0^G(B_1) & \xrightarrow{\varphi_*\times\psi_*} & K_0^G(A_2)\times K_0^G(B_2) \\ \downarrow & & \downarrow \\ K_0^G(A_1\otimes B_1) & \xrightarrow{(\varphi\otimes\psi)_*} & K_0^G(A_2\otimes B_2) \end{array}$$

where $\varphi: A_1 \longrightarrow A_2$ and $\psi: B_1 \longrightarrow B_2$ are equivariant unital homomorphisms of G-algebras.

Let H be a finite dimensional representation space of G, and suppose that $p \in L(V)\otimes A$ and $q \in L(W)\otimes B$ for finite dimensional representation spaces V and W of G. Then $[H][p]$ is represented by $1_H\otimes p \in L(H)\otimes L(V)\otimes A$ and $[H][q]$ is represented by $1_H\otimes q \in L(H)\otimes L(W)\otimes B$. It is now clear that the images of both of the pairs $([H][p],[q])$ and $([p],[H][q])$ are represented by $1_H\otimes p\otimes q \in L(H\otimes V\otimes W)\otimes A\otimes B$, where $p\otimes q$ is regarded as an element of $L(V)\otimes L(W)\otimes A\otimes B$. Therefore the pairing defined in the previous paragraph is $R(G)$-bilinear and defines an $R(G)$-module homomorphism

$$\omega(A,B): K_0^G(A)\otimes_{R(G)}K_0^G(B) \longrightarrow K_0^G(A\otimes B).$$

Associativity for ω as defined so far is obvious.

Now consider the case in which A is not unital. We have split exact sequences

$$0 \longrightarrow K_0^G(A) \longrightarrow K_0^G(A^+) \longrightarrow K_0^G(\mathbf{C}) \longrightarrow 0$$

and

$$0 \longrightarrow K_0^G(A\otimes B) \longrightarrow K_0^G(A^+\otimes B) \longrightarrow K_0^G(B) \longrightarrow 0.$$

We tensor the first of these over $R(G)$ with $K_0^G(B)$. Since the sequence was split exact, it remains split exact, and we obtain a commutative diagram with exact rows:

$$\begin{array}{ccccccccc} 0 & \longrightarrow & K_0^G(A)\otimes_{R(G)}K_0^G(B) & \longrightarrow & K_0^G(A^+)\otimes_{R(G)}K_0^G(B) & \longrightarrow & K_0^G(B) & \longrightarrow & 0 \\ & & & & \downarrow \omega(A^+,B) & & \downarrow \omega(\mathbf{C},B) & & \\ 0 & \longrightarrow & K_0^G(A\otimes B) & \longrightarrow & K_0^G(A^+\otimes B) & \longrightarrow & K_0^G(B) & \longrightarrow & 0 \end{array}$$

The map $\omega(\mathbf{C},B)$ is easily seen to be the canonical isomorphism $R(G)\otimes_{R(G)}K_0^G(B) \simeq K_0^G(B)$. There is therefore a unique $R(G)$-module homomorphism

$$\omega(A,B): K_0^G(A)\otimes_{R(G)}K_0^G(B) \longrightarrow K_0^G(A\otimes B)$$

which makes the left hand square also commutative. We now extend the definition of ω to the case in which B is also nonunital by the same method. The extended pairing is clearly still natural for pairs of maps and is associative.

Next, we extend the pairing to K_1^G. It is notationally convenient to take the suspension SA to be $C_0(\mathbf{R})\otimes A$, rather than $C_0((0,1))\otimes A$. Recall from definition 2.8.1 that $K_1^G(A)$ is defined to be $K_0^G(SA)$. Furthermore, there is a natural suspension isomorphism $s_A : K_0^G(A) \longrightarrow K_0^G(S^2A)$, which by proposition 2.8.6 and the definition of ω is given by $s_A(\lambda)=\omega(C_0(\mathbf{R}^2),A)(\eta_0\otimes\lambda)$ for a certain element η_0 of $K_0^G(C_0(\mathbf{R}^2))$. We now define $\omega(A,B)$ on the other summands $K_i^G(A)\otimes K_j^G(B)$ of $K_*^G(A)\otimes K_*^G(B)$, where i and j are not both zero, by the following composites:

$$K_1^G(A)\otimes_{R(G)}K_0^G(B) \xrightarrow{\cong} K_0^G(SA)\otimes_{R(G)}K_0^G(B)$$

$$\xrightarrow{\omega(SA,B)} K_0^G(SA\otimes B) \xrightarrow{\cong} K_0^G(S(A\otimes B)) \xrightarrow{\cong} K_1^G(A\otimes B),$$

$$K_0^G(A)\otimes_{R(G)}K_1^G(B) \xrightarrow{\cong} K_0^G(A)\otimes_{R(G)}K_0^G(SB)$$

$$\xrightarrow{\omega(A,SB)} K_0^G(A\otimes SB) \xrightarrow{\cong} K_0^G(S(A\otimes B)) \xrightarrow{\cong} K_1^G(A\otimes B),$$

and

$$K_1^G(A)\otimes_{R(G)}K_1^G(B) \xrightarrow{\cong} K_0^G(SA)\otimes_{R(G)}K_0^G(SB)$$

$$\xrightarrow{-\omega(SA,SB)} K_0^G(SA\otimes SB) \xrightarrow{\cong} K_0^G(S^2(A\otimes B)) \xrightarrow{s_{A\otimes B}^{-1}} K_0^G(A\otimes B).$$

In the last of these, the minus sign is important. We further identify $SA\otimes SB$ with $C_0(\mathbf{R})\otimes A\otimes C_0(\mathbf{R})\otimes B$, and $S^2(A\otimes B)$ with $C_0(\mathbf{R})\otimes C_0(\mathbf{R})\otimes A\otimes B$. The map from $SA\otimes SB$ to $S^2(A,B)$ is then given by

$$f_1\otimes a\otimes f_2\otimes b \longrightarrow f_1\otimes f_2\otimes a\otimes b$$

for $f_1,f_2\in C_0(\mathbf{R})$, $a\in A$, and $b\in B$. It is important not to reverse the order of f_1 and f_2 here, since by lemma 6.1.2, the map $f_1\otimes f_2 \longrightarrow f_2\otimes f_1$ is multiplication by -1 on K-theory.

Since the isomorphisms $K_0^G(SA)\simeq K_1^G(A)$ and $K_1^G(SA)\simeq K_0^G(A)$ are natural, it is clear that α as extended is still natural for pairs of maps. We next prove that α respects suspensions, that is, that the following diagrams commute, where indices are taken mod 2:

$$\begin{array}{ccc} K_i^G(A)\otimes_{R(G)}K_j^G(B) & \xrightarrow{\omega(A,B)} & K_{i+j}^G(A\otimes B) \\ \downarrow\simeq & & \downarrow\simeq \\ K_{i+1}^G(SA)\otimes_{R(G)}K_j^G(B) & \xrightarrow{\omega(SA,B)} & K_{i+j+1}^G(SA\otimes B), \end{array} \qquad (*)$$

and the corresponding diagram in which the suspension is applied to B rather than to A. We will actually prove this only for (*), the other cases being essentially the same. There are four cases to be considered, corresponding to the two possible values of i and j. Notice that we now use $\omega(A,B)$ to denote the full pairing $K_*^G(A)\otimes_{R(G)}K_*^G(B)\longrightarrow K_*^G(A\otimes B)$.

Case 1. $i=1, j=0$. Then the commutativity of (*) is exactly the definition of $\omega(A,B)\colon K_1^G(A)\otimes_{R(G)}K_0^G(B)\longrightarrow K_1^G(A\otimes B)$.

Case 2. $i=1, j=1$. We consider the following diagram:

$$\begin{array}{ccc} K_1^G(A)\otimes_{R(G)}K_1^G(B) & \xrightarrow{\omega(A,B)} & K_0^G(A\otimes B) \\ \downarrow\simeq & & \downarrow\simeq \\ K_0^G(SA)\otimes_{R(G)}K_0^G(SB) & \xrightarrow{-\omega(SA,SB)} & K_0^G(SA\otimes SB) \\ \downarrow\simeq & & \downarrow\simeq \\ K_0^G(SA)\otimes_{R(G)}K_1^G(B) & \xrightarrow{-\omega(SA,B)} & K_1^G(SA\otimes B). \end{array}$$

The isomorphism from $K_0^G(SA\otimes SB)$ to $K_1^G(SA\otimes B)$ is the composite of φ_* and the isomorphism $K_0^G(S(SA\otimes B))\simeq K_1^G(SA\otimes B)$, where

$$\varphi\colon C_0(\mathbf{R})\otimes A\otimes C_0(\mathbf{R})\otimes B\longrightarrow C_0(\mathbf{R})\otimes C_0(\mathbf{R})\otimes A\otimes B$$

is defined by

$$\varphi(f_1\otimes a\otimes f_2\otimes b) = f_2\otimes f_1\otimes a\otimes b\ .$$

(Note that φ reverses the order of f_1 and f_2.) The top rectangle commutes by the definition of $\omega(A,B)$ and the bottom one commutes by the definition of $\omega(SA,B)$. However, the composite

$$K_0^G(A\otimes B)\longrightarrow K_0^G(SA\otimes SB)\xrightarrow{\varphi_*}K_0^G(S^2A\otimes B)\longrightarrow K_1^G(SA\otimes B)$$

is the negative of the usual isomorphism $K_0^G(A\otimes B)\simeq K_1^G(SA\otimes B)$, by the previous lemma. Therefore the diagram (*) commutes for $i=j=1$.

Case 3. $i=0, j=0$. Here, we expand the diagram (*) as follows:

$$\begin{array}{ccc}
K_0^G(A)\otimes_{R(G)}K_0^G(B) & \xrightarrow{\omega(A,B)} & K_0^G(A\otimes B) \\
\simeq\downarrow s_A\otimes(id_B)_* & & \simeq\downarrow s_{A\otimes B} \\
K_0^G(S^2A)\otimes_{R(G)}K_0^G(B) & \xrightarrow{\omega(S^2A,B)} & K_0^G(S^2A\otimes B) \\
\simeq\downarrow & & \simeq\downarrow \\
K_1^G(SA)\otimes_{R(G)}K_0^G(B) & \xrightarrow{\omega(SA,B)} & K_1^G(SA\otimes B).
\end{array}$$

The top rectangle commutes by the associativity of ω, since $s_A(\lambda)=\omega(C_0(\mathbf{R}^2),A)(\eta_0\otimes\lambda)$ for a certain element $\eta_0\in K_0^G(C_0(\mathbf{R}^2))$, and the bottom rectangle commutes by the definition of $\omega(SA,B)$. This time the composite of the vertical maps on the right is the correct map.

Case 4. $i=0, j=1$. For this case, we expand (*) as follows:

$$\begin{array}{ccc}
K_0^G(A)\otimes_{R(G)}K_1^G(B) & \xrightarrow{\omega(A,B)} & K_1^G(A\otimes B) \\
\downarrow\simeq & & \downarrow\simeq \\
K_0^G(A)\otimes_{R(G)}K_0^G(SB) & \xrightarrow{\omega(A,SB)} & K_0^G(A\otimes SB) \\
\downarrow\simeq & & \downarrow\simeq \\
K_1^G(SA)\otimes_{R(G)}K_0^G(SB) & \xrightarrow{\omega(SA,SB)} & K_1^G(SA\otimes SB) \\
\downarrow\simeq & & \downarrow\simeq \\
K_1^G(SA)\otimes_{R(G)}K_1^G(B) & \xrightarrow{\omega(SA,B)} & K_0^G(SA\otimes B).
\end{array}$$

The top rectangle commutes by the definition of $\omega(A,SB)$, the middle one commutes by case 3, and the bottom commutes by an argument similar to that of case 2. (There is a minus sign in the definition of $\omega(SA,B)$, which is cancelled out because the isomorphism $K_1^G(SA\otimes SB)\simeq K_0^G(SA\otimes B)$ used in the definition of $\omega(SA,B)$ is the negative of the one needed to produce the correct right vertical map in (*).)

We have thus shown that the diagrams (*) commute for all i,j. As mentioned earlier, the arguments for the corresponding diagrams in which the suspensions are on B rather than A are similar. The results for these cases can also be derived from those already obtained by the following commutativity principle: the diagram

$$\begin{array}{ccc}
K_i^G(A)\otimes_{R(G)}K_j^G(B) & \xrightarrow{\omega(A,B)} & K_{i+j}^G(A\otimes B) \\
\downarrow\simeq & & \downarrow\simeq \\
K_j^G(B)\otimes_{R(G)}K_i^G(A) & \xrightarrow{\omega(B,A)} & K_{i+j}^G(B\otimes A)
\end{array}$$

in which the vertical maps are the flip maps, commutes up to the sign $(-1)^{ij}$. This is obvious for $i=j=0$, and follows from the definition of ω and lemma 6.1.2 for the other cases.

It now follows that ω also respects connecting homomorphisms, because, by lemma 2.8.10, the connecting homomorphisms are composites of suspension isomorphisms and maps of the form φ_* or $(\varphi_*)^{-1}$. The associativity of ω is essentially obvious.

Q.E.D.

6.1.4 Remark. The sign convention referred to in [95] is in the choice of the periodicity isomorphism s_A. We could just as easily have defined $s_A(\lambda)=\omega(C_0(\mathbf{R}^2),A)(-\eta_0\otimes\lambda)$. (Note that η_0 and λ are in K_0^G, where ω is canonically determined.) This has the effect of reversing the sign of the isomorphism $K_1^G(SA)\simeq K_0^G(A)$. However, the only bad effect of carelessness with the signs, such as leaving out the minus sign in the definition of

$$\omega(A,B): K_1^G(A)\otimes_{R(G)}K_1^G(B)\longrightarrow K_0^G(A\otimes B),$$

would be that various diagrams would commute only up to sign, which is not a serious problem for the uses we make of this pairing.

6.1.5 Remark. For separable algebras A and B, we could have defined ω as follows: identify $K_i^G(A)$ with $KK_G^i(\mathbf{C},A)$ as in theorem 3.6.12, do the same for B, and set $\omega(\eta\otimes\lambda)=\eta\otimes_{\mathbf{C}}\lambda$. This ω satisfies (4) of proposition 6.1.3 by the form of the isomorphism $K_0^G(A)\simeq KK_G^0(\mathbf{C},A)$ as obtained in theorem 3.6.12, and properties (1) through (3) follow from corollary 3.6.3.

6.1.6 Corollary (to proposition 6.1.3). Let G be a compact Lie group, and let P be a prime ideal in $R(G)$. Then there is a pairing (of degree 0)

$$\omega_P: K_*^G(A)_P\otimes_{R(G)_P}K_*^G(B)_P\longrightarrow K_*^G(A\otimes B)_P$$

for G-algebras A and B, such that ω_P is a natural $R(G)_P$-module homomorphism. Furthermore, ω_P respects suspensions and connecting homomorphisms, and is associative.

Proof. Localize the pairing in 6.1.3 at P, noting that tensor products commute with localization by [5], proposition 3.7.

Q.E.D.

We will need two further properties of the pairing of proposition 6.1.3, which are given in the next two propositions.

6.1.7 Proposition. Let G be a finite group, and let (G,A,α) and (G,B,β) be G-algebras such that one of them is nuclear and the action α is trivial. Then there are natural isomorphisms

$$K_*^G(A)\otimes_{R(G)}K_*^G(B)\simeq K_*(A)\otimes K_*(C^*(G,B)) \quad \text{and} \quad K_*^G(A\otimes B)\simeq K_*(A\otimes C^*(G,B)).$$

These isomorphisms identify the equivariant pairing $\omega(A,B)$ with the nonequivariant pairing $\omega(A,C^*(G,B))$.

This proposition surely holds for general compact groups, but the proof becomes messier, and we only need it for finite groups.

Proof of proposition 6.1.7. We have $K_*^G(A) \simeq K_*(A) \otimes R(G)$ by remark 2.8.5, and thus we have natural isomorphisms

$$K_*^G(A) \otimes_{R(G)} K_*^G(B) \simeq K_*(A) \otimes R(G) \otimes_{R(G)} K_*^G(B)$$
$$\simeq K_*(A) \otimes K_*^G(B) \simeq K_*(A) \otimes K_*(C^*(G,B)),$$

where the last step is from theorem 2.8.3 (7). Furthermore, $C^*(G, A \otimes B) \simeq A \otimes C^*(G,B)$ since G acts trivially on A, and since one of A and B is nuclear. Therefore

$$K_*^G(A \otimes B) \simeq K_*(C^*(G, A \otimes B)) \simeq K_*(A \otimes C^*(G,B)).$$

Now we must identify the pairing. Since both the equivariant and nonequivariant pairings are natural and respect suspensions, it is sufficient to consider only K_0 and K_0^G, and to consider only unital A and B. It is also convenient to consider finitely generated projective modules instead of projections.

Let E be a finitely generated projective (G,A)-module, and let F be a finitely generated projective (G,B)-module. Then the algebraic tensor product $E \odot F$ is a finitely generated projective module over the algebraic tensor product $A \odot B$, with a compatible action of G. In fact, if E is a summand in $V \otimes A$ and F is a summand in $W \otimes B$ for finite dimensional representation spaces V and W of G (see proposition 2.2.5), then $E \odot F$ is a summand in $V \otimes W \otimes (A \odot B)$. It therefore has a completion $E \otimes F$ which is a finitely generated projective $(G, A \otimes B)$-module, and one can check that the completion does not depend on the choice of embeddings in free modules. (See the discussion at the beginning of chapter VII of [46].) The pairing ω then clearly satisfies $\omega([E] \otimes [F]) = [E \otimes F]$.

Now consider the following diagram.

$$\begin{array}{ccc} K_0^G(A) \otimes_{R(G)} K_0^G(B) & \xrightarrow{\omega_0} & K_0^G(A \otimes B) \\ \downarrow \simeq & & \downarrow = \\ K_0(A) \otimes_{\mathbf{Z}} K_0^G(B) & \xrightarrow{\omega_1} & K_0^G(A \otimes B) \\ \downarrow \simeq & & \downarrow \simeq \\ K_0(A) \otimes_{\mathbf{Z}} K_0(C^*(G,B)) & \xrightarrow{\omega_2} & K_0^G(A \otimes C^*(G,B)). \end{array} \qquad (*)$$

Here $\omega_0 = \omega(A,B)$ is the equivariant pairing, and $\omega_2 = \omega(A, C^*(G,B))$ is the nonequivariant pairing of [95], which is obtained by letting G be trivial in proposition 6.1.3. Also ω_1 is the pairing defined as follows: if E is a finitely generated projective A-module, and (G,F,λ) is a finitely generated projective (G,B)-module, then $\omega_1([E] \otimes [F]) = [E \otimes F]$, where $E \otimes F$ has the G-action $g \longrightarrow id_E \otimes \lambda_g$, which makes it a finitely generated projective $(G, A \otimes B)$-module. Finally, the vertical maps in (*) come from the isomorphisms in the first paragraph of the proof. The proposition will therefore be proved if we can show that the diagram (*) commutes.

We now need to use the specific form of the isomorphism $K_0^G(B) \simeq K_0(C^*(G,B))$ of Julg's Theorem 2.6.1. In the case at hand, G is finite. Therefore $C^*(G,B) = L^1(G,B)$ as a *-algebra (both are the set of all functions from G to B); only the norms are different. Furthermore, $C^*(G,B)$ is again unital, and the isomorphism $K_0^G(B) \simeq K_0(C^*(G,B))$ is given simply by regarding a finitely generated projective (G,B)-module as a finitely generated projective $C^*(G,B)$-module, using the integrated form action of $L^1(G,B)$ given in lemma 2.2.3. (There is no need to complete E with respect to some other topology, which is the effect of the isomorphism $K_0(L^1(G,B)) \simeq K_0(C^*(G,B))$ in the general case. This is why the proof is simpler for finite groups.) If E is a finitely generated projective (G,B)-module, we will write $\widetilde{E}$ for E regarded as a $C^*(G,B)$-module. We also use the same notation for $(G,A\otimes B)$-modules.

Let E be a finitely generated projective A-module, and let (G,F,λ) be a finitely generated projective (G,B)-module. Let E_0 denote E regarded as a finitely generated projective (G,A)-module with the trivial action of G. Then $[E]\otimes[F]$ is an element of the middle left group of (*), namely $K_0(A)\otimes_{\mathbf{Z}} K_0^G(B)$. Its image in the upper left group, $K_0^G(A)\otimes_{R(G)} K_0^G(B)$, is $[E_0]\otimes[F]$. Since

$$\omega_0([E_0]\otimes[F]) = [E_0\otimes F] = \omega_1([E]\otimes[F]),$$

the upper rectangle commutes. Since $(E\otimes F)^{\sim} = E\otimes\widetilde{F}$, the lower rectangle also commutes. Therefore the whole diagram commutes, as required. Q.E.D.

6.1.8 Proposition. The pairing ω respects exterior equivalence. That is, if G is a compact group, α^0 and α^1 are exterior equivalent actions of G on a C^*-algebra A, and (G,B,β) is a G-algebra, then there is a commutative diagram

$$\begin{array}{ccc} K_*^G(A,\alpha^0)\otimes_{R(G)}K_*^G(B) & \xrightarrow{\omega(A,\alpha^0,B)} & K_*^G(A\otimes B,\alpha^0\otimes\beta) \\ \downarrow\simeq & & \downarrow\simeq \\ K_*^G(A,\alpha^1)\otimes_{R(G)}K_*^G(B) & \xrightarrow{\omega(A,\alpha^1,B)} & K_*^G(A\otimes B,\alpha^1\otimes\beta)\,. \end{array} \qquad (*)$$

Here $K_*^G(A,\alpha^0)$ is the equivariant K-theory of A for the action α^0, and similarly for the other groups. The vertical isomorphisms are given by the exterior equivalence of α^0 with α^1, as in theorem 2.8.3 (5), and by the corresponding exterior equivalence of $\alpha^0\otimes\beta$ with $\alpha^1\otimes\beta$.

Proof. The exterior equivalence of α^0 and α^1 means that there is a cocycle $g \longrightarrow u_g$ from G to the unitary group of $M(A)$ such that $\alpha_g^1(a)=u_g\,\alpha_g^0(a)u_g^*$ for $g\in G$. (See definition 2.7.1.) The corresponding exterior equivalence of $\alpha^0\otimes\beta$ with $\alpha^1\otimes\beta$ is then given by the cocycle $g \longrightarrow u_g\otimes 1$. (Note that $u_g\otimes 1$ certainly defines an element of the unitary group of $M(A\otimes B)$.) Since the pairing is natural, it is sufficient to exhibit the isomorphism $K_*^G(A,\alpha^0)\simeq K_*^G(A,\alpha^1)$ as a composite of induced maps φ_* for equivariant homomorphisms φ, in a natural way.

By [79], 8.11.2, there is an action $\alpha : G \longrightarrow Aut(M_2(A))$ such that the maps

$$\varphi^i : (G,A,\alpha^i) \longrightarrow (G,M_2(A),\alpha)$$

defined by

$$\varphi^0(a) = \begin{bmatrix} a & 0 \\ 0 & 0 \end{bmatrix} \quad \text{and} \quad \varphi^1(a) = \begin{bmatrix} 0 & 0 \\ 0 & a \end{bmatrix}$$

are equivariant. It is given by

$$\alpha_g \begin{bmatrix} a & b \\ c & d \end{bmatrix} = \begin{bmatrix} \alpha_g^0(a) & \alpha_g^0(b)u_g^* \\ u_g \alpha_g^0(c) & \alpha_g^1(d) \end{bmatrix}.$$

This construction is clearly natural; in particular, $\alpha\otimes\beta$ and $\varphi^i\otimes id_B$ satisfy the analogous conditions for the exterior equivalence of $\alpha^0\otimes\beta$ and $\alpha^1\otimes\beta$. We now only have to show that φ_*^1 is invertible and that $(\varphi_*^1)^{-1}\circ\varphi_*^0$ is the usual isomorphism $K_*^G(A,\alpha^0)\simeq K_*^G(A,\alpha^1)$.

Set $v_g = \begin{bmatrix} u_g & 0 \\ 0 & 1 \end{bmatrix}$ for $g\in G$. Then one readily checks that $g \longrightarrow v_g$ defines an exterior equivalence between the actions α and $id_{M_2}\otimes\alpha^1$ on $M_2(A)$. Now consider the diagram

$$(A,\alpha^0) \xrightarrow{\varphi^0} (M_2(A),\alpha) \xleftarrow{\varphi^1} (A,\alpha^1). \qquad (**)$$

If we apply the exterior equivalence $g \longrightarrow u_g$ to (A,α^0), the exterior equivalence $g \longrightarrow v_g$ to $(M_2(A),\alpha)$, and the trivial exterior equivalence $g \longrightarrow 1$ to (A,α^1), then the maps φ^0 and φ^1 remain equivariant. In order to distinguish them on K-theory, we will rename them ψ^0 and ψ^1. Thus, the result of applying the exterior equivalences above to the diagram (**) is the diagram

$$(A,\alpha^1) \xrightarrow{\psi^0} (M_2(A), id_{M_2}\otimes\alpha^1) \xleftarrow{\psi^1} (A,\alpha^1).$$

By naturality of the isomorphism of equivariant K-theory for exterior equivalent actions, we obtain the following commutative diagram, in which $f : K_*^G(A,\alpha^0)\longrightarrow K_*^G(A,\alpha^1)$ is the isomorphism of theorem 2.8.3 (5):

$$\begin{array}{ccccc} K_*^G(A,\alpha^0) & \xrightarrow{\varphi_*^0} & K_*^G(M_2(A),\alpha) & \xleftarrow{\varphi_*^1} & K_*^G(A,\alpha^1) \\ f\downarrow\simeq & & \downarrow\simeq & & \downarrow= \\ K_*^G(A,\alpha^1) & \xrightarrow{\psi_*^0} & K_*^G(M_2(A),id_{M_2}\otimes\alpha^1) & \xleftarrow{\psi_*^1} & K_*^G(A,\alpha^1). \end{array}$$

In the bottom row, $\psi^0(a) = w\psi^1(a)w^*$ for $a\in A$, where $w = \begin{bmatrix} 0 & 1 \\ 1 & 0 \end{bmatrix}$. Since w is invariant for the action $id_{M_2}\otimes\alpha^1$ of G, it follows that $\psi_*^0 = \psi_*^1$. A diagram chase now shows that $\varphi_*^0 = \varphi_*^1\cdot f$. We now compute ψ_*^1 using theorem 2.8.3 (7). Taking crossed products in the lower right part of the diagram shows that ψ_*^1 is given on the ordinary K-theory of the crossed products by the map $\bar{\psi}_*^1$, where

$$\bar{\psi}^1 : C^*(G,A,\alpha^1) \longrightarrow M_2(C^*(G,A,\alpha^1))$$

is defined by $\bar{\psi}^1(a) = \begin{bmatrix} 0 & 0 \\ 0 & a \end{bmatrix}$. Since $\bar{\psi}_*^1$ is an isomorphism (see, for example, theorem

2.8.3 (4)), the map ψ_*^1 is also, and therefore so is φ_*^1. We now have $(\varphi_*^1)^{-1}\cdot\varphi_*^0 = f$, as desired. This completes the proof that the diagram (*) commutes. Q.E.D.

6.2. The Künneth Formula: Some Special Cases

6.2.1 Notation. Throughout this section, G is a finite *cyclic* group, and $P \subset R(G)$ is a prime ideal with support G. (That is, P is not the inverse image of a prime ideal in $R(H)$ for any subgroup H of G. See the discussion preceding the Localization Theorem 1.2.4.) To simplify notation, we let $R = R(G)_P$, and we set $h_i(A) = K_i^G(A)_P$ and $h_*(A) = K_*^G(A)_P$, for any G-algebra A. We denote the pairing of corollary 6.1.6 by ω_P.

The purpose of this section is to prove the analogs of some of the results in section 2 of [95] for the homology theory h_*. Thus, we are going to prove, under certain technical hypotheses on A, that $\omega_P(A,B)$ is an isomorphism if $h_*(B)$ is a projective R-module. This means of course that

$$\omega_P : h_*(A) \otimes_R h_*(B) \longrightarrow h_*(A \otimes B)$$

is an isomorphism. We start off with an important fact about the structure of R.

6.2.2 Proposition. Let n be the order of G, and let $\zeta \in \mathbf{C}$ be a primitive nth root of unity. Let $\mathbf{Z}[\zeta]$ be the subring of $\mathbf{C}$ generated by $\mathbf{Z}$ and ζ. Then there is a prime ideal $Q \subset \mathbf{Z}[\zeta]$ such that R is isomorphic to the localized ring $\mathbf{Z}[\zeta]_Q$.

Proof. Let τ generate $\widehat{G}$. Let $\varphi : \mathbf{Z}[x] \longrightarrow R(G)$ be the unique ring homomorphism from the polynomial ring $\mathbf{Z}[x]$ in the indeterminate x to $R(G)$ such that $\varphi(x) = \tau$. Then φ is surjective, and its kernel is $I = \langle x^n - 1\rangle$, the ideal generated by $x^n - 1$. Therefore $\varphi^{-1}[P]$ is a prime ideal in $\mathbf{Z}[x]$ which contains $x^n - 1$. Write

$$x^n - 1 = \prod_{d|n} f_d(x), \qquad (*)$$

where $d\,|\,n$ means d divides n, and $f_d(x)$ is the minimal polynomial over $\mathbf{Q}$ for the primitive d-th root of unity $\zeta^{n/d}$. We assume that $f_d(x)$ is normalized so it has leading coefficient equal to 1. Then $f_d(x)$ is well known to have integer coefficients. (See [64], page 206.)

Suppose $f_d(x) \in \varphi^{-1}[P]$ for some $d \neq n$. Then $x^d - 1 \in \varphi^{-1}[P]$, because $f_d(x)$ is a factor of $x^d - 1$. Consequently $\tau^d - 1 \in P$. Let H be the subgroup of G of order d; then $\langle \tau^d - 1\rangle$ is exactly the kernel of the restriction homomorphism $R(G) \longrightarrow R(H)$. Thus if $\tau^d - 1 \in P$ then the support of P is contained in H, which contradicts the assumption that the support of P is G. Therefore $f_d(x) \notin \varphi^{-1}[P]$ for all $d \neq n$. Since $\varphi^{-1}[P]$ is a prime ideal and $x^n - 1 \in \varphi^{-1}[P]$, it follows from (*) that $f_n(x) \in \varphi^{-1}[P]$.

The ideal $I_{\varphi^{-1}[P]}$ in $\mathbf{Z}[x]_{\varphi^{-1}[P]}$ is generated by $x^n - 1$. Since $f_d(x) \notin \varphi^{-1}[P]$ for $d \neq n$, the elements $f_d(x)$ are invertible in $\mathbf{Z}[x]_{\varphi^{-1}[P]}$ and therefore $I_{\varphi^{-1}[P]}$ is generated by $f_n(x)$, which is the minimal polynomial of ζ over $\mathbf{Q}$. Let J be the ideal in $\mathbf{Z}[x]$ generated by $f_n(x)$. Then we have $J_{\varphi^{-1}[P]} = I_{\varphi^{-1}[P]}$. Therefore

$$R = R(G)_P \simeq (\mathbf{Z}[x]/I)_P \simeq \mathbf{Z}[x]_{\varphi^{-1}[P]}/I_{\varphi^{-1}[P]}$$
$$= \mathbf{Z}[x]_{\varphi^{-1}[P]}/J_{\varphi^{-1}[P]} \simeq (\mathbf{Z}[x]/J)_Q,$$

where Q is the image of $\varphi^{-1}[P]$ in $\mathbf{Z}[x]/J$. (Note that $\varphi^{-1}[P]$ contains J.) Now $\mathbf{Z}[x]/J$ is isomorphic to $\mathbf{Z}[\zeta]$, and Q is a prime ideal in $\mathbf{Z}[x]/J$. So $R \simeq \mathbf{Z}[\zeta]_Q$, as desired.

Q.E.D.

In the rest of this section, we prove that $\omega_P(A,B)$ is an isomorphism for $h_*(B)$ projective and for two special classes of G-algebras A. We then prove a series of lemmas showing how to enlarge the class of algebras A for which $\omega_P(A,B)$ is an isomorphism, and at the end of the section we apply the lemmas to show that, if $h_*(B)$ is projective, then $\omega_P(A,B)$ is an isomorphism for all type I G-algebras A. We assume throughout that B is nuclear; we could, however, assume instead that all algebras occurring as A are nuclear.

For the next lemma, recall that the Künneth formula in [95] (theorem 4.1) is proved to hold in case one of the algebras is in the category **N** defined as follows. **N** is the smallest category of separable nuclear C^*-algebras such that:

(1) **N** contains all separable type I algebras.

(2) If

$$0 \longrightarrow I \longrightarrow A \longrightarrow B \longrightarrow 0$$

is exact and two of the algebras are in **N**, then so is the third.

(3) **N** is closed under the formation of direct limits.

(4) **N** is closed under the formation of crossed products by **Z** and **R**.

(5) If A and B are stably isomorphic ($K \otimes A \simeq K \otimes B$, where K is the algebra of compact operators on a separable Hilbert space), and $A \in \mathbf{N}$, then $B \in \mathbf{N}$.

6.2.3 Lemma. Let $A \in \mathbf{N}$, let G act trivially on A, and let B be a nuclear G-algebra such that $h_*(B)$ is projective. Then $\omega_P(A,B)$ is an isomorphism.

Proof. Let ω_0 denote the nonequivariant pairing $\omega(A, C^*(G,B))$. We then have the following diagram, in which the rectangle commutes by proposition 6.1.7 and the right hand column is exact by the ordinary Künneth formula ([95], theorem 4.1):

$$\begin{array}{ccc} & & 0 \\ & & \downarrow \\ K^G_*(A)\otimes_{R(G)}K^G_*(B) & \xrightarrow{\simeq} & K_*(A)\otimes_{\mathbf{Z}}K_*(C^*(G,B)) \\ \omega(A,B)\ \downarrow & & \downarrow\ \omega_0 \\ K^G_*(A\otimes B) & \xrightarrow{\simeq} & K_*(A\otimes C^*(G,B)) \\ & & \downarrow \\ & & Tor^{\mathbf{Z}}(K_*(A),(C^*(G,B))) \\ & & \downarrow \\ & & 0 \end{array}$$

In the right hand column, the first and last terms are $R(G)$-modules because $K_*(C^*(G,B))$ is, and the middle term is an $R(G)$-module because it is $K_*(C^*(G,A\otimes B))$. The naturality of the Künneth formula then implies that the maps in the right hand column are $R(G)$-module homomorphisms. We want to show that $\omega_P(A,B)$ is an isomorphism; since localization is exact, this will be the case if we can show that $Tor^{\mathbf{Z}}(K_*(A),K_*(C^*(G,B)))_P = 0$.

We simplify the notation. Let $M = K_*(A)$, which is an abelian group. Let $N = K_*(C^*(G,B))$, which is an $R(G)$-module such that N_P is a projective R-module. (Recall that $R = R(G)_P$.) Then we have to show that $Tor^{\mathbf{Z}}(M,N)_P = 0$.

Let

$$0 \longrightarrow X \longrightarrow Y \longrightarrow M \longrightarrow 0$$

be a free resolution of M over $\mathbf{Z}$. (That is, the sequence is exact and X and Y are free abelian groups.) Then there is an exact sequence (see [47], section 8.3):

$$0 \longrightarrow Tor^{\mathbf{Z}}(M,N) \longrightarrow X\otimes_{\mathbf{Z}}N \longrightarrow Y\otimes_{\mathbf{Z}}N \longrightarrow M\otimes_{\mathbf{Z}}N \longrightarrow 0 .$$

Since N is an $R(G)$-module, and the sequence is natural, it is a sequence of $R(G)$-modules. We localize this sequence at P. We have

$$(X\otimes_{\mathbf{Z}}N)_P \simeq X\otimes_{\mathbf{Z}}N\otimes_{R(G)}R(G)_P \simeq X\otimes_{\mathbf{Z}}N_P ,$$

and similarly with X replaced by Y, so the sequence becomes

$$0 \longrightarrow Tor^{\mathbf{Z}}(M,N)_P \longrightarrow X\otimes_{\mathbf{Z}}N_P \longrightarrow Y\otimes_{\mathbf{Z}}N_P \longrightarrow M\otimes_{\mathbf{Z}}N_P \longrightarrow 0 .$$

Using the same free resolution to compute $Tor^{\mathbf{Z}}(M,N_P)$, we obtain the exact sequence

$$0 \longrightarrow Tor^{\mathbf{Z}}(M,N_P) \longrightarrow X\otimes_{\mathbf{Z}}N_P \longrightarrow Y\otimes_{\mathbf{Z}}N_P \longrightarrow M\otimes_{\mathbf{Z}}N_P \longrightarrow 0 .$$

Therefore $Tor^{\mathbf{Z}}(M,N)_P \simeq Tor^{\mathbf{Z}}(M,N_P)$. Now N_P is a projective module over R by hypothesis. It follows from proposition 6.2.2 that R is a subring of $\mathbf{C}$, and hence is torsion-free over $\mathbf{Z}$. Therefore N_P, being a projective R-module, is also torsion-free over $\mathbf{Z}$, and $Tor^{\mathbf{Z}}(M,N)_P = Tor^{\mathbf{Z}}(M,N_P) = 0$, as desired. Q.E.D.

We point out that we made no essential use of the special hypotheses on G and P until the second to the last sentence of the above proof, where we appealed to proposition 6.2.2. The rest of the argument generalizes to arbitrary compact groups G and arbitrary prime ideals P in $R(G)$.

We now consider the second special class of algebras promised earlier. In this case, we do not have to assume that $h_*(B)$ is projective.

6.2.4 Proposition. Let A be a separable G-algebra, and suppose that G acts on $Prim(A)$ with no fixed points. Let B be any nuclear G-algebra. Then $\omega_P(A,B)$ is an isomorphism.

Proof. It follows from proposition 4.3.7 that $h_*(A) = h_*(A \otimes B) = 0$. Therefore the domain and codomain of $\omega_P(A,B)$ are both zero. Q.E.D.

Notice that the previous lemma applies to algebras that need not be in the category **N** considered in [95].

We now show how to extend the category of algebras A such that $\omega_P(A,B)$ is an isomorphism whenever $h_*(B)$ is projective.

6.2.5 Lemma. (Compare [95], proposition 2.6.) Let A be a G-algebra and let I be a G-invariant ideal in A. Let B be a nuclear G-algebra such that $h_*(B)$ is projective. If ω_P is an isomorphism for two of the pairs (I,B), (A,B), and $(A/I,B)$, then it is for the third as well. If the sequence

$$0 \longrightarrow I \longrightarrow A \longrightarrow A/I \longrightarrow 0$$

is equivariantly split, and $\omega_P(A,B)$ is an isomorphism, then $\omega_P(I,B)$ and $\omega_P(A/I,B)$ are isomorphisms.

Proof. We have a diagram:

$$
\begin{array}{ccc}
\vdots & & \vdots \\
\downarrow & & \downarrow \\
(h_i(I)\otimes_R h_0(B))\oplus(h_{1-i}(I)\otimes_R h_1(B)) & \xrightarrow{\omega_P(I,B)} & h_i(I\otimes B) \\
\downarrow & & \downarrow \\
(h_i(A)\otimes_R h_0(B))\oplus(h_{1-i}(A)\otimes_R h_1(B)) & \xrightarrow{\omega_P(A,B)} & h_i(A\otimes B) \\
\downarrow & & \downarrow \\
(h_i(A/I)\otimes_R h_0(B))\oplus(h_{1-i}(A)\otimes_R h_1(B)) & \xrightarrow{\omega_P(A/I,B)} & h_i((A/I)\otimes B) \\
\downarrow & & \downarrow \\
(h_{1-i}(I)\otimes_R h_0(B))\oplus(h_i(I)\otimes_R h_1(B)) & \xrightarrow{\omega_P(I,B)} & h_{1-i}(I\otimes B) \\
\downarrow & & \downarrow \\
\vdots & & \vdots
\end{array}
\qquad (*)
$$

It is commutative because the pairing ω_P is natural and respects connecting homomorphisms. The right hand column is exact because it is obtained by tensoring the long exact sequence

$$\cdots \longrightarrow h_i(I) \longrightarrow h_i(A) \longrightarrow h_i(A/I) \longrightarrow h_{1-i}(I) \longrightarrow \cdots$$

over R with the projective, hence flat, R-module $h_*(B)$. The left hand column is the long exact sequence associated to the exact sequence

$$0 \longrightarrow I\otimes B \longrightarrow A\otimes B \longrightarrow (A/I)\otimes B \longrightarrow 0. \qquad (**)$$

By assumption, two out of every three horizontal maps are isomorphisms. By the Five Lemma, the third is as well.

If the sequence

$$0 \longrightarrow I \longrightarrow A \longrightarrow A/I \longrightarrow 0$$

is split, then so is the sequence of tensor products (**), and the vertical sequences in the diagram (*) break up into split short exact sequences. We obtain a diagram with split exact columns:

$$\begin{array}{ccc}
0 & & 0 \\
\downarrow & & \downarrow \\
(h_i(I)\otimes_R h_0(B))\oplus(h_{1-i}(I)\otimes_R h_1(B)) & \xrightarrow{\omega_P(I,B)} & h_i(I\otimes B) \\
\downarrow & & \downarrow \\
(h_i(A)\otimes_R h_0(B))\oplus(h_{1-i}(A)\otimes_R h_1(B)) & \xrightarrow{\omega_P(A,B)} & h_i(A\otimes B) \\
\mu \uparrow \; \downarrow & & \downarrow \; \uparrow \nu \\
(h_i(A/I)\otimes_R h_0(B))\oplus(h_{1-i}(A/I)\otimes_R h_1(B)) & \xrightarrow{\omega_P(A/I,B)} & h_i((A/I)\otimes B) \\
\downarrow & & \downarrow \\
0 & & 0
\end{array}$$

Since the product is natural, we find that $\nu\cdot\omega_P(A/I,B) = \omega_P(A,B)\cdot\mu$. We are assuming that $\omega_P(A,B)$ is an isomorphism, and a diagram chase now shows that $\omega_P(A/I,B)$ is an isomorphism. (This conclusion does not follow without the maps μ and ν.) It now follows that $\omega_P(I,B)$ is also an isomorphism. Q.E.D.

6.2.6 Lemma. (Compare [95], proposition 2.4.) Let (A_λ) be a directed system of G-algebras, and let B be a nuclear G-algebra. If $\omega_P(A_\lambda,B)$ is an isomorphism for all λ, then $\omega_P(\varinjlim A_\lambda, B)$ is an isomorphism. (Here we don't need to assume that $h_*(B)$ is projective.)

Proof. We have $(\varinjlim A_\lambda)\otimes B \simeq \varinjlim(A_\lambda\otimes B)$ since B is nuclear. Furthermore, h_* commutes with direct limits. Therefore there is a commutative diagram

$$\begin{array}{ccc}
h_*(\varinjlim A_\lambda)\otimes_R h_*(B) & \xrightarrow{\omega_P(\varinjlim A_\lambda, B)} & h_*((\varinjlim A_\lambda)\otimes B) \\
\downarrow\simeq & & \downarrow\simeq \\
(\varinjlim h_*(A_\lambda))\otimes_R h_*(B) & & h_*(\varinjlim(A_\lambda\otimes B)) \\
\downarrow\simeq & & \downarrow\simeq \\
\varinjlim(h_*(A_\lambda)\otimes_R h_*(B)) & \xrightarrow{\varinjlim \omega_P(A_\lambda, B)} & \varinjlim h_*(A_\lambda\otimes B)
\end{array}$$

Since each $\omega_P(A_\lambda,B)$ is an isomorphism, so is their direct limit. Therefore $\omega_P(\varinjlim A_\lambda, B)$ is an isomorphism. Q.E.D.

6.2.7 Corollary. Let A be a G-algebra with a G-invariant composition series $(A_\lambda)_{\lambda\le\kappa}$, such that $A_0=0$. Let B be a nuclear G-algebra such that $h_*(B)$ is projective. If $\omega_P(A_{\lambda+1}/A_\lambda, B)$ is an isomorphism for all λ, then $\omega_P(A,B)$ is an isomorphism.

Proof. We prove that $\omega_P(A_\lambda,B)$ is an isomorphism by transfinite induction on λ. For $\lambda=0$ it is trivial, since $A_0=0$. If it is known for λ then it follows for $\lambda+1$ by lemma 6.2.5. If it is known for all $\lambda<\mu$, where μ is a limit ordinal, then it follows for μ by lemma 6.2.6. Q.E.D.

6.2.8 Lemma. Let A be a C^*-algebra, and let α^0 and α^1 be two exterior equivalent actions of G on A. Let (G,B,β) be a nuclear G-algebra. If $\omega_P(A,B)$ is an isomorphism for the action α^0, then $\omega_P(A,B)$ is an isomorphism for the action α^1.

Proof. Write $K_*^G(A,\alpha^0)$ for the equivariant K-theory of A for the action α^0, etc. By proposition 6.1.8, there is a commutative diagram

$$\begin{array}{ccc} K_*^G(A,\alpha^0)\otimes_{R(G)}K_*^G(B) & \xrightarrow{\omega(A,\alpha^0,B)} & K_*^G(A\otimes B,\alpha^0\otimes\beta) \\ \downarrow\simeq & & \downarrow\simeq \\ K_*^G(A,\alpha^1)\otimes_{R(G)}K_*^G(B) & \xrightarrow{\omega(A,\alpha^1,B)} & K_*^G(A\otimes B,\alpha^1\otimes\beta). \end{array}$$

Localize this diagram at P. Then the top horizontal map becomes an isomorphism. Therefore so does the bottom horizontal map. Q.E.D.

6.2.9 Lemma. Let A and B be G-algebras with B nuclear. Let K be the algebra of compact operators on some Hilbert space, with some action γ of G. Then $\omega_P(A,B)$ is an isomorphism if and only if $\omega_P(K\otimes A,B)$ is an isomorphism.

Proof. First assume that the action on K is trivial. By theorem 2.8.3 (4) and the definition of ω, the isomorphism $K_*^G(A)\simeq K_*^G(K\otimes A)$ is given by $\eta \longrightarrow \omega(K,A)([p]\otimes\eta)$, where p is any one dimensional projection in K. The associativity of the product now implies that $\omega_P(A,B)$ is an isomorphism if and only if $\omega_P(K\otimes A,B)$ is, by localizing the commutative diagram

$$\begin{array}{ccc} K_*^G(K)\otimes_{R(G)}K_*^G(A)\otimes_{R(G)}K_*^G(B) & \longrightarrow & K_*^G(K)\otimes_{R(G)}K_*^G(A\otimes B) \\ \downarrow & & \downarrow \\ K_*^G(K\otimes A)\otimes_{R(G)}K_*^G(B) & \longrightarrow & K_*^G(K\otimes A\otimes B). \end{array}$$

Now suppose the action γ of G on K is not trivial. Then, since G is cyclic, γ is inner by lemma 4.2.12. Consequently, it is exterior equivalent to the trivial action ι. Therefore $\gamma\otimes\alpha$ is exterior equivalent to $\iota\otimes\alpha$, where α is the action of G on A. The result now follows from the previous lemma and the first paragraph of the proof. Q.E.D.

We now use these results to prove that if A is separable type I, then $\omega_P(A,B)$ is an isomorphism whenever $h_*(B)$ is projective. The proof requires a lemma.

6.2.10 Lemma. Let H be a Hilbert space, let $A=K(H)\otimes C_0(X)$, where X is a second countable locally compact Hausdorff space, and let $\alpha: G\longrightarrow Aut(A)$ be an action of G which is trivial on $Prim(A)$ (which, of course, is just X). Let B be a nuclear G-

algebra such that $h_*(B)$ is projective. Then $\omega_P(A,B)$ is an isomorphism.

Proof. In [79], 8.7.4, it is shown that every automorphism of $K(H)$ is inner. We adapt the proof given there to show that the action α is "locally inner" in an appropriate sense.

Let g generate G, and identify A with $C_0(X,K(H))$. Since G acts trivially on $Prim(A)$, we have $\alpha_g[I]=I$ for every ideal I of A. Taking

$$I = I_x = \{f \in C_0(X,K(H)) : f(x) = 0\}$$

for a fixed $x \in X$, we obtain an automorphism α_g^x of $A/I_x \simeq K(H)$. Clearly $\alpha_g(f)(x) = \alpha_g^x(f(x))$ for any $f \in C_0(X,K(H))$.

Fix $x_0 \in X$. Let p be a one-dimensional projection in $K(H)$. Then $\alpha_g^{x_0}(p)$ is also a one-dimensional projection. Therefore there is a unitary $u_{x_0} \in L(H)$ such that $u_{x_0} p u_{x_0}^* = \alpha_g^{x_0}(p)$. Furthermore, $x \longrightarrow \alpha_g^x(p)$ is continuous, since α_g is an automorphism of $C_0(X,K(H)) = A$. Therefore there is a neighborhood U of x_0 such that $\|\alpha_g^x(p) - \alpha_g^{x_0}(p)\| < \frac{1}{2}$ for $x \in U$. Lemma 2.5.3 (2) provides a unitary $v_x \in L(H)$ such that $v_x \alpha_g^x(p) v_x^* = \alpha_g^{x_0}(p)$. Recall that $v_x = t(t^*t)^{-\frac{1}{2}}$, where

$$t = \alpha_g^x(p)\alpha_g^{x_0}(p) + (1 - \alpha_g^x(p))(1 - \alpha_g^{x_0}(p)).$$

Therefore $x \longrightarrow v_x$ is continuous and $v_{x_0} = 1$. Define $u_x = v_x^* u_{x_0}$ for $x \in U$. Then $u_x p u_x^* = \alpha_g^x(p)$. Now set $\beta_g^x(a) = u_x^* \alpha_g^x(a) u_x$ for $x \in U$, $a \in K(H)$, and $g \in G$. It follows that β_g^x is an automorphism of $K(H)$ such that $a \longrightarrow \beta_g^x(a)$ is continuous for $a \in K(H)$, and such that $\beta_g^x(p) = p$.

Let ξ be a unit vector in pH. It is shown in the proof of [79], 8.7.4, that the definition $w_x(a\xi) = \beta_g^x(a)\xi$ for $a \in K(H)$ defines a unitary operator $w_x \in L(H)$ such that $w_x a w_x^* = \beta_g^x(a)$ for all $a \in K(H)$. Since $x \longrightarrow \beta_g^x(a)$ is continuous, it follows from the definition that $x \longrightarrow w_x \xi$ is continuous for all $\xi \in H$. Consequently $x \longrightarrow w_x$ is a strong operator continuous function from U to $L(H)$. Therefore so is $x \longrightarrow u_x w_x$. Furthermore, $u_x w_x a (u_x w_x)^* = \alpha_g^x(a)$.

Let n be the order of G. Then $(\alpha_g^x)^n = id_{K(H)}$ for $x \in U$. Therefore $(u_x w_x)^n$ is a scalar multiple of the identity, that is, $(u_x w_x)^n = \lambda_x$ for some $\lambda_x \in \mathbf{C}$ with $|\lambda_x| = 1$. Furthermore, $x \longrightarrow \lambda_x$ is continuous. Let U_0 be a neighborhood of x_0 that is sufficiently small that there is a continuous branch of $\lambda_x^{-1/n}$ defined on U_0, and set $s_x = \lambda_x^{-1/n} u_x w_x$. Then for $x \in U_0$, we have $s_x^n = 1$, and for $f \in C_0(U_0,K(H))$, we have

$$s_x f(x) s_x^* = \alpha_g^x(f(x)) = \alpha_g(f)(x)$$

for all $x \in U_0$. The function $x \longrightarrow s_x$ is clearly in the multiplier algebra of $C_0(U_0,K(H))$, and, since $s_x^n = 1$, we conclude that $\alpha_{(\cdot)}|_{C_0(U_0,K(H))}$ is inner. To summarize our conclusions: under the hypotheses of the lemma, there is a nonzero G-invariant ideal $I = C_0(U_0,K(H))$ in A such that $\alpha_{(\cdot)}|_I$ is inner.

We now claim that $\omega_P(I,B)$ is an isomorphism. By lemma 6.2.8, it is sufficient to show that $\omega_P(I,B)$ is an isomorphism when I has the trivial action of G. By lemma 6.2.9, this will be the case if $\omega_P(C_0(U_0),B)$ is an isomorphism, since $I = K(H)\otimes C_0(U_0)$. But $\omega_P(C_0(U_0), B)$ is an isomorphism by lemma 6.2.3.

By repeating the process for A/I rather than A, we can construct a composition series (I_λ) for A such that the actions on all the factors $I_{\lambda+1}/I_\lambda$ have the same form as the action on I. Therefore $\omega_P(A,B)$ is an isomorphism by corollary 6.2.7.

Q.E.D.

6.2.11 Proposition. (Compare [95], theorem 2.13.) Let (G,A,α) be a separable type I G-algebra, and let B be a nuclear G-algebra such that $h_*(B)$ is projective. Then $\omega_P(A,B)$ is an isomorphism.

Proof. By [29], 4.4.5, the primitive ideal space of any type I C^*-algebra contains a dense Hausdorff open subset. Since primitive ideal spaces are Baire spaces ([29], 3.4.13), and since G is finite, the intersection of the translates under elements of G of a dense open subset is again dense and open. Thus, the primitive ideal space of a type I G-algebra contains a dense Hausdorff G-invariant open subset. (We will prove in chapter 8 that this actually holds for general compact metrizable groups.) We can therefore construct a G-invariant composition series (I_λ) for A such that each $I_{\lambda+1}/I_\lambda$ has Hausdorff primitive ideal space. By corollary 6.2.7, we may therefore reduce to the case in which $Prim(A)$ is Hausdorff.

Let

$$X = \{x \in Prim(A) : gx = x \text{ for all } g \in G\}.$$

Then X is a closed G-invariant subset of $Prim(A)$, and its complement U is open. Let I be the ideal in A corresponding to U. Then $Prim(I)$ is Hausdorff and G has no fixed points in $Prim(I)$, so by proposition 6.2.4, $\omega_P(I,B)$ is an isomorphism. By lemma 6.2.5, $\omega_P(A,B)$ will be an isomorphism if we can show that $\omega_P(A/I,B)$ is an isomorphism.

By Lemma 6.2.9, $\omega_P(A/I,B)$ is an isomorphism if and only if $\omega_P(K\otimes(A/I),B)$ is an isomorphism, where K is the algebra of compact operators on a Hilbert space H. We may therefore assume that A/I is stable, that is, $K\otimes(A/I) \simeq A/I$. It follows ([29],4.5.5) that A/I has a composition series for which the factors are continuous trace algebras. (Since G acts trivially on $Prim(A/I)$, all ideals are G-invariant.) Each stable separable continuous trace algebra is the algebra of continuous sections of a locally trivial continuous field (see [90], beginning of section 6), and therefore has a composition series in which the factors are the sets of continuous sections of trivial continuous fields, that is, have the form $K(H)\otimes C_0(X)$ for appropriate H and X. By the previous lemma, $\omega_P(K(H)\otimes C_0(X),B)$ is an isomorphism, since G acts trivially on the primitive ideal space. Several applications of corollary 6.2.7 now show that $\omega_P(K\otimes(A/I),B)$ is an isomorphism, as needed. Q.E.D.

6.2.12 Remark. We have assumed throughout this section that all algebras appearing in the second position in $\omega_P(\cdot,\cdot)$ are nuclear. It is, however, clear that the results on ω_P all continue to hold if this assumption is replaced by the assumption that all algebras appearing in the first position are nuclear, since the only use made of nuclearity is to guarantee the uniqueness of tensor products.

6.3. The Equivariant K-Theory of Crossed Products by Z, R, and F_n

We are still missing the analogs of propositions 2.5 and 2.7 in [95], which are used there to show that the category **N** of algebras for which the Künneth formula holds is closed under formation of crossed products by **Z** and **R**. We need the analogs of the Connes isomorphism theorem ([20]) and the Pimsner-Voiculescu exact sequences ([82] and [83]) for equivariant K-theory in order to proceed. They follow very easily from the nonequivariant versions. We start with:

6.3.1 Proposition. Let (G,A,α) be a G-algebra, where G is an arbitrary group. Let $\beta : H \longrightarrow Aut(A)$ be an action of a locally compact group H on A which commutes with α, that is, $\alpha_g \cdot \beta_h = \beta_h \cdot \alpha_g$ for $g \in G$ and $h \in H$. Then the crossed product $C^*(H,A,\beta)$ and the reduced crossed product $C_r^*(H,A,\beta)$ ([79], 7.7.4) are G-algebras in a natural way.

Proof. Define $\lambda : G \longrightarrow Aut(L^1(H,A,\beta))$ by $\lambda_g(f)(h) = \alpha_g(f(h))$ for $f \in L^1(H,A,\beta)$, $g \in G$, and $h \in H$. Using the fact that α_g commutes with β_h, it is trivial to check that λ_g is a $*$-automorphism of $L^1(H,A,\beta)$, and it is obvious that $g \longrightarrow \lambda_g$ is a homomorphism.

We want to show that $g \longrightarrow \lambda_g(f)$ is continuous for $f \in L^1(H,A,\beta)$. First, assume that $f \in C_c(H,A)$. Then the function $(g,h) \longrightarrow \alpha_g(f(h))$ from $G \times H$ to A is jointly continuous. Since f has compact support, a standard argument shows that $g \longrightarrow \lambda_g(f)$ is continuous for the supremum norm on $C_c(H,A)$, that is, if (g_n) is a net converging to $g \in G$, then $\|\lambda_{g_n}(f) - \lambda_g(f)\|_\infty \longrightarrow 0$. Again using the fact that f has compact support, we conclude that if $g_n \longrightarrow g$ then $\|\lambda_{g_n}(f) - \lambda_g(f)\|_1 \longrightarrow 0$. Since $C_c(H,A)$ is dense in $L^1(H,A,\beta)$, and since $\|\lambda_g\| = 1$ for all $g \in G$, it follows that $g \longrightarrow \lambda_g(f)$ is continuous for all $f \in L^1(H,A,\beta)$. Thus, λ is continuous.

It follows immediately that λ extends to a continuous action of G on $C^*(H,A,\beta)$. Furthermore, λ_g is easily seen to preserve the norm which defines $C_r^*(H,A,\beta)$, since if σ is the regular representation of $L^1(H,A,\beta)$ defined by the representation π of A ([79], 7.7.1), then $\sigma \cdot \lambda_g$ is the regular representation defined by $\pi \cdot \alpha_g$. So λ also extends to a continuous action on $C_r^*(H,A,\beta)$.

Naturality is obvious. Q.E.D.

Throughout this section, if (G,A,α) is a G-algebra and β is an action of H on A which commutes with α, then $C^*(H,A,\beta)$ and $C_r^*(H,A,\beta)$ will be assumed to carry the action of this proposition.

6.3.2 Theorem (Connes isomorphism). Let G be a compact group, let (G,A,α) be a G-algebra, and let $\beta : \mathbf{R} \longrightarrow Aut(A)$ be a continuous action of $\mathbf{R}$ commuting with α. Then there is a natural isomorphism of $R(G)$-modules $K_i^G(C^*(\mathbf{R},A,\beta)) \simeq K_{1-i}^G(A)$.

Proof. We obviously have $C^*(G,C^*(\mathbf{R},A)) \simeq C^*(\mathbf{R},C^*(G,A))$. By the ordinary Connes isomorphism ([20]), there is a natural isomorphism

$$K_i(C^*(\mathbf{R}, C^*(G,A))) \simeq K_{1-i}(C^*(G,A)).$$

By theorem 2.8.3 (7), there is therefore a natural isomorphism of abelian groups $K_i^G(C^*(\mathbf{R},A)) \simeq K_{1-i}^G(A)$. Theorem 2.8.3 (7) also implies that the $R(G)$-module structure on equivariant K-theory is determined by certain homomorphisms of C^*-algebras, so the naturality of this isomorphism implies that it is an $R(G)$-module homomorphism. Q.E.D.

The same method of proof yields the following theorems.

6.3.3 Theorem (Pimsner-Voiculescu sequence). Let G be a compact group, let (G,A,α) be a G-algebra, and let $\beta : \mathbf{Z} \longrightarrow Aut(A)$ be an action of $\mathbf{Z}$ which commutes with α. Then there is a natural 6 term exact sequence of $R(G)$-modules

$$\begin{array}{ccccc} K_0^G(A) & \xrightarrow{id-\beta(-1)_*} & K_0^G(A) & \longrightarrow & K_0^G(C^*(\mathbf{Z},A,\beta)) \\ \uparrow & & & & \downarrow \\ K_1^G(C^*(\mathbf{Z},A,\beta)) & \longleftarrow & K_1^G(A) & \xleftarrow{id-\beta(-1)_*} & K_1^G(A). \end{array}$$

Proof. We repeat the argument from the proof of theorem 6.3.2, using theorem 2.4 of [82] in place of [20]. Q.E.D.

6.3.4 Theorem (Pimsner-Voiculescu sequence for F_n). Let G be a compact group, let (G,A,α) be a G-algebra, and let $\beta : F_n \longrightarrow Aut(A)$ be an action of the free group F_n on n generators which commutes with α. Then there is a natural 6 term exact sequence

$$\begin{array}{ccccc} (K_0^G(A))^n & \longrightarrow & K_0^G(A) & \longrightarrow & K_0^G(C_r^*(F_n,A,\beta)) \\ \downarrow & & & & \downarrow \\ K_1^G(C_r^*(F_n,A,\beta)) & \longleftarrow & K_1^G(A) & \longleftarrow & (K_1^G(A))^n. \end{array}$$

Proof. This is the same as the previous theorem, this time using [83], theorem 3.5. Q.E.D.

6.3.5 Corollary. If A is separable, then there is an exact sequence as in the previous theorem with $C_r^*(F_n,A,\beta)$ replaced by $C^*(F_n,A,\beta)$.

Proof. We have natural isomorphisms

$$K_*^G(C_r^*(F_n,A,\beta)) \simeq K_*(C_r^*(F_n,C^*(G,A)))$$
$$\simeq K_*(C^*(F_n,C^*(G,A))) \simeq K_*^G(C^*(F_n,A,\beta)).$$

Here the first and last isomorphisms come from the C^*-algebra isomorphisms

$$C^*(G,C_r^*(F_n,A)) \simeq C_r^*(G\times F_n,A) \simeq C_r^*(F_n,C^*(G,A))$$

and similarly with C^* in place of C_r^* (note that G, being compact, is amenable), and the middle isomorphism is the K-amenability of F_n, as in [24], theorem 2.1 and the example following definition 2.2. (The middle isomorphism is the step which requires separability.) Q.E.D.

We digress to consider the implications of these results for K-freeness of actions on crossed products. Let G be a compact Lie group, let (G,A,α) be a G-algebra, and let $\beta: H \longrightarrow Aut(A)$ be an action commuting with α, where H is one of $\mathbf{R}, \mathbf{Z}$, or F_n. It is clear from the theorems above that if P is a prime ideal in $R(G)$, and $K_*^G(A)_P = 0$, then $K_*^G(C_r^*(H,A,\beta))_P = 0$. Thus, if α has locally discrete K-theory, then so does the action of G on $C_r^*(H,A,\beta)$. It is tempting to think that a similar statement might hold for K-freeness. As we now show, that is in fact not the case.

6.3.6 Example. Let $B = \mathbf{C}\oplus\mathbf{C}$, let $G = \mathbf{Z}/2\mathbf{Z}$, and define $\lambda: G \longrightarrow Aut(B)$ by sending the nontrivial element g of G to the automorphism $(a,b) \longrightarrow (b,a)$ of B. Then λ is a totally K-free action.

Further define $\beta: \mathbf{Z} \longrightarrow Aut(B)$ by $\beta_n = \lambda_g$ for n odd and $\beta_n = id_B$ for n even. Clearly β commutes with λ. Let $C = C^*(\mathbf{Z},B,\beta)$, and let $\mu: G \longrightarrow Aut(C)$ be the induced action of G on the crossed product. We are going to show that μ is not K-free. To do this, we will show that (G,C,μ) is isomorphic to the G-algebra (G,A,α) considered in example 4.1.7. It was shown there that α is not K-free. Recall that $A = M_2\otimes C(S^1)$, and that $\alpha_g(a) = waw^*$ for $a\in A$, where w is the unitary given by

$$w(\exp(it)) = \begin{bmatrix} 0 & \exp(\frac{1}{2}it) \\ \exp(-\frac{1}{2}it) & 0 \end{bmatrix}$$

for $t\in[0,2\pi)$. (As pointed out in example 4.1.7, w is not in A, but waw^* is in A for $a\in A$. We have changed the notation slightly from example 4.1.7.)

It follows immediately from corollary 2.10 of [38] that $C \simeq M_2\otimes C(S^1)$. (The groups G and H which appear there are taken to be $\mathbf{Z}$ and $2\mathbf{Z}$ respectively.) It is, however, not obvious what this isomorphism does to the action μ. We therefore define a map $\varphi: C \longrightarrow M_2\otimes C(S^1)$ as follows. We regard B as a subalgebra of $C = C^*(\mathbf{Z},B,\beta)$ in the obvious way, namely $x\in B$ is identified with the function in $L^1(\mathbf{Z},B,\beta)$ taking the value x at 0 and taking the value 0 elsewhere. Further, let $\delta_1 \in L^1(\mathbf{Z},B,\beta)$ be the function taking the value 1 at 1 and 0 elsewhere. Thus, δ_1 is a unitary element of $C^*(\mathbf{Z},B,\beta)$ such that $\delta_1 x \delta_1^* = \beta_1(x)$ for $x\in B$. Now set $\varphi(\delta_1) = \begin{bmatrix} 0 & z \\ 1 & 0 \end{bmatrix}$

and $\varphi(a,b) = \begin{bmatrix} a & 0 \\ 0 & b \end{bmatrix}$, where z is the identity function on S^1, $z(e^{it}) = e^{it}$, and a and b are regarded as constant functions on S^1. It is easily seen that $\varphi(1) = 1$, that $\varphi(\delta_1)$ is unitary, and that $\varphi(\delta_1)\varphi(a,b)\varphi(\delta_1)^* = \varphi(\beta_1(a,b))$. By the definition of the crossed product, φ defines a *-homomorphism from C to $M_2 \otimes C(S^1)$.

The elements

$$\begin{bmatrix} 1 & 0 \\ 0 & 0 \end{bmatrix} = \varphi(1,0), \quad \begin{bmatrix} 0 & 0 \\ 0 & 1 \end{bmatrix} = \varphi(0,1), \quad \begin{bmatrix} 0 & 0 \\ 1 & 0 \end{bmatrix} = \varphi(\delta_1)\varphi(1,0), \quad \text{and} \quad \begin{bmatrix} z & 0 \\ 0 & z \end{bmatrix} = \varphi(\delta_1)^2$$

are all in the range of φ. Since they generate $M_2 \otimes C(S^1)$ as a C^*-algebra, it follows that φ is surjective. As was mentioned above, C is isomorphic to $M_2 \otimes C(S^1)$. We therefore have $Prim(C) \simeq S^1$. Since φ defines an isomorphism from $C/\,Ker\,\varphi$ to $M_2 \otimes C(S^1)$, it follows that $S^1 - Prim(Ker\,\varphi)$ is homeomorphic to S^1. This can only happen if $Prim(Ker\,\varphi) = \emptyset$, that is, if $Ker\,\varphi = 0$. Therefore φ is injective, and is thus an isomorphism. An easy computation now shows that φ intertwines the action μ on C and the action α on $M_2 \otimes C(S^1)$. So φ is an equivariant isomorphism from (G,C,μ) to (G,A,α), as desired.

6.3.7 Remark. Let (G,A,α) be a G-algebra, where G is a compact Lie group, and let $\beta : \mathbf{Z} \longrightarrow Aut(A)$ be the trivial action. Then $C^*(\mathbf{Z},A,\beta)$ is just $A \otimes C(S^1)$, and the induced action of G on it is $g \longrightarrow \alpha_g \otimes id_{C(S^1)}$. Thus, if α is (totally) K-free then in this case the action of G on $C^*(\mathbf{Z},A,\beta)$ *is* (totally) K-free, by proposition 4.3.2. If β is an inner action, then a similar result holds, since β is exterior equivalent to the trivial action.

At the other extreme, we have the following somewhat more interesting result:

6.3.8 Theorem. Let G be a compact Lie group, let (G,A,α) be a G-algebra, and let $\beta : \mathbf{Z} \longrightarrow Aut(A)$ be an action commuting with α. Assume that the strong Connes spectrum $\widetilde{\Gamma}(\beta)$ ([58]) is all of the dual group S^1 of $\mathbf{Z}$. If α is (totally) K-free, then so is the induced action of G on $C^*(\mathbf{Z},A,\beta)$.

The proof of this theorem requires the well-known correspondence between ideals in a C^*-algebra which are invariant under the action of a locally compact abelian group and ideals in the crossed product which are invariant for the dual action. This correspondence is implicit in [79], 7.9.6 and 7.9.7, and also in section 2 of [70]. However, we have been unable to find a proof in the literature, so we prove this result here.

6.3.9. Proposition. Let H be a locally compact abelian group, and let (H,A,α) be an H-algebra. Then the map which assigns to each H-invariant ideal I in A the ideal $C^*(H,I)$ is a one-to-one correspondence between the H-invariant ideals of A and the $\widehat{H}$-invariant ideals of $C^*(H,A)$.

Proof. It is trivial to see that if I is H-invariant, then $C^*(H,I)$ is $\widehat{H}$-invariant. Let c_1 be the map defined in the statement of the proposition. Also let c_2 be the corresponding map from $\widehat{H}$-invariant ideals J of $C^*(H,A)$ to H-invariant ideals $C^*(\widehat{H},J)$ of $C^*(\widehat{H},C^*(H,A))$. By Takai duality ([79], 7.9.3), if I is an H-invariant ideal of A, then $c_2 \cdot c_1(I) = C^*(\widehat{H},C^*(H,I))$ is equivariantly isomorphic to $K(L^2(H))\otimes I$, where $L^2(H)$ carries the left regular representation of H. This isomorphism is in fact the restriction of the isomorphism $C^*(\widehat{H},C^*(H,A)) \simeq K(L^2(H))\otimes A$. Since $K(L^2(H))$ is nuclear and simple, proposition 4.2.9 implies that the correspondence $I \longrightarrow K(L^2(H))\otimes I$ is bijective from H-invariant ideals in A to H-invariant ideals in $K(L^2(H))\otimes A$. Therefore $c_2 \cdot c_1$ is bijective. Letting c_3 be the corresponding map from H-invariant ideals in the double crossed product to $\widehat{H}$-invariant ideals in the triple crossed product, the same argument proves that $c_3 \cdot c_2$ is bijective. It follows that c_2, and hence also c_1, is bijective. Q.E.D.

Proof of theorem 6.3.8. We consider K-freeness first. Let I be a G-invariant ideal in $C^*(\mathbf{Z},A,\beta)$. By lemma 3.4 in [58], I is also invariant under the dual action $\widehat{\beta}$. Therefore $I = C^*(\mathbf{Z},J,\beta)$ for some β-invariant ideal J of A, by the previous proposition. Since $\mathbf{Z}$ is discrete, the ideal J is in fact the set of fixed points in I for $\widehat{\beta}$, and is therefore also G-invariant. By assumption, if P is any prime ideal in $R(G)$ which does not contain $I(G)$, then $K_*^G(J)_P = 0$. From theorem 6.3.3, we conclude that $K_*^G(I)_P = 0$. Thus, the induced action on $C^*(\mathbf{Z},A,\beta)$ is K-free.

If α is totally K-free, then applying this argument to $\alpha|_H$ for all closed subgroups H of G shows that the action of G on $C^*(\mathbf{Z},A,\beta)$ is totally K-free. Q.E.D.

Of course, a similar result holds for actions of $\mathbf{R}$. However, it is not clear what happens for either full or reduced crossed products by F_n.

We now return to our original reason for considering the theorems at the beginning of this section. We therefore again let G be a finite cyclic group, $P \subset R(G)$ a prime ideal with support G, $R = R(G)_P$, and $h_*(A) = K_*^G(A)_P$.

6.3.10 Lemma. (Compare [95], proposition 2.5.) Let A and B be G-algebras, with B nuclear. Let λ be a continuous action of $\mathbf{R}$ on A which commutes with the action of G on A. If $\omega_P(A,B)$ is an isomorphism, then so is $\omega_P(C^*(\mathbf{R},A),B)$.

Proof. We want to show that the diagram

$$\begin{array}{ccc} K_*^G(A)\otimes_{R(G)}K_*^G(B) & \xrightarrow{\omega} & K_*^G(A\otimes B) \\ \gamma \downarrow\simeq & & \delta \downarrow\simeq \\ K_*^G(C^*(\mathbf{R},A))\otimes_{R(G)}K_*^G(B) & \xrightarrow{\omega} & K_*^G(C^*(\mathbf{R},A)\otimes B) \end{array}$$

is commutative. The vertical maps γ and δ of course come from theorem 6.3.2 and the fact that $C^*(\mathbf{R},A\otimes B) \simeq C^*(\mathbf{R},A)\otimes B$, where $\mathbf{R}$ acts trivially on B. Because the pairing ω is natural and respects suspensions, it is sufficient to consider the pairing on $K_0^G(B)$ for B unital. Let $p \in L(V)\otimes B$ be a G-invariant projection; then proposition 6.1.3 implies that $\omega(\eta\otimes[p]) = \varphi_*(\eta)$ for $\eta \in K_*^G(A)$, where φ is the map from A to

$L(V)\otimes A\otimes B$ given by composing $a \longrightarrow a\otimes p$ with the canonical isomorphism $A\otimes L(V)\otimes B \simeq L(V)\otimes A\otimes B$. (Here we identify $K_*^G(L(V)\otimes A\otimes B)$ with $K_*^G(A\otimes B)$, as in theorem 2.8.3 (4).) Let $\bar{\varphi}$ be the corresponding map from $C^*(\mathbf{R},A)$ to $L(V)\otimes C^*(\mathbf{R},A)\otimes B$. Then

$$\omega(\eta\otimes\gamma([p])) = \bar{\varphi}_*\cdot\gamma([p]) = \delta\cdot\varphi_*([p]) = \delta(\omega(\eta\otimes[p]))$$

for $\eta\in K_*^G(A)$, by naturality of the Connes isomorphism. This shows that the diagram commutes.

We now localize this diagram at P. The upper horizontal map becomes an isomorphism; therefore so does the lower one. Q.E.D.

6.3.11 Lemma. (Compare [95], proposition 2.7.) Let A and B be G-algebras, with B nuclear and $h_*(B)$ projective. Let λ be an action of $\mathbf{Z}$ on A commuting with the action of G on A. If $\omega_P(A,B)$ is an isomorphism, then so is $\omega_P(C^*(\mathbf{Z},A),B)$.

Proof. Let $D=C^*(\mathbf{Z},A)$. Consider the diagram:

$$\begin{array}{ccc}
\vdots & & \vdots \\
\downarrow & & \downarrow \\
(h_i(A)\otimes_R h_0(B))\oplus(h_{1-i}(A)\otimes_R h_1(B)) & \xrightarrow{\omega_P(A,B)} & h_i(A\otimes B) \\
\downarrow & & \downarrow \\
(h_i(A)\otimes_R h_0(B))\oplus(h_{1-i}(A)\otimes_R h_1(B)) & \xrightarrow{\omega_P(A,B)} & h_i(A\otimes B) \\
\downarrow & & \downarrow \\
(h_i(D)\otimes_R h_0(B))\oplus(h_{1-i}(D)\otimes_R h_1(B)) & \xrightarrow{\omega_P(D,B)} & h_i(D\otimes B) \\
\downarrow & & \downarrow \\
(h_{1-i}(A)\otimes_R h_0(B))\oplus(h_i(A)\otimes_R h_1(B)) & \xrightarrow{\omega_P(A,B)} & h_{1-i}(A\otimes B) \\
\downarrow & & \downarrow \\
\vdots & & \vdots
\end{array}$$

The vertical maps come from the Pimsner-Voiculescu exact sequence, theorem 6.3.3. The diagram commutes by an argument analogous to the one in the previous lemma. The right hand column is exact since $C^*(\mathbf{Z},A\otimes B)\simeq C^*(\mathbf{Z},A)\otimes B$ when $\mathbf{Z}$ acts trivially on B. The left hand column is exact because it is obtained by tensoring an exact sequence with the projective R-module $h_*(B)$. By hypothesis, the maps $\omega_P(A,B)$ are isomorphisms. Therefore, by the Five Lemma, $\omega_P(D,B)$ is an isomorphism. Q.E.D.

6.3.12 Lemma. Let A and B be G-algebras, with B nuclear and $h_*(B)$ projective. Let λ be an action of F_n on A commuting with the action of G on A. If $\omega_P(A,B)$ is an isomorphism, then so is $\omega_P(C_r^*(F_n,A),B)$, and, if A is separable, so is $\omega_P(C^*(F_n,A),B)$.

Proof. The proof is essentially the same as for actions of **Z**, using theorem 6.3.4 and corollary 6.3.5 instead of theorem 6.3.3. Q.E.D.

6.3.13 Remark. Just as in section 6.2, the results on $\omega_P(\cdot,\cdot)$ continue to hold if, instead of assuming that the second variable is nuclear, we assume that all algebras appearing in the first position are nuclear. In lemmas 6.3.10 and 6.3.11 this means assuming that A rather than B is nuclear, since crossed products of nuclear C^*-algebras by amenable groups are nuclear. In lemma 6.3.12, this means assuming that both A and $C_r^*(F_n, A)$ (or $C^*(F_n, A)$) are nuclear, which is unlikely but by no means impossible.

6.4. The Künneth Formula: The General Case

The category of G-algebras for which we obtain a theorem is given in the following definition.

6.4.1 Definition. Let G be a finite cyclic group. Let $\mathbf{N}_0^G$ be the category of all G-algebras A such that for every prime ideal $P \subset R(G)$ with support G and every nuclear G-algebra B such that $K_*^G(B)_P$ is a projective $R(G)_P$-module, the pairing $\omega_P(A, B)$ is an isomorphism.

The results of the previous two sections show that there are many algebras in $\mathbf{N}_0^G$. For instance, $\mathbf{N}_0^G$ contains all separable type I G-algebras and is closed under the formation of extensions, equivariant direct limits, certain crossed products, etc. We will postpone the exact statement because there is one more operation, not yet discussed, under which $\mathbf{N}_0^G$ is closed, and the proof of this depends on the Künneth formula itself.

As Schochet does in [95], we need a geometric realization theorem in order to obtain the full Künneth theorem for the category $\mathbf{N}_0^G$. This is where the homological dimension of the ring of coefficients enters. It is defined as follows. First, we recall that the projective dimension of a module over a ring is the length of the shortest projective resolution of the module ([66], chapter 18). The homological dimension of the ring is then the supremum of the projective dimensions of all modules over the ring ([66], chapter 18, where it is called the global dimension). It turns out that a ring has homological dimension 1 if and only if every submodule of a projective module is projective. (This will become clear in the course of the proof of the next lemma.) The ring of integers has homological dimension 1, since in fact every subgroup of a free abelian group is free abelian. This fact plays a crucial role in Schochet's geometric realization, and thus it should be no surprise that we need the following lemma.

6.4.2 Lemma. Let G be a finite cyclic group and let $P \subset R(G)$ be a prime ideal with support G. Then every submodule of a projective $R(G)_P$-module is projective.

Proof. Let $R = R(G)_P$. By proposition 6.2.2, R is isomorphic to a localization $\mathbf{Z}[\zeta]_Q$, where ζ is a primitive root of unity in $\mathbf{C}$, and Q is a prime ideal in $\mathbf{Z}[\zeta]$. By [65], chapter 4, theorem 4, $\mathbf{Z}[\zeta]$ is the ring of algebraic integers in the algebraic number field $\mathbf{Q}[\zeta]$. From [65], chapter 1, section 6, we conclude that $\mathbf{Z}[\zeta]$ is a Dedekind domain, and by [65], chapter 1, section 7, we find that $\mathbf{Z}[\zeta]_Q$ is a discrete valuation ring. By [5], remark after theorem 11.23, it follows that $\mathbf{Z}[\zeta]_Q$ is a regular local ring of dimension 1. By [66], theorem 42, it follows that $R \simeq \mathbf{Z}[\zeta]_Q$ has homological (or global) dimension 1. This implies ([66], chapter 18, lemma 2 and the definition immediately following) that $Ext_R^2(M,N) = 0$ for all R-modules M and N.

Now let T be a projective R-module, and let S be a submodule. The long exact sequence for Ext_R shows that for all R-modules N, we have

$$Ext_R^1(S,N) = Ext_R^2(T/S,N) = 0 .$$

(We have $Ext_R^n(T,N) = 0$ for all $n > 0$, since T is projective.) Therefore S is projective by [66], chapter 18, lemma 1.

Q.E.D.

6.4.3 Remark. No such result holds for $R(G)$ itself for the groups we are considering. If $G = S^1$, the circle group, it is in fact true that the homological dimension of $R(G)$ is not too large (it is 2); however, the homological dimension of $R(\mathbf{Z}/2\mathbf{Z})$ is infinite. One can check this as follows. Let I be a maximal ideal containing the augmentation ideal $I(\mathbf{Z}/2\mathbf{Z})$. Then it can be shown that $R(\mathbf{Z}/2\mathbf{Z})_I$ is not an integral domain. By [66], theorem 36, it is not regular, so by [66], theorem 45, it has infinite homological dimension. Now [66], chapter 18, lemma 5, implies that $R(\mathbf{Z}/2\mathbf{Z})$ also has infinite homological dimension.

6.4.4 Lemma. (Compare [95], lemma 3.1.) Let G be a compact group, and let (G,B,β) be a unital G-algebra. Then there is a Hilbert space H with a representation of G, a commutative G-algebra F, and an equivariant inclusion $\varphi : F \longrightarrow K(H) \otimes B$ such that $K_*^G(F)$ is a free $R(G)$-module and $\varphi_* : K_*^G(F) \longrightarrow K_*^G(B)$ is surjective. (Here $K(H)$ has the inner G-action given by the representation of G on H, and we identify $K_*^G(K(H) \otimes B)$ with $K_*^G(B)$.)

Proof. Let $\{p_k\}_{k \in D_0}$ be a family of G-invariant projections $p_k \in L(V_k) \otimes B$ for finite dimensional representation spaces V_k of G, such that $\{[p_k]\}_{k \in D_0}$ generates $K_0^G(B)$ as an $R(G)$-module. Let $\{u_k\}_{k \in D_0}$ be a set of G-invariant unitaries $u_k \in L(W_k) \otimes B$ for finite dimensional representation spaces W_k of G, such that $\{[u_k]\}_{k \in D_1}$ generates $K_1^G(B)$ as an $R(G)$-module, and $[u_k] \neq 0$ for any k. (See theorem 2.8.8. We actually only proved theorem 2.8.8 for finite abelian groups, so that our proof is complete only in that case. But in fact we only use this lemma for finite cyclic groups. If $K_1^G(B) = 0$, take $D_1 = \varnothing$.) Let

$$H = \left[\bigoplus_{k \in D_0} V_k \right] \oplus \left[\bigoplus_{k \in D_1} W_k \right].$$

Then we can regard the direct sum

$$F_0 = \left[\bigoplus_{k \in D_0} L(V_k) \otimes B \right] \oplus \left[\bigoplus_{k \in D_1} (L(W_k) \otimes B) \right]$$

as a subalgebra of $K(H) \otimes B$. For each $k \in D_0$, let q_k be the image of p_k under the inclusion of $L(V_k) \otimes B$ in F_0, and for $k \in D_1$, let v_k be the image of $u_k - 1$ under the inclusion of $L(W_k) \otimes B$ in F_0. Let F be the C^*-subalgebra of F_0 generated by the q_k and the v_k, and let φ be the inclusion of F in $K(H) \otimes B$.

It is clear that F is commutative, and that G acts trivially on F. Since the C^*-algebra generated by q_k is $\mathbf{C}$ and the C^*-algebra generated by v_k is $C_0(\mathbf{R})$, we see that $K_*(F)$ is a free abelian group. (To see that v_k generates a C^*-algebra isomorphic to $C_0(\mathbf{R})$, we observe that u_k is unitary, and its spectrum must be all of S^1 – otherwise we could use the functional calculus to connect u_k to 1 by a continuous path of G-invariant unitaries, and we would have $[u_k] = 0$. Therefore the C^*-algebra generated by $u_k - 1$ is $C_0(sp(u) - \{1\})$, which is isomorphic to $C_0(\mathbf{R})$. The C^*-algebra generated by the image v_k of $u_k - 1$ in F_0 is the same, because the map is injective.) Since G acts trivially on F, we find that $K_*^G(F) \simeq R(G) \otimes K_*(F)$ is a free $R(G)$-module. Finally, it is clear that $\varphi_*([q_k]) = [p_k]$, while $v_k + 1$ is a unitary element in F^+ such that $\varphi^+(v_k + 1)$ is a unitary in $(K(H) \otimes B)^+$ defining the same class in $K_1^G(B)$ as u_k. It follows that φ_* is surjective. Q.E.D.

6.4.5 Proposition. (Geometric realization; compare [95], section 3.) Let G be a finite cyclic group, let P be a prime ideal in $R(G)$ with support G, and let $h_* = K_*^G(\cdot)_P$. For any G-algebra B there is an equivariant exact sequence

$$0 \longrightarrow K \otimes SB \longrightarrow C \longrightarrow F \longrightarrow 0,$$

where K is the algebra of compact operators with an inner action of G, such that application of the functor h_* yields a projective resolution

$$0 \longrightarrow h_*(C) \longrightarrow h_*(F) \longrightarrow h_*(B) \longrightarrow 0$$

of $h_*(B)$. (That is, $h_*(C)$ and $h_*(F)$ are projective $R(G)_P$-modules, and the sequence is exact.) If B is nuclear, then so are C and F.

Proof. Let $\varphi: F \longrightarrow K \otimes B$ be the equivariant homomorphism constructed in the previous lemma. Recall that $C(K \otimes B) = C_0((0,1]) \otimes (K \otimes B)$ is the cone over $K \otimes B$, and that the mapping cone of φ is given by

$$C\varphi = \{(a, f) \in F \oplus C(K \otimes B) : \varphi(a) = f(1)\},$$

where $C(K \otimes B)$ is viewed as functions from $(0,1]$ into $K \otimes B$.

Define $\lambda : C\varphi \longrightarrow F$ by $\lambda(a,f) = a$, and define $\mu : C\varphi \longrightarrow C(K\otimes B)$ by $\mu(a,f) = f$. Then we have an equivariant commutative diagram with exact rows

$$\begin{array}{ccccccccc} 0 & \longrightarrow & S(K\otimes B) & \longrightarrow & C\varphi & \xrightarrow{\lambda} & F & \longrightarrow & 0 \\ & & \downarrow = & & \downarrow \mu & & \downarrow \varphi & & \\ 0 & \longrightarrow & S(K\otimes B) & \longrightarrow & C(K\otimes B) & \longrightarrow & K\otimes B & \longrightarrow & 0, \end{array}$$

where $S(K\otimes B)$ is $C_0((0,1))\otimes K\otimes B$ and can therefore be regarded as an ideal in $C\varphi$ and in $C(K\otimes B)$. We apply h_* to obtain the following commutative diagram with exact rows, in which $h_*(C(K\otimes B)) = 0$ because $C(K\otimes B)$ is equivariantly contractible:

$$\begin{array}{ccccccccc} h_{1-i}(F) & \xrightarrow{\partial_{1-i}} & h_i(S(K\otimes B)) & \longrightarrow & h_i(C\varphi) & \xrightarrow{\lambda_*} & h_i(F) & \xrightarrow{\partial_i} & h_{1-i}(S(K\otimes B)) \\ \downarrow \varphi_* & & \downarrow = & & & & \downarrow \varphi_* & & \downarrow = \\ h_{1-i}(K\otimes B) & \xrightarrow{\simeq} & h_i(S(K\otimes B)) & \longrightarrow & 0 & \longrightarrow & h_i(K\otimes B) & \xrightarrow{\simeq} & h_{1-i}(S(K\otimes B)). \end{array}$$

Since φ_* is surjective by construction, we find that ∂_{1-i} and ∂_i are surjective, so that λ_* is injective. Furthermore, $Ker\ (\varphi_*) = Ker\ (\partial_i)$. Therefore, with $C = C\varphi$, the sequence

$$0 \longrightarrow h_*(C) \xrightarrow{\lambda_*} h_*(F) \xrightarrow{\varphi_*} h_*(K\otimes B) \longrightarrow 0$$

is exact. Since $h_*(F)$ is a free $R(G)_P$-module, $h_*(C)$ is projective by lemma 6.4.2. It remains only to identify $h_*(K\otimes B)$ with $h_*(B)$. Q.E.D.

6.4.6 Theorem. (Künneth formula; compare [95], theorem 4.1.) Let G be a finite cyclic group, let P be a prime ideal in $R(G)$ with support G, let $R = R(G)_P$, and let $h_* = K_*^G(\cdot)_P$. Then for G-algebras A and B with $A \in \mathbf{N}_0^G$ and B nuclear, there is a natural short exact sequence

$$0 \longrightarrow h_*(A)\otimes_R h_*(B) \xrightarrow{\omega_P} h_*(A\otimes B) \xrightarrow{\varepsilon} Tor^R(h_*(A), h_*(B)) \longrightarrow 0,$$

where ω_P has degree 0 and ε has degree 1.

Proof. The proof is essentially the same as the proof of theorem 4.1 in [95], except that tensor products must be taken over R instead of $\mathbf{Z}$, $Tor^{\mathbf{Z}}$ must be replaced by Tor^R, and we use projective R-modules in place of free or torsion-free abelian groups. We omit the details. Q.E.D.

We now return to the question of how large the category $\mathbf{N}_0^G$ really is. Before stating the theorem, we should point out that we know of no examples of G-algebras which are not in $\mathbf{N}_0^G$.

6.4.7 Theorem. Let G be a finite cyclic group. Then the category $\mathbf{N}_0^G$ satisfies the following properties.

(1) $\mathbf{N}_0^G$ contains all separable type I G-algebras.

(2) $\mathbf{N}_0^G$ contains all separable G-algebras A such that the action of G on *Prim*(A) has no fixed points.

(3) If

$$0 \longrightarrow I \longrightarrow A \longrightarrow B \longrightarrow 0$$

is an equivariant exact sequence such that two of the three algebras are in $\mathbf{N}_0^G$, then so is the third. If the sequence is equivariantly split, and A is in $\mathbf{N}_0^G$, then so are I and B.

(4) If A and B are stably isomorphic as G-algebras, and $A \in \mathbf{N}_0^G$, then $B \in \mathbf{N}_0^G$.

(5) If (G,A,α) and (G,A,β) are exterior equivalent G-algebras, and $(G,A,\alpha) \in \mathbf{N}_0^G$, then $(G,A,\beta) \in \mathbf{N}_0^G$.

(6) $\mathbf{N}_0^G$ is closed under the formation of equivariant direct limits.

(7) $\mathbf{N}_0^G$ is closed under the formation of crossed products by actions of $\mathbf{R}$ and $\mathbf{Z}$ which commute with the action of G, of reduced crossed products by actions of F_n which commute with the action of G, and of full crossed products by actions of F_n on separable algebras which commute with the action of G.

(8) If H is a proper subgroup of G, and A is an H-algebra, then $C(G \times_H A) \in \mathbf{N}_0^G$.

(9) If $A, B \in \mathbf{N}_0^G$ and B is nuclear, then $A \otimes B \in \mathbf{N}_0^G$.

Before proving this theorem, some remarks are in order. First, if A is in the category $\mathbf{N}$ of [95], and G acts trivially on A, then $A \in \mathbf{N}_0^G$ by lemma 6.2.3. But is it in fact easily seen that conditions (1) through (7) of the theorem imply that $(G,A) \in \mathbf{N}_0^G$ without appealing to lemma 6.2.3.

Secondly, it can be shown without too much difficulty that the smallest category $\mathbf{M}$ satisfying (1) through (8) is already closed under the formation of tensor products when one of the algebras is nuclear. The proof goes by showing that if B is nuclear and in $\mathbf{M}$, then the category of G-algebras A such that $A \otimes B$ is in $\mathbf{M}$ contains $\mathbf{C}$ and satisfies conditions (2) through (8). The arguments of section 6.2, and of section 2 of [95], essentially show that any category of G-algebras containing $\mathbf{C}$ and satisfying (2) through (7) must contain all separable type I G-algebras. (Notice that this argument shows that the category $\mathbf{N}$ of [95] is also already closed under formation of tensor products.)

Finally, we point out that $\mathbf{N}_0^G$ also contains all $(G, C_0(X))$-algebras A such that the action of G on X has no fixed points. Indeed, remarks 4.6.2 and 4.6.3 imply that for any nuclear G-algebra B, we have $K_*^G(A)_P = K_*^G(A \otimes B)_P = 0$. However, this gives little added generality, since G must act on *Prim*(A) with no fixed points, so that if A is separable then A is covered by part (2).

Proof of theorem 6.4.7. Parts (1) through (6) are, in order, proposition 6.2.11, proposition 6.2.4, lemma 6.2.5, lemma 6.2.9, lemma 6.2.8, and lemma 6.2.6. Part (7) is lemmas 6.3.10, 6.3.11, and 6.3.12. To prove part (8), let H be a proper subgroup of G

and let A be an H-algebra. Then for any G-algebra B, we have

$$C(G\times_H A)\otimes B\simeq C(G\times_H(A\otimes B))$$

by lemma 5.1.7, so that if P is a prime ideal with support G, then

$$K_*^G(C(G\times_H A))_P = K_*^G(C(G\times_H A)\otimes B)_P = 0$$

by lemma 4.3.6. So $\omega_P(C(G\times_H A), B)$ is an isomorphism.

It remains to prove part (9). Let $P\subset R(G)$ be a prime ideal with support G, and as usual let $R = R(G)_P$ and $h_* = K_*^G(\cdot)_P$. Let $A, B\in \mathbf{N}_0^G$, and assume that B is nuclear. Let C be any nuclear G-algebra such that $h_*(C)$ is projective. Then we are trying to show that $\omega_P(A\otimes B, C)$ is an isomorphism. Consider the following diagram:

$$\begin{array}{ccc}
0 & & 0 \\
\downarrow & & \downarrow \\
h_*(A)\otimes h_*(B)\otimes h_*(C) & \xrightarrow{id_{h_*(A)}\otimes\omega_P(B,C)} & h_*(A)\otimes h_*(B\otimes C) \\
\downarrow & & \downarrow \\
h_*(A\otimes B)\otimes h_*(C) & \xrightarrow{\omega_P(A\otimes B,\, C)} & h_*(A\otimes B\otimes C) \\
\downarrow & & \downarrow \\
Tor^R(h_*(A), h_*(B))\otimes h_*(C) & \xrightarrow{\eta} & Tor^R(h_*(A), h_*(B\otimes C)) \\
\downarrow & & \downarrow \\
0 & & 0
\end{array}$$

Here the left hand column is the Künneth formula for $h_*(A\otimes B)$, tensored with $h_*(C)$. Since $h_*(C)$ is projective, it is exact. The right hand column is the Künneth formula for $h_*(A\otimes(B\otimes C))$; note that theorem 6.4.6 shows that it is exact because $B\otimes C$ is nuclear. The top square is commutative by the associativity of the pairing ω, and $id_{h_*(A)}\otimes\omega_P(B,C)$ is an isomorphism because $B\in\mathbf{N}_0^G$. To define η, let us first recall ([47], section IV.11) that if M and N are R-modules, then $Tor^R(M,N)$ is computed by taking a projective resolution for M, tensoring it over R with N, and forming the homology of the resulting complex. It follows that if X is a free R-module, then $Tor^R(M, N\otimes_R X)$ is naturally isomorphic to $Tor^R(M,N)\otimes_R X$. Therefore the same holds for any direct summand in a free R-module, and in particular for the projective R-module $h_*(C)$. Thus, there is a natural isomorphism

$$Tor^R(h_*(A),\, h_*(B))\otimes h_*(C) \simeq Tor^R(h_*(A),\, h_*(B)\otimes h_*(C)).$$

The map η is the composite of this map with $Tor^R(h_*(A),\, \omega_P(B,C))$. Since $\omega_P(B,C)$ is an isomorphism, so is η. Furthermore, the lower square of the diagram commutes by the naturality of the Künneth formula; we omit the details. It now follows from the Five Lemma that $\omega_P(A\otimes B,\, C)$ is also an isomorphism, as desired. Q.E.D.

6.4.8 Remark. The Künneth formula of theorem 6.4.6 also holds for arbitrary G-algebras (G,B,β) provided that (G,A,α) is required to be in the category of *nuclear*

G-algebras such that for any prime ideal P with support G and any not necessarily nuclear G-algebra (G,B,β) with $K_*^G(B)_P$ projective, the pairing $\omega_P(A,B)$ is an isomorphism. Theorem 6.4.7 also holds for this category, with the following small modifications: in parts (2) and (8), the algebra A must be nuclear, and in part (7) the crossed products by F_n are only in the category if they are still nuclear. Of course, in part (9) both algebras are now nuclear by assumption. The proofs for parts (1) through (7) are the appropriate modifications of the relevant results from sections 6.2 and 6.3, as discussed in remarks 6.2.12 and 6.3.13. (Note that the category of nuclear C^*-algebras is closed under all the operations which appear except for full and reduced crossed products by F_n.) The proofs for parts (8) and (9) are exactly as in the proof of theorem 6.4.7.

6.5 The Categories $\mathbf{N}^G$ and $\mathbf{N}^G_{nuc}$

It is now time to consider exactly which G-algebras the theorem on K-freeness of actions on tensor products is going to work for. In the first place, the fact that the Künneth formula holds only for equivariant K-theory localized at certain prime ideals in the representation rings of cyclic groups means that we will have to restrict ourselves to G-algebras (G,A,α) such that $(H,A,\alpha|_H)\in\mathbf{N}_0^H$ for all cyclic subgroups H of G. The category of all such G-algebras satisfies most of the conditions of theorem 6.4.7. (In condition (2) the action on $Prim(A)$ must be required to be free, and condition (8) must also be modified.) A more serious difficulty arises because of the need to look at arbitrary invariant ideals. The category $\mathbf{N}_0^H$ is not obviously closed under passage to invariant ideals, and this leads to further restrictions on the algebras which may be considered. The following definition is appropriate.

6.5.1 Definition. Let G be a finite group. Let $\mathbf{N}_1^G$ be the category of all G-algebras (G,A,α) such that $(H,A,\alpha|_H)\in\mathbf{N}_0^H$ for every cyclic subgroup H of G. Further, let $\mathbf{N}^G$ be the category of all G-algebras (G,A,α) such that $(G,I,\alpha_{(\cdot)}|_I)\in\mathbf{N}_1^G$ for every G-invariant ideal I of A.

When applying this definition, remember that by convention a cyclic subgroup H of G has more than one element. The purpose of this convention is to rule out the possibility of having a prime ideal in $R(H)$ with support H contain $I(H)$, since we do not want to localize at such ideals. Note that property (2′) in the next lemma would not hold without the convention. Also note that, according to this convention, if G is the one element group then $\mathbf{N}^G$ is the category of all C^*-algebras.

As the definition is written, when determining if something is in $\mathbf{N}^G$ we restrict to invariant ideals before restricting to cyclic subgroups. However, these operations could also be done in the other order, and this is in effect proved in lemma 6.5.3. First, however, we need some properties of the category $\mathbf{N}_1^G$.

6.5.2 Lemma. If G is a finite group, the category $\mathbf{N}_1^G$ satisfies properties (1), (3) through (7), and (9) of theorem 6.4.7, and the following substitutes for properties (2) and (8) of that theorem:

(2') $\mathbf{N}_1^G$ contains all separable G-algebras A such that the action of G on $Prim(A)$ is free.

(8') If H is a subgroup of G, and $A \in \mathbf{N}_1^H$, then $C(G\times_H A) \in \mathbf{N}_1^G$.

Proof. These properties are all trivial consequences of theorem 6.4.7, except for property (8'), which we now prove. The proof is similar to arguments used in the proof of theorem 4.3.4, but is simpler because G is assumed to be discrete. Let $(H,A,\alpha) \in \mathbf{N}_1^H$, and let $(G, C(G\times_H A), \beta)$ be the corresponding twisted product algebra. Let G_0 be a cyclic subgroup of G, and let $G_0\times H$ act on G via the formula $(g_0, h)g = g_0 g h^{-1}$ for $g \in G$, $g_0 \in G_0$, and $h \in H$. Let X be the orbit space $G/(G_0\times H)$, which is a finite set with the discrete topology, and let $\pi: G \longrightarrow X$ be the quotient map. For each $x \in X$ there is an ideal $J_x \subset C(G\times_H A)$ consisting of all functions in $C(G\times_H A)$ which vanish off the subset $\pi^{-1}[x]$ of G. The ideals J_x are G_0-invariant, and we have $C(G\times_H A) = \bigoplus_{x\in X} J_x$. So it is sufficient to prove that $J_x \in \mathbf{N}_0^{G_0}$.

Now $\pi^{-1}[x] = G_0 g H$ for some $g \in G$. Let $H_0 = \{g_0 \in G_0 : g_0 g H = gH\}$. Then H_0 is a subgroup of G_0, and it is easily seen, using proposition 2.9.6, that $J_x = C(G_0\times_{H_0} I)$, where I is the set of functions in $C(G\times_H A)$ which vanish off gH. The map $f \longrightarrow f(g)$ defines an isomorphism of C^*-algebras $I \simeq A$, and a calculation shows that this map intertwines the action $\beta|_{H_0}|_I$ of H_0 on I with the action γ of H_0 on A defined by $\gamma_{h_0} = \alpha_{g^{-1}h_0 g}$ for $h_0 \in H_0$. Note that $g^{-1}H_0 g \subset H$, so this makes sense. Thus, the systems $(H_0, I, \beta|_{H_0}|_I)$ and $(g^{-1}H_0 g, A, \alpha|_{g^{-1}H_0 g})$ are isomorphic. There are now two cases. If $H_0 \neq G_0$, then $J_x = C(G_0\times_{H_0} I) \in \mathbf{N}_0^{G_0}$ by theorem 6.4.7 (8). Otherwise, we have $H_0 = G_0$. Then J_x is isomorphic to the $g^{-1}G_0 g$-algebra A, and $g^{-1}G_0 g$ is a cyclic subgroup of H. Since $A \in \mathbf{N}_1^H$, it follows from the definition of $\mathbf{N}_1^H$ that $J_x \in \mathbf{N}_0^{G_0}$ in this case also. This completes the proof that $C(G\times_H A) \in \mathbf{N}_0^G$. Q.E.D.

6.5.3 Lemma. Let G be a finite group, let H be a subgroup of G, and let $(G,A,\alpha) \in \mathbf{N}^G$. Then $(H, A, \alpha|_H) \in \mathbf{N}^H$.

Proof. It follows immediately from the definition that if $(G,A,\alpha) \in \mathbf{N}_1^G$ then $(H, A, \alpha|_H) \in \mathbf{N}_1^H$. We must therefore show that if (G,A,α) is a G-algebra such that $(H, I, \alpha|_H|_I) \in \mathbf{N}_1^H$ for all G-invariant ideals I of A, then $(H, I, \alpha|_H|_I) \in \mathbf{N}_1^H$ for all H-invariant ideals I of A. The proof uses the Ideal Decomposition Lemma 5.3.3. We use the notation introduced there, except that, for e the identity of the group, we write I_e instead of $I_{\{e\}}$. We make free use of the fact (lemma 5.3.2) that sums and intersections of ideals distribute over each other.

The proof uses two nested induction arguments. We start out by proving, by downward induction on subsets S of G containing e, that $(H, I_S \cap I, \alpha|_H|_{I_S \cap I}) \in \mathbf{N}_1^H$ for all H-invariant ideals I of A. This is certainly true for $S=G$, because $I_G \cap I = I_G$ is G-invariant. So let S be a subset of G such that $e \in S$, and suppose $(H, I_T \cap I, \alpha|_H|_{I_T \cap I}) \in \mathbf{N}_1^H$ for all subsets T of G which properly contain S. We now use an induction argument on the size of a set T disjoint from S to show that $\sum_{g \in T} I_{S \cup \{g\}} \cap I$, together with its H-action, is in $\mathbf{N}_1^H$. For a one element set T, this is just the outer induction hypothesis, so suppose it holds for some T, and let $k \in G$, $k \notin T \cup S$. Let $L = I_{S \cup \{k\}} \cap I$. Then (4b) of the Ideal Decomposition Lemma 5.3.3 shows that

$$(I_{S \cup \{k\}} \cap I) \cap \left[\sum_{g \in T} I_{S \cup \{g\}} \cap I \right] = \sum_{g \in T} L_{S \cup \{g\}} \cap L .$$

This ideal, equipped with its H-action, is in $\mathbf{N}_1^H$ by the inner induction assumption applied to L. Since $I_{S \cup \{g\}} \cap I$ is also in $\mathbf{N}_1^H$, we conclude that the quotient

$$(I_{S \cup \{k\}} \cap I) / \left[\sum_{g \in T} L_{S \cup \{g\}} \cap L \right]$$

is in $\mathbf{N}_1^H$. Using the isomorphism $(M+N)/N \simeq M/M \cap N$, we obtain an exact sequence

$$0 \longrightarrow \sum_{g \in T} I_{S \cup \{g\}} \cap I \longrightarrow \sum_{g \in T \cup \{k\}} I_{S \cup \{g\}} \cap I \longrightarrow (I_{S \cup \{k\}} \cap I) / \left[\sum_{g \in T} L_{S \cup \{g\}} \cap L \right] \longrightarrow 0 .$$

Consequently, $\sum_{g \in T \cup \{k\}} I_{S \cup \{g\}} \cap I$ is in $\mathbf{N}_1^H$. By induction we can now conclude that $(H, I_{\bar{S}} \cap I, \alpha|_H|_{I_{\bar{S}} \cap I}) \in \mathbf{N}_1^H$.

The quotient $I_S / I_{\bar{S}}$ is in $\mathbf{N}_1^H$, because I_S and $I_{\bar{S}}$ are both G-invariant ideals in A. Now (5) of the Ideal Decomposition Lemma implies that $(I_S \cap I)/(I_{\bar{S}} \cap I)$ is H-equivariantly isomorphic to a direct summand of $I_S / I_{\bar{S}}$, and is therefore also in $\mathbf{N}_1^H$, because by lemma 6.5.2, $\mathbf{N}_1^H$ satisfies the second part of property (3) in theorem 6.4.7. Therefore $(H, I_S \cap I, \alpha|_H|_{I_S \cap I}) \in \mathbf{N}_1^H$, as desired. This completes the outer induction step.

We therefore conclude that $(H, I_e \cap I, \alpha|_H|_{I_e \cap I}) \in \mathbf{N}_1^H$. Since $I_e \cap I = I$, this completes the proof.

Q.E.D.

We are now almost in a position to prove a descriptive theorem for the category $\mathbf{N}^G$ analogous to theorem 6.4.7 for the category $\mathbf{N}_0^H$. Of course, some of the statements in theorem 6.4.7 will have to be weakened. We still, however, need to do some more work for one of the statements which does not need to be weakened, namely the one involving tensor products. We will show that if two G-algebras (G,A,α) and (G,B,β) are in $\mathbf{N}^G$, and if one of them is nuclear, then their tensor product is in $\mathbf{N}^G$.

It is clear from what we have already proved in theorem 6.4.7 that if I and J are G-invariant ideals in A and B respectively, and if γ is the obvious action of G on $I \otimes J$, then $(G, I \otimes J, \gamma) \in \mathbf{N}_1^G$. Unfortunately, many G-invariant ideals in $A \otimes B$ are not of this form, or even sums of ideals of this form. For example, if $G = \mathbf{Z}/2\mathbf{Z}$, $A = B = \mathbf{C} \oplus \mathbf{C}$, and $\alpha = \beta$ sends the nontrivial element of G to the automorphism which interchanges the two summands, then the ideal

$$\left[(\mathbf{C} \oplus 0) \otimes (\mathbf{C} \oplus 0)\right] \oplus \left[(0 \oplus \mathbf{C}) \otimes (0 \oplus \mathbf{C})\right]$$

is G-invariant but contains no nonzero G-invariant ideal of the form $I \otimes J$.

In order to handle ideals which are not tensor products, we will need to use the Ideal Decomposition Lemma again. We will need essentially the same argument in the next section as well, where we will need to prove, under appropriate assumptions on A and B, that all G-invariant ideals in $A \otimes B$ have locally discrete K-theory. We therefore state the required argument as a separate lemma. We will need to use the fact (proposition 4.2.9) that if A and B are C^*-algebras, with one of them nuclear, then $Prim(A \otimes B) \simeq Prim(A) \times Prim(B)$. In particular, if an ideal in $A \otimes B$ has the form $I \otimes J$ for I an ideal in A and J an ideal in B, we will call it "rectangular", since the corresponding open subset of $Prim(A) \times Prim(B)$ is a rectangle.

6.5.4 Lemma. Let G be a finite group. Let (P) be a property which a G-algebra may or may not have, and suppose that:

(1) In any short exact sequence of G-algebras, if two out of the three terms have the property (P), then so does the third.

(2) If D is a G-algebra with an increasing transfinite sequence $(I_\lambda)_{\lambda < \kappa}$ of G-invariant ideals whose union is dense in D, and if each I_λ has (P), then D has (P).

Further, let A and B be G-algebras, one of which is nuclear, and assume that:

(3) Every G-invariant rectangular ideal in $A \otimes B$ has (P).

(4) If H is a proper subgroup of G, and I is an H-invariant ideal in $A \otimes B$, then the G-algebra $C(G \times_H I)$ has (P).

Then every G-invariant ideal in $A \otimes B$ has (P).

Proof. We start with an easy consequence of condition (1), namely:

(5) Let I and J be G-invariant ideals in $A \otimes B$. If I, J, and $I \cap J$ have (P), then so does $I + J$.

The proof of this is just like lemma 5.3.4: consideration of the short exact sequence

$$0 \longrightarrow I \cap J \longrightarrow I \longrightarrow I/(I \cap J) \longrightarrow 0$$

shows that $I/(I \cap J)$ has (P), and the isomorphism $I/(I \cap J) \simeq (I + J)/J$ together with the short exact sequence

$$0 \longrightarrow J \longrightarrow I + J \longrightarrow (I + J)/J \longrightarrow 0$$

now shows that $I+J$ has (P), as desired.

The remainder of the proof is divided into 4 steps.

Step 1. Let J be any G-invariant ideal in $A \otimes B$. Then for any ideal I in $A \otimes B$, if $J \cap I_G$ has (P), then $J \cap I_e$ has (P).

Proof of step 1. As usual with arguments involving the Ideal Decomposition Lemma, we use two nested inductions. In the outer one we prove, by downward induction on subsets S of G containing e, that $J \cap I_S$ has (P), the case $S=G$ being the hypothesis. So let S be a subset of G such that $e \in S$, and assume that $J \cap I_T$ has (P) for all ideals I in $A \otimes B$ and all T properly containing S. Now we show that

$$\sum_{g \in T} J \cap I_{S \cup \{g\}} \text{ has } (P) \tag{*}$$

for all ideals I such that $J \cap I_G$ has (P) and all nonempty subsets T of G disjoint from S, by upward induction on the size of T. The case in which T has one element is part of the previous induction hypothesis. So assume (*) holds for a set T disjoint from S, and let $k \notin S \cup T$. Let $L = I \cap I_{S \cup \{k\}}$. Then $L_G = I_G \cap I_{S \cup \{k\}} = I_G$, using (6) of the Ideal Decomposition Lemma and the relation $I_G \subset I_{S \cup \{k\}}$. Since $J \cap I_G$ has (P) by assumption, we can apply the induction hypothesis to L. We find that

$$J \cap \sum_{g \in T} L_{S \cup \{g\}} = (J \cap I_{S \cup \{k\}}) \cap \Big(\sum_{g \in T} J \cap I_{S \cup \{g\}}\Big) \text{ has } (P), \tag{**}$$

where the equality follows from (4a) of the Ideal Decomposition Lemma. Furthermore, the ideal $J \cap I_{S \cup \{k\}}$ has (P). By property (5) at the beginning of the proof, this, together with (*) and (**), implies that

$$\sum_{g \in T \cup \{k\}} J \cap I_{S \cup \{g\}}$$

has (P). This completes the inner induction, and we conclude, taking $T=G-S$, that $J \cap I_{\bar{S}}$ has (P).

By (6) of the Ideal Decomposition Lemma, $J \cap I_S = (J \cap I)_S$ and $J \cap I_{\bar{S}} = (J \cap I)_{\bar{S}}$. Therefore, by (5) of that lemma, $(J \cap I_S)/(J \cap I_{\bar{S}}) \simeq C(G \times_H M)$ for a certain proper subgroup H of G and a certain H-invariant ideal M in $(J \cap I_S)/(J \cap I_{\bar{S}})$. Let L be the inverse image of M in $A \otimes B$ under the quotient map from $J \cap I_S$ to $(J \cap I_S)/(J \cap I_{\bar{S}})$. Then L and $J \cap I_{\bar{S}}$ are H-invariant ideals in $A \otimes B$, so that the algebras $C(G \times_H L)$ and $C(G \times_H (J \cap I_{\bar{S}}))$ have (P) by property (4). By property (1), their quotient, which is $C(G \times_H M)$, also has (P). Applying (1) to the short exact sequence

$$0 \longrightarrow J \cap I_{\bar{S}} \longrightarrow J \cap I_S \longrightarrow C(G \times_H M) \longrightarrow 0$$

now enables us to conclude that $J \cap I_S$ has (P). This completes the outer induction.

It now follows that $J \cap I_e$ has (P), as desired, and step 1 is proved.

Step 2. Let $I_1, \ldots, I_n$ be rectangular ideals in $A \otimes B$. Then $(I_1)_e \cap \cdots \cap (I_n)_e$ has (P).

Proof of step 2. We use induction on n. For the case $n=1$, notice that $(I_1)_G$ is an intersection of rectangular ideals. Since rectangular ideals in $A \otimes B$ correspond to open rectangles in $Prim(A)\times Prim(B)$, and since the intersection of rectangles is a rectangle, we conclude that $(I_1)_G$ is a G-invariant rectangular ideal. Therefore $(I_1)_G$ has (P) by condition (3). By step 1, with $J=A\otimes B$, we get that $(I_1)_e$ has (P).

Now suppose we know that $\bigcap_{i=1}^{n}(I_i)_e$ has (P) for any n rectangular ideals $I_1,\dots,I_n$, and let I_{n+1} be a rectangular ideal. Then $I_i \cap (I_{n+1})_G$ is an intersection of rectangular ideals and is hence rectangular. Therefore

$$\left[\bigcap_{i=1}^{n}(I_i)_e\right]\cap(I_{n+1})_G=\bigcap_{i=1}^{n}((I_i)_e\cap(I_{n+1})_G)$$

has (P) by the induction hypothesis, because $(I_i\cap(I_{n+1})_G)_e=(I_i)_e\cap(I_{n+1})_G$. Since $\bigcap_{i=1}^{n}(I_i)_e$ is G-invariant, step 1 implies that $\bigcap_{i=1}^{n+1}(I_i)_e$ has (P), as desired. This completes the induction and proves step 2.

Step 3. Let $(J_\lambda)_{\lambda<\kappa}$ be a set of G-invariant ideals indexed by the set of ordinal numbers $\lambda<\kappa$, such that each J_λ has the form $J_\lambda=\bigcap_{i=1}^{n(\lambda)}(I_{\lambda,i})_e$ for rectangular ideals $I_{\lambda,i}$. Let $L_\lambda=\overline{\sum_{\mu<\lambda} J_\mu}$. Then L_λ has (P) for all λ.

Proof of step 3. We prove this by induction on λ, for all such sequences (J_λ) simultaneously. The case $\lambda=0$ is step 2. So assume the result is true for all $\lambda<\eta$, where η is some ordinal.

Case 1. Suppose η is a successor of some other ordinal λ, that is, $\eta=\lambda+1$. Then L_λ has (P) by assumption. Also J_λ has (P) by step 2. Furthermore, each $J_\lambda\cap J_\mu$ is an intersection of finitely many ideals of the form I_e, namely

$$J_\lambda\cap J_\mu=\bigcap_{i=1}^{n(\lambda)}(I_{\lambda,i})_e\cap\bigcap_{i=1}^{n(\mu)}(I_{\mu,i})_e\,.$$

By the induction assumption,

$$J_\lambda\cap L_\lambda=\overline{\sum_{\mu<\lambda}J_\lambda\cap J_\mu}$$

has (P). Since J_λ and L_λ have (P), property (5) at the beginning of the proof shows that $L_\eta=J_\lambda+L_\lambda$ has (P).

Case 2. Suppose η is a limit ordinal. Then

$$L_\eta=\overline{\sum_{\lambda<\eta}J_\lambda}=\overline{\bigcup_{\lambda<\eta}L_\lambda},$$

which has (P) by condition (2) and the induction hypothesis. This completes the proof of step 3.

Step 4. Proof of the lemma. Let L be any G-invariant ideal in $A\otimes B$. Let $U\subset Prim(A)\times Prim(B)$ be the corresponding open subset. By the definition of the product topology, U is the union of a set $(U_\lambda)_{\lambda<\kappa}$ of open rectangles. Let I_λ be the rectangular ideal corresponding to U_λ, so that $L=\overline{\sum_{\lambda<\kappa}I_\lambda}$. Further, let $J_\lambda=(I_\lambda)_e$, which is a G-invariant ideal containing I_λ and contained in L. Then also $L=\overline{\sum_{\lambda<\kappa} J_\lambda}$. Step 3 now implies that L has (P), as desired. Q.E.D.

Notice that in step 4, only the special case of step 3 in which each $n(\lambda)=1$ was used. However, the proof of step 3 by induction required the more general case in which $n(\lambda)$ is an arbitrary positive integer.

In the next theorem, the numbering of the first 9 items corresponds to the numbering of the analogous properties in theorem 6.4.7.

6.5.5 Theorem. Let G be a finite group. Then the category $\mathbf{N}^G$ satisfies the following properties.

(1) $\mathbf{N}^G$ contains all separable type I G-algebras.

(2) $\mathbf{N}^G$ contains all separable G-algebras A such that the action of G on $Prim(A)$ is free.

(3) If

$$0 \longrightarrow I \longrightarrow A \longrightarrow B \longrightarrow 0$$

is an equivariant exact sequence, then A is in $\mathbf{N}^G$ if and only if both I and B are.

(4) If A and B are stably isomorphic as G-algebras, and $A \in \mathbf{N}^G$, then $B \in \mathbf{N}^G$.

(5) If (G,A,α) and (G,B,β) are exterior equivalent G-algebras, and $(G,A,\alpha) \in \mathbf{N}^G$, then $(G,A,\beta) \in \mathbf{N}^G$.

(6a) If A is a G-algebra which has a composition series $(A_\lambda)_{\lambda\leq\kappa}$ such that each $A_{\lambda+1}/A_\lambda$ is in $\mathbf{N}^G$, then A is in $\mathbf{N}^G$.

(6b) If A is an AF algebra, and α is an action of G on A such that there is an increasing sequence A_n of G-invariant finite dimensional subalgebras with $\overline{\bigcup_{n=1}^{\infty} A_n}=A$, then $(G,A,\alpha) \in \mathbf{N}^G$.

(7) Let H be either $\mathbf{R}$ or $\mathbf{Z}$, let $(G,A,\alpha) \in \mathbf{N}^G$, and let β be an action of H on A which commutes with α. If the strong Connes spectrum $\tilde{\Gamma}(\beta)$ is all of $\hat{H}$, then $C^*(H,A,\beta) \in \mathbf{N}^G$.

(8) If H is a subgroup of G, and $(H,A,\alpha) \in \mathbf{N}^H$, then $C(G \times_H A) \in \mathbf{N}^G$.

(9) If $A, B \in \mathbf{N}^G$ and B is nuclear, then $A \otimes B \in \mathbf{N}^G$.

(10) If $(G,A,\alpha) \in \mathbf{N}^G$ and H is a subgroup of G, then $(H,A,\alpha|_H) \in \mathbf{N}^H$.

As after theorem 6.4.7, some remarks are in order. The reasons for the modifications of properties (2), (3), and (8) in theorem 6.4.7 are fairly evident. We are unable to prove properties (6) and (7) of that theorem because of lack of control over ideals in direct limits and in general crossed products. Properties (6a), (6b) and (7) above are what we can prove. However, it is certainly possible for a direct limit or a crossed product not covered by these properties to be in $\mathbf{N}^G$, and for this to be provable by showing that for every G-invariant ideal I and every cyclic subgroup H of G, the H-algebra (H,I) is in the smallest category closed under all the operation in theorem 6.4.7. This is this case, for instance, for the algebra in example 6.3.6. We should also mention that a version of the Ideal Decomposition Lemma that worked for the circle group S^1 would enable us to prove that $\mathbf{N}^G$ is closed under arbitrary crossed products by commuting actions of $\mathbf{Z}$. Indeed, for $A \in \mathbf{N}^G$ we would have $C^*(\mathbf{Z}, A) \in \mathbf{N}^{G \times S^1}$, since ideals invariant under the dual action are again crossed products, and we would then be able to get $C^*(\mathbf{Z},A) \in \mathbf{N}^G$ by lemma 6.5.3.

Proof of theorem 6.5.5. Part (10) follows immediately from lemma 6.5.3. The proofs of parts (1), (2), (4), and (5) follow from the corresponding parts of lemma 6.5.2 and obvious relations among ideals, such as the fact that an ideal in a type I algebra is again type I, that exterior equivalent actions have the same invariant ideals, etc. Part (3) follows from part (3) of lemma 6.5.2 and the definition of $\mathbf{N}^G$, and part (6a) follows from part (3), part (6) of lemma 6.5.2, and a standard transfinite induction argument. For part (6b), we use the fact that finite dimensional G-algebras are in $\mathbf{N}^G$ (by part (1)), and that if $A = \overline{\bigcup_{n=0}^{\infty} A_n}$ is as in (6b) and I is a (G-invariant) ideal in A, then $I = \overline{\bigcup_{n=0}^{\infty} (I \cap A_n)}$. (This last fact follows by applying [12], theorem 3.3, to the unitizations. We do not assume that AF algebras are unital.)

To prove (7), observe that the assumption implies that every ideal I in $C^*(H,A,\beta)$ is $\hat{H}$-invariant, by lemma 3.4 in [58]. Therefore $I = C^*(H,J,\beta)$ for some H-invariant ideal J in A, by proposition 6.3.9. If I is G-invariant, then $C^*(\hat{H},I,\hat{\beta}) = K(L^2(H)) \otimes J$ is also G-invariant, so that J is G-invariant. Part (7) now follows from part (7) of lemma 6.5.2. Also, part (8) follows from part (8') of lemma 6.5.2 and the fact that every G-invariant ideal I in $C(G \times_H A)$ has the form $C(G \times_H J)$ for some H-invariant ideal J in A. (This fact was proved at the beginning of the proof of theorem 4.3.4.)

It remains to prove part (9). We will actually prove that if (9) holds for all proper subgroups of a finite group G, then (9) holds for G as well. An induction argument then finishes the proof, since, if G is the one element group, then $\mathbf{N}^G$ is the category of all C^*-algebras.

Accordingly, assume (9) holds for all proper subgroups H of G, and let A and B be G-algebras in $\mathbf{N}^G$. Say that a G-algebra has the property (P) if it is in $\mathbf{N}_1^G$. We will use the previous lemma, so we need to verify conditions (1) through (4) of its hypotheses. Condition (1) is just (3) of lemma 6.5.2, and condition (2) is a special case of (6) of lemma 6.5.2. Condition (3) follows from the definition of $\mathbf{N}^G$ and (9) of lemma 6.5.2, while condition (4) follows from the hypothesis on subgroups, (8') of lemma 6.5.2, and part (10) of the present theorem, which we have already proved. The previous lemma now implies that all G-invariant ideals of $A \otimes B$ are in $\mathbf{N}_1^G$, that is, that $A \otimes B$ is in $\mathbf{N}^G$. Q.E.D.

The category $\mathbf{N}^G$ contains many non-nuclear C^*-algebras, for instance the algebras $C_r^*(F_n)$ for $n\geq 2$. If we want to consider K-freeness of tensor products of a given G-algebra, on which the action of G is totally K-free, with arbitrary, not necessarily nuclear, G-algebras, we will therefore have to restrict ourselves to the somewhat smaller category given in the following definition.

6.5.6 Definition. Let G be a finite group. Then $\mathbf{N}_{nuc}^G$ is the full subcategory of nuclear C^*-algebras (G,A,α) such that, for every G-invariant ideal I of A, every cyclic subgroup H of G, every prime ideal $P\subset R(H)$ with support H, and every G-algebra B such that $K_*^H(B)_P$ is a projective $R(H)_P$-module, the pairing

$$\omega_P(A,B): K_*^H(A)_P \otimes_{R(H)_P} K_*^H(B)_P \longrightarrow K_*^H(A\otimes B)_P$$

is an isomorphism.

6.5.7 Theorem. Let G be a finite group. Then the category $\mathbf{N}_{nuc}^G$ satisfies all of the properties of theorem 6.5.5, except that in condition (2), the algebras must be required to be nuclear, and in condition (8), (H,A,α) must be required to be in $\mathbf{N}_{nuc}^H$.

Proof. Remark 6.4.8, together with the fact that all of the operations under which $\mathbf{N}_{nuc}^G$ is claimed to be closed preserve nuclearity, shows that the proofs of theorem 6.5.5 and the lemmas leading up to it apply equally well to the category $\mathbf{N}_{nuc}^G$. Q.E.D.

6.6. K-Freeness of Actions on Tensor Products

Let G be a cyclic group of prime order, and let (G,A,α) and (G,B,β) be G-algebras such that (G,A,α) is in $\mathbf{N}^G$, B is nuclear, and one of the actions α and β is K-free. Then we can conclude from our Künneth formula that if $P\subset R(G)$ is a prime ideal not containing $I(G)$, then $K_*^G(I\otimes J)_P = 0$ for all G-invariant ideals I of A and J of B. Lemma 6.5.4 can then be used to show that the diagonal action $\alpha\otimes\beta$ is K-free. However, we want our results on tensor products to apply to groups other than $\mathbf{Z}/p\mathbf{Z}$ for p prime. To accomplish this, we must somehow get around the restriction, needed for the Künneth formula, that the prime ideal considered have support G. Our method works only for some groups, and requires that we use induction arguments on subgroups. Therefore we need to consider totally K-free actions. The first lemma will be needed to reduce the case of a prime ideal not containing $I(G)$ to the case of a prime ideal with support G.

6.6.1 Lemma. Let G be a compact abelian group, and let H be a closed subgroup. Let M be an $R(H)$-module, and regard M as an $R(G)$-module via the restriction Res_H^G. Let Q be a prime ideal in $R(H)$, and let $P = (Res_H^G)^{-1}[Q]$. Then $M_Q = 0$ as an $R(H)$-module if and only if $M_P = 0$ as an $R(G)$-module.

Proof. We first observe that the restriction $Res_H^G: R(G) \longrightarrow R(H)$ is surjective. Since $R(G)$ is isomorphic to the group ring $\mathbf{Z}[\widehat{G}]$ and similarly $R(H) \simeq \mathbf{Z}[\widehat{H}]$ in a canonical way, this follows immediately from the fact that $\widehat{G} \longrightarrow \widehat{H}$ is surjective.

Now let $m \in M$. Then the image of m in M_P is zero if and only if there is $r \in R(G)$, $r \notin P$, such that $rm = 0$. The image of m in M_Q is zero if and only if there is $s \in R(H)$, $s \notin Q$, such that $sm = 0$. Given such an r, let $s = Res_H^G(r)$, while given such an s, use the surjectivity of Res_H^G to choose r such that $Res_H^G(r) = s$. So $M_P = 0$ if and only if $M_Q = 0$. Q.E.D.

We need one more lemma before proceeding to the proof of the theorem on tensor products of action.

6.6.2 Lemma. Let p be a prime number, and let P be a prime ideal in $R(\mathbf{Z}/p^n\mathbf{Z})$ such that P does not contain $I(\mathbf{Z}/p^n\mathbf{Z})$. Let τ be a generator of $(\mathbf{Z}/p^n\mathbf{Z})^\wedge$. Then P contains $f(\tau)$ for exactly one irreducible factor $f(x)$ of the polynomial $1 - x^{p^n}$ in $\mathbf{Z}[x]$.

Proof. Let $h(x)$ be the polynomial $h(x) = 1 + x + \cdots + x^{p-1}$. Then

$$1 - x^{p^n} = (1-x)h(x)h(x^p)\cdot \cdots \cdot h(x^{p^{n-1}}).$$

It is well known that each factor $h(x^{p^k})$ is irreducible over $\mathbf{Q}$. (It is the minimal polynomial of the primitive p^kth roots of unity.)

We claim that the remainder on dividing $h(x^{p^k})$ by $h(x)$ is p. Indeed, $(1-x)h(x) = 1 - x^p$, so that $h(x)$ divides $x^r - x^{r+p}$ for any r. Therefore $h(x)$ divides $x^r - x^s$ whenever p divides $r - s$. It follows that $h(x)$ divides the polynomial

$$(1 - x^{p^k}) + (1 - x^{2p^k}) + \cdots + (1 - x^{(p-1)p^k}),$$

which is exactly $p - h(x^{p^k})$. Therefore the remainder on dividing $h(x^{p^k})$ by $h(x)$ is p, as claimed. Furthermore, if $l < k$, then the remainder on dividing $h(x^{p^k})$ by $h(x^{p^l})$ is also p, as can be seen by replacing x by x^{p^l}.

Since P is a prime ideal, the equation

$$(1-\tau)h(\tau)h(\tau^p)\cdot \cdots \cdot h(\tau^{p^{n-1}}) = 1 - \tau^{p^n} = 0,$$

together with the fact that $1 - \tau \notin P$, implies that $h(\tau^{p^k}) \in P$ for some k. (The element $1-\tau$ is not in P because $1-\tau$ generates the augmentation ideal.) Now suppose that $h(\tau^{p^k}) \in P$ for more than one value of k. Let l be the smallest number such that $h(\tau^{p^l}) \in P$, and let k be the next smallest number such that $h(\tau^{p^k}) \in P$. By the result of the previous paragraph, there is a polynomial $g(x)$ such that $h(x^{p^l})g(x) + p = h(x^{p^k})$. If we evaluate this expression at $x = \tau$, we obtain $p \in P$. But the major part of the proof of theorem 5.2.6 was devoted to showing that if G is a finite p-group and $P \subset R(G)$ is a prime ideal which does not contain $I(G)$, then $p \notin P$. We have therefore obtained a contradiction. Q.E.D.

6.6.3 Tensor Product Theorem. Let G be a finite p-group. Let (G,A,α) and (G,B,β) be G-algebras, with $(G,A,\alpha) \in \mathbf{N}^G$ and B nuclear. If either α or β is totally K-free, then so is the diagonal action $\alpha\otimes\beta$.

Proof. We first consider the case $G=\mathbf{Z}/p^n\mathbf{Z}$, and we prove the result by induction on n. For $n=0$ the assertion to be proved is vacuous, so assume that it is known to hold for all $k<n$. We then prove it for n.

Let $P\subset R(G)$ be a prime ideal not containing $I(G)$, and let $S\subset G$ be the support of P. There are then two cases.

Case 1. $S=G$. We use lemma 6.5.4. Let the property (P) used in that lemma be defined by the equation $K_*^G(A)_P=0$. Then the long exact sequence associated to an equivariant short exact sequence shows that hypothesis (1) of the lemma holds. To check hypothesis (2), let $A=\overline{\bigcup_{\lambda<\kappa} I_\lambda}$. Then $K_*^G(A)_P=\varinjlim_{\lambda<\kappa} K_*^G(I_\lambda)_P$, so if each I_λ has (P) then so does A. For hypothesis (3), let I and J be any G-invariant ideals of A and B respectively. By the definition of the category $\mathbf{N}^G$, we have $(G,I,\alpha_{(\cdot)}|_I)\in\mathbf{N}_0^G$. Furthermore, J is nuclear. The Künneth formula, theorem 6.4.6, therefore yields an exact sequence

$$0\longrightarrow K_*^G(I)_P\otimes_R K_*^G(J)_P\longrightarrow K_*^G(I\otimes J)_P\longrightarrow Tor^R(K_*^G(I)_P,K_*^G(J)_P)\longrightarrow 0,$$

where $R=R(G)_P$. By assumption, $K_*^G(I)_P=0$ (if α is totally K-free), or $K_*^G(J)_P=0$ (if β is totally K-free). Therefore $K_*^G(I\otimes J)_P=0$. We have thus shown that an arbitrary G-invariant rectangular ideal in $A\otimes B$ has (P), and hypothesis (3) is verified. For hypothesis (4), let H be a proper subgroup of G and let L be an H-invariant ideal in $A\otimes B$. Then the action of H on L has locally discrete K-theory by the induction hypothesis, and therefore the action of G on $C(G\times_H L)$ has locally discrete K-theory by theorem 4.3.4. In particular, $K_*^G(C(G\times_H L))_P=0$, that is, $C(G\times_H L)$ has (P). Lemma 6.5.4 now implies that $K_*^G(L)_P=0$ for all G-invariant ideals L of $A\otimes B$.

Case 2. $S\neq G$. By lemma 6.5.2, we have $(A,S,\alpha|_S)\in\mathbf{N}^S$. Therefore the induction hypothesis applies to this S-algebra. Let $Q\subset R(S)$ be the prime ideal such that $(Res_S^G)^{-1}[Q]=P$. (There is such an ideal because S is the support of P.) For any G-invariant ideal L in $A\otimes B$, consider the composition

$$K_*^G(L)_P\xrightarrow{Res_S^G}K_*^S(L)_P\xrightarrow{Ind_S^G}K_*^G(L)_P. \qquad (*)$$

By proposition 5.1.3 (2), this composition is multiplication by the element $Ind_S^G(1)\in R(G)$. Let p^k be the order of S, and let p^r be the index of S in G. (Thus $k+r=n$.) Let τ be a generator of $\widehat{G}$; then one easily checks that $Ind_S^G(1)=h(\tau)$, where

$$h(x)=1+x^{p^k}+x^{2p^k}+\cdots+x^{(p^r-1)p^k}.$$

Now $(1-x^{p^k})h(x)=1-x^{p^n}$, and by the previous lemma, $f(\tau)\in P$ for exactly one irreducible factor $f(x)$ of $1-x^{p^n}$. Since $1-\tau^{p^k}$ is in the kernel of Res_S^G, we have $1-\tau^{p^k}\in P$. Since P is a prime ideal, the irreducible factor $f(x)$ divides $1-x^{p^k}$, and so does not divide $h(x)$. (The polynomial $1-x^{p^n}$ has no multiple roots.) It follows that $h(\tau)\notin P$. Therefore multiplication by $Ind_S^G(1)$ on $K_*^G(L)_P$ is invertible. Now $(*)$ shows that this map factors through $K_*^S(L)_P$. By lemma 6.5.1, we have $K_*^S(L)_P = K_*^S(L)_Q$, which is zero by the induction hypothesis. Therefore $K_*^G(L)_P = 0$.

Combining cases 1 and 2, we have found that for any G-invariant ideal L in $A \otimes B$ and any prime ideal $P \subset R(G)$ not containing $I(G)$, we have $K_*^G(L)_P = 0$. Therefore $\alpha \otimes \beta$ is K-free. Combining this with the induction hypothesis, which implies that $\alpha \otimes \beta |_H$ is K-free for any proper subgroup H of G, we conclude that $\alpha \otimes \beta$ is totally K-free. The induction is now complete.

Now consider the general case, in which G is an arbitrary finite p-group. Suppose α is totally K-free. Then $\alpha|_S$ is totally K-free for every cyclic subgroup S of G. It follows from what we have done that $(\alpha \otimes \beta)|_S$ is totally K-free for every cyclic subgroup S of G. Theorem 5.2.6 now implies that $\alpha \otimes \beta$ is totally K-free. A similar argument implies that $\alpha \otimes \beta$ is totally K-free if β is totally K-free. Q.E.D.

If we instead require A to be nuclear, then we obtain:

6.6.4 Theorem. Let G be a finite p-group. Let (G,A,α) and (G,B,β) be G-algebras, and assume that $(G,A,\alpha) \in \mathbf{N}_{nuc}^G$. If either α or β is totally K-free, then so is the diagonal action $\alpha \otimes \beta$.

Proof. The only change that needs to be made in the proof of the previous theorem is to substitute for theorem 6.4.6 the part of remark 6.4.8 concerning the Künneth formula. Q.E.D.

6.6.5 Remark. We point out that, in the previous two theorems, one does not need the full Künneth formula to prove that if β is totally K-free then $\alpha \otimes \beta$ is totally K-free. Indeed, if $(G,A,\alpha) \in \mathbf{N}^G$, then $(H, I, \alpha|_H|_I) \in \mathbf{N}_0^H$ for every cyclic subgroup H of G and every H-invariant ideal I of A. If now J is an H-invariant ideal in B, and $P \subset R(H)$ is a prime ideal with support H, then $K_*^H(J)_P = 0$, which is certainly a projective $R(H)_P$-module. Therefore $\omega_P(I,J)$ is an isomorphism, and we conclude that

$$K_*^H(I \otimes J)_P \simeq K_*^H(I) \otimes_{R(H)_P} K_*^H(J)_P = 0 .$$

One might therefore hope that the theorems could be proved in this case without such a restrictive condition on G, since the homological dimension plays no obvious role. The idea might be to define a category $\mathbf{M}_0^G$ of G-algebras (G,A,α) such that, whenever (G,B,β) is a nuclear G-algebra with locally discrete K-theory, then the diagonal action $\alpha \otimes \beta$ has locally discrete K-theory. The analogs of many of the results proved in section 6.3 for the category $\mathbf{N}_0^H$ do indeed hold. Unfortunately, however, it is not clear how to show that all type I G-algebras are in such a category. The stumbling block is G-algebras of the form (G, M_n, α), where the action α is given by conjugation by a projective unitary representation which does not lift to a unitary representation. Note that the method used in proposition 6.2.11, and the lemmas leading up to it, to handle general type I G-algebras depended on considering prime ideals with support G, and the method for circumventing this restriction required that G be a p-group.

One would really hope that theorem 6.6.3 would hold without any of the restrictive condition except perhaps that one of the algebras be nuclear. We have no counterexamples for the possible weaker sets of hypotheses, but the only other way we see to get a theorem is to start with the assumption that one of the actions is KK-free. We then obtain a similar result — without using anything from sections 1 through 4 of this chapter. The substitute for the Künneth formula is the fact that there is a unital ring homomorphism from $KK_G^0(A,A)$ to $KK_G^0(A\otimes B,A\otimes B)$, which is one of the basic properties of KK-theory. (See proposition 3.5.2.) We thus have:

6.6.6 Theorem. Let G be an arbitrary finite group, and let (G,A,α) and (G,B,β) be separable G-algebras such that A is nuclear. If α is KK-free, then the diagonal action $\alpha\otimes\beta$ is totally K-free.

Proof. We prove that $(\alpha\otimes\beta)|_H$ is totally K-free for all subgroups H of G, by induction on H. The result is vacuous for $H=\{e\}$. So let H_0 be a fixed subgroup, and assume the result is true for all proper subgroups H of H_0. We are going to use lemma 6.5.4 to show that the action of H_0 on L has locally discrete K-theory for all H_0-invariant ideals L of $A\otimes B$. This, together with the induction hypothesis, will imply that $(\alpha\otimes\beta)|_{H_0}$ is totally K-free, and will complete the induction and prove the theorem. Accordingly, let (P) be the property of having locally discrete K-theory. Part (1) of the hypothesis of lemma 6.5.4 is satisfied by lemma 4.1.4, part (2) is lemma 4.2.14, and part (4) follows from theorem 4.3.4 and the induction hypothesis. It remains to verify part (3).

Let I and J be H_0-invariant ideals in A and B respectively. By proposition 3.5.2, there is a unital $R(H_0)$-algebra homomorphism $\tau_J : KK_{H_0}^0(I,I) \longrightarrow KK_{H_0}^0(I\otimes J,I\otimes J)$. By the KK-Subgroup Theorem 5.3.6, $\alpha|_{H_0}$ is KK-free. Therefore, for each prime ideal $P\subset R(H_0)$ not containing $I(H_0)$, the image of 1_I in $KK_{H_0}^0(I,I)_P$ is zero, and it follows that the image of $1_{I\otimes J}=\tau_J(1_I)$ in $KK_{H_0}^0(I\otimes J,I\otimes J)_P$ is zero. So $KK_{H_0}^0(I\otimes J,I\otimes J)_P$ is a unital ring in which $1=0$, and is therefore the zero ring. Thus $(\alpha\otimes\beta)|_{I\otimes J}$ has discrete KK-theory, and therefore has locally discrete K-theory by proposition 4.4.6. We have now shown that all rectangular H_0-invariant ideals have (P), as required.

Q.E.D.

It would of course be nice to be able to conclude that $\alpha\otimes\beta$ is KK-free under the hypotheses of this theorem. Unfortunately, the property of having discrete KK-theory does not satisfy part (2) of the hypothesis of lemma 6.5.4. (See example 4.4.12.) It is not clear how to get around this problem.

Recall that at the end of section 4.2 we discussed what additional conditions one could require of an equivariant homomorphism $\varphi : A \longrightarrow B$ in order to conclude that if the action of G on A is totally K-free, then so is the action of G on B. (This is the analog of the fact that if $f : X \longrightarrow Y$ is an equivariant map of G-spaces, and if G acts freely on Y, then G acts freely on X.) We can now prove the following result in this direction.

6.6.7 Theorem. Let G be a finite group, and let $\varphi : A \longrightarrow B$ be an equivariant homomorphism of G-algebras (G,A,α) and (G,B,β). Assume that B is generated as a C^*-algebra by the image $\varphi[A]$ and the relative commutant of $\varphi[A]$ in B. If either

(1) A is separable nuclear, B is separable, and α is KK-free, or

(2) G is a p-group, $(G,A,\alpha) \in \mathbf{N}^G_{nuc}$, and α is totally K-free,

then β is totally K-free.

Proof. Let C be the relative commutant of $\varphi[A]$ in B, and let $\gamma = \beta_{(\cdot)}|_C$. Since the images in B of A under φ and of C under the identity map commute, there is a unique homomorphism ψ from the maximal tensor product $A \otimes_{\max} C$ to B such that $\psi(a \otimes c) = \varphi(a)c$ for $a \in A$ and $c \in C$. (See [104], proposition IV.4.7.) Clearly ψ is equivariant, and the hypotheses of the theorem imply that ψ is surjective. Since A is nuclear (under either hypothesis (1) or hypothesis (2)), there is in fact only one possible C^* tensor product. Thus, $A \otimes_{\max} C = A \otimes C$. Under hypothesis (1), C must also be separable, and the diagonal action $\alpha \otimes \gamma$ is totally K-free by theorem 6.6.6. Under hypothesis (2), $\alpha \otimes \gamma$ is totally K-free by theorem 6.6.4. In either case, B is equivariantly isomorphic to $(A \otimes C)/ Ker(\psi)$, and therefore β is totally K-free by proposition 4.2.6. Q.E.D.

We point out that the possible improvements in the tensor product theorems discussed in remark 6.6.5 would not be useful in improving this theorem, because we have total K-freeness in the wrong factor. We also point out that this theorem implies special cases of several of the results in chapter 4, such as theorem 4.6.1 and its corollaries for finite groups.

There is no hope of obtaining a tensor product theorem for K-freeness, as opposed to total K-freeness, as the following proposition shows.

6.6.8 Proposition. Let G be a compact Lie group, and let (G,A,α) be a G-algebra. For each closed subgroup H of G, let β_H be the action of G on $C(G/H)$ coming from the standard action of G on G/H by translation. Then α is totally K-free if and only if $\alpha \otimes \beta_H$ is K-free for every closed subgroup H of G.

Proof. We prove that $\alpha \otimes \beta_H$ is K-free if and only if $\alpha|_H$ is K-free. Since $C(G/H) \simeq C(G \times_H \mathbf{C})$, lemma 5.1.7 shows that $A \otimes C(G/H) \simeq C(G \times_H A)$ as G-algebras. The result now follows from theorem 4.3.4. Q.E.D.

The actual situation is even worse than is suggested by this proposition, as is shown by the following example of G-algebras (G,A,α) and (G,B,β) such that both α and β are K-free but $\alpha \otimes \beta$ is not.

6.6.9 Example. Let $G=(\mathbf{Z}/2\mathbf{Z})^3$, and let G be generated by the elements g, h, and k of order 2. Let $A=B=M_2\oplus M_2$. Define elements $u, v\in M_2$ by

$$u=\begin{pmatrix}1 & 0\\ 0 & -1\end{pmatrix} \quad \text{and} \quad v=\begin{pmatrix}0 & 1\\ 1 & 0\end{pmatrix}.$$

Define an action $\alpha: G\longrightarrow Aut(A)$ by

$$\alpha_g(a,b)=(uau^*, ubu^*), \quad \alpha_h(a,b)=(vav^*, vbv^*), \quad \text{and} \quad \alpha_k(a,b)=(b,a),$$

for $a,b\in M_2$. (Compare with example 4.2.3.) Let $\varphi: G\longrightarrow G$ be the automorphism of G such that $\varphi(g)=h$, $\varphi(h)=k$, and $\varphi(k)=g$. Define $\beta: G\longrightarrow Aut(B)$ by $\beta_t=\alpha_{\varphi(t)}$ for $t\in G$.

Let H be the subgroup of G generated by g and h. By proposition 2.9.6, (G,A,α) is isomorphic to the G-algebra $C(G\times_H M_2)$ induced by the action of H on M_2. Since the action of H on M_2 is K-free by example 4.2.3, we conclude that α is K-free by theorem 4.3.4. Clearly β is therefore also K-free.

Now consider the diagonal action $\alpha\otimes\beta$ on $A\otimes B$. We have $A\otimes B\simeq M_4\oplus M_4\oplus M_4\oplus M_4$, where the subgroup $\{e,h,k,hk\}$ permutes the summands transitively and the element g sends each summand to itself. By proposition 2.9.6, we have $A\otimes B\simeq C(G\times_{H_0}M_4)$, where $H_0=\{e,g\}$. The action of H_0 on M_4 is inner by lemma 4.2.12, and therefore does not have locally discrete K-theory by lemma 4.1.5. Consequently $\alpha\otimes\beta$ does not have locally discrete K-theory, by theorem 4.3.4 again.

This example also shows that the Künneth formula, or even anything like it, fails dramatically for G-algebras. Indeed, for an appropriate prime ideal $P\subset R(G)$, we have $K_*^G(A)_P=K_*^G(B)_P=0$ but $K_*^G(A\otimes B)_P\neq 0$.

Chapter 7

K-Freeness, Saturation, and the Strong Connes Spectrum

The purpose of this chapter is to study the relation between K-freeness and other forms of freeness of group actions on C^*-algebras. One of these other forms of freeness is the concept of saturation of an action, due to Marc Rieffel (unpublished). If G is a compact group, then a continuous action α of G on a C^*-algebra A is saturated if a suitable completion of A (in a different norm) is in a natural way an $A^G - C^*(G,A)$ (Morita) equivalence bimodule, as defined in [87]. (This concept is closely related to the concept of "large spectral subspaces" introduced in [77] for actions of the circle group.) Since separable Morita equivalent C^*-algebras have isomorphic K-groups, this concept is also related to the question of whether there is an isomorphism $K_*^G(A) \simeq K_*(A^G)$. (Compare condition (2) of theorem 1.1.1).

The other condition we consider is that the strong Connes spectrum $\widetilde{\Gamma}(\alpha)$ of an action α of an abelian group G on a C^*-algebra be the entire dual group $\widehat{G}$. The Connes spectrum $\Gamma(\alpha)$ and its variants were used by Olesen and Pedersen in [71], [72], and [73] (see also chapter 8 of [79]) to study primeness and simplicity of crossed products and outerness of automorphisms of C^*-algebras. The strong Connes spectrum $\widetilde{\Gamma}(\alpha)$ was later introduced by Kishimoto in [58] for similar reasons. The condition that $\widetilde{\Gamma}(\alpha)$ equal $\widehat{G}$ is more restrictive than the condition that $\Gamma(\alpha)$ equal $\widehat{G}$, and is more suitable for our purposes. The strong Connes spectrum is actually closely related to saturation, as will become apparent.

Easy examples show that even a trivial action can be K-free. (Consider the Cuntz algebra $\mathbf{O}_2$.) In order to prove that K-freeness implies any of these other properties, we therefore need to assume that the K-theory of the C^*-algebra involved is sufficiently nontrivial. For this purpose, we introduce K-visibility of C^*-algebras and several related concepts. We are then able to show, for example, that a K-free action α of a finite group G on a highly K-visible C^*-algebra A satisfies $\widetilde{\Gamma}(\alpha) = \widehat{G}$. The hypothesis that G be finite is shown by an example to be necessary. The converse of this result is also false.

We do not consider totally K-free actions in this chapter. The explanation for this is that the other conditions considered also do not behave well on passage to subgroups. This is illustrated in example 4.2.3, where it was shown that the Connes spectrum of a certain action $\alpha : G \longrightarrow Aut(A)$ is all of $\widehat{G}$, but that $\Gamma(\alpha|_H)$ is trivial for nontrivial subgroups H of G. One can check that in fact $\widetilde{\Gamma}(\alpha) = \widehat{G}$, while $\widetilde{\Gamma}(\alpha|_H)$, being smaller than $\Gamma(\alpha|_H)$, is also trivial.

This chapter is organized as follows. In the first section, we define and present the basic properties of saturated actions. We show that saturation satisfies at least some of the properties that one would expect from a concept of freeness of group actions on C^*-algebras. The results are mostly due to Rieffel (unpublished). In section 2, we introduce the concept of hereditary saturation. We say that an action is hereditarily saturated if the induced actions on invariant hereditary subalgebras are all saturated. We then prove some of the properties of this condition, including the result that an action α of a compact abelian group G is hereditarily saturated if and only if $\widetilde{\Gamma}(\alpha) = \widehat{G}$. In the third section, we define K-visibility and several variants. We then prove some lemmas about subquotients of C^*-algebras which are needed to apply these concepts. Section 4 is devoted to the proof of the theorem mentioned above, a related theorem, and several examples. In section 5 we prove that there are lots of K-visible algebras, to which the results of section 4 are applicable. Finally, in the last section we consider the relationship between $K_*(A^G)$ and $K_*^G(A)$ for a K-free action of G on A, in light of the results of section 4.

7.1. Saturated Actions

Given a G-algebra (G,A,α), where G is compact, we want to make A into an $A^G - C^*(G,A,\alpha)$ bimodule. We will then be able to say that α is saturated if this bimodule is a Morita equivalence bimodule (see [87]). We will need the following lemma, which is the analog of proposition 4.10 of [84] for our situation. (Here we are considering functions with values in a C^*-algebra, but the subgroup H which appears in [84] is trivial here.)

7.1.1 Lemma. Let G be a compact group, and let (G,A,α) be a G-algebra. Regard $C(G,A)$ as a subalgebra (not closed) of $L^1(G,A,\alpha)$, so that multiplication is taken to be convolution. For $r, s \in C(G,A)$, we have

$$(rss^*r^*)(e) \leq \| s \|^2 (rr^*)(e),$$

where e is the identity in G, and the norm is that of $C^*(G,A,\alpha)$.

Proof. Fix $r \in C(G,A)$. Let φ be a positive linear functional on A, and define, for $k \in G$, a functional $\Psi(k)$ on A by

$$\Psi(k)(a) = \varphi\left[\int_G r(h)\alpha_h(a)r(hk)^* dh\right].$$

This formula is motivated as follows: a calculation shows that, for $f \in C(G,A)$, we have

$$\varphi((rfr^*)(e)) = \int_G \Psi(g)(f(g))\,dg. \qquad (*)$$

Now r is a continuous function on a compact set, and is hence uniformly continuous. Therefore, for $\varepsilon > 0$ and for k_1 sufficiently close to k_2, we have $\|r(hk_1)^* - r(hk_2)^*\| < \varepsilon$ for all $h \in G$, whence

$$\| (\Psi(k_1) - \Psi(k_2))(a) \| < \| \varphi \| \; \| a \| \; \| r \|_\infty \varepsilon.$$

It follows that Ψ is a continuous function from G to the dual A^* of A. Furthermore, the relation

$$((rf)(rf)^*)(e) = \int_G (rf)(g)(rf)(g)^* dg \geq 0$$

for $f \in C(G,A)$ implies that

$$\int_G \Psi(g)((ff^*)(g)) = \varphi(rff^*r^*)(e) \geq 0$$

for all $f \in C(G,A)$, where the first step follows from (*). By [79], 7.6.8, the formula

$$\psi(f) = \int_G \Psi(g)(f(g))\, dg$$

defines a positive linear functional on $C^*(G,A,\alpha)$. Therefore, for $s \in C^*(G,A,\alpha)$, we have

$$\psi(ss^*) \leq \| s \|^2 \psi(1).$$

(If necessary, we adjoin an identity to $C^*(G,A,\alpha)$.) In view of (*), this says that for $s \in C(G,A)$, we have

$$\varphi((rss^*r^*)(e)) \leq \| s \|^2 \varphi((rr^*)(e)).$$

This equation holds for an arbitrary positive functional φ on A. Therefore,

$$(rss^*r^*)(e) \leq \| s \|^2 (rr^*)(e),$$

as desired. Q.E.D.

The function $x \longrightarrow \tilde{x}$ in the following definition will be used throughout this section.

7.1.2 Definition. Let G be a compact group, and let (G,A,α) be a G-algebra. For $x \in A$, define $\tilde{x}$ to be the element of $C^*(G,A,\alpha)$ defined in $L^1(G,A)$ by $\tilde{x}(g) = \alpha_g(x)$ for $g \in G$.

This function is not a homomorphism, although it becomes one when restricted to A^G. It is clearly linear, and does satisfy further good properties, as we will see in the proof of lemma 7.1.7.

We are now ready to define the necessary bimodule structure. The left action of A^G on A will be the obvious one, and the right action of $L^1(G,A,\alpha)$ on A will be just the integrated form representation, which was defined for finitely generated projective (G,A,α)-modules in lemma 2.2.3. Unfortunately, this action of $L^1(G,A,\alpha)$ on A does not in general extend to an action of $C^*(G,A,\alpha)$. We will therefore have to complete A in a new norm, derived from the inner products we define.

7.1.3 Proposition. Let G be a compact group, and let (G,A,α) be a G-algebra. For $a \in A^G$, $x, y \in A$, and $f \in L^1(G,A)$, we make the following definitions:

ax is the usual product of a and x as elements of A,

$xf = \int_G \alpha_g^{-1}(xf(g))\, dg$,

$\langle x, y \rangle_{A^G} = \int_G \alpha_g(xy^*)\, dg$,

and

$<x,y>_{C^*(G,A)}$ is the function $g \longrightarrow x^*\alpha_g(y)$.

With these operations, an appropriate completion $\bar{A}$ of A is an $A^G - C^*(G,A,\alpha)$ bimodule which may fail to be a Morita equivalence bimodule ([87], definition on page 287) only in that the range of $<\cdot,\cdot>_{C^*(G,A)}$ need not be dense in $C^*(G,A)$.

Proof. We first observe that $<x,y>_{A^G} = (\tilde{x}\,\tilde{y}^*)(e)$ and $<x,y>_{C^*(G,A)} = (\tilde{x}^*\tilde{y})$ for $x,y \in A$. It is now straightforward to check that the definitions above make A into an $A^G - C^*(G,A,\alpha)$ bimodule with inner products satisfying the conditions preceding the definition on page 287 of [87], except that the right multiplication is only defined for elements in the dense subalgebra $L^1(G,A)$ of $C^*(G,A,\alpha)$, and the equation

$$<x,yf>_{C^*(G,A)} = <x,y>_{C^*(G,A)}f$$

only holds for $f \in L^1(G,A)$. The condition (1) of the definition, namely $(<x,y>_{A^G})z = x<y,z>_{C^*(G,A)}$, is also easily seen to hold — both sides are equal to $\int_G \alpha_g^{-1}(xy^*)z\,dg$.

We now check that the condition (2) of the definition holds. The only part requiring any effort is to show that, for $a \in A^G$, $x \in A$, and $f \in L^1(G,A)$, we have

$$<ax,ax>_{C^*(G,A)} \le \|a\|^2 <x,x>_{C^*(G,A)} \text{ and } <xf,xf>_{A^G} \le \|f\|^2 <x,x>_{A^G}, \quad (*)$$

where in both cases the inequalities are as positive elements of the C^*-algebras to which these elements belong. The first one is easy: let δ_e be the point mass measure at the identity in G, so that $a\delta_e$ is an element of the multiplier algebra of $C^*(G,A)$. (See [79], 7.6.2.) Then

$$<ax,ax>_{C^*(G,A)} = \tilde{x}^*(a\delta_e)^*(a\delta_e)\tilde{x} \le \|a\delta_e\|^2\tilde{x}^*\tilde{x} = \|a\|^2 <x,x>_{C^*(G,A)},$$

where we use the fact that $a \in A^G$ in verifying the first step. For the other part of (*), let $f \in L^1(G,A)$ and $x \in A$. Then

$$<xf,xf>_{A^G} = \int_G \alpha_g \left[\left(\int_G (\alpha_h^{-1}(xf(h))\,dh \right) \left(\int_G \alpha_k^{-1}(xf(k))^*\,dk \right) \right] dg .$$

An ugly calculation shows that, if f is continuous, then this expression is equal to $(\tilde{x}ff^*\tilde{x}^*)(e)$; we omit the details. By lemma 7.1.1,

$$(\tilde{x}ff^*\tilde{x}^*)(e) \le \|f\|^2(\tilde{x}\tilde{x}^*)(e) = \|f\|^2 <x,x>_{A^G},$$

as desired. By continuity, the relation $<xf,xf>_{A^G} \le \|f\|^2 <x,x>_{A^G}$ now holds for all $f \in L^1(G,A)$, not just for f continuous.

Define a new norm on A by setting $\|x\|_2 = \|<x,x>_{A^G}\|^{1/2}$. (For a proof that this is really a norm, see [84], proposition 2.10, or [76], proposition 2.3.) Let $\bar{A}$ be the completion of A in this norm. Then $<\cdot,\cdot>_{A^G}$ extends to an A^G-valued inner product on $\bar{A}$ by continuity. Furthermore,

$$\| <x,y>_{C^*(G,A)} \| \le \| <x,x>_{C^*(G,A)} \|^{1/2} \| <y,y>_{C^*(G,A)} \|^{1/2}$$
$$= \| <x,x>_{A^G} \|^{1/2} \| <y,y>_{A^G} \|^{1/2} = \| x \|_2 \| y \|_2,$$

where the inequality follows from [84], proposition 2.9, or [76], proposition 2.3, and the first equality follows from [85], proposition 3.1. (That proposition clearly applies even if the ranges of the inner products are not dense.) Therefore $<\cdot,\cdot>_{C^*(G,A)}$ also extends by continuity to a $C^*(G,A)$-valued inner product on $\bar{A}$. The relations (*) now imply that the actions of A^G and $L^1(G,A)$ on A can be extended by continuity to actions of A^G and $C^*(G,A)$ on $\bar{A}$, such that (*) continues to hold. We then find that $(<x,y>_{A^G})z = x<y,z>_{C^*(G,A)}$ for all $x,y,z \in \bar{A}$. Thus, all the properties of an $A^G - C^*(G,A)$ equivalence bimodule except property (3) of the definition in [87] have been verified for $\bar{A}$.

Condition (3) specifies that the linear span of the range of $<\cdot,\cdot>_{A^G}$ should be dense in A^G and that the linear span of the range of $<\cdot,\cdot>_{C^*(G,A)}$ should be dense in $C^*(G,A)$. We now prove the first of these; the second is not part of the conclusion of our proposition. Let $x \in A^G$, and let (e_λ) be an approximate identity for A^G. Then

$$<x,e_\lambda> = \int_G \alpha_g(xe_\lambda^*)\, dg = xe_\lambda^*,$$

which converges to x. So the range of $<\cdot,\cdot>_{A^G}$ is dense in A^G, as desired. Q.E.D.

7.1.4 Definition (Rieffel). Let G be a compact group, and let (G,A,α) be a G-algebra. We say α is saturated if the bimodule $\bar{A}$ of the previous proposition is in fact an equivalence bimodule.

The term "saturated" comes from a relation between saturated actions of abelian groups and saturation of a certain Banach *-algebraic bundle in the sense of J.M.G. Fell [35]. This relation will become clearer below, after theorem 7.1.15.

7.1.5 Corollary (of the proof of proposition 7.1.3). An action α of a compact group G on a C^*-algebra A is saturated if and only if the elements $\tilde{x}^*\tilde{y}$, for $x,y \in A$, span a dense subspace of $C^*(G,A,\alpha)$.

Proof. We have $<x,y>_{C^*(G,A)} = \tilde{x}^*\tilde{y}$ for $x,y \in A$. These elements span a dense subspace of $C^*(G,A)$ if and only if the elements $<x,y>_{C^*(G,A)}$ for $x,y \in \bar{A}$ do, because A is dense in $\bar{A}$. This is the case if and only if $\bar{A}$ is an equivalence bimodule, because, by proposition 7.1.3, all the other properties of an equivalence bimodule are already known to hold. Q.E.D.

7.1.6 Proposition. Let G be a compact group. For each G-algebra (G,A,α), there is a natural homomorphism of abelian groups from $K_*(A^G)$ to $K_*^G(A)$. If A is unital, it is defined on K_0 by sending the class of a projection $p \in M_n(A^G)$ to the class of p

regarded as a G-invariant projection in $L(\mathbf{C}^n)\otimes A$, where G acts trivially on $\mathbf{C}^n$.

Proof. It is obvious that the definition given in the statement of the proposition defines a natural homomorphism of abelian groups from $K_0(A^G)$ to $K_0^G(A)$ for unital G-algebras A. By considering the unitization, it can be extended to nonunital algebras, and by considering suspensions, it can be extended to K_1. Q.E.D.

7.1.7 Lemma. Let G be a compact group, and let (G,A,α) be a G-algebra. Define a map $j : A^G \longrightarrow C^*(G,A,\alpha)$ by $j(x)=\tilde{x}$, where $\tilde{x}$ is as defined in 7.1.2. Under the identification of $K_*(C^*(G,A,\alpha))$ with $K_*^G(A)$ via the isomorphism of theorem 2.8.3 (7), j_* becomes the natural map from $K_*(A^G)$ to $K_*^G(A)$ of the previous proposition.

Proof. It is easy to check that j is in fact a homomorphism. By naturality, it is sufficient to consider K_0 and unital algebras A. The isomorphism from $K_0^G(A)$ to $K_0(C^*(G,A))$ was constructed in two stages (see the proof of Julg's theorem 2.6.1): we first assigned to a finitely generated projective (G,A,α)-module (G,E,λ) the $L^1(G,A)$-module E with the integrated form action, and then we considered the inclusion from $L^1(G,A)$ into $C^*(G,A)$. Since the range of j is contained in $L^1(G,A)$, it is sufficient to consider only the first step. That is, if $p \in M_n(A^G)$ is a projection, it is sufficient to show that $\tilde{p}\,[L^1(G,A)^n\,]$ is isomorphic as an $L^1(G,A)$-module to $p\,[A^n]$, where $\tilde{p}$ is the obvious element in $M_n\,(L^1(G,A))$.

We first consider the case $n=1$ and $p=1$. Thus, we want to find an isomorphism as $L^1(G,A)$-modules from A to $qL^1(G,A)$, where $q=\tilde{1}$ is the function taking the constant value 1. Define $\varphi : A \longrightarrow L^1(G,A)$ by $\varphi(a)=\tilde{a}$. Then a simple calculation shows that $q\,\varphi(a)=\varphi(a)$, so that $\varphi(a)\in qL^1(G,A)$. Clearly φ is injective. Also, the image of φ is all of $qL^1(G,A)$, because one can easily show that if $f \in L^1(G,A)$ then $qf=\varphi(a)$ with $a=\int_G \alpha_h\,(f\,(h^{-1}))\,dh$. Next, we check that φ is a right $L^1\,(G,A)$-module homomorphism. If $a \in A$ and $f \in L^1(G,A)$, then

$$(\varphi\,(a)f\,)(g)=\int_G \alpha_h\,(a)\,\alpha_h\,(f\,(h^{-1}g\,))\,dh$$
$$=\int_G \alpha_{gh^{-1}}(af\,(h))\,dh=\alpha_g\,(af\,)=\varphi(af\,)\,(g\,),$$

where the second step follows from the change of variables h to gh^{-1}.

In order to deal with the general case, we introduce left A^G-module structures on A and on $L^1(G,A)$, defined by the inclusion homomorphism from A^G to A and the homomorphism $a \longrightarrow \tilde{a}$ from A^G to $L^1(G,A)$. Then A is an $A^G-L^1(G,A)$ bimodule; the bimodule structure is exactly the one of proposition 7.1.3. A trivial calculation shows that $L^1(G,A)$ is also an $A^G-L^1(G,A)$ bimodule, and it is easily seen that φ is a bimodule homomorphism. In particular, $qL^1(G,A)$ is an $A^G-L^1(G,A)$ subbimodule of $L^1(G,A)$ which is isomorphic to A via φ. We can now extend φ to an isomorphism of the $M_n\,(A^G)-L^1(G,A)$ bimodules A^n and $(q\,L^1(G,A))^n$. Since φ is an isomorphism of bimodules, it restricts to an isomorphism of right $L^1(G,A)$-modules

$$p\,[A^n]\simeq\tilde{p}\,[(q\,L^1(G,A))^n\,]=\tilde{p}\,[L^1(G,A)^n\,],$$

for any projection $p \in M_n(A^G)$, as desired. Q.E.D.

7.1.8 Proposition. Let G be a second countable compact group, and let (G,A,α) be a separable G-algebra. If α is saturated, then the natural map $K_*(A^G) \longrightarrow K_*^G(A)$ is an isomorphism.

Proof. By the previous lemma, it suffices to show that j_* is an isomorphism, where $j(x) = \tilde{x}$. Since j is injective, it is sufficient to prove that the inclusion of $j[A^G]$ in $C^*(G,A)$ induces an isomorphism on K-theory. To do this, it is, by proposition 2.9.1, sufficient to prove that $j[A^G]$ is a full hereditary subalgebra in $C^*(G,A)$. Now by [89], we have $j[A^G] = qC^*(G,A)q$, where $q = \tilde{1}$ is the projection in $M(C^*(G,A))$ given by the function in $L^1(G,A)$ which takes the constant value 1. (Note that $q \in C^*(G,A)$ if A is unital.) It is therefore clear that $j[A^G]$ is a hereditary subalgebra, and it only remains to show that it is full.

Let L be the closed ideal in $C^*(G,A)$ generated by $j[A^G]$. Let (e_λ) be an approximate identity for A contained in A^G. (It is well known that G-invariant approximate identities exist when G is compact: take any approximate identity and average it over G.) Then for $a, b \in A$, we have

$$<a, b>_{C^*(G,A)} = \lim_\lambda <e_\lambda a, e_\lambda b>_{C^*(G,A)} = \lim_\lambda <e_\lambda \tilde{a}, e_\lambda \tilde{b}>_{C^*(G,A)}.$$

Here $e_\lambda \tilde{a}$ is the product of the elements $e_\lambda \in A$ and $\tilde{a} \in L^1(G,A)$, using the $L^1(G,A)$-module structure on A defined in proposition 7.1.2. (It is easy to check that $b\tilde{a} = ba$ for $a \in A$ and $b \in A^G$, where ba is the product of b and a as elements of A.) We therefore have

$$\begin{aligned} <a, b>_{C^*(G,A)} &= \lim_\lambda \tilde{a}^* <e_\lambda, e_\lambda>_{C^*(G,A)} \tilde{b} \\ &= \lim_\lambda \tilde{a}^* \tilde{e}_\lambda^* \tilde{e}_\lambda \tilde{b} = \lim_\lambda \tilde{a}^* j(e_\lambda^* e_\lambda) \tilde{b} \in L. \end{aligned}$$

Since α is saturated, corollary 7.1.5 implies that the elements $<a, b>_{C^*(G,A)} = \tilde{a}^* \tilde{b}$ span a dense subspace of $C^*(G,A)$. Therefore $L = C^*(G,A)$, and we conclude that $j[A^G]$ is full in $C^*(G,A)$, as desired. Q.E.D.

The proposition just proved shows that saturated actions satisfy the analog of condition (2) of theorem 1.1.1. In particular, a saturated action of a compact Lie group on a separable commutative unital C^*-algebra is free. This can, however, be proved directly and in greater generality, as will be seen below.

We now prove some other general properties of saturated actions. These are analogous to some of the properties of K-freeness, and thus provide additional justification for considering saturation to be related to freeness of an action. (We will demonstrate one of its disadvantages in the next section.) The results proved here will not be used in the proof of the main results of this chapter; some of their consequences will, however, be needed in the next chapter.

The next lemma provides a more convenient way of verifying that an action is saturated.

7.1.9 Lemma (Rieffel). Let G be a compact group, and let (G,A,α) be a G-algebra. Then α is saturated if and only if the functions $g \longrightarrow a\alpha_g(b)$, in $L^1(G,A,\alpha)$, for $a,b \in A$, span a dense subspace of $C^*(G,A,\alpha)$.

Proof. Since $(\tilde{a}^*(b^*)^\sim)(g) = a^*\alpha_g(b^*)$, this follows immediately from corollary 7.1.5. Q.E.D.

7.1.10 Proposition (Rieffel). Let G be a compact group, and let

$$0 \longrightarrow I \longrightarrow A \stackrel{\pi}{\longrightarrow} B \longrightarrow 0$$

be an equivariant exact sequence. Let ε, α, and β be the actions of G on I, A, and B respectively. If ε and β are saturated, then so is α.

Proof. Let I_0 be the linear span in $C^*(G,I)$ of the functions $g \longrightarrow a\varepsilon_g(b)$ for $a,b \in I$. Similarly define A_0 and B_0.

By proposition 2.8.2, the sequence

$$0 \longrightarrow C^*(G,I) \longrightarrow C^*(G,A) \stackrel{\sigma}{\longrightarrow} C^*(G,B) \longrightarrow 0$$

is exact. Here σ is the map induced by π. Clearly $I_0 \subset A_0$. We claim that $\sigma[A_0] = B_0$. Indeed, if $a,b \in B$ then there are $x,y \in A$ such that $\pi(x) = a$ and $\pi(y) = b$. Thus, $\pi(x\alpha_g(y)) = a\beta_g(b)$, and this proves the claim. By assumption (using lemma 7.1.9), I_0 and B_0 are dense in $C^*(G,I)$ and in $C^*(G,B)$ respectively. Routine arguments now show that A_0 is dense in $C^*(G,A)$. Another application of lemma 7.1.9 now finishes the proof. Q.E.D.

7.1.11 Proposition (Rieffel). Let G be a compact group, let (G,A,α) be a G-algebra, and let I be a G-invariant ideal in A. If α is saturated, then so is the induced action on A/I.

Proof. Similar to the proof of the previous proposition, and omitted. Q.E.D.

7.1.12 Proposition (Rieffel). Let G be a compact group, and let $(G, C_0(X), \alpha)$ be a commutative G-algebra. If α is saturated, then the action of G on X is free.

Proof. Assume α is saturated, and let $x \in X$. Then the orbit Gx of x is a compact subset of X, and therefore the G-algebra $(G, C(Gx))$ is a quotient algebra of $(G,C_0(X))$. By the previous proposition, the action of G on $C(Gx)$ is saturated. Consequently, $\mathbf{C} = C(Gx)^G$ is strongly Morita equivalent to $C^*(G,C(Gx))$.

Let H be the stabilizer of x. Then Gx is equivariantly homeomorphic to G/H, and therefore $C^*(G,C(Gx)) \simeq C^*(G,C(G/H))$. By [88], the algebra $C^*(G,C(G/H))$ is strongly Morita equivalent to $C^*(H)$. (This is situation 5 of [88].) Since strong

Morita equivalence is transitive (this is essentially theorem 5.9 in [84]), it follows that **C** is strongly Morita equivalent to $C^*(H)$. By theorem 3.1 of [85], it follows that $C^*(H)$ is simple. Since H is a compact group, this can only happen if $H=\{e\}$. We have thus shown that the stabilizer of every point in X is trivial, so the action of G on X is free. Q.E.D.

7.1.13 Proposition. Let G be a compact group, and let $(G,A_\lambda,\alpha^{(\lambda)})_{\lambda\in D}$ be a directed system of G-algebras. Let (G,A,α) be the direct limit $\varinjlim_\lambda (G,A_\lambda,\alpha^{(\lambda)})$. If each $\alpha^{(\lambda)}$ is saturated, then so is α.

Proof. Let $B_\lambda \subset C^*(G,A_\lambda)$ be the linear span of all functions $g \longrightarrow a\alpha_g^{(\lambda)}(b)$, for $a,b\in A_\lambda$. Similarly let $B\subset C^*(G,A)$ be the linear span of all functions $g \longrightarrow a\alpha_g(b)$, for $a,b\in A$. Let $\varphi_\lambda : C^*(G,A_\lambda) \longrightarrow C^*(G,A)$ be the canonical map. Then $\varphi_\lambda[B_\lambda]\subset B$. Since B_λ is dense in $C^*(G,A_\lambda)$ for all λ by lemma 7.1.9, and since $\bigcup_\lambda \varphi_\lambda[C^*(G,A_\lambda)]$ is dense in $C^*(G,A)$, it follows that B is dense in $C^*(G,A)$. Therefore α is saturated by lemma 7.1.9. Q.E.D.

7.1.14 Corollary. Let G be a compact group, and let (G,A,α) be a G-algebra which has a G-invariant composition series $(I_\lambda)_{\lambda\leq\kappa}$ such that the induced actions on the composition factors $I_{\lambda+1}/I_\lambda$ are all saturated. Then α is saturated.

Proof. This follows immediately from the previous proposition and proposition 7.1.10. Q.E.D.

To end this section, we prove an important characterization of saturated actions of abelian groups. We should mention that Marc Rieffel has an analogous but more complicated result for nonabelian groups.

7.1.15 Theorem (Rieffel). Let G be a compact abelian group, and let (G,A,α) be a G-algebra. For $\tau\in\widehat{G}$, let

$$A_\tau=\{a\in A : \alpha_g(a)=\tau(g)a \text{ for all } g\in G\}.$$

Then α is saturated if and only if $\overline{A_\tau^* A_\tau}=A^G$ for every $\tau\in\widehat{G}$. (Here $\overline{A_\tau^* A_\tau}$ is the closed linear span of all a^*b for $a,b\in A_\tau$).

We can now explain the connection between saturated actions and J.M.G. Fell's saturated bundles. The collection $\{A_\tau\}_{\tau\in\widehat{G}}$ is easily shown to define a Banach $*$-algebraic bundle over the discrete group $\widehat{G}$. (See [35], section 11.) The crucial property here is that $A_\sigma A_\tau\subset A_{\sigma\tau}$ and $A_\tau^*=A_{\tau^{-1}}$ for $\sigma,\tau\in\widehat{G}$. This bundle is saturated in the sense of section 11 of [35] exactly when $\overline{A_{\tau^{-1}}A_\tau}=A_1$ for all $\tau\in\widehat{G}$, where 1 is the identity in $\widehat{G}$. But A_1 is just A^G.

Proof of theorem 7.1.15. Suppose $\overline{A_\tau^* A_\tau} = A^G$ for all $\tau \in \hat{G}$. Let V be the closed linear span in $C(G,A)$ with respect to $\| \ \|_\infty$ of all functions $g \longrightarrow a\alpha_g(b)$ for $a, b \in A$. We are going to show that $V = C(G,A)$. Since $\| \ \|_\infty$ is larger than $\| \ \|_1$, which is in turn larger than the norm of $C^*(G,A)$, and since $C(G,A)$ is dense in $L^1(G,A)$ for $\| \ \|_1$, lemma 7.1.9 will then imply that α is saturated.

Let $a \in A$, let $\varepsilon > 0$, let $\tau \in \hat{G}$, and choose $e \in A^G$ such that $\| ae - a \| < \varepsilon$. (This can be done because A has a G-invariant approximate identity.) Further choose $x_1, \ldots, x_n, y_1, \ldots, y_n \in A_\tau$ such that

$$\| e - \sum_{k=1}^{n} x_k^* y_k \| < \varepsilon \| a \|^{-1}.$$

For $g \in G$, set

$$f(g) = \sum_{k=1}^{n} x_k^* \alpha_g(y_k) = \tau(g) \sum_{k=1}^{n} x_k^* y_k .$$

Then we have

$$\| \tau(g)a - f(g) \| \leq \| \tau(g)(a - ae) \| + \| a \| \, \| \tau(g)e - f(g) \| < 2\varepsilon .$$

Since $\varepsilon > 0$ is arbitrary, it follows that the functions $g \longrightarrow \tau(g)a$ are in V for $a \in A$ and $\tau \in \hat{G}$. Now $C(G)$ is the closed linear span of the elements of $\hat{G}$ for $\| \ \|_\infty$, so we conclude that $g \longrightarrow f(g)a$ is in V for any $f \in C(G)$. So $C(G,A) \subset V$, and α is saturated.

For the converse, suppose α is saturated. Then the elements $\tilde{a}^* \tilde{b}$ for $a, b \in A$ span a dense subspace of $C^*(G,A)$, by corollary 7.1.5. Let $\varepsilon > 0$, let $\tau \in \hat{G}$, and let $x \in A^G$. Then there are $a_1, \ldots, a_n, b_1, \ldots, b_n \in A$ such that

$$\| \sum_{k=1}^{n} \tilde{a}_k^* \tilde{b}_k - \hat{\alpha}_\tau(\tilde{x}) \| < \varepsilon ,$$

where $\hat{\alpha}$ is the dual action of $\hat{G}$ on $C^*(G,A)$. Let $p = \tilde{1}$ be the constant function on G taking the value 1; it is at least an element of the multiplier algebra $M(C^*(G,A))$. Furthermore, p is a projection, and, for $b \in A$ and $g \in G$, we have

$$(\hat{\alpha}_\tau^{-1}(\tilde{b})p)(g) = \int_G \tau(h)^{-1} \alpha_h(b)\, dh .$$

The right hand side of this equation does not depend on g; we call it b_τ and note that it is in A_τ. A similar computation shows that, for $a \in A$ and $g \in G$ we also have $(p\,\hat{\alpha}_\tau^{-1}(\tilde{a}^*))(g) = \tau(g)^{-1} a_\tau^*$. We now obtain

$$(p\,\hat{\alpha}_\tau^{-1}(\tilde{a}^*\tilde{b})p)(g) = a_\tau^* b_\tau = (a_\tau^* b_\tau)^\sim (g),$$

since $a_\tau^* b_\tau \in A^G$. Combining this with the fact that $y \longrightarrow \tilde{y}$ is an injective homomorphism from A^G to $C^*(G,A)$ whose range lies in $pC^*(G,A)p$, we obtain

$$\| \sum_{k=1}^{n} (a_k)_\tau^* (b_k)_\tau - x \| = \| \sum_k ((a_k)_\tau^* (b_k)_\tau)^\sim - \tilde{x} \| = \| p\,(\sum_k \hat{\alpha}_\tau^{-1}(a_k^* b_k) - \tilde{x})p \|$$

$$\leq \| \sum_k \hat{\alpha}_\tau^{-1}(a_k^* b_k) - \tilde{x} \| = \| \sum_k a_k^* b_k - \hat{\alpha}_\tau(\tilde{x}) \| < \varepsilon .$$

Since $\varepsilon > 0$ is arbitrary, we have shown that $A^G \subset \overline{A_\tau^* A_\tau}$. We clearly have $A_\tau^* A_\tau \subset A^G$, so we obtain $\overline{A_\tau^* A_\tau} = A^G$, as desired. Q.E.D.

This theorem essentially says that the action α is saturated if and only if $\widetilde{Sp}(\alpha) = \widehat{G}$, where $\widetilde{Sp}(\alpha)$ is as defined by Kishimoto in [58]. The calculation necessary to show the equivalence of $\widetilde{Sp}(\alpha) = \widehat{G}$ and $\overline{A_\tau^* A_\tau} = A^G$ for all $\tau \in \widehat{G}$ will be carried out in the proof of theorem 7.2.7 below.

7.2. Hereditary Saturation and the Strong Connes Spectrum

Saturation is not a strong enough condition to rule out many inner actions, as the following result shows. Before stating this result, let us recall the definition of the Arveson spectrum $Sp(\alpha)$ ([79], 8.1.6) of a continuous action α of an abelian group G on a Banach space A, in the special case in which the group is compact. (The definition is much simpler in this case.) Let A_τ be as in the statement of theorem 7.1.15. Then $Sp(\alpha)$ is the subset of the dual group $\widehat{G}$ given by $Sp(\alpha) = \{\tau \in \widehat{G} : A_\tau \neq 0\}$.

7.2.1 Proposition (Rieffel). Let G be a compact abelian group, and let u be a unitary representation of G on a Hilbert space H. Let $\alpha : G \longrightarrow Aut(K(H))$ be the corresponding inner action, given by $\alpha_g(a) = u_g a u_g^*$. Then α is saturated if and only if the Arveson spectrum satisfies $Sp(u) = \widehat{G}$.

Proof. We have $H = \bigoplus_{\tau \in \widehat{G}} H_\tau$. It is easily seen that $K(H)_\tau$ consists of those operators which map H_σ into $H_{\tau\sigma}$ for every $\sigma \in \widehat{G}$. Thus,

$$K(H)_\tau = \overline{\bigoplus_{\sigma \in \widehat{G}} K(H_\sigma, H_{\tau\sigma})}.$$

In particular,

$$K(H)^G = \overline{\bigoplus_{\sigma \in \widehat{G}} K(H_\sigma)}.$$

It is clear that $\overline{K(H_\sigma, H_\tau) K(H_\rho, H_\sigma)} = K(H_\rho, H_\tau)$ if $H_\sigma \neq 0$ (one need only consider rank 1 operators), and that $\overline{K(H_\sigma, H_\tau) K(H_\rho, H_\sigma)} = 0$ if $H_\sigma = 0$. It follows that $\overline{K(H)_\tau^* K(H)_\tau} = K(H)^G$ for all τ if $H_\tau \neq 0$ for all τ. Conversely, if some $H_\tau = 0$, then choose σ such that $H_\sigma \neq 0$. Then one easily checks that $K(H_\sigma) \not\subset \overline{K(H)_\rho^* K(H)_\rho}$, where $\rho = \tau\sigma^{-1}$. Q.E.D.

To rule out this sort of action, we extend Rieffel's definition as follows.

7.2.2 Definition. Let G be a compact group, and let (G, A, α) be a G-algebra. Then α is hereditarily saturated if for every nonzero G-invariant hereditary subalgebra B of A, the restricted action $\alpha_{(\cdot)}|_B$ is saturated.

The actions considered in the previous proposition are never hereditarily saturated, since if one of them is saturated, that implies by the proposition that there is a G-invariant projection $p \neq 0$ such that G acts trivially on pH. Therefore G acts trivially on the nonzero hereditary subalgebra $pK(H)p$, and the action is not hereditarily saturated.

Using the properties of saturation proved in the previous section, we obtain the following properties of hereditary saturation. They will be needed in chapter 8.

7.2.3 Proposition. Let G be compact group. Let

$$0 \longrightarrow I \longrightarrow A \xrightarrow{\pi} B \longrightarrow 0$$

be an equivariant exact sequence of G-algebras. Then the action on A is hereditarily saturated if and only if the actions on I and B are.

Proof. Regard I as an ideal in A. Let the actions on I and B be hereditarily saturated, and let C be a G-invariant hereditary subalgebra of A. Then there is an equivariant exact sequence

$$0 \longrightarrow C \cap I \longrightarrow C \longrightarrow \pi[C] \longrightarrow 0.$$

Here $C \cap I$ is a G-invariant hereditary subalgebra of I, and $\pi[C]$ is a G-invariant hereditary subalgebra of B. Therefore the action on C is saturated by proposition 7.1.10, and the action on A is hereditarily saturated.

For the converse, let the action on A be hereditarily saturated. Since every G-invariant hereditary subalgebra of I is also a hereditary subalgebra of A, the action on I is certainly hereditarily saturated. If C is a G-invariant hereditary subalgebra of B, then $\pi^{-1}[C]$ is a G-invariant hereditary subalgebra of A, and $C \simeq \pi^{-1}[C]/I$. Then the action on C is saturated by proposition 7.1.11, so the action on B is hereditarily saturated. Q.E.D.

7.2.4 Proposition. Let G be a compact group, and let (G,A,α) be a G-algebra with a G-invariant composition series $(I_\lambda)_{\lambda\le\kappa}$. If the induced actions on the factors $I_{\lambda+1}/I_\lambda$ are all hereditarily saturated, then so is α.

Proof. Let B be a G-invariant hereditary subalgebra of A. Then it is readily verified that $(I_\lambda \cap B)_{\lambda\le\kappa}$ is a G-invariant composition series for B, such that the factors

$$(I_{\lambda+1} \cap B)/(I_\lambda \cap B) \simeq ((I_{\lambda+1} \cap B) + I_\lambda)/I_\lambda$$

are G-invariant hereditary subalgebras of the factors $I_{\lambda+1}/I_\lambda$. From corollary 7.1.14, we conclude that the action on B is saturated. Q.E.D.

We next consider $(G,C_0(X))$-algebras. (See definition 4.5.1.) In section 4.6, it was shown that if a compact Lie group G acts freely on X, and if (G,A,α) is a separable $(G,C_0(X))$-algebra, then α is K-free. We want to prove a similar result for hereditary saturation. First, we prove a lemma.

7.2.5 Lemma. Let G be a compact group, and let X be a locally compact Hausdorff free G-space. Then the functions $(g,x) \longrightarrow f_1(x)f_2(g^{-1}x)$, for $f_1, f_2 \in C_0(X)$, span a dense subspace of $C_0(G\times X)$.

Proof. The linear span of the functions $(g,x) \longrightarrow f_1(x)f_2(g^{-1}x)$ is clearly a subalgebra of $C_0(G\times X)$ and is closed under complex conjugation. We want to apply the Stone-Weierstrass theorem to conclude that its closure is all of $C_0(G\times X)$; we need only prove that the functions $(g,x) \longrightarrow f_1(x)f_2(g^{-1}x)$ separate the points of $G\times X$. Let (g,x) and (h,y) be distinct points in $G\times X$. Then there are two cases.

Case 1. Assume $x \neq y$. Choose $f_1 \in C_0(X)$ such that $f_1(x) \neq f_1(y)$. Choose $f_2 \in C_0(X)$ such that $f_2(z) = 1$ for all z in some compact subset of X containing the orbits Gx and Gy. Then

$$f_1(x)f_2(g^{-1}x) = f_1(x) \neq f_1(y) = f_1(y)f_2(h^{-1}y).$$

Case 2. Assume $x = y$. Then $g \neq h$. Choose $f_1 \in C_0(X)$ such that $f_1(x) = 1$. Choose $f_2 \in C_0(X)$ such that $f_2(g^{-1}x) \neq f_2(h^{-1}x)$. (This can be done since $g^{-1}x \neq h^{-1}x$.) Then, since $x = y$, we have

$$f_1(x)f_2(g^{-1}x) = f_2(g^{-1}x) \neq f_2(h^{-1}x) = f_1(y)f_2(h^{-1}y).$$

Q.E.D.

We can now conclude that if X is a free G-space, then the action of G on $C_0(X)$ is saturated, by using lemma 7.1.9. This result is also a corollary of the following theorem.

7.2.6 Theorem. Let G be a compact group, and let X be a locally compact Hausdorff free G-space. Let $(G, C_0(X), \beta)$ be the corresponding G-algebra. Let (G, A, α) be a $(G, C_0(X), \beta)$-algebra. Then α is hereditarily saturated.

Proof. We first show that α is saturated. We will show that the functions $g \longrightarrow r(g)a$ for $r \in C(G)$ and $a \in A$ can be approximated in $\|\ \|_\infty$ on $C(G,A)$ by linear combinations of functions of the form $g \longrightarrow a_k\alpha_g(b_k)$. This will show that the functions $g \longrightarrow a\alpha_g(b)$, for $a, b \in A$, span a $\|\ \|_\infty$-dense subspace of $C(G,A)$. Since $\|\ \|_\infty$ is larger than $\|\ \|_1$, which is in turn larger than the norm on $C^*(G,A)$, and since $C(G,A)$ is dense in $L^1(G,A)$ for $\|\ \|_1$, we will have verified the hypotheses of lemma 7.1.9, and will have thus shown that α is saturated.

Let $r \in C(G)$, let $a \in A$, and let $\varepsilon > 0$. Let (e_λ) be a G-invariant approximate identity for A, and choose λ such that $\| a - ae_\lambda \| < \frac{1}{3}\varepsilon(\| r \|_\infty + 1)^{-1}$. Further choose $f \in C_0(X)$ such that $0 \le f \le 1$ and $\| a - fa \| < \frac{1}{3}\varepsilon(\| r \|_\infty + 1)^{-1}$. (Such a function exists by condition (2) in the definition of a $C_0(X)$-algebra.) Then the function $(g,x) \longrightarrow r(g)f(x)$ is in $C_0(G\times X)$, so by the previous lemma there are $f_{11}, \ldots, f_{1n}, f_{21}, \ldots, f_{2n} \in C_0(X)$ such that

$$| r(g)f(x) - \sum_1^n f_{1k}(x)f_{2k}(g^{-1}x) | < \min(1, \tfrac{1}{3}\varepsilon \| a \|^{-1})$$

for all $g \in G$ and $x \in X$. Set $a_k = f_{1k} a$ and $b_k = f_{2k} e_\lambda$. Then

$$\| r(g)a - \sum_1^n a_k \alpha_g(b_k) \|$$

$$\leq \| r(g)a - r(g)fa \| + \| r(g)fa - \left[\sum_1^n f_{1k} \beta_g(f_{2k}) \right] a e_\lambda \|$$

$$\leq \| r \|_\infty \| a - fa \| + \| r(g)f - \sum_1^n f_{1k} \beta_g(f_{2k}) \| \, \| a \|$$

$$+ \| \sum_1^n f_{1k} \beta_g(f_{2k}) \|_\infty \| a - ae_\lambda \| ,$$

which is less than ε by the choice of f, f_{ik}, and e_λ, using the fact that

$$\| \sum_1^k f_{1k} \beta_g(f_{2k}) \| < sup\{ |r(g)f(x)| : g \in G, x \in X \} + 1 \leq \| r \|_\infty + 1 .$$

We thus conclude that α is saturated.

Now let B be any G-invariant hereditary subalgebra of A. If $b \in B$ and $f \in C_0(X)$, then

$$(fb)(fb)^* = |f|^2 bb^* \leq \| f \|_\infty^2 bb^* \in B ,$$

so that $(fb)(fb)^* \in B$. For similar reasons, $(fb)^*(fb) \in B$, and we conclude that $fb \in B$ because B is hereditary. (See [79], 1.5.2.) Therefore $(G, B, \alpha_{(\cdot)}|_B)$ is also a $(G, C_0(X))$-algebra, and so $\alpha_{(\cdot)}|_B$ is saturated by the previous argument. Thus α is hereditarily saturated. Q.E.D.

We are now going to investigate the relationship between hereditary saturation and the strong Connes spectrum, defined on page 70 of [58]. We therefore recall the definition, and also the definition of the Connes spectrum ([79], 8.8.2), in the special case in which the abelian group involved is compact. Thus, let G be a compact abelian group, and let (G,A,α) be a G-algebra. As mentioned above, the Arveson spectrum of α is $Sp(\alpha) = \{\tau \in \hat{G} : A_\tau \neq 0\}$. Then the Connes spectrum of α is

$$\Gamma(\alpha) = \bigcap_B Sp(\alpha_{(\cdot)}|_B),$$

where B ranges over all nonzero G-invariant hereditary subalgebras of A. It is in fact a subgroup of $\hat{G}$ ([79], 8.8.4). The strong spectrum ([58], page 70) is given by $\tilde{Sp}(\alpha) = \{\tau \in \hat{G} : \overline{A_\tau^* A A_\tau} = A\}$. In analogy with the Connes spectrum, the strong Connes spectrum is defined by

$$\tilde{\Gamma}(\alpha) = \bigcap_B \tilde{Sp}(\alpha_{(\cdot)}|_B),$$

where again B ranges over all nonzero G-invariant hereditary subalgebras of A. It is clear from the alternate characterization of $\tilde{\Gamma}(\alpha)$ given in lemma 3.4 of [58] that $\tilde{\Gamma}(\alpha)$ is at least a subsemigroup of $\hat{G}$. If we compare theorem 7.1.15 with the definition of $\tilde{\Gamma}(\alpha)$, we obtain the following theorem.

7.2.7 Theorem. Let G be a compact abelian group, and let (G,A,α) be a G-algebra. Then α is hereditarily saturated if and only if $\tilde{\Gamma}(\alpha) = \hat{G}$.

Proof. It is clearly sufficient to show that $\tilde{Sp}(\alpha) = \hat{G}$ if and only if α is saturated. By theorem 7.1.15, α is saturated if and only if $\overline{A_\tau^* A_\tau} = A^G$ for all $\tau \in \hat{G}$, while by the definition of $\tilde{Sp}(\alpha)$, we have $\tilde{Sp}(\alpha) = \hat{G}$ if and only if $\overline{A_\tau^* A A_\tau} = A^G$ for all $\tau \in \hat{G}$. Therefore we need only prove that $\overline{A_\tau^* A_\tau} = A^G$ if and only if $\overline{A_\tau^* A A_\tau} = A$.

If $\overline{A_\tau^* A_\tau} = A^G$, then we have

$$A = \overline{A^G A A^G} = \overline{A_\tau^*(A_\tau A A_\tau^*)A_\tau} \subset \overline{A_\tau^* A A_\tau},$$

so $A = \overline{A_\tau^* A A_\tau}$. (The first step follows from the existence of a G-invariant approximate identity in A.) For the converse, let P be the standard conditional expectation from A to A^G, given by $P(x) = \int_G \alpha_g(x)dg$. Then P is continuous, and for $a, b \in A_\tau$, we have

$$\begin{aligned} P(a^* x b) &= \int_G \alpha_g(a^*)\alpha_g(x)\alpha_g(b)\,dg \\ &= \int_G (\tau(g)^{-1}a^*)\alpha_g(x)(\tau(g)b)\,dg = a^* P(x) b\,. \end{aligned}$$

Therefore, by applying P, we obtain $(\overline{A_\tau^* A A_\tau})^G = \overline{A_\tau^* A^G A_\tau}$. Consequently, if $\overline{A_\tau^* A A_\tau} = A$, then

$$A^G = (\overline{A_\tau^* A A_\tau})^G = \overline{A_\tau^* A^G A_\tau} \subset \overline{A_\tau^* A_\tau} \subset A^G\,.$$

Q.E.D.

Specializing further, we obtain the following result:

7.2.8 Theorem. Let G be a finite abelian group, and let (G,A,α) be a G-algebra. Then the following are equivalent:

(1) The action α is hereditarily saturated.

(2) $\tilde{\Gamma}(\alpha) = \hat{G}$.

(3) For every G-invariant ideal I of A such that $I \neq A$, we have $\Gamma(\alpha^I) = \hat{G}$, where α^I is the induced action of G on A/I.

(4) $\hat{G}$ acts trivially on *Prim* $(C^*(G,A,\alpha))$.

If A is separable, then these conditions imply:

(5) For every G-invariant hereditary subalgebra B of A, the natural map $K_*(B^G) \longrightarrow K_*^G(B)$ is an isomorphism.

Proof. (1)$\Longleftrightarrow$(2) is the previous theorem.

(2)$\Longleftrightarrow$(3) is proposition 4.1 in [60].

(2)$\Longleftrightarrow$(4) is lemma 3.4 in [58].

(1) $\Longrightarrow$(5) is proposition 7.1.8. Q.E.D.

We have included condition (3) to show how our concepts and the strong Connes spectrum are related to the perhaps more familiar Connes spectrum. The chain of implications we are really interested in is (4)$\Longrightarrow$(2)$\Longrightarrow$(1)$\Longrightarrow$(5), since it will turn out later that the condition we are actually able to check is (4).

The converse of the implication (5)$\Longrightarrow$(1) is false, as is shown by the following example.

7.2.9 Example. Let $G = \mathbf{Z}/2\mathbf{Z}$ and let $A = \mathbf{O}_2$ be the Cuntz algebra [21]. Let $\alpha : G \longrightarrow Aut(A)$ be the trivial action. Then $K_*^G(A) = R(G) \otimes K_*(\mathbf{O}_2) = 0$, since $K_*(\mathbf{O}_2) = 0$ by [23]. Furthermore, if B is any hereditary subalgebra of A, then B is full because A is simple. Since $\mathbf{O}_2$ is separable, the inclusion of B in A is an isomorphism on K-theory, by proposition 2.9.1. So $K_*(B) = 0$ also, and therefore $K_*^G(B) = 0$. We conclude that for every G-invariant hereditary subalgebra B, the map $K_*(B^G) \longrightarrow K_*^G(B)$ is an isomorphism, since $B^G = B$, and thus both K-groups are zero. However, α is certainly not even saturated, let alone hereditarily saturated. We also point out that α is totally K-free.

7.3. Some Properties of Subquotients of C^*-Algebras

We want to find a condition on a C^*-algebra A such that a K-free action on A must be hereditarily saturated. The last example suggests that one should require that the K-theory of A be sufficiently nontrivial, and in fact conditions of this sort do work, at least for finite abelian groups. Let us call an algebra of the form I/J, where I and J are ideals of A with $J \subset I$ and $J \neq I$, a subquotient of A. (Thus, we do not admit the zero algebra as a subquotient.) If A is a G-algebra, and I and J are G-invariant, then we call I/J a G-invariant subquotient of A. The conditions which we need are contained in the following definition. We give several versions because we need stronger conditions for some groups than for others.

7.3.1 Definition. Let A be a C^*-algebra. Then we say A is K-visible if every subquotient of A has in turn a subquotient C such that $K_*(C)$ contains an element of infinite order. We say A is p-weakly K-visible, where $p \in \mathbf{Z}$ is a fixed prime number, if every subquotient of A has in turn a subquotient C such that $K_*(C)$ has a non-zero element whose order is not a power of p. Finally, we say A is highly K-visible if every subquotient of A has in turn a subquotient C such that $K_0(C)$ contains a totally positive element, where $\eta \in K_0(C)$ is called totally positive if $\eta = [p]$ where p is a projection in some $M_n(C)$, and $-[p]$ is not represented by any projection in any $M_k(C)$. (Notice that this condition makes sense even if C is not unital, since a projection p in a matrix algebra over C always defines an element of $K_0(C)$.)

7.3.2 Remark. A highly K-visible C^*-algebra is K-visible, and a K-visible C^*-algebra is p-weakly K-visible for any prime p. Indeed, the second assertion is obvious, and the first follows from the fact that a totally positive element $[p] \in K_0(C)$ must have infinite order. (If $[p]$ has order n, then $-[p]$ is represented by a direct sum of $n-1$

copies of p.)

In section 7.5 we will prove that there are many (highly) K-visible C^*-algebras. For the moment, we will content ourselves with exhibiting just three classes of highly K-visible C^*-algebras, and then we will go on to show how K-visibility and its relatives are used. Recall that a C^*-algebra A is called finite, or directly finite, if whenever $a \in A$ satisfies $a^*a \le aa^*$, then $a^*a = aa^*$, and that A is called stably finite if $M_n(A)$ is finite for all n. (See [43].) Also recall that an AF algebra is a C^*-algebra which is a direct limit of finite dimensional C^*-algebras. (See [12]. However, we make no assumption about units.)

7.3.3 Proposition. Type I, AF, and unital stably finite simple C^*-algebras are all highly K-visible.

Proof. Type I algebras: Since every subquotient of a type I algebra is again type I, it is enough to show that every type I algebra A has a subquotient C such that $K_0(C)$ contains a totally positive element. By [29], 4.4.5, there is a nonempty Hausdorff open subset U in $Prim(A)$. Let I be the corresponding ideal in A. Let x be any point in U, and let J be the ideal corresponding to the open subset $U-\{x\}$ of $Prim(A)$. Since A is type I, we have $I/J \simeq K(H)$ for some Hilbert space H. Since the class of any nonzero projection in $K(H)$ is totally positive, this shows that type I algebras are highly K-visible.

AF algebras: Since every subquotient of an AF algebra is again AF (apply [12], 1.5 and theorem 3.3, to the unitizations), it is sufficient to prove that if A is an AF algebra, then $K_0(A)$ has a totally positive element. This follows immediately from the standard facts about the K-theory of AF algebras. (See [31]. This is essentially just the fact that as an ordered group, $K_0(A)$ is a direct limit of groups $\mathbb{Z}^n$.)

Unital stably finite simple algebras: Let A be a unital stably finite simple C^*-algebra. Then A has no nontrivial subquotients, so it is enough to show that $K_0(A)$ has a totally positive element. We claim that the class of the identity in A will do. So suppose not; then $-[1]=[p]$ for some projection p in some matrix algebra $M_n(A)$. Therefore there is a projection q in some matrix algebra $M_m(A)$, such that, with $k=m+1+n$, the projections $q\oplus 0$ and $q\oplus 1\oplus p$ are Murray-von Neumann equivalent in $M_k(A)$. But if $u\in M_k(A)$ satisfies $u^*u = q\oplus 0$ and $uu^* = q\oplus 1\oplus p$, then we have $u^*u < uu^*$, contradicting the assumption that $M_k(A)$ is finite. So $[1]$ is totally positive, and A is highly K-visible. Q.E.D.

7.3.4 Corollary. Let G be a finite group, and let (G,A,α) be a G-algebra for which A is G-simple, stably finite, and unital. Then A is highly K-visible.

Proof. G-simple of course means that there are no nontrivial G-invariant ideals. It follows from [73], lemma 2.1, that G acts transitively on $Prim(A)$, and that $Prim(A)$ is therefore homeomorphic to a coset space of G. Consequently $Prim(A)$ is discrete, and A is a direct sum of simple algebras. The assumption on A implies that each of them must be stably finite and unital, and hence highly K-visible. Now the direct sum of highly K-visible algebras is obviously also highly K-visible. Q.E.D.

In order to use K-visibility and related properties, we need some properties of subquotients of C^*-algebras. In particular, we will have to work with the subsets of the primitive ideal space of a C^*-algebra A which correspond to subquotients of A. These turn out to be the locally closed subsets of $Prim(A)$. Recall that a subset S of a topological space X is called locally closed if $S = E \cap U$ where E is a closed subset and U is an open subset of X. Since $\bar{S} \subset E$, we have $S \subset \bar{S} \cap U \subset E \cap U = S$, so the closed subset E may be taken to be the closure of S. Thus, a subset of X is locally closed if and only if it is an open subset of its closure.

We also temporarily introduce the notation $I(U)$ for the ideal in a C^*-algebra corresponding to an open subset U of its primitive ideal space.

7.3.5 Lemma. Let A be a C^*-algebra. Then the primitive ideal space of a subquotient of A can be canonically identified with a locally closed subset of $Prim(A)$. Conversely, given a nonempty locally closed subset S of of $Prim(A)$, there are ideals I and J of A with $J \subset I$ such that $Prim(I/J) = S$, and the subquotient I/J is independent of the particular choices of I and J up to canonical isomorphism.

Proof. Let C be a subquotient of A, and write $C = I/J$. Then $Prim(C)$ can be canonically identified with $Prim(I) - Prim(J)$, which is equal to $Prim(I) \cap (Prim(A) - Prim(J))$ and is thus a locally closed subset of $Prim(A)$.

For the converse, let $S \subset Prim(A)$ be locally closed. Then $S = \bar{S} \cap U$ for some open subset U of $Prim(A)$. Set $V = U \cap (Prim(A) - \bar{S})$, which is an open subset of $Prim(A)$. Then $S = U - V$, so $S = Prim(I(U)/I(V))$, as required.

Now let U_0 and V_0 be open subsets of $Prim(A)$ such that $V_0 \subset U_0$ and $U_0 - V_0 = S$. We must show that there is a canonical isomorphism $I(U_0)/I(V_0) \simeq I(U)/I(V)$. The first step is to show that $V_0 = U_0 \cap (Prim(A) - \bar{S})$. We certainly have $V_0 \subset U_0$, and we also have $V_0 \subset Prim(A) - \bar{S}$, since V_0 is an open subset of $Prim(A)$ which is disjoint from S. Thus $V_0 \subset U_0 \cap (Prim(A) - \bar{S})$. The reverse inclusion also holds, since the equation $S = U_0 - V_0$ implies that already $U_0 \cap (Prim(A) - S) \subset V_0$.

Let $Y = U \cap U_0$, and let $Z = V \cap V_0$. Since $V = U \cap (Prim(A) - \bar{S})$ and $V_0 = U_0 \cap (Prim(A) - \bar{S})$, we have $Z = Y \cap (Prim(A) - \bar{S})$. Since also $S \subset Y$, we furthermore have $Y - Z = S$. It is therefore sufficient to show that there is a canonical isomorphism $I(U)/I(V) \simeq I(Y)/I(Z)$, since the same reasoning will then imply that there is also a canonical isomorphism $I(U_0)/I(V_0) \simeq I(Y)/I(Z)$. Now

$$V \cup Y = V \cup (U \cap U_0) = (V \cup U) \cap (V \cup U_0) = U \cap (V \cup S \cup V_0) = U.$$

A similar calculation shows that $V \cap Y = Z$. Therefore there is a canonical isomorphism

$$I(U)/I(V) = (I(V) + I(Y))/I(V) \simeq I(Y)/(I(Y) \cap I(V)) = I(Y)/I(Z),$$

as desired. Q.E.D.

Note that, because of this lemma, we can speak of *the* subquotient of A corresponding to a locally closed subset $S \subset Prim(A)$. We next consider the question of G-invariance of the subquotient corresponding to a G-invariant locally closed subset. We need the following lemma.

7.3.6 Lemma. Let G be any group, and let X be a not necessarily Hausdorff G-space. If $S \subset X$ is a G-invariant locally closed subset, then $S = \bar{S} \cap U$ for some G-invariant open subset U of X.

Proof. Since S is locally closed, there is an open subset U_0 of X such that $S = \bar{S} \cap U_0$. Let $U = \bigcup_{g \in G} U_0$. Then U is an open G-invariant subset of X. Obviously $S \subset \bar{S} \cap U$. To show the reverse inclusion, let $x \in \bar{S} \cap U$. Then $x \in gU_0$ for some $g \in G$. Consequently

$$x \in \bar{S} \cap gU_0 = g\,(\bar{S} \cap U_0) = gS = S,$$

where the first equality follows from the G-invariance of $\bar{S}$. So $S = \bar{S} \cap U$. Q.E.D.

7.3.7 Corollary. Let G be an arbitrary group, and let (G,A,α) be a G-algebra. If S is a G-invariant locally closed subset of $Prim(A)$, then S corresponds to a G-invariant subquotient.

Proof. Let $S = E \cap U$, with E closed, U open, and both sets G-invariant. Then S corresponds to the G-invariant subquotient $I(U)/I(V)$, where $V = U \cap (Prim(A) - E)$ is G-invariant. Q.E.D.

We will need some further properties of locally closed sets. These are presumably known, but we prove them here for completeness. We say that subsets S and T of a topological space X are separated if there are disjoint open sets U and V in X such that $S \subset U$ and $T \subset V$.

7.3.8 Lemma. Let X be any topological space. Then:

(1) The intersection of two locally closed sets is locally closed.

(2) A locally closed subset of a locally closed set is locally closed.

(3) The union of two separated locally closed sets is locally closed.

Proof. (1) Obvious.

(2) Let S be a locally closed subset of X, and let T be a locally closed subset of S. Then $T = F_0 \cap V_0$ where F_0 is a closed subset of S and V_0 is an open subset of S. Therefore $F_0 = S \cap F$ and $V_0 = S \cap V$, where F is a closed subset of X and V is an open subset of X. Also $S = E \cap U$ where E is closed and U is open. Then $T = (E \cap F) \cap (U \cap V)$, which is a locally closed subset of X.

(3) Let $S = \bar{S} \cap U$ and $T = \bar{T} \cap V$ be locally closed subsets of X, where U and V are open. Further let Y and Z be disjoint open subsets of X such that $S \subset Y$ and $T \subset Z$. Then $\bar{S} \cap Z = \bar{T} \cap Y = \varnothing$. Let $W = (U \cap Y) \cup (V \cap Z)$. Then

$$\begin{aligned}(\overline{S \cup T}) \cap W &= (\bar{S} \cup \bar{T}) \cap ((U \cap Y) \cup (V \cap Z)) \\ &= (\bar{S} \cap U \cap Y) \cup (\bar{S} \cap V \cap Z) \cup (\bar{T} \cap U \cap Y) \cup (\bar{T} \cap V \cap Z) \\ &= (S \cap Y) \cup (T \cap Z) = S \cup T.\end{aligned}$$

(Note that the two middle terms of the third expression are both equal to the empty set.) So $S \cup T$ is locally closed. Q.E.D.

7.3.9 Remark. The union of two arbitrary locally closed sets is in general not locally closed, even in a compact metric space. Let $X = [-2, 2] \times [-2, 2]$, let $U = (-1, 1) \times (-1, 1)$, and let $E = \{(1, 1)\}$. Then U is open, E is closed, E and U are disjoint, but $E \cup U$ is not locally closed. This is easily seen using the following alternate characterization of locally closed sets: a subset S of X is locally closed if and only if for every $x \in S$ there is a neighborhood U of x in X such that $S \cap U$ is closed in U.

7.3.10 Corollary (of lemma 7.3.8). Let A be a C^*-algebra, and let C and D be subquotients of A corresponding to separated subsets of $Prim(A)$. Then $C \oplus D$ is also a subquotient of A.

Proof. Obvious from lemma 7.3.5 and lemma 7.3.8 (3). Q.E.D.

The fact that this result does not hold for arbitrary disjoint subsets of $Prim(A)$ will be something of an inconvenience. Therefore we will need the following lemma, which will of course be applied to primitive ideal spaces.

7.3.11 Lemma. Let G be a finite group, and let X be a T_0 G-space. Let $x \in X$. Then there are locally closed subsets S_y of X for $y \in Gx$, such that $y \in S_y \subset \overline{Gx}$, the sets S_y are pairwise separated in $\overline{Gx}$, and the set $S = \bigcup_{y \in Gx} S_y$ is a G-invariant locally closed subset of X.

This result is of course trivial in case X is a Hausdorff space, since S_y may be taken to be $\{y\}$. Unfortunately, however, primitive ideal spaces of C^*-algebras need not be Hausdorff.

Proof of lemma 7.3.11. Since G is finite, we can apply lemma 2.1 of [73] to conclude that Gx is homeomorphic to G/G_x, where Gx has the relative topology as a subset of X. Therefore Gx is discrete. Consequently, there is an open subset V_0 of $\overline{Gx}$ such that $V_0 \cap Gx = \{x\}$. Let $V = \bigcap_{g \in G_x} gV_0$, which is a G_x-invariant open subset of $\overline{Gx}$ such that $x \in V$. Let $g \in G$ and suppose that $g \notin G_x$. Then

$$gV \cap V \cap Gx = \{gx\} \cap \{x\} = \varnothing.$$

Thus, $gV \cap V$ is an open subset of $\overline{Gx}$ whose intersection with Gx is empty. It is easy to see that we must therefore have $gV \cap V = \emptyset$. Let R be a system of left coset representatives of G_x in G. Then from the above we conclude that the sets gV for $g \in R$ are pairwise disjoint. Furthermore, if $gG_x = hG_x$, then $g^{-1}h \in G_x$, and $gV = g(g^{-1}hV) = hV$, since V is G_x-invariant. Let $S = \bigcup_{g \in G} gV$, which is the disjoint union of the sets gV for $g \in R$, and set $S_{gx} = gV$. (This is well defined since we have shown that if $gx = hx$ then $gV = hV$.) The sets S_y are locally closed by lemma 7.3.8 (2), since they are open subsets of the closed set $\overline{Gx}$. Similarly S is locally closed. The sets S_y are pairwise separated in $\overline{Gx}$ because they are pairwise disjoint open subsets of $\overline{Gx}$. Q.E.D.

The first application of this lemma is to relate K-visibility to actions of finite groups, as follows.

7.3.12 Lemma. Let G be a finite group, and let (G,A,α) be a G-algebra. If A is highly K-visible (respectively, p-weakly K-visible for some prime p), then every G-invariant subquotient of A has in turn a G-invariant subquotient C such that $K_0(C)$ contains a totally positive element (respectively, $K_*(C)$ has an element whose order is not a power of p).

Thus, if A is highly K-visible, then one can consider only G-invariant subquotients when applying the definition.

Proof of lemma 7.3.12. Let B be a G-invariant subquotient of A. Choose $x \in \mathit{Prim}(B)$ such that G_x is minimal among the stabilizers of all points of $\mathit{Prim}(B)$. (This can certainly be done, since G is finite.) Let S and S_y for $y \in Gx$ be as in the previous lemma. For $g \notin G_x$, we have $gS_x \cap S_x = \emptyset$. Consequently, the stabilizer of every point of S_x is contained in G_x, and, by the choice of x, is therefore equal to G_x.

Since S_x is locally closed in $\mathit{Prim}(B)$, it corresponds to a subquotient D of B. By assumption, there is a subquotient E of D such that $K_0(E)$ has a totally positive element η (respectively, $K_*(E)$ has an element η whose order is not a power of p). The subquotient E corresponds to a locally closed subset T_0 of S_x. Since G_x acts trivially on T_0, it follows that the sets T_0 and gT_0 are either equal (in case $g \in G_x$) or are separated in $\overline{Gx}$ (in case $g \notin G_x$). (If $g \notin G_x$, then S_x and S_{gx} are separated in $\overline{Gx}$ by the choice of S. Since $T_0 \subset S_x$ and $gT_0 \subset S_{gx}$, it follows that T_0 and gT_0 are also separated in $\overline{Gx}$.) Therefore the G-invariant set $T = \bigcup_{g \in G} gT_0$ is the union of finitely many pairwise separated locally closed subsets of $\overline{Gx}$. It follows from lemma 7.3.8 (3) that T is a locally closed subset of $\overline{Gx}$, and hence of $\mathit{Prim}(B)$ by lemma 7.3.8 (2).

Let R be a system of left coset representatives of G_x in G. Then T is the disjoint union of the sets gT_0 for $g \in R$. If C is the subquotient of B corresponding to T, then it follows from corollary 7.3.10 that $C = \bigoplus_{g \in R} \alpha_g[E]$. (Here, if I/J is a subquotient of a

G-algebra, we write $\alpha_g[I/J]$ for the subquotient $\alpha_g[I]/\alpha_g[J]$.) The algebra C is a G-invariant subquotient of B, and $K_*(C) \simeq \bigoplus_{g\in R} K_*(E)$. Clearly the element $(\eta, 0, \ldots, 0)$ of $K_*(C)$ is a totally positive element in $K_0(C)$ (respectively, an element of $K_*(C)$ whose order is not a power of p). Q.E.D.

7.4. The Main Theorems

There are actually two versions of the main theorem, since it turns out that one needs a stronger version of K-visibility for more complicated groups. We first prove a lemma which is applicable to both versions.

7.4.1 Lemma. Let G be a finite abelian group, and let (G,A,α) be a G-algebra such that $\widetilde{\Gamma}(\alpha) \neq \widehat{G}$. Suppose that A is highly K-visible (respectively, p-weakly K-visible for some prime p). Then there exists a proper subgroup H of $\widehat{G}$, a G-invariant subquotient D of A, a $\widehat{G}$-invariant subquotient C of $C^*(G,A)$, and an H-invariant subquotient Q of C, such that:

(1) $C \simeq C^*(G,D)$ as $\widehat{G}$-algebras.

(2) $C \simeq C(\widehat{G} \times_H Q)$ as $\widehat{G}$-algebras.

(3) $K_0(D)$ has a totally positive element (respectively, $K_*(D)$ has an element whose order is not a power of p).

Proof. By theorem 7.2.8, the action of $\widehat{G}$ on $Prim(C^*(G,A))$, obtained from the dual action $\widehat{\alpha}$ of $\widehat{G}$ on $C^*(G,A)$, is not trivial. Therefore there is $x \in Prim(C^*(G,A))$ such that $\widehat{G}_x \neq \widehat{G}$. We let $H = \widehat{G}_x$ and $Z = \overline{\widehat{G}x}$. An application of lemma 7.3.11 yields locally closed sets $S_y \subset Z$ for $y \in \widehat{G}x$, such that the sets S_y are pairwise separated in Z, and $S = \bigcup_{y \in \widehat{G}x} S_y$ is a locally closed G-invariant subset of Z.

Let C_0 be the $\widehat{G}$-invariant subquotient of $C^*(G,A)$ corresponding to S. The correspondence between $\widehat{G}$-invariant ideals of $C^*(G,A)$ and G-invariant ideals of A (proved in proposition 6.3.9) is easily seen to imply that $C_0 \simeq C^*(G,D_0)$ for some G-invariant subquotient D_0 of A. By lemma 7.3.12, there is a G-invariant subquotient D of D_0 such that $K_0(D)$ contains a totally positive element (respectively, an element whose order is not a power of p). Let $C = C^*(G,D)$, which is a $\widehat{G}$-invariant subquotient of C_0.

The set S_x is open in S, and therefore corresponds to an ideal Q_0 in C_0. Let R be a system of coset representatives for $H = \widehat{G}_x$ in $\widehat{G}$. Then S is the disjoint union of the sets $S_{\tau x} = \tau S_x$ for $\tau \in R$. Since these sets are pairwise separated in Z, repeated application of corollary 7.3.10 shows that C_0 is the direct sum, as τ ranges over R, of the subquotients $\widehat{\alpha}_\tau[Q_0]$ which correspond to τS_x. That is, $C_0 = \bigoplus_{\tau \in R} \widehat{\alpha}_\tau[Q_0]$. Now let

Q be the subquotient of Q_0 which corresponds to the intersection of S_x with the locally closed set corresponding to C. Then Q is H-invariant by corollary 7.3.7, since S_x and C are H-invariant. Just as above, we find that $C = \bigoplus_{\tau \in R} \hat{\alpha}_\tau[Q]$. Therefore $C \simeq C(\hat{G} \times_H Q)$ as $\hat{G}$-algebras, by proposition 2.9.6. Q.E.D.

7.4.2 Theorem. Let G be a cyclic group of prime order p, and let (G,A,α) be a G-algebra. If A is p-weakly K-visible, and α is K-free, then $\widetilde{\Gamma}(\alpha) = \hat{G}$.

Proof. Assume that $\widetilde{\Gamma}(\alpha) \neq \hat{G}$. Let D, C, Q, and H be as in the previous lemma, as applied to p-weakly K-visible algebras. (Thus, $K_*(D)$ contains an element whose order is not a power of p.) Then H is the trivial subgroup, since $\hat{G}$ has prime order. In particular, $C = \bigoplus_{\tau \in \hat{G}} \hat{\alpha}_\tau[Q]$.

Let $l^2(G)$ carry the left regular representation of G, and let $K(l^2(G))$ carry the associated inner action of G. By Takai duality ([79], 7.9.3), we have $C^*(\hat{G}, C^*(G,D)) \simeq K(l^2(G)) \otimes D$. Therefore there are isomorphisms

$$K_*(Q) \simeq K_*^{\hat{G}}(C) \simeq K_*(C^*(\hat{G},C)) \simeq K_*(K(l^2(G)) \otimes D) \simeq K_*(D),$$

where the first step is proposition 2.9.4, the second step is theorem 2.8.3 (7), the third step is Takai duality, and the fourth step is stability, theorem 2.8.3 (4). Since $K_*(D)$ has an element whose order is not a power of p, it follows that $K_*(Q)$ has an element η whose order is not a power of p.

Since $C = \bigoplus_{\tau \in \hat{G}} \hat{\alpha}_\tau[Q]$, we have

$$K_*^G(D) \simeq K_*(C) = \bigoplus_{\tau \in \hat{G}} \tau K_*(Q).$$

(Recall from theorem 2.8.3 (7) that multiplication by τ on $K_*(C^*(G,D))$ is $(\hat{\alpha}_\tau)_*$.) We can identify η with the element $(\eta, 0, \ldots, 0)$ of $K_*(C)$. Let τ be a generator of $\hat{G}$. Then we find that the elements $\eta, \tau\eta, \ldots, \tau^{p-1}\eta$ are all independent over $\mathbf{Z}$ in the sense that if $n_0, \ldots, n_{p-1} \in \mathbf{Z}$ are such that $\sum_{r=0}^{p-1} n_r \tau^r \eta = 0$, then $n_r \tau^r \eta = 0$ for each r. (The elements $\tau^r \eta$ are all in different direct summands of $K_*(C)$.)

We are going to show that $(1-\tau)^n \eta$ is not zero for any n. For $0 \leq r \leq p-1$ and $n > 0$ let

$$c_{n,r} = \sum_{0 \leq j \leq n,\, j = r \bmod p} (-1)^j \binom{n}{j},$$

where $\binom{n}{j}$ is the binomial coefficient $\frac{n!}{j!(n-j)!}$. Then

$$(1-\tau)^n \eta = \sum_{r=0}^{p-1} c_{n,r} \tau^r \eta.$$

If $(1-\tau)^n \eta = 0$, then $c_{n,r} \tau^r \eta = 0$ for all r. The order of $\tau^r \eta$ is obviously the same as the order of η, so it follows that $c_{n,r}$ is a multiple of the order of η for all r. (If η has infinite order, then we must have $c_{n,r} = 0$ for all r.) If q is any prime such that $q \neq p$

and q divides the order of η (take any $q \neq p$ if η has infinite order), then it follows that $c_{n,r} = 0 \bmod q$ for all r. (Such a prime q exists, because, by the choice of η, the order of η is not a power of p.) Now let F be the field with q elements, and let ω be an element of its algebraic closure such that $1 + \omega + \cdots + \omega^{p-1} = 0$. Since $p \neq q$, the element 1 does not satisfy this equation, and so $\omega \neq 1$. However, $\omega^p = 1$, so

$$0 \neq (1-\omega)^n = \sum_{r=0}^{p-1} c_{n,r}\, \omega^r = 0,$$

where the last equality follows from the fact that $c_{n,r} = 0 \bmod q$ for all r. This contradiction shows that $(1-\tau)^n \eta \neq 0$ for all n.

We conclude that D does not have locally discrete K-theory, since $1 - \tau \in I(G)$. Since D is a G-invariant subquotient of A, this contradicts the K-freeness of α. Therefore the assumption made at the beginning of this proof, that $\widetilde{\Gamma}(\alpha) \neq \widehat{G}$, is untenable. Thus $\widetilde{\Gamma}(\alpha) = \widehat{G}$. Q.E.D.

Similar methods yield the other version of the main theorem, which deals with actions of finite abelian groups. The proof is complicated by the fact that nontrivial subgroups of the dual group must be considered as possible stabilizer subgroups for primitive ideals in the crossed product. This difference also makes it necessary to start with a stronger hypothesis.

7.4.3 Theorem. Let G be a finite abelian group, and let (G,A,α) be a G-algebra. If A is highly K-visible, and α is K-free, then $\widetilde{\Gamma}(\alpha) = \widehat{G}$.

Proof. Again, suppose that $\widetilde{\Gamma}(\alpha) \neq \widehat{G}$, and let H, D, C, and Q be as in lemma 7.4.1, as applied to highly K-visible C^*-algebras. Thus, we assume that $K_0(D)$ contains a totally positive element η_0. In particular, there is $n > 0$ and a projection $p_0 \in M_n(D)$ such that $\eta_0 = [p_0]$.

Let $l^2(G)$ carry the left regular representation u of G, and let $K(l^2(G))$ have the associated inner action λ of G. Let e_g be the projection from $l^2(G)$ onto the standard basis vector corresponding to g, that is, if $\xi \in l^2(G)$ then

$$e_g \xi(h) = \begin{cases} \xi(h) & h = g \\ 0 & h \neq g. \end{cases}$$

Then one checks that $\lambda_g(e_h) = u_g e_h u_g^* = e_{gh}$. Define $p \in K(l^2(G)) \otimes M_n(D)$ by

$$p = \sum_{g \in G} e_g \otimes \alpha_g(p_0).$$

Since the summands are orthogonal, p is a projection in $K(l^2(G)) \otimes M_n(D)$, and it is easily seen to be G-invariant. (Actually, $[p] = Ind_{\{e\}}^{G}(\eta_0)$ in the notation of proposition 5.1.3. However, we need the concrete form of it given here.)

Let $\eta \in K_0^G(D)$ be the element $\eta = [p]$. We claim that $m\eta \notin I(G)K_0^G(D)$ for any integer $m > 0$. To see this, let $r : K_0^G(D) \longrightarrow K_0(D)$ be the restriction map $Res_{\{e\}}^{G}$ of proposition 5.1.3. (This is just the obvious homomorphism defined by regarding G-

invariant projections simply as projections.) Then clearly $r(\mu)=0$ for $\mu \in I(G)K_0^G(D)$. (If q is a G-invariant projection in some $L(V)\otimes D^+$, and W_0 and W_1 are representation spaces of G of the same dimension, then $1_{W_0}\otimes q$ and $1_{W_1}\otimes q$ become equivalent when the action of G is ignored.) Consequently, if $m\eta \in I(G)K_0^G(D)$, then $m[p]=0$ in $K_0(D)$. Therefore in $K_0(D)$ we have

$$-[p_0] = (m-1)[p_0] + m \sum_{g\in G-\{e\}} [\alpha_g(p_0)],$$

which contradicts the assumption that $[p_0]$ is totally positive. This proves the claim.

By construction, we have $C^*(G,D) \simeq C(\hat{G}\times_H Q)$ as $\hat{G}$-algebras. Now $C(\hat{G}\times_H Q)$ is isomorphic to a direct sum $\bigoplus_{y\in \hat{G}/H} Q_y$, where each Q_y is isomorphic to Q as a C^*-algebra. If β denotes the action of $\hat{G}$ on this direct sum, then $\beta_\tau[Q_y]=Q_{\tau y}$ for $\tau\in\hat{G}$ and $y\in \hat{G}/H$. We have

$$K_0^G(D) \simeq K_0(C^*(G,D)) \simeq K_0(C(\hat{G}\times_H Q)) \simeq \bigoplus_{y\in\hat{G}/H} K_0(Q_y),$$

where the $R(G)$-module structures on $K_0(C^*(G,D))$ and $K_0(C(\hat{G}\times_H Q))$ are determined by the action of $\hat{G}$ on the algebras. Therefore $M=\bigoplus_{y\in\hat{G}/H} K_0(Q_y)$ is an $R(G)$-module, and we have $\tau\cdot K_0(Q_y)=K_0(Q_{\tau y})$ for $\tau\in\hat{G}$ and $y\in\hat{G}/H$.

Since $\eta\in K_0^G(D)$, we can identify η with its image in M. We then obtain $\eta=\sum_{y\in\hat{G}/H}\eta_y$, where $\eta_y\in K_0(Q_y)$. We now claim that there is $z\in\hat{G}/H$ such that $m\eta_z\notin I(G)M$ for all $m>0$. Suppose not; then for all y there is $m_y>0$ such that $m_y\eta_y\in I(G)M$. Let $m=\prod_{y\in\hat{G}/H} m_y$. (This is a finite product.) Then $m\eta_y\in I(G)M$ for all y, whence $m\eta\in I(G)M$, a contradiction. Therefore the claim is proved. Choose z as in the claim. Since $I(G)M$ is a submodule of M, we actually have $m\eta_z\notin I(G)M$ for all $m\neq 0$.

Choose $\tau\in\hat{G}$ such that $\tau\notin H$. Let τ^l be the smallest positive power of τ which is in H. Fix $n>0$. Then $(1-\tau)^n\eta_z=\sum_{r=0}^{l-1}\lambda_r$, where

$$\lambda_r = \sum_{0\le j\le n,\, j=r \bmod l} (-1)^j \binom{n}{j} \tau^j \eta_z .$$

Here $\binom{n}{j}$ is the binomial coefficient as usual. Now $\lambda_r\in K_0(Q_{\tau^r z})$. Since $M=\bigoplus_{y\in\hat{G}/H}K_0(Q_y)$, and the elements $\tau^r z$ of $\hat{G}/H$ are distinct for $0\le r<l$, we conclude that $(1-\tau)^n\eta_z=0$ if and only if $\lambda_r=0$ for all r. Define

$$c_r = \sum_{0\le j\le n,\, j=r \bmod l} (-1)^j \binom{n}{j}.$$

Then if ζ is a primitive l-th root of unity, we have $0\neq(1-\zeta)^n=\sum_{r=0}^{l-1}c_r\zeta^r$.

Therefore there is r such that $c_r \neq 0$. We have

$$\lambda_r - c_r \tau^r \eta_z = \sum_{0 \le j \le n,\, j = r \bmod l} (-1)^j \binom{n}{j} (\tau^j - \tau^r)\eta_z \in I(G)M.$$

Now $c_r\eta_z \notin I(G)M$ by the choice of z, so $c_r\tau^r\eta_z \notin I(G)M$ because τ^r is invertible in $R(G)$. Therefore $\lambda_r \notin I(G)M$, and in particular $\lambda_r \neq 0$. We conclude that $(1-\tau)^n\eta_z \neq 0$. Since this is true for all $n > 0$, we have found an element of M which is not annihilated by any power of $1-\tau$.

Since $M \simeq K_0^G(D)$ as an $R(G)$-module, it follows that the action of G on D does not have locally discrete K-theory, which contradicts the assumption that the action α of G on A is K-free. Thus the assumption $\tilde{\Gamma}(\alpha) \neq \hat{G}$ leads to a contradiction, and we conclude that $\tilde{\Gamma}(\alpha) = \hat{G}$. Q.E.D.

7.4.4 Corollary. Let G be a finite abelian group, and let (G,A,α) be a G-algebra such that A is unital, stably finite, and G-simple. If α is K-free then $C^*(G,A,\alpha)$ is simple.

Proof. By corollary 7.3.4, A is highly K-visible. Therefore $\tilde{\Gamma}(\alpha) = \hat{G}$ by the previous theorem. Theorem 3.5 of [58] now implies that $C^*(G,A,\alpha)$ is simple. Q.E.D.

7.4.5 Remark. It is evident from the proof of the previous theorem that the assumption we really needed to make is that every G-invariant subquotient of A has in turn a G-invariant subquotient C such that $K_*^G(C)/I(G)K_*^G(C)$ has an element of infinite order. Actually, the existence of an element of order relatively prime to the order of G would do. It is tempting to call this condition K^G-visibility, and to then say that if α is a K-free action of a finite abelian group G on a C^*-algebra A such that (G,A,α) is K^G-visible, then $\tilde{\Gamma}(\alpha) = \hat{G}$. However, K^G-visibility seems to be much harder to verify than the other forms of K-visibility. Indeed, the only way I know to show that a G-algebra (G,A,α) is K^G-visible is to show that A is highly K-visible, and then to use the argument in the first part of the proof of theorem 7.4.3.

We next present an example which shows that one really can use K-theory to prove that a crossed product is simple. This example is not new, being in fact a concrete case of a class of examples considered in section 4 of [58]. (Our methods are not applicable to the more general situation considered there.)

7.4.6 Example. Let $\mathbf{O}_3$ be the Cuntz algebra [21], that is, $\mathbf{O}_3$ is the universal C^*-algebra of three isometries s_0, s_1, and s_2 such that $s_0s_0^* + s_1s_1^* + s_2s_2^* = 1$. By the universal property of $\mathbf{O}_3$, there is a unique automorphism α of $\mathbf{O}_3$ such that $\alpha(s_0) = s_1$, $\alpha(s_1) = s_2$, and $\alpha(s_2) = s_0$. This automorphism has order 3 and therefore defines an action of $G = \mathbf{Z}/3\mathbf{Z}$ on $\mathbf{O}_3$. We also call the action α.

We first want to show that α has locally discrete K-theory. Define an action λ of G on M_3 as follows. We identify M_3 with $L(l^2(\hat{G}))$, define $u_g\xi(\tau) = \tau(g)\xi(\tau)$ for $\xi \in l^2(\hat{G})$, $g \in G$, and $\tau \in \hat{G}$, and define $\lambda_g(a) = u_g a u_g^*$. (The representation u is a disguised form of the regular representation.) Write λ for λ_h, where h is a fixed generator of G. Let p_0 be the orthogonal projection onto the function $\xi_0 \in l^2(\hat{G})$ defined by

$$\xi_0(\tau) = \begin{cases} 1 & \tau = 1 \\ 0 & \tau \neq 1, \end{cases}$$

and let p_1 be the orthogonal projection onto the function

$$\xi_1(\tau) = \begin{cases} 1 & \tau = \sigma \\ 0 & \tau \neq \sigma, \end{cases}$$

where σ is a fixed generator of $\hat{G}$. Define $\varphi : \mathbf{O}_3 \longrightarrow M_3 \otimes \mathbf{O}_3$ by $\varphi(a) = p_0 \otimes a$, and define $\psi : \mathbf{O}_3 \longrightarrow M_3 \otimes \mathbf{O}_3$ by $\psi(a) = p_1 \otimes a$. Since p_0 and p_1 are G-invariant, it is clear that φ and ψ are equivariant. In fact, it follows from lemma 2.7.5 that the induced maps φ_* and ψ_* on equivariant K-theory are isomorphisms, and that $\varphi_*^{-1} \circ \psi_*(\eta) = \sigma\eta$ for $\eta \in K_*^G(\mathbf{O}_3)$. We are next going to prove that φ and ψ are G-homotopic. As a consequence, we will obtain $\sigma\eta = \eta$ for $\eta \in K_*^G(\mathbf{O}_3)$, or $(1-\sigma)K_*^G(\mathbf{O}_3) = 0$. Since $1-\sigma$ generates $I(G)$, it will follow that the given action on $\mathbf{O}_3$ has discrete K-theory.

Let $t \longrightarrow p_t$ for $t \in [0,1]$ be a continuous path of projections in M_3 such that p_0 and p_1 are the projections defined above. (Of course, we cannot take p_t to be G-invariant.) Set $q_j = s_j s_j^*$ for $j = 0, 1, 2$. Then the q_j are orthogonal projections whose sum is 1. Let

$$e_t = p_t \otimes q_0 + \lambda(p_t) \otimes q_1 + \lambda^2(p_t) \otimes q_2,$$

which is an orthogonal sum

$$e_t = \sum_{j=0}^{2} (\lambda \otimes \alpha)^j (p_t \otimes q_0),$$

and is hence a G-invariant projection. We have $e_0 = p_0$ and $e_1 = p_1$, and $t \longrightarrow e_t$ is continuous. Recall from lemma 2.5.3 (2) that G-invariant projections p and q satisfying $\| p - q \| < ½$ are unitarily equivalent via a G-invariant unitary. The construction given there shows that the unitary can be made to depend continuously on p and q. Using the uniform continuity of $t \longrightarrow p_t \otimes 1$ and $t \longrightarrow e_t$, it is now easy to construct a continuous function $t \longrightarrow u_t$ from $[0,1]$ to the unitary group of $M_3 \otimes \mathbf{O}_3$ such that $u_0 = 1$ and $u_t(p_t \otimes 1)u_t^* = e_t$ for all t. (Note that u_t is not G-invariant, and it is not obvious that we can choose $u_1 = 1$.) Now $u_1(p_1 \otimes 1)u_1 = e_1 = p_1 \otimes 1$, so u_1 commutes with $p_1 \otimes 1$. Therefore $u_1 = v + w$, where

$$v \in (p_1 \otimes 1)(M_3 \otimes \mathbf{O}_3)(p_1 \otimes 1) \simeq \mathbf{O}_3 \quad \text{and} \quad w \in (1 - p_1 \otimes 1)(M_3 \otimes \mathbf{O}_3)(1 - p_1 \otimes 1)$$

are unitaries. By [23], corollary 3.12, the unitary group of $\mathbf{O}_3$ is connected, so there is a continuous path $t \longrightarrow v_t$ for $t \in [1,2]$ of unitaries in $\mathbf{O}_3$ such that $v_1 = v$ and $v_2 = 1$. Identifying $\mathbf{O}_3$ with $(p_1 \otimes 1)(M_3 \otimes \mathbf{O}_3)(p_1 \otimes 1)$, set $u_t = v_t + w$ for $t \in (1,2]$. Further set $p_t = p_1$ and $e_t = e_1$ for $t \in (1,2]$.

Define a partial isometry $x_t \in M_3 \otimes \mathbf{O}_3$ by $x_t = (p_t \otimes s_0)u_t^*$. Then

$$(\lambda \otimes \alpha)^j (x_t x_t^*) = \lambda^j (p_t) \otimes q_j,$$

and

$$(\lambda\otimes\alpha)^j\,(x_t^*x_t) = (\lambda\otimes\alpha)^j\,(u_t(p_t\otimes 1)u_t^*) = (\lambda\otimes\alpha)^j\,(e_t) = e_t\,.$$

Since $e_t = \sum_{j=0}^{2} \lambda^j\,(p_t)\otimes q_j$, the assignment $s_j \longrightarrow (\lambda\otimes\alpha)^j\,(x_t)$ determines a G-homotopy of homomorphisms $\varphi_t : \mathbf{O}_3 \longrightarrow M_3\otimes\mathbf{O}_3$ for $t\in[0,2]$, such that $\varphi_t(1) = e_t$. By definition $\varphi_0 = \varphi$, and since

$$(p_2\otimes 1)u_2(p_2\otimes 1) = p_2\otimes 1 = p_1\otimes 1\,,$$

we have $\varphi_2 = \psi$. So φ and ψ are G-homotopic, as desired, and it follows that the action α of G on $\mathbf{O}_3$ has locally discrete K-theory.

Since $\mathbf{O}_3$ is simple, it now follows that α is K-free. Also, $K_0(\mathbf{O}_3) \simeq \mathbf{Z}/2\mathbf{Z}$ by [23]. This, together with the simplicity of $\mathbf{O}_3$, implies that $\mathbf{O}_3$ is 3-weakly K-visible. Therefore theorem 7.4.2 implies that $\widetilde{\Gamma}(\alpha) = \widehat{G}$. By theorem 3.5 of [58], we find that $C^*(\mathbf{Z}/3\mathbf{Z}, \mathbf{O}_3, \alpha)$ is simple.

7.4.7 Remark. The method of the previous example is certainly not the recommended way to prove that $C^*(\mathbf{Z}/3\mathbf{Z}, \mathbf{O}_3, \alpha)$ is simple. Indeed, I believe (but have not confirmed) that the existence of the homotopy constructed there should directly imply that $\widetilde{\Gamma}(\alpha) = \widehat{G}$, without considering K-theory. (Similar homotopies will appear in the main theorem on K-freeness of actions on AF algebras, in chapter 9.) However, this example is interesting because it does show what can be done with K-freeness, and also because it provides an example of an action on a Cuntz algebra which is K-free for nontrivial reasons.

The results of this section unfortunately do not generalize to infinite compact Lie groups. It is clear that the dual group $\widehat{G}$ must be compact for the proof of theorem 7.4.3 to work, and we now show by an example that this theorem is false for $G = S^1$. It turns out that the action in our example is even totally K-free.

7.4.8 Example. Let $A = A_\vartheta$ be the rotation algebra corresponding to the irrational number ϑ. (See [86].) That is, $A = C^*(\mathbf{Z}, C(S^1), \rho)$, where ρ is the action of $\mathbf{Z}$ defined by $\rho_1(f)(\lambda) = f(\exp(-2\pi i\vartheta)\lambda)$ for $f\in C(S^1)$ and $\lambda\in S^1$. It will be useful to write $A = C^*(z, u)$ where z and u are two unitaries satisfying the relation $uzu^* = \omega z$, with $\omega = \exp(-2\pi i\vartheta)$. Here, we identifiy z with the function on S^1 given by $z(\lambda) = \lambda$, and u with the function in $L^1(\mathbf{Z}, C(S^1))$ whose value at 1 is the constant function 1 and whose values elsewhere on $\mathbf{Z}$ are all zero.

Let $G = S^1$, and let $\alpha : G \longrightarrow Aut(A)$ be the dual action $\alpha = \widehat{\rho}$. Thus, we have $\alpha_g(z) = z$ and $\alpha_g(u) = gu$ for $g\in S^1$. We will show that α is K-free but that $\widetilde{\Gamma}(\alpha) = 0$. The computation of $\widetilde{\Gamma}(\alpha)$ is straightforward: by lemma 3.2 of [58] we have

$$\widetilde{\Gamma}(\alpha) = \{n\in\mathbf{Z} : \rho_n[I]\subset I \text{ for all closed ideals } I\subset C(S^1)\},$$

which is obviously zero. In fact, the (in general larger) Connes spectrum $\Gamma(\alpha)$ is also zero. Indeed, we have

$$\Gamma(\alpha) = \{ n \in \mathbf{Z} : \rho_n[I] \cap I \neq 0 \text{ for all nonzero closed ideals } I \subset C(S^1) \},$$

by [79], 8.11.7, and this set is also obviously equal to $\{0\}$.

We now verify K-freeness. By Takai duality, $C^*(G,A,\alpha) \simeq K(l^2(\mathbf{Z})) \otimes C(S^1)$, and the isomorphism is equivariant for the action $\hat{\alpha}$ of $\mathbf{Z}$ on $C^*(G,A,\alpha)$ and the diagonal action $\gamma \otimes \rho$ on $K(l^2(\mathbf{Z})) \otimes C(S^1)$, where γ is an appropriate inner action (coming from the left regular representation of $\mathbf{Z}$). Consequently $K_*^G(A) \simeq K_*(C(S^1))$, and furthermore this isomorphism identifies $(\hat{\alpha}_n)_*$ with $(\rho_n)_*$ for $n \in \mathbf{Z}$. Since any rotation on S^1 is homotopic to the identity, we see that $(\rho_n)_*$ is the identity on $K_*(C(S^1))$. It follows that the dual action $\hat{\alpha}$ is trivial on K-theory, whence $I(G) K_*^G(A) = 0$. Since A is simple, this implies that α is K-free.

Since A has a faithful trace, it is certainly stably finite. Since A is simple and unital, it is highly K-visible, and we have shown that theorem 7.4.3 does not hold for $G = S^1$.

As we now show, this situation cannot be repaired by requiring total K-freeness. Indeed, the action α of this example is even totally K-free. To prove this, let H be an arbitrary proper subgroup of S^1; then we must show that $\alpha|_H$ has locally discrete K-theory. Now $H \simeq \mathbf{Z}/n\mathbf{Z}$ for some integer $n > 1$, and H is generated by the element $\zeta = \exp(\frac{2\pi i}{n}) \in S^1$. Let us write β for $\alpha|_H$. Then we must compute the K-theory of the crossed product $C^*(H,A,\beta)$. This algebra is the universal C^*-algebra generated by three unitaries z, u, and s satisfying the relations

$$uzu^* = \omega z, \quad szs^* = s, \quad sus^* = \zeta u, \quad \text{and} \quad s^n = 1. \tag{*}$$

Observe that A^H is generated by z and u^n. Since these unitaries satisfy $u^n z (u^n)^* = \omega^n z$, we must have $A^H = A_{n\vartheta}$, the rotation algebra for the irrational number $n\vartheta$. Define elements $\varphi(z)$, $\varphi(u)$, and $\varphi(s)$ in $M_n(A^H)$ by

$$\varphi(z) = \begin{pmatrix} z & & & & \\ & \omega z & & 0 & \\ & & \cdot & & \\ & & & \cdot & \\ & 0 & & & \cdot \\ & & & & \omega^{n-1} z \end{pmatrix},$$

$$\varphi(u) = \begin{pmatrix} 0 & 1 & & & \\ & 0 & 1 & & 0 \\ & & & \cdot & \\ & 0 & & & \cdot \\ & & 0 & & 1 \\ u^n & & & & 0 \end{pmatrix},$$

and

$$\varphi(s) = \begin{bmatrix} 1 & & & & \\ & \zeta^{-1} & & 0 & \\ & & \cdot & & \\ & & & \cdot & \\ & 0 & & \cdot & \\ & & & & \zeta^{-n+1} \end{bmatrix}.$$

It is easily shown that the unitaries $\varphi(z)$, $\varphi(u)$, and $\varphi(s)$ satisfy the relations (*). Therefore there is a homomorphism $\varphi : C^*(H,A,\beta) \longrightarrow M_n(A^H)$ taking the specified values on z, u and s.

We now show that φ is surjective. Regarding the diagonal matrices with entries in $C(S^1)$ as functions on the disjoint union X of n copies of S^1, we see that $\varphi(z)$ and $\varphi(s)$ separate the points of X. The Stone-Weierstrass theorem now implies that all diagonal matrices with entries in $C(S^1)$ are in the range of φ. The diagonal matrix with diagonal entries all equal to u^n is also in the range of φ, since it is $\varphi(u)^n$. It now follows that all diagonal matrices are in the range of φ. Finally, if e_{ij} denotes the ij matrix unit, then for every i, e_{ii} is in the range of φ because it is diagonal, so, for $i < n$, the matrix unit $e_{i,i+1} = e_{ii}\,\varphi(u)$ is in the range of φ. It is now easily seen that φ is surjective.

Now consider the map $j : A^H \longrightarrow C^*(H,A,\beta)$ of lemma 7.1.7. Here, we have $j(a) = \sum_{k=0}^{n-1} as^k$ for $a \in A^H$, whence $\varphi \cdot j(a)$ is the matrix whose 1,1 entry is a and whose other entries are zero. Consequently $(\varphi \cdot j)_*$ is an isomorphism from $K_*(A^H)$ to $K_*(M_n(A^H))$. Let τ be the generator of $\widehat{H}$ such that $\tau(\zeta) = \zeta$. Then for $a \in A^H$ it is easy to see that $(\varphi \cdot \widehat{\beta}_\tau \cdot j)(a)$ is the matrix whose 2,2 entry is a and whose remaining entries are zero. We therefore clearly have $(\varphi \cdot \widehat{\beta}_\tau \cdot j)_* = (\varphi \cdot j)_*$. Now β is saturated, since A_{τ^k} contains the unitary u^k for every k. By proposition 7.1.8, the map j_* is therefore an isomorphism. Since $(\varphi \cdot j)_*$ is an isomorphism, so is φ_*. It follows that $(\widehat{\beta}_\tau)_*$ is the identity, and, since $1 - \tau$ generates $I(H)$, that $I(H)\, K_*^H(A) = 0$. Since A is simple, we conclude that β is K-free. This shows that α is totally K-free.

We point out that it is now easy to show that φ is an isomorphism. Indeed, β is K-free, A is highly K-visible, and H is finite abelian, so $C^*(H,A,\beta)$ is simple by corollary 7.4.4. So φ can have no kernel. In fact, however, there is a map analogous to φ in the case that $C(S^1)$ is replaced by an arbitrary **Z**-algebra, and we presume that it is always an isomorphism, independently of any considerations involving K-freeness or K-visibility.

7.5. The Existence of K-Visible C^*-Algebras

In proposition 7.3.3, we proved that type I, AF, and unital stably finite simple C^*-algebras are all highly K-visible. Obviously we would like to be able to apply the theorems in the previous section to more algebras than those. Therefore we show in this section that there are in fact more (highly, p-weakly) K-visible C^*-algebras.

7.5.1 Lemma. Let A be a (highly, p-weakly) K-visible C^*-algebra. Then any subquotient of A is also (highly, p-weakly) K-visible.

Proof. Obvious from the definition. Q.E.D.

7.5.2 Lemma. Let

$$0 \longrightarrow I \longrightarrow A \longrightarrow B \longrightarrow 0$$

be an exact sequence of C^*-algebras. Then A is (highly, p-weakly) K-visible if and only if I and B are.

Proof. If A is (highly, p-weakly) K-visible, then I and B are also, by the previous lemma. For the converse, let I and B be (highly, p-weakly) K-visible, and let $Q = L/M$ be a subquotient of A. Then there is an exact sequence

$$0 \longrightarrow (L \cap I)/(M \cap I) \longrightarrow Q \longrightarrow (L+I)/(M+I) \longrightarrow 0. \qquad (*)$$

(This is most easily checked by considering subsets of $Prim(A)$.) Since by convention Q is not zero, one of the other two algebras in (*) is not zero. If $(L \cap I)/(M \cap I)$ is not zero, then it is a subquotient of I, and if $(L+I)/(M+I)$ is not zero, then it is a subquotient of $B = A/I$. In either case, the nonzero algebra has a subquotient C such that $K_*(C)$ has an element of infinite order (respectively, $K_0(C)$ has a totally positive element, $K_*(C)$ has an element of order not a power of p). Since C is a subquotient of Q, we have shown that A is (highly, p-weakly) K-visible. Q.E.D.

7.5.3 Lemma. Let A be a C^*-algebra with a composition series $(I_\lambda)_{\lambda \leq \kappa}$, such that the composition factors $I_{\lambda+1}/I_\lambda$ are (highly, p-weakly) K-visible for all λ. Then A is (highly, p-weakly) K-visible.

Proof. We prove by induction on λ that I_λ is (highly, p-weakly) K-visible. Of course we have $I_0 = 0$ by definition, so I_1 is (highly, p-weakly) K-visible by assumption. Now assume I_μ is (highly, p-weakly) K-visible for all $\mu < \lambda$.

Case 1. Assume λ is a successor ordinal, that is, $\lambda = \mu+1$ for some μ. Then I_λ is (highly, p-weakly) K-visible because I_μ and I_λ/I_μ are, by the previous lemma.

Case 2. Assume λ is a limit ordinal. Then $I_\lambda = \overline{\bigcup_{\mu<\lambda} I_\mu}$. Let L/M be a subquotient of I_λ. Since $L \neq M$, there is $\mu < \lambda$ such that $L \cap I_\mu \neq M \cap I_\mu$. Then $Q = (L \cap I_\mu)/(M \cap I_\mu)$ is a subquotient of both I_μ and L/M. Since Q is a subquotient of I_μ, it has in turn a subquotient C such that $K_*(C)$ has an element of infinite order (respectively, $K_0(C)$

has a totally positive element, $K_*(C)$ has an element whose order is not a power of p). Since C is also a subquotient of L/M, this shows that I_λ is (highly, p-weakly) K-visible, and thus completes the induction. Q.E.D.

7.5.4 Lemma. Let A and B be stably isomorphic C^*-algebras. Then A is (highly, p-weakly) K-visible if and only if B is.

Proof. It is clearly enough to show that A is (highly, p-weakly) K-visible if and only if $K \otimes A$ is, where K is the algebra of compact operators on an arbitrary Hilbert space. By proposition 4.2.9, there is a one-to-one correspondence between ideals of A and ideals of $K \otimes A$, given by $I \longrightarrow K \otimes I$. Therefore a similar result holds for subquotients, and it is consequently sufficient to show that $K_*(A)$ has an element of infinite order (respectively, $K_0(A)$ has a totally positive element, $K_*(A)$ has an element of order not a power of p) if and only if $K \otimes A$ does. Since $K_*(K \otimes A) \simeq K_*(A)$, the only condition needing further proof is the condition that $K_0(A)$ have a totally positive element. However, it is clear from the construction of the isomorphism between $K_0(A^+)$ and $K_0(K \otimes A^+)$ in proposition 2.7.5 (the group here is trivial) that the classes in $K_0(A^+)$ represented by projections are the same as those in $K_0(K \otimes A^+)$ which are represented by projections. Since $K_0(A)$ is a subgroup of $K_0(A^+)$, the same holds for it as well. In particular, $K_0(A)$ contains a totally positive element if and only if $K_0(K \otimes A)$ does. Q.E.D.

7.5.5 Corollary. Let A be separable and (highly, p-weakly) K-visible. Then so is any hereditary subalgebra of A.

Proof. Let B be a hereditary subalgebra of A. Then by [15], B is stably isomorphic to the ideal it generates in A. The conclusion now follows from lemmas 7.5.4 and 7.5.2. Q.E.D.

7.5.6 Lemma. Let A and B be K-visible nuclear C^*-algebras, and suppose that A together with all its ideals is in the category **N** of [95]. (The definition of **N** is reproduced just before lemma 6.2.3.) Then $A \otimes B$ is K-visible.

Proof. We first show that every subquotient of $A \otimes B$ which has the form $C \otimes D$, for C and D subquotients of A and B respectively, has in turn a subquotient N such that $K_*(N)$ has an element of infinite order. Let P be a subquotient of C such that $K_*(P)$ has an element of infinite order, and let Q be a subquotient of D such that $K_*(Q)$ has an element of infinite order. Then P is a subquotient of A, and therefore has the form L/M, where L and M are ideals in A. By assumption, $L, M \in \mathbf{N}$, so that also $P \in \mathbf{N}$. By the Künneth formula ([95], theorem 4.1), there is a graded exact sequence

$$0 \longrightarrow K_*(P) \otimes K_*(Q) \longrightarrow K_*(P \otimes Q) \longrightarrow \mathrm{Tor}(K_*(P), K_*(Q)) \longrightarrow 0.$$

Since $K_*(P)$ and $K_*(Q)$ both have elements of infinite order, so does $K_*(P) \otimes K_*(Q)$. (This can be seen by looking at $K_*(P)$ and $K_*(Q)$ modulo their torsion subgroups.)

Therefore $K_*(P\otimes Q)$ has an element of infinite order. Since $P\otimes Q$ is a subquotient of $C\otimes D$, this case is done.

Now consider the general case. By proposition 4.2.9, we have

$$Prim(A\otimes B) \simeq Prim(A)\times Prim(B).$$

Let Q be any subquotient of $A\otimes B$, and let R be the corresponding locally closed subset of $Prim(A)\times Prim(B)$. Write $R=E\cap U$, where E is closed and U is open. Let $(x,y)\in R$. Then there are open sets V in $Prim(A)$ and W in $Prim(B)$ such that $x\in V$, $y\in W$, and $V\times W\subset U$. Let S be the closure of $\{x\}$ in V, which is a closed subset of V and hence a locally closed subset of $Prim(A)$. Similarly, the closure T of $\{y\}$ in W is a locally closed subset of $Prim(B)$. Furthermore, $S\times T$ is the closure of $\{(x,y)\}$ in $V\times W$, and is therefore contained in $E\cap U=R$. Let C be the subquotient of A corresponding to S, and let D be the subquotient of B corresponding to T. Then $C\otimes D$ is a subquotient of Q, and by the special case considered in the first part of the proof, $C\otimes D$ has a subquotient N such that $K_*(N)$ has an element of infinite order. Since N is a subquotient of Q, this shows that $A\otimes B$ is K-visible. Q.E.D.

7.5.7 Remark. This lemma is the real reason for introducing K-visibility. (K-visibility does not figure in any of the theorems in the previous section.) The analogous result for p-weak K-visibility is clearly false: let p, q and r be three distinct prime numbers. Then the Cuntz algebras $\mathbf{O}_{q+1}$ and $\mathbf{O}_{r+1}$ are in $\mathbf{N}$ and satisfy $K_*(\mathbf{O}_{q+1})\simeq \mathbf{Z}/q\mathbf{Z}$ and $K_*(\mathbf{O}_{r+1})\simeq \mathbf{Z}/r\mathbf{Z}$. (See [23].) Therefore $\mathbf{O}_{q+1}$ and $\mathbf{O}_{r+1}$ are p-weakly K-visible. However,

$$\mathbf{Z}/q\mathbf{Z}\otimes\mathbf{Z}/r\mathbf{Z}=Tor(\mathbf{Z}/q\mathbf{Z},\mathbf{Z}/r\mathbf{Z})=0,$$

so by the Künneth formula ([95], theorem 4.1), $K_*(\mathbf{O}_{q+1}\otimes\mathbf{O}_{r+1})=0$. Since $\mathbf{O}_{q+1}\otimes\mathbf{O}_{r+1}$ is simple (by proposition 4.2.9), it is is not p-weakly K-visible. The result analogous to lemma 7.5.6 for highly K-visible C^*-algebras seems to depend on the following question, which I have not been able to answer.

7.5.8 Question. Let A and B be (nuclear) C^*-algebras, and let $p\in M_n(A)$, and $q\in M_n(B)$ be projections such that $[p]$ is a totally positive element in $K_0(A)$ and $[q]$ is a totally positive element in $K_0(B)$. Does it follow that $[p\otimes q]$ is a totally positive element in $K_0(A\otimes B)$?

We do, however, have the following result.

7.5.9 Lemma. Let A be highly K-visible and let B be either AF or type I. Then $A\otimes B$ is highly K-visible.

Proof. We first consider the case in which B is type I. By [29], 4.5.6, B has a composition series $(I_\lambda)_{\lambda\le\kappa}$ such that $Prim(I_{\lambda+1}/I_\lambda)$ is Hausdorff for all λ. Since $(A\otimes I_\lambda)_{\lambda\le\kappa}$ is a composition series for $A\otimes B$, it suffices by lemma 7.5.3 to prove that $A\otimes B$ is highly K-visible whenever B is a type I algebra with a Hausdorff primitive ideal space. By the argument in the second part of the proof of lemma 7.5.6, it is sufficient to show that each subquotient of the form $C \otimes D$ has a subquotient N such that $K_0(N)$ has a totally positive element. Since A is highly K-visible, C has a subquotient P such that $K_0(P)$ has a totally positive element. Since D is type I with Hausdorff primitive ideal space, D has a quotient $Q \simeq K(H)$ for some Hilbert space H. Therefore $K_0(P \otimes Q) \simeq K_0(P \otimes K(H))$ has a totally positive element by lemma 7.5.4, and we conclude that $A \otimes B$ is highly K-visible.

Now let B be an AF algebra. As before, it is sufficient to prove that every subquotient of $A \otimes B$ of the form $C \otimes D$ has in turn a subquotient N such that $K_0(N)$ has a totally positive element. By hypothesis, C has a subquotient P such that $K_0(P)$ has a totally positive element $\eta = [p]$ for some projection $p \in M_k(P)$. We will show that $K_0(P \otimes D)$ has a totally positive element.

Since D is a subquotient of an AF algebra, it is again AF. (Apply [12], 1.5 and theorem 3.3, to appropriate unitizations.) Therefore $D = \overline{\bigcup_{n=1}^{\infty} D_n}$, where the algebras D_n are finite dimensional. Let $q \in D$ be any nonzero projection. Then $[q]$ is totally positive. (See [31] for facts about the K-theory of AF algebras.) We may take $q \in D_m$ for some m. Since D_n is finite dimensional, and $[q]$ is totally positive in $K_0(D_n)$ for $n \geq m$, it is clear that $[p \otimes q]$ is totally positive in $K_0(P \otimes D_n)$ for $n \geq m$. Suppose there were a projection $r \in M_l(P \otimes D)$ for some l such that $-[p \otimes q] = [r]$. Since

$$M_l(P \otimes D) = \overline{\bigcup_{k=1}^{\infty} M_l(P \otimes D_n)},$$

there is $j \geq m$ and $a \in M_l(P \otimes D_j)$ such that $a^* = a$ and $\| r - a \| < \frac{1}{4}$. An application of the continuous functional calculus, using a function f such that

$$f(t) = \begin{cases} 0 & t \in [-\frac{1}{4}, \frac{1}{4}] \\ 1 & t \in [\frac{3}{4}, \frac{5}{4}], \end{cases}$$

yields a projection $s = f(a)$ in $M_l(P \otimes D_j)$ such that $\| r - s \| < \frac{1}{2}$. Then $[r] = [s]$ in $M_l(P \otimes D_j)$ by lemma 2.5.3 (2). (We cannot directly apply the fact that $K_0(P \otimes D) = \varinjlim K_0(P \otimes D_n)$, since we do not know that P has a unit.) Now $[s] + [p \otimes q] = 0$ in $K_0(P \otimes D)$, so the isomorphism $K_0(P \otimes D) \simeq \varinjlim K_0(P \otimes D_n)$ implies that $[s] + [p \otimes q] = 0$ in $K_0(P \otimes D_n)$ for some $n \geq j$. This contradicts the fact that $[p \otimes q]$ is totally positive in $K_0(P \otimes D_n)$. So $[p \otimes q]$ is totally positive in $K_0(P \otimes D)$, and we have shown that $A \otimes B$ is highly K-visible. Q.E.D.

7.5.10 Remark. Similar arguments show that if A is p-weakly K-visible and B is either type I or AF, then $A \otimes B$ is p-weakly K-visible. We omit the proofs.

To sum up the results of this section, we have:

7.5.11 Theorem. The categories of highly K-visible, K-visible, and p-weakly K-visible C^*-algebras satisfy the following properties.

(1) They contain all type I, AF, and G-simple unital stably finite C^*-algebras (for any finite group G).

(2) They are closed under the formation of subquotients, extensions, and hereditary subalgebras.

(3) If an algebra A is in one of these categories, then so is any algebra stably isomorphic to A.

(4) If an algebra A is in one of these categories, then so is $A \otimes B$ for any type I or AF algebra B.

In addition, the category of K-visible C^*-algebras satisfies the following property:

(5) If A and B are K-visible and nuclear, and A together with all its ideals is in the category **N** of [95], then $A \otimes B$ is K-visible.

7.6. The K-Theory of the Fixed Point Algebra

In theorem 1.1.1, it was shown that if X is a compact G-space, then the conditions that the action of G on X be free, that X have discrete K-theory, and that the map $K^*(X/G) \longrightarrow K^*_G(X)$ be an isomorphism, are all equivalent. This suggests that one consider the relationship between K-freeness and the properties in the following definition.

7.6.1 Definition. Let G be a compact group, and let (G,A,α) be a G-algebra. We say that α is K-saturated if the map $K_*(A^G) \longrightarrow K^G_*(A)$ of proposition 7.1.6 is an isomorphism. We further say that α is hereditarily K-saturated if $\alpha_{(\cdot)}|_B$ is K-saturated for every nonzero G-invariant hereditary subalgebra B of A.

We immediately obtain:

7.6.2 Proposition. Let (G,A,α) be a G-algebra, where G is compact and second countable and A is separable. If α is (hereditarily) saturated then α is (hereditarily) K-saturated.

Proof. This follows from proposition 7.1.8. Q.E.D.

As a corollary, we obtain:

7.6.3 Theorem. Let (G,A,α) be a separable G-algebra. Assume that either G is finite abelian and A is highly K-visible, or that G is cyclic of prime order p and A is p-weakly K-visible. If α is K-free then α is K-saturated.

Proof. This is immediate from theorems 7.2.8, 7.4.2, 7.4.3, and the preceding proposition. Q.E.D.

The converse of theorem 7.6.3 is false. In fact, in example 9.3.5 we will produce an action of $\mathbf{Z}/2\mathbf{Z}$ on a separable C^*-algebra which is hereditarily saturated but which does not even have locally discrete K-theory. Thus the analog of the implication (2) $\Longrightarrow$ (6) in theorem 1.1.1 is false. We will, however, show in chapter 8 that there is a converse of theorem 7.6.3 for actions of cyclic groups on type I algebras. We will need the next proposition, which is the analog of proposition 7.2.3. We first prove a lemma.

7.6.4 Lemma. Let G be a compact group, and let

$$0 \longrightarrow I \longrightarrow A \longrightarrow B \longrightarrow 0$$

be an equivariant exact sequence of G-algebras. If the actions on two of the three algebras I, A, and B are K-saturated, then so is the action on the third.

Proof. Using the compactness of G, it is easy to see that the sequence

$$0 \longrightarrow I^G \longrightarrow A^G \longrightarrow B^G \longrightarrow 0$$

of fixed point algebras is exact. The naturality of the map $K_*(A^G) \longrightarrow K_*^G(A)$ of proposition 7.1.6 now shows that there is a commutative diagram with exact rows:

$$\begin{array}{ccccccccccc} \cdots \longrightarrow & K_i(I^G) & \longrightarrow & K_i(A^G) & \longrightarrow & K_i(B^G) & \longrightarrow & K_{1-i}(I^G) & \longrightarrow & K_{1-i}(A^G) & \longrightarrow \cdots \\ & \downarrow & & \downarrow & & \downarrow & & \downarrow & & \downarrow & \\ \cdots \longrightarrow & K_i^G(I) & \longrightarrow & K_i^G(A) & \longrightarrow & K_i^G(B) & \longrightarrow & K_{1-i}^G(I) & \longrightarrow & K_{1-i}^G(A) & \longrightarrow \cdots \end{array}$$

By assumption, two out of every three vertical arrows in this diagram are isomorphisms. By the Five Lemma, the third is as well. Q.E.D.

7.6.5 Proposition. Let G be a compact group and let

$$0 \longrightarrow I \longrightarrow A \xrightarrow{\pi} B \longrightarrow 0$$

be an equivariant exact sequence of G-algebras. Then the action on A is hereditarily K-saturated if and only if the actions on I and B are.

Proof. Regard I as a G-invariant ideal in A. Let the actions on I and B be hereditarily K-saturated, and let C be a G-invariant hereditary subalgebra of A. Then, exactly as in the proof of proposition 7.2.3, there is an exact sequence

$$0 \longrightarrow C \cap I \longrightarrow C \longrightarrow \pi[C] \longrightarrow 0,$$

in which $C \cap I$ is a G-invariant hereditary subalgebra of I and $\pi[C]$ is a G-invariant hereditary subalgebra of B. By hypothesis, the actions on $C \cap I$ and $\pi[C]$ are K-saturated. Therefore so is the action on C, by the previous lemma. Thus, the action on A is hereditarily K-saturated.

Conversely, let the action on A be hereditarily K-saturated. Since every G-invariant hereditary subalgebra of I is also a G-invariant hereditary subalgebra of A, it is clear that the action on I is hereditarily K-saturated. Now let C be any G-invariant hereditary subalgebra of B. Then $\pi^{-1}[C]$ is a G-invariant hereditary subalgebra of A, and there is an equivariant exact sequence

$$0 \longrightarrow I \longrightarrow \pi^{-1}[C] \longrightarrow C \longrightarrow 0.$$

By the previous lemma, the action on C is K-saturated, and we have thus shown that the action on B is hereditarily K-saturated. Q.E.D.

We will not develop the further properties of hereditary K-saturation here, although we presume that analogs of many of the properties of K-freeness proved in chapter 4 hold.

One would hope that K-freeness of an action would imply K-saturation without the additional hypotheses of theorem 7.6.3, since we are dealing with the question of whether one K-theoretic property of freeness of group actions implies another. Unfortunately, theorem 7.6.3 fails without the K-visibility assumption, as the following example shows.

7.6.6 Example. Let $G = \mathbf{Z}/2\mathbf{Z}$, and let $A = \mathbf{O}_3$ be the Cuntz algebra (yet again!). Let $\alpha: G \longrightarrow Aut(A)$ be the trivial action. In [23], it is shown that $K_*(A) \simeq \mathbf{Z}/2\mathbf{Z}$, so that

$$K_*^G(A) \simeq R(G) \otimes \mathbf{Z}/2\mathbf{Z} \simeq \mathbf{Z}/2\mathbf{Z} \times \mathbf{Z}/2\mathbf{Z}.$$

Since $A^G = A$, we see that the map $K_*(A^G) \longrightarrow K_*^G(A)$ is not an isomorphism. However, α has locally discrete K-theory by lemma 4.1.5, and since G and A are both simple, this implies that α is totally K-free.

Nevertheless, something can be said, at least for the case $G = \mathbf{Z}/2\mathbf{Z}$.

7.6.7 Theorem. Let $(\mathbf{Z}/2\mathbf{Z}, A, \alpha)$ be a separable $\mathbf{Z}/2\mathbf{Z}$-algebra, and suppose α is K-free. Then the map $K_*(B^{\mathbf{Z}/2\mathbf{Z}}) \longrightarrow K_*^{\mathbf{Z}/2\mathbf{Z}}(B)$ is an isomorphism up to 2-torsion for every $\mathbf{Z}/2\mathbf{Z}$-invariant hereditary subalgebra B of A. That is, α is "hereditarily K-saturated up to 2-torsion".

Proof. Let $G = \mathbf{Z}/2\mathbf{Z}$. Since the restriction of a K-free action on a separable C^*-algebra to an invariant hereditary subalgebra is also K-free by proposition 4.2.13, we need only show that $K_*(A^G) \longrightarrow K_*^G(A)$ is an isomorphism up to 2-torsion. This is the same as showing that

$$K_*(A^G) \otimes_{\mathbf{Z}} \mathbf{Z}[1/2] \longrightarrow K_*^G(A) \otimes_{\mathbf{Z}} \mathbf{Z}[1/2]$$

is an isomorphism. Equivalently, we must show that

$$S^{-1} K_*(A^G) \longrightarrow S^{-1} K_*^G(A)$$

is an isomorphism, where S is the multiplicative system in $\mathbf{Z}$ consisting of all powers of 2.

Let τ be the nontrivial element of $\widehat{G}$, so that as a Banach space we have $A = A^G \oplus A_\tau$, where

$$A_\tau = \{a \in A : \alpha_g(a) = \tau(g)a \text{ for all } g \in G\}$$

as usual. Let $J = \overline{A_\tau A_\tau^*}$, which is an ideal in A^G, and let $I = J + A_\tau$, which is an ideal in A. (We have

$$A^G(J + A_\tau) \subset J + A_\tau \quad \text{and} \quad A_\tau(J + A_\tau) \subset A_\tau + A_\tau^2 \subset A_\tau + J,$$

and similarly on the other side.) We clearly have $A/I \simeq A^G/J$. Furthermore, the action of G on A/I is trivial. Since α is K-free, this action also has locally discrete K-theory, and we conclude from lemma 4.1.5 that $K_*(A^G/J)$ is a 2-torsion group. In particular, $S^{-1}K_*(A^G/J) = 0$.

By [78], there is an exact sequence of abelian groups

$$\begin{array}{ccccc} K_0(A^G) & \xrightarrow{j_0} & K_0(C^*(G,A)) & \longrightarrow & K_0(A^G/J) \\ \uparrow & & & & \downarrow \\ K_1(A^G/J) & \longleftarrow & K_1(C^*(G,A)) & \xleftarrow{j_1} & K_1(A^G), \end{array}$$

where j_0 and j_1 correspond to the maps $K_0(A^G) \longrightarrow K_0^G(A)$ and $K_1(A^G) \longrightarrow K_1^G(A)$ under the isomorphism $K_*(C^*(G,A)) \simeq K_*^G(A)$. If we apply the localization functor S^{-1} to this sequence, it remains exact. Since $S^{-1}K_*(A^G/J) = 0$, it follows that $S^{-1}j_0$ and $S^{-1}j_1$ are isomorphisms. Q.E.D.

7.6.8 Conjecture. A similar result should hold for K-free actions of $\mathbf{Z}/p\mathbf{Z}$ for any prime p. That is, the map $K_*(B^{\mathbf{Z}/p\mathbf{Z}}) \longrightarrow K_*^{\mathbf{Z}/p\mathbf{Z}}(B)$ should be an isomorphism up to p-torsion for every invariant hereditary subalgebra B.

The method of proof of theorem 7.6.7 also yields the following result.

7.6.9 Theorem. Let $(\mathbf{Z}/2\mathbf{Z}, A, \alpha)$ be a separable $\mathbf{Z}/2\mathbf{Z}$-algebra. Assume that A satisfies the following weak form of K-visibility: any subquotient of A has in turn a subquotient Q such that $K_*(Q) \neq 0$. If α is hereditarily K-saturated, then α is hereditarily saturated.

Proof. Let $G = \mathbf{Z}/2\mathbf{Z}$. Since the hypotheses hold for any G-invariant hereditary subalgebra of A, it is sufficient to prove that α is saturated. (The proofs of lemmas 7.5.2 and 7.5.4, and therefore also of corollary 7.5.5, clearly apply to the form of K-visibility considered here just as well as to the stronger forms considered in section 5. Thus, every nonzero hereditary subalgebra of A satisfies the K-visibility hypothesis of the theorem.) Now let I be as in the proof of theorem 7.6.7. Then theorem 7.1.15 implies that α is saturated if and only if $I = A$.

Suppose $I \neq A$. As in the proof of theorem 7.6.7, I is a G-invariant ideal of A such that G acts trivially on A/I. By assumption, there is a subquotient Q of A/I

such that $K_*(Q) \neq 0$. Since G acts trivially on Q, we have $K_*^G(Q) \simeq K_*(Q) \otimes R(G)$ by remark 2.8.5. Since $G = \mathbf{Z}/2\mathbf{Z}$, we have $R(G) = \mathbf{Z} + \mathbf{Z}\tau$ where τ is the nontrivial element of $\widehat{G}$, and therefore, as an abelian group, $K_*^G(Q) \simeq K_*(Q) \oplus K_*(Q)$. Furthermore, $Q^G = Q$, and the natural map from $K_*(Q^G)$ to $K_*^G(Q)$ easily seen to be given by $\eta \longrightarrow \eta \otimes 1$ when $K_*^G(Q)$ is identified with $K_*(Q) \otimes R(G)$, and therefore by $\eta \longrightarrow (\eta, 0)$ when $K_*^G(Q)$ is identified with $K_*(Q) \oplus K_*(Q)$. This map is not an isomorphism, because $K_*(Q) \neq 0$. From proposition 7.6.5 we conclude that α is not hereditarily K-saturated. (Proposition 7.6.5 implies that the induced action on any G-invariant subquotient of an algebra with a hereditarily K-saturated action of G is again hereditarily K-saturated.) Thus, we have a contradiction, and we must conclude that α is in fact saturated, as desired. Q.E.D.

Note that the proof does *not* show that if α is K-saturated then α is saturated. We also presume that this theorem holds in greater generality:

7.6.10 Conjecture. In theorem 7.6.9, the group $\mathbf{Z}/2\mathbf{Z}$ can be replaced by an arbitrary finite group.

We will show in the next chapter that a hereditarily K-saturated action of a finite cyclic group or of the circle group S^1 on a type I algebra is necessarily hereditarily saturated.

7.6.11 Remark. The proof of theorem 7.6.9 can also be adapted to give a shorter proof of theorem 7.4.2 for the case in which $G = \mathbf{Z}/2\mathbf{Z}$ and A is separable. We want to show that if $G = \mathbf{Z}/2\mathbf{Z}$ and $\alpha : G \longrightarrow Aut(A)$ is a K-free action on a separable 2-weakly K-visible algebra A, then α is hereditarily saturated. Again, the hypotheses hold for any G-invariant hereditary subalgebra of A as well, and it is therefore sufficient to show that α is saturated. Just as in the proof of theorem 7.6.9, it is enough to show that the ideal I considered there is all of A. If not, then A/I has a subquotient Q such that $K_*(Q)$ contains an element whose order is not a power of 2. Since G acts trivially on Q, lemma 4.1.5 implies that this action does not have locally discrete K-theory, contradicting the K-freeness of α. This completes the proof.

Chapter 8

Type I Algebras

In this chapter, we apply the theory constructed in earlier chapters to the special case of type I algebras. We obtain much stronger results than for the general case. The most striking result is that an action of a compact Lie group on a separable type I algebra is totally K-free if and only if the corresponding action of G on the primitive ideal space is free. We also find that a K-free action of a finite cyclic group, or of the circle group S^1, on a separable type I algebra, is actually totally K-free. We are furthermore able to show that for actions of finite abelian groups on separable type I algebras, K-freeness is in fact equivalent to hereditary saturation. Of course, not everything that works for spaces works for type I algebras: recall that the algebra in example 4.2.3, which led to the introduction of total K-freeness, was M_2.

The first section of this chapter is devoted to deriving consequences of free action on the primitive ideal space. The primitive ideal spaces of type I algebras have special properties which allow us to drop the restriction, used in the proof of theorem 4.3.8, that the group be finite. In section 2, we show that several conditions imply that the action on the primitive ideal space is free, in the special case of actions of finite cyclic groups and S^1 on type I algebras. As a corollary, we show that if the action of a compact Lie group is totally K-free, then it is free on the primitive ideal space. Finally, the last section examines the relationship between K-freeness and hereditary saturation when these conditions do not imply that the action on the primitive ideal space is free. We then summarize the results of this chapter in theorem 8.3.7.

8.1. Free Action on the Primitive Ideal Space

In his book on C^*-algebras, Dixmer proves ([29], 4.4.5) that the primitive ideal space of every type I C^*-algebra A (called postliminal there) contains a dense Hausdorff open subset. One can then proceed to construct a composition series $(I_\lambda)_{\lambda \leq \kappa}$ for A such that all of the composition factors $I_{\lambda+1}/I_\lambda$ have Hausdorff primitive ideal spaces. If A is now a G-algebra for a finite group G, then the fact that $Prim(A)$ is a Baire space ([29], 3.4.13) can be used to construct a dense Hausdorff open G-invariant subset U as follows: choose any dense Hausdorff open subset V, and take U to be the intersection of the finitely many translates of V under G. This was in fact done in the proof of proposition 6.2.11. In this chapter, we consider arbitrary compact Lie groups, for which this proof breaks down. Nevertheless, the result still holds.

8.1.1 Lemma. Let G be a compact metrizable group, and let X be a G-space which is a Baire space and contains a dense Hausdorff open subset. Then X contains a G-invariant dense Hausdorff open subset.

Proof. We first construct a nonempty G-invariant Hausdorff open subset. Let U be a dense Hausdorff open subset of X, and let S be a countable dense subset of G. (Note that a compact metrizable space is separable.) Since X is a Baire space, we have $\bigcap_{g\in S} g^{-1}U \neq \emptyset$. Let x_0 be an element of this intersection. Then for $g \in S$, we have $gx_0 \in U$. Consequently $U \cap Gx_0$ is dense in Gx_0.

Define $\varphi : G \times U \longrightarrow X$ by $\varphi(g,x) = gx$. Then φ is continuous, and equivariant if we give $G \times U$ the G-action $g \cdot (h,x) = (gh,x)$ for $g,h \in G$ and $x \in U$. The set $\varphi^{-1}[U]$ is an open subset of $G \times U$ which contains (e,x_0), where e is the identity of G. The translates $g \cdot \varphi^{-1}[U]$ obviously cover the compact set $G\times\{x_0\}$, so that there are $g_1,\ldots,g_n \in G$ such that

$$G\times\{x_0\} \subset \bigcup_{k=1}^{n} g_k \cdot \varphi^{-1}[U].$$

By a standard argument involving the compactness of G, there is in fact an open set $W \subset U$ such that $x_0 \in W$ and

$$G\times W \subset \bigcup_{k=1}^{n} g_k \cdot \varphi^{-1}[U].$$

Let

$$V = W \cap \left(\bigcap_{k=1}^{n} g_k \; U \right).$$

Then V is a Hausdorff open subset of X, since V is contained in U. We claim that $V \neq \emptyset$. We will in fact prove that $V \cap Gx_0 \neq \emptyset$. Now

$$V \cap Gx_0 = (W \cap Gx_0) \cap \bigcap_{k=1}^{n} (g_k \, U \cap Gx_0).$$

By the first paragraph of the proof, $U \cap Gx_0$ is dense in Gx_0. Therefore $g_k\, U \cap Gx_0 = g_k\,(U \cap Gx_0)$ is also dense in Gx_0. Since G is compact metrizable, it is a Baire space, hence its continuous image Gx_0 is also a Baire space. The sets $g_k\, U \cap Gx_0$ are clearly open in Gx_0, so we find that $\bigcap_{k=1}^{n} (g_k\, U \cap Gx_0)$ is a dense open subset of Gx_0. Therefore its intersection with the nonempty open subset $W \cap Gx_0$ is nonempty. This proves that $V \cap Gx_0 \neq \emptyset$, as claimed.

We now claim that GV is a nonempty G-invariant Hausdorff open subset of X. The only part of the claim that is not obvious is that GV is Hausdorff. We must therefore prove that two points $x,y \in GV$ with $x \neq y$ can be separated. We first consider the special case of points $x \in V$ and $y \in GV$, with $x \neq y$. Then $y = gz$ for some $z \in V$ and $g \in G$. We then have $(g,z) \in G\times W$, so that there is some k such that

$(g,z) \in g_k \cdot \varphi^{-1}[U]$. Consequently $y \in g_k U$. Since $V \subset g_k U$ by the definition of V, we have $x, y \in g_k U$. Since $g_k U$ is open and Hausdorff, x and y have disjoint neighborhoods. Now let $x, y \in GV$ be arbitrary. Then there is $h \in G$ such that $h^{-1}x \in V$. Also, $h^{-1}y \in GV$ because GV is G-invariant. By the special case considered above, there are disjoint neighborhoods Z_1 of $h^{-1}x$ and Z_2 of $h^{-1}y$ respectively. Therefore hZ_1 and hZ_2 are disjoint neighborhoods of x and y. Since x and y are arbitrary points in GV, we have shown that GV is Hausdorff, as claimed.

We now show that there is a G-invariant *dense* Hausdorff open subset of X. Let $\{U_\lambda\}_{\lambda \in I}$ be a maximal family of disjoint G-invariant Hausdorff open subsets, which exists by Zorn's lemma, and let U be the union of the U_λ for $\lambda \in I$. Then U is a G-invariant Hausdorff open subset. Suppose that U is not dense. Then $X - \overline{U}$ is a G-invariant open subset of X. The set $X - \overline{U}$ is easily seen to be a Baire space, and to contain a dense Hausdorff open subset. (Take its intersection with a dense Hausdorff open subset of X.) The first part of the proof now shows that $X - \overline{U}$ contains a G-invariant Hausdorff open subset, contradicting the maximality of the family $\{U_\lambda\}_{\lambda \in I}$. So U is dense. Q.E.D.

8.1.2 Corollary. Let G be a compact metrizable group, and let (G,A,α) be a type I G-algebra. Then A has a G-invariant composition series $(I_\lambda)_{\lambda \le \kappa}$ such that each factor $I_{\lambda+1}/I_\lambda$ has a Hausdorff primitive ideal space.

Proof. Set $I_0 = 0$, and construct I_λ by transfinite induction. Suppose I_μ has been constructed for all $\mu < \lambda$. If λ is a limit ordinal, we set $I_\lambda = \overline{\bigcup_{\mu<\lambda} I_\mu}$, which is clearly a G-invariant ideal in A. Otherwise, λ is a successor ordinal, that is, $\lambda = \mu + 1$ for some ordinal μ. Let $X = \mathit{Prim}(A/I_\mu)$. Since A/I_μ is type I, X is a Baire space by [29], 3.4.13, and has a dense Hausdorff open subset by [29], 4.4.5. By the lemma, X has a G-invariant dense Hausdorff open subset U. Let J be the ideal in A/I_μ corresponding to U. Then J is G-invariant. Let I_λ be the inverse image of J in A, which is also G-invariant and satisfies $I_\lambda / I_\mu \simeq J$, which has Hausdorff primitive ideal space. This complete the inductive step of the construction.

The composition series must terminate somewhere, since each factor $I_{\lambda+1}/I_\lambda$ corresponds to a nonempty subset of $\mathit{Prim}(A)$. Q.E.D.

To apply this corollary, we need the following proposition.

8.1.3 Proposition. Let G be a group, and let (G,A,α) be a G-algebra such that $\mathit{Prim}(A)$ is Hausdorff. Then (G,A,α) is a $(G,C_0(\mathit{Prim}(A)))$-algebra, as in definition 4.5.1, in a canonical manner.

Proof. Let $X = \mathit{Prim}(A)$. Then X is a locally compact Hausdorff G-space. By the Dauns-Hofmann theorem ([79], 4.4.8), the center of the multiplier algebra $M(A)$ can be canonically identified with the space $C_b(X)$ of bounded continuous complex-valued functions on X. In particular, if $f \in C_0(X)$ and $a \in A$, then we have a product

$fa \in A$. It is clear that all of the conditions defining a $(G,C_0(X))$-algebra are satisfied, except possibly for the unit conditions. Thus, we must show that if (f_λ) is an approximate identity for $C_0(X)$ and $a \in A$, then $f_\lambda a \longrightarrow a$ is norm.

For each point $x \in X$, choose a representation π_x of A whose kernel is x. Then the product fa for $f \in C_0(X)$ and $a \in A$ is characterized by the relation $\pi_x(fa) = f(x)\pi_x(a)$ for all $x \in X$. If $a \in A$, then the function $x \longrightarrow \| \pi_x(a) \|$ is continuous by [29], 3.3.9. Furthermore, by [29], 3.3.7, the subset $\{x \in X : \| \pi_x(a) \| \geq \varepsilon\}$ is compact for every $\varepsilon > 0$. Therefore $x \longrightarrow \| \pi_x(a) \|$ is a continuous function on X vanishing at infinity. Consequently, if (f_λ) is an approximate identity for $C_0(X)$, we have

$$\| a - f_\lambda a \| = \sup_{x \in X} \| \pi_x(a) - \pi_x(f_\lambda a) \| = \sup_{x \in X} \Big(| 1 - f_\lambda(x) | \, \| \pi_x(x)(a) \| \Big),$$

which converges to zero. So $f_\lambda a \longrightarrow a$, and A is a $(G,C_0(X))$-algebra. Q.E.D.

We are now in a position to prove the main results of this section.

8.1.4 Theorem. Let G be a compact Lie group, and let (G,A,α) be a separable type I G-algebra. If G acts freely on $Prim(A)$, then α is totally K-free.

Proof. By corollary 8.1.2, the algebra A has a G-invariant composition series $(I_\lambda)_{\lambda \leq \kappa}$ such that each $I_{\lambda+1}/I_\lambda$ has a Hausdorff primitive ideal space. By proposition 4.2.15, it is sufficient to prove that the action on each $I_{\lambda+1}/I_\lambda$ is totally K-free. Let $X = Prim(I_{\lambda+1}/I_\lambda)$. Then X is a subset of $Prim(A)$, and therefore G acts freely on X. By the previous proposition, $I_{\lambda+1}/I_\lambda$ is a $(G,C_0(X))$-algebra, where $C_0(X)$ is separable. By theorem 4.6.1, the action of G on $I_{\lambda+1}/I_\lambda$ is totally K-free. Q.E.D.

8.1.5 Remark. As pointed out in remark 4.6.3, separability is not really necessary for the results on K-freeness of actions on $C(X)$- and $C_0(X)$-algebras. Therefore we don't really need to assume separability here either.

8.1.6 Theorem. Let G be a compact metrizable group, and let (G,A,α) be a type I G-algebra. If G acts freely on $Prim(A)$, then α is hereditarily saturated.

Proof. As in the proof of theorem 8.1.4, we use corollary 8.1.2 to reduce to the case in which $Prim(A)$ is Hausdorff; we replace the appeal to proposition 4.2.15 by an appeal to proposition 7.2.4. The result now follows from proposition 8.1.3, using theorem 7.2.6 instead of theorem 4.6.1. Q.E.D.

Theorems 8.1.4 and 8.1.6 are not surprising. Of more interest is the converse of theorem 8.1.4, and a partial converse of theorem 8.1.6, which will be proved in the next section.

The situation for KK-freeness is not so nice, as is shown by example 4.4.12, in which we exhibited a free action of the circle group on a locally compact space X such that the corresponding action on $C_0(X)$ is not KK-free. Thus, it is not surprising that the best we can do is the following proposition. The separability hypothesis appearing in it is necessary for the proof, unlike in theorem 8.1.4.

8.1.7 Proposition. Let G be a compact Lie group, and let (G,A,α) be a separable type I G-algebra. Assume that A has a finite G-invariant composition series $(I_k)_{k\leq n}$ such that $Prim(I_{k+1}/I_k)$ is compact for all $k<n$. If G acts freely on $Prim(A)$, then α is KK-free.

Proof. Let $X_k = Prim(I_{k+1}/I_k)$. Then by proposition 8.1.3, the subquotient I_{k+1}/I_k is a $(G,C(X_k))$-algebra. Since X_k is a subset of $Prim(A)$, we see that G acts freely on X_k, and therefore, by theorem 4.5.8, the action of G on I_{k+1}/I_k is KK-free. (This is where separability is needed.) Now the action on $I_0=0$ is KK-free, and if the action on I_k is KK-free, then an application of proposition 4.4.8 to the exact sequence

$$0 \longrightarrow I_k \longrightarrow I_{k+1} \longrightarrow I_{k+1}/I_k \longrightarrow 0$$

shows that the action on I_{k+1} is KK-free. So, by induction, the action on $I_n = A$ is KK-free. Q.E.D.

8.2. Actions of Cyclic Groups

The first step toward the converse of theorem 8.1.4 is the following lemma.

8.2.1 Lemma. Let G be a finite cyclic group, and let (G,A,α) be a type I G-algebra. If G does not act freely on $Prim(A)$, then there is a G-invariant subquotient of A which is equivariantly isomorphic to $C(G\times_{G_0}K(H))$ for a subgroup G_0 of G with more than one element, some Hilbert space H, and some inner action of G_0 on $K(H)$.

Proof. Since the action of G on $Prim(A)$ is not free, there is $x\in Prim(A)$ such that the stabilizer G_x is not $\{e\}$. Let $(I_\lambda)_{\lambda\leq\kappa}$ be a G-invariant composition series for A such that $Prim(I_{\lambda+1}/I_\lambda)$ is Hausdorff for all λ, which exists by corollary 8.1.2. Since

$$Prim(A) = \bigcup_{\lambda+1\leq\kappa} (Prim(I_{\lambda+1}) - Prim(I_\lambda)),$$

there is λ such that $x\in Prim(I_{\lambda+1}) - Prim(I_\lambda)$. Now $Prim(I_{\lambda+1}/I_\lambda)$ is a Hausdorff space, and homeomorphic to $Prim(I_{\lambda+1}) - Prim(I_\lambda)$. Therefore the orbit Gx can be considered as a discrete, and therefore closed, subset of $Prim(I_{\lambda+1}/I_\lambda)$. Let J be the ideal in $I_{\lambda+1}/I_\lambda$ corresponding to the complement of Gx in $Prim(I_{\lambda+1}/I_\lambda)$. Then $B=(I_{\lambda+1}/I_\lambda)/J$ is a G-invariant subquotient of A whose primitive ideal space is Gx. Let I be the G_x-invariant ideal in B corresponding to $\{x\}$. Since $Gx=\{gx : g\in R\}$, where R is a system of left coset representatives for G_x in G, we have $B=\bigoplus_{g\in R}\alpha_g[I]$. Proposition 2.9.6 now implies that $B\simeq C(G\times_{G_x}I)$ as G-algebras. Since I is simple and

A is type I, we have $I \simeq K(H)$, the space of compact operators on a Hilbert space H. Since G_x is finite cyclic, its action on $K(H)$ is inner by lemma 4.2.12. Q.E.D.

The fact that the action on $K(H)$ is inner in this lemma is crucial, because it will enable us to apply the stability of K-theory, theorem 2.8.3 (4). There is one other compact Lie group such that any action of any closed subgroup on $K(H)$ is necessarily inner, namely the circle group S^1. Therefore we have:

8.2.2 Lemma. Let (S^1,A,α) be a type I S^1-algebra. If the action of S^1 on $Prim(A)$ is not free, then there is an S^1-invariant subquotient of A which is equivariantly isomorphic to $C(S^1 \times_G K(H))$ for some closed subgroup G of S^1 with more than one element, some Hilbert space H, and some inner action of G on $K(H)$.

Proof. Choose $x \in Prim(A)$ such that the stabilizer G of x is not trivial. Then G is a closed subgroup of S^1, and must therefore be either S^1 or a finite cyclic subgroup. Furthermore, $S^1 x$ is a compact subset of $Prim(A)$. The argument used in the proof of lemma 8.2.1 now produces a subquotient B of A such that $Prim(B) \simeq S^1 x$ and is Hausdorff. Let β be the induced action of S^1 on B. We want to prove that B has the correct form, and there are two cases.

Case 1. $G = S^1$. Then $B \simeq K(H)$ for some Hilbert space H. Since all automorphisms of $K(H)$ are of the form $a \longrightarrow uau^*$ for some unitary $u \in L(H)$, and since the automorphism $a \longrightarrow uau^*$ is trivial if u is a multiple of the identity, it follows that the automorphism group of $K(H)$ is the projective unitary group $PU(H)$, which is the quotient of the unitary group of H by the multiples of the identity. It is easily checked that the topology of pointwise convergence in norm on $K(H)$ corresponds to the topology on $PU(H)$ obtained from the strong operator topology on the unitary group of H. Thus, β defines a strong operator continuous projective unitary representation $z \longrightarrow w(z)$ of S^1 on H. There is a corresponding projective unitary representation of $\mathbf{R}$ defined by $t \longrightarrow w(e^{it})$. By [69], theorem 5 and the remarks following its proof, there is a unitary representation $t \longrightarrow u(t)$ of $\mathbf{R}$ on H such that the image of $u(t)$ in $PU(H)$ is $w(e^{it})$. In particular, $u(2\pi)$ must be a scalar multiple of the identity, that is, $u(2\pi) = \lambda \cdot 1$ with $|\lambda| = 1$. Choose $r \in \mathbf{R}$ such that $e^{2\pi i r} = \lambda$, and set $v(t) = e^{-irt} u(t)$. Then $v(2\pi) = 1$, so we actually have a unitary representation of S^1, and $\beta_{exp(it)}(a) = v(t) a v(t)^*$. Therefore β is an inner action, as desired.

Case 2. $G \neq S^1$. Then G is a finite cyclic subgroup, generated by the element $g_0 = e^{it_0}$, where t_0 is the smallest positive number such that $e^{it_0} \in G$. Let I be the ideal in B corresponding to the complement of the point x in $Prim(B) = S^1 x$. Then $B/I \simeq K(H)$ for some Hilbert space H. Let $\pi : B \longrightarrow K(H)$ be the quotient map. Define $\varphi : B \longrightarrow C(S^1, K(H))$ by $\varphi(a)(g) = \pi(\beta_g^{-1}(a))$. Since I is a G-invariant ideal, $K(H)$ carries an induced action γ of G. It is then easily checked that φ is in fact an equivariant homomorphism from B to $C(S^1 \times_G K(H))$. The map φ is clearly injective, because if $\varphi(a) = 0$ then $\sigma(a) = 0$ for every irreducible representation σ of B.

We want to show that φ is surjective. We identify $C(S^1 \times_G K(H))$ with the set of continuous sections of a continuous field ([29], 10.3.1) of C^*-algebras on the space S^1/G as follows: the fiber over every point is $K(H)$, and the continuous sections are the continuous functions f from $[0, t_0)$ into $K(H)$ such that $\lim_{t \to t_0} f(t) = \gamma_{g_0}^{-1}(f(0))$, where $[0, t_0)$ is identified with S^1/G in the obvious way. Since γ_{g_0} is an automorphism, it is clear that we have in fact defined a continuous field. Since S^1/G is canonically homeomorphic to $Prim(B)$, we can apply [29], 4.2.5, to conclude that if $s_1, s_2 \in [0, t_0)$ with $s_1 \neq s_2$, and $a_1, a_2 \in K(H)$, there is $c \in B$ such that the image in $K(H)$ of c under the representation of B corresponding to s_i is a_i. This means that the section over $[0, t_0)$ corresponding to $\varphi(c)$ takes the value a_i at s_i for $i = 1, 2$. By [29], 11.5.3, it follows that all continuous sections over $[0, t_0)$ correspond to sections of the form $\varphi(c)$ for $c \in B$. Therefore φ is surjective.

We have now shown φ is an isomorphism. Since γ is an action of a finite cyclic group on $K(H)$, it is inner by lemma 4.2.12, and thus B has the desired form. Q.E.D.

Using these two lemmas, we now obtain the following:

8.2.3 Theorem. Let G be either a finite cyclic group or S^1. Let (G, A, α) be a type I G-algebra. Then any one of the following conditions implies that the action of G on $Prim(A)$ is free.

(1) The action α is K-free.

(2) The action α is hereditarily saturated.

(3) The action α is hereditarily K-saturated.

Proof. It is clear from proposition 4.2.2 that if α is K-free, then so is the induced action on any G-invariant subquotient. Similarly, if α is hereditarily (K-)saturated, then so is the induced action on any G-invariant subquotient, by propositions 7.2.3 and 7.6.5.

To prove the theorem, it is now sufficient to assume that the action of G on $Prim(A)$ is not free, and to produce a G-invariant subquotient B of A such that none of the properties (1), (2), or (3) holds. We will take B to be the subquotient $C(G \times_{G_0} K(H))$ provided by lemma 8.2.1 if G is finite cyclic, and by lemma 8.2.2 if $G = S^1$. Here G_0 is a closed subgroup of G with more than one element, and the action γ of G_0 on $K(H)$ is inner, that is, there is a unitary representation u of G_0 on H such that $\gamma_g(a) = u_g a u_g^*$ for $g \in G_0$ and $a \in K(H)$.

To show that condition (1) fails, we observe that

$$K_*^G(C(G \times_{G_0} K(H))) \simeq K_*^{G_0}(K(H)) \simeq R(G_0),$$

where the first isomorphism follows from proposition 2.9.4 and the second one follows from theorem 2.8.3 (4) and the fact that γ is inner. Taking a prime ideal $P \subset R(G)$ whose support is G_0, we see that the action on B does not even have locally discrete K-theory.

In order to show that (2) and (3) fail, we construct an appropriate hereditary subalgebra of B. Since G_0 is abelian, there is a one dimensional subspace H_0 of H which is invariant for the representation u. Let p be the projection onto H_0. Then p is a G_0-invariant projection in $K(H)$. Therefore the function $q : G \longrightarrow K(H)$, defined by $q(g) = p$ for $g \in G$, is a G-invariant projection in $C(G \times_{G_0} K(H))$. It is readily verified that the hereditary subalgebra $C = qC(G \times_{G_0} K(H))q$ is isomorphic to $C(G \times_{G_0} \mathbf{C})$, which is just $C(G/G_0)$. Since G does not act freely on G/G_0, we find that $K_*(C^G) \longrightarrow K_*^G(C)$ is not an isomorphism by theorem 1.1.1, and the action on C is not saturated by corollary 7.1.12. Thus (2) and (3) also fail for this subquotient.

Q.E.D.

8.2.4 Corollary. Let G be either finite cyclic or equal to S^1. Let α be a K-free action of G on a separable type I algebra. Then α is totally K-free.

Proof. By the previous theorem, G acts freely on $Prim(A)$. By theorem 8.1.4, α is therefore totally K-free. Q.E.D.

Another corollary of theorem 8.2.3 is the following theorem, which is the converse of theorem 8.1.4.

8.2.5 Theorem. Let G be a compact Lie group, and let (G, A, α) be a type I G-algebra. If α is totally K-free, then G acts freely on $Prim(A)$.

Proof. Suppose G does not act freely on $Prim(A)$. Then there is a point $x \in Prim(A)$ such that $G_x \neq \{e\}$. The subgroup G_x must be closed. Since G is a compact Lie group, G_x contains an element $g \neq e$ of finite order. Let H be the cyclic subgroup generated by g. Then $\alpha|_H$ is K-free, but H does not act freely on $Prim(A)$, since $hx = x$ for $h \in H$. This contradicts theorem 8.2.3. Q.E.D.

8.3. The Relationship Between K-Freeness and Hereditary Saturation

The restrictions on the group G considered in theorem 8.2.3 are necessary, as the following example shows.

8.3.1 Example. Let $G = \mathbf{Z}/2\mathbf{Z} \times \mathbf{Z}/2\mathbf{Z}$, let $A = M_2$, and let α be the action of example 4.2.3. It was shown there that α is K-free. It was further shown that $\Gamma(\alpha) = \widehat{G}$. Since A is simple, it follows from theorem 7.2.8 that α is hereditarily saturated, and also that $K_*(B^G) \longrightarrow K_*^G(B)$ is an isomorphism for every G-invariant hereditary subalgebra of A. Thus, all of the conditions (1), (2), and (3) of theorem 8.2.3 hold, but G does not act freely on $Prim(A)$.

Nevertheless, something can be said about more general groups. We need a lemma.

8.3.2 Lemma. Let A be a separable type I algebra, and let φ be an automorphism of A which is trivial on $Prim(A)$. Then for all $\eta \in K_*(A)$ there is n such that $(id - \varphi_*)^n(\eta) = 0$.

Proof. If A is not unital, then the hypotheses also hold for the extension of φ to A^+. Furthermore, if the conclusion holds for A^+, then it holds for A. We may therefore assume that A is unital. Define an action $\alpha : \mathbf{Z} \longrightarrow Aut(A)$ by $\alpha_n = \varphi^n$ for $n \in \mathbf{Z}$, and let $B = C^*(\mathbf{Z}, A, \alpha)$. By [58], lemma 3.2, we have

$$\widetilde{\Gamma}(\widehat{\alpha}) = \{g \in \mathbf{Z} : \alpha_g[I] \subset I \text{ for all ideals } I \text{ of } A\},$$

that is, $\widetilde{\Gamma}(\widehat{\alpha}) = \mathbf{Z}$. Therefore $\widehat{\alpha}$ is hereditarily saturated by theorem 7.2.7.

We want to apply theorem 8.2.3 (2), so we need to know that B is type I. We will prove this by showing that the image of every irreducible representation π of B contains the compact operators. By [79], 7.6.6, the representation π is the integrated form of a covariant representation (u, σ) of $(\mathbf{Z}, A, \alpha)$ on a Hilbert space H. Since A is unital and $\mathbf{Z}$ is discrete, we can identify A with a subalgebra of B and $\mathbf{Z}$ with a subgroup of the unitary group of B, and obtain the representations u and σ by restriction. In particular, the image of π contains the image of σ. We are going to show that σ is irreducible.

By [61], the automorphism φ is universally weakly inner, that is, the induced automorphism φ'' on the second dual A'' of A has the form $\varphi''(a) = waw^*$ for some unitary $w \in A''$. The representation σ extends to a normal representation σ'' of A'' on H by [79], 3.7.7, and we have, for $a \in A$,

$$u_1 \sigma(a) u_1^* = \sigma(\alpha_1(a)) = \sigma(\varphi(a)) = \sigma''(w)\sigma(a)\sigma''(w)^*.$$

Therefore, with $v = \sigma''(w)$, we have $v^* u_1 \in \sigma[A]'$, the commutant of $\sigma[A]$. Since the equation $u_1\sigma(a)u_1^* = v\,\sigma(a)v^*$ holds for all $a \in A$, and since $\sigma[A''] = \sigma[A]''$, where $\sigma[A]''$ is the double commutant of $\sigma[A]$, we have $u_1 v u_1^* = vvv^* = v$. Therefore v commutes with u_1. Similarly, v^* commutes with u_1. It follows that $v^* u_1$ also commutes with u_1. Consequently $v^* u_1 \in \pi[B]'$, and the irreducibility of π implies that u_1 is a scalar multiple of v. So $u_1 \in \sigma[A]''$. Since $\pi[B]$ is generated by u_1 and $\sigma[A]$, we conclude $\pi[B]'' = \sigma[A]''$. Therefore the irreducibility of π implies the irreducibility of σ.

Now σ is an irreducible representation of the type I algebra A. Consequently $\sigma[A]$ contains the compact operators. Since $\pi[B]$ contains $\sigma[A]$, we conclude that $\pi[B]$ also contains the compact operators. Because π is an arbitrary irreducible representation of B, we conclude that B is type I.

We now apply theorem 8.2.3 (2), and conclude that the action of $S^1 = \widehat{\mathbf{Z}}$ on $Prim(B)$ induced by $\widehat{\alpha}$ is free. By theorem 8.1.4, the action $\widehat{\alpha}$ is K-free. By proposition 4.1.3, for every $\tau \in \widehat{S}^1$ and every $\eta \in K_*(C^*(S^1, B))$ there is n such that $[id - (((\widehat{\alpha})^\frown)_\tau)_*]^n \eta = 0$. By Takai duality ([79], 7.9.3), the $\mathbf{Z}$-algebra $(\mathbf{Z}, C^*(S^1, B), (\widehat{\alpha})^\frown)$ is isomorphic to the $\mathbf{Z}$-algebra $(\mathbf{Z}, K(l^2(\mathbf{Z})) \otimes A, \lambda \otimes \alpha)$, where λ is the inner action on $K(l^2(\mathbf{Z}))$ obtained from the regular representation of $\mathbf{Z}$ on $l^2(\mathbf{Z})$. This implies that for every $\eta \in K_*(K(l^2(\mathbf{Z})) \otimes A)$ there is n such that $(id - (\lambda_1 \otimes \alpha_1)_*)^n(\eta) = 0$. Now λ_1 is an inner automorphism of $K(l^2(\mathbf{Z}))$, that is, it is given by conjugation by a unitary operator on $l^2(\mathbf{Z})$. Since the unitary group of a separable infinite dimensional Hilbert space is connected, it follows that λ_1 is homotopic to the identity. Hence $\lambda_1 \otimes \alpha_1$ is homotopic to $id_{K(l^2(\mathbf{Z}))} \otimes \alpha_1$. It follows that $(\lambda_1 \otimes \alpha_1)_* = (id_{K(l^2(\mathbf{Z}))} \otimes \alpha_1)_*$. From the stability of K-theory, theorem 2.8.3 (4), we now conclude that for all $\eta \in K_*(A)$ there is n such that $(id - (\alpha_1)_*)^n \eta = 0$. Since $\alpha_1 = \varphi$, this completes the proof. Q.E.D.

The following corollary may be of independent interest.

8.3.3 Corollary. Let φ be a universally weakly inner automorphism of a separable type I algebra A. Then for every $\eta \in K_*(A)$ there is n such that $(id - \varphi_*)^n \eta = 0$.

Proof. The fact that φ is universally weakly inner also implies that φ acts trivially on *Prim*(A), again by [61]. Q.E.D.

Of course, if φ is actually inner, we can choose $n = 1$.

8.3.4 Question. Can the assumption that A is type I in this corollary be dropped?

A positive answer to this question would enable one to conclude, for example, that if α is an action of a compact abelian Lie group whose dual action is universally weakly inner, then α is K-free. As further evidence for a positive answer, we point out that corollary 2.3 of [59] implies that a universally weakly inner automorphism φ of a separable simple C^*-algebra A is necessarily inner (in $(M(A))$. Therefore $\varphi_* = id_{K_*(A)}$. We note, however, that lemma 8.3.2 does not hold for AF algebras, as will be shown in example 9.3.5.

We now return to consideration of type I algebras.

8.3.5 Theorem. Let G be a compact abelian Lie group, and let (G,A,α) be a separable type I G-algebra. If α is hereditarily saturated, then α is K-free.

Proof. By theorem 7.2.8, the dual action $\hat{\alpha}$ induces the trivial action on *Prim*$(C^*(G,A,\alpha))$. It follows from [103], theorem 6.1, that $C^*(G,A,\alpha)$ is type I. For $\tau \in \hat{G}$ and $\eta \in K_*(C^*(G,A))$, we conclude from lemma 8.3.2 that there is n such that $(id - (\hat{\alpha}_\tau)_*)^n(\eta) = 0$. By proposition 4.1.3, the action α has locally discrete K-theory.

Now let I be any G-invariant ideal of A. Then I is also type I, and $\alpha_{(\cdot)}|_I$ is also hereditarily saturated, by proposition 7.2.3. Therefore $\alpha_{(\cdot)}|_I$ has locally discrete K-theory. So α is K-free. Q.E.D.

8.3.6 Theorem. Let G be a finite abelian group, and let (G,A,α) be a separable type I G-algebra. Then the following are equivalent:

(1) The action α is K-free.

(2) The dual action $\hat{\alpha}$ induces the trivial action on *Prim*$(C^*(G,A,\alpha))$.

(3) The action α is hereditarily saturated.

Proof. (2)$\Longleftrightarrow$(3) is theorem 7.2.8. (1)$\Longrightarrow$(2) follows from theorems 7.4.3 and 7.2.8, since A is highly K-visible by proposition 7.3.3. (3)$\Longrightarrow$(1) follows from the previous theorem. Q.E.D.

We summarize our results on separable type I algebras in the following theorem.

8.3.7 Theorem. Consider the following conditions on an action α of a compact Lie group G on separable type I algebra A.

(1) G acts freely on $Prim(A)$.

(2) The action α is totally K-free.

(3) The action α is K-free.

(4) The action α is hereditarily saturated.

(5) The action α is hereditarily K-saturated.

If G is finite cyclic, or $G = S^1$, then all five conditions are equivalent. If G is finite abelian, the following implications hold:

$$(1) \Longleftrightarrow (2) \Longrightarrow (3) \Longleftrightarrow (4) \Longrightarrow (5).$$

The implication (2)$\Longrightarrow$(3) cannot be reversed. If G is an arbitrary compact Lie group, then the following implications hold:

$$(1) \Longleftrightarrow (2) \Longrightarrow (3) \text{ and } (2) \Longrightarrow (4) \Longrightarrow (5).$$

We point out that condition (5) says nothing about Morita equivalence.

8.3.8 Question. Are the conditions (3), (4), and (5) equivalent for general compact Lie groups? I know of no counterexamples.

The following example shows that it will not do to replace condition (5) with the weaker condition that α be K-saturated.

8.3.9 Example. Let n be a positive integer. Near the beginning of section 6 of [90], there is a construction of a continuous trace C^*-algebra A whose primitive ideal space is the 3-sphere S^3 and whose K-theory is given by $K_0(A) = 0$ and $K_1(A) = \mathbf{Z}/n\mathbf{Z}$. Now let $n = 1$. Then $K_*(A) = 0$. If G is a compact Lie group acting trivially on A, then $A^G = A$ and $K_*^G(A) \simeq R(G) \otimes K_*(A) = 0$. Therefore the map from $K_*(A^G)$ to $K_*^G(A)$ must be an isomorphism. However, G certainly does not act freely on $Prim(A)$.

Chapter 9

AF Algebras

We now turn our attention to AF algebras. We recall ([12]; [31], section 2) that a C^*-algebra A is called AF if it is a direct limit of a sequence of finite dimensional C^*-algebras. Equivalently,

$$A = \overline{\bigcup_{n=0}^{\infty} A_n},$$

where A_n is finite dimensional, and $A_n \subset A_{n+1}$, for all n. (We do not insist that A or the maps from A_n to A_{n+1} be unital.) If α is an action of a compact group G on A, then it is an open question whether the subalgebras A_n can be chosen to be G-invariant. (The answer is not even known for $G = \mathbf{Z}/2\mathbf{Z}$.) In order to be able to prove things about K-freeness of actions on AF algebras, we will therefore assume that the subalgebras A_n can be chosen to be G-invariant. We call such an action α an AF action. (We point out that proposition 7.3.3 and theorem 7.4.3 together show that if (G,A,α) is a G-algebra, such that A is AF, G is finite abelian, and α is K-free, then $\widetilde{\Gamma}(\alpha) = \widehat{G}$. This result does *not* require that α be an AF action.)

As in chapter 8, the restriction on the type of algebra considered yields stronger results than in the general case. Here, however, the results are somewhat different, and the proofs are more difficult. It is easy to show, using the fact that the K_1 group of an AF algebra is zero, that an AF action of a finite group which has locally discrete K-theory is actually K-free. It is much more difficult to prove the main theorem of this chapter, which gives a number of equivalent conditions for an AF action of a finite abelian group to be totally K-free. As part of this theorem, we prove that a totally K-free AF action of a finite abelian group must be KK-free; we even prove the existence of certain G-homotopies of homomorphisms which immediately imply KK-freeness.

We start out in section 1 by proving some elementary properties of K-freeness of AF actions of finite groups. In the second section, we examine the consequences of K-freeness for a special kind of AF action, namely the locally representable actions of [45]. For these, we find that K-freeness implies total K-freeness, and is characterized by a certain factorization property. Section 3 is devoted to the construction of a number of examples, all but one of which are product type actions on UHF algebras. A number of examples promised in earlier sections are found here. Using the results of the first two sections, it is usually easy to show that they have the properties claimed. In the fourth section, we state the main theorem on actions of finite abelian groups on AF algebras, and prove the easy parts of it. We also give a partial generalization to the nonabelian case. The next two sections are each devoted to one of the two hard implications in the main theorem. In the final section, we show how to compute the equivariant KK-theory of AF algebras, and prove a result needed for the proof of the nonabelian case of the main theorem.

9.1. Elementary Properties of K-Free Actions on AF Algebras

9.1.1 Definition. Let G be a group, and let A be an AF algebra. An action $\alpha: G \longrightarrow Aut(A)$ is called an AF action if there exists an increasing sequence $(A_n)_{n\geq 0}$ of finite dimensional subalgebras of A such that $A = \overline{\bigcup_{n=0}^{\infty} A_n}$ and such that $\alpha_g[A_n] = A_n$ for all n and for all $g \in G$. An AF G-algebra is a G-algebra (G,A,α) such that A is an AF algebra and α is an AF action.

As mentioned in the introduction, it is not known whether every action of a finite group on an AF algebra is AF. Since the methods of this chapter work only for AF actions, we will consider only such actions.

The following proposition is of course well known. It is in fact one of the reasons that one would like to be able to show that all actions of finite groups on AF algebras are AF.

9.1.2 Proposition. Let G be a finite group, and let (G,A,α) be an AF G-algebra. Then $C^*(G,A,\alpha)$ is an AF algebra. If G is abelian, then the dual action $\hat{\alpha}$ is also AF.

Proof. Write $A = \overline{\bigcup_{n=0}^{\infty} A_n}$, where (A_n) is an increasing sequence of G-invariant finite dimensional subalgebras. Then

$$C^*(G,A,\alpha) = \overline{\bigcup_{n=0}^{\infty} C^*(G,A_n,\alpha_{(\cdot)}|_{A_n})},$$

and the algebras $C^*(G,A_n,\alpha_{(\cdot)}|_{A_n})$ form an increasing sequence of finite dimensional subalgebras, which are invariant for the dual action in case G is abelian. Therefore $C^*(G,A,\alpha)$ is AF, and the dual action is AF if G is abelian. Q.E.D.

9.1.3 Corollary. Let G be a finite group, and let (G,A,α) be an AF G-algebra. Then $K_1^G(A) = 0$ and $K_0^G(A)$ is torsion-free as an abelian group.

Proof. We have $K_*^G(A) \simeq K_*(C^*(G,A,\alpha))$ by theorem 2.8.3 (7). It is well known ([31], corollary 8.2 and section 3) that K_0 of an AF algebra is torsion-free and also ([31], proof of corollary 9.2) that K_1 of an AF algebra is zero. Q.E.D.

9.1.4 Proposition. Let G be a finite group, and let (G,A,α) be an AF G-algebra. Then α has locally discrete K-theory if and only if α is K-free.

Proof. We only have to show that if α has locally discrete K-theory, then α is K-free. So assume α has locally discrete K-theory, and let I be a G-invariant ideal in A. Let β be the induced action of G on the quotient $B = A/I$, and let γ be of G the action $\alpha_{(\cdot)}|_I$ on I. It is well known that I and A/I are AF algebras, but we must also show that the actions γ and β are AF.

Since α is AF, there is an increasing sequence (A_n) of G-invariant finite dimensional subalgebras such that $A = \overline{\bigcup_{n=0}^{\infty} A_n}$. The proof that A/I is AF ([12], 1.5 for the unital case; the nonunital case is the same) consists of looking at the images of the algebras A_n in A/I. Since these are clearly finite dimensional and G-invariant, and since their union is dense in A/I, we have shown that β is AF. The proof that I is AF consists of showing that

$$I = I \cap \left(\overline{\bigcup_{n=0}^{\infty} A_n} \right) = \overline{\bigcup_{n=0}^{\infty} \left[I \cap A_n \right]}.$$

(See [12], 3.3 for the unital case; the nonunital case follows immediately from the unital case.) Since I and A_n are G-invariant, so is $I \cap A_n$. Therefore γ is AF.

In the long exact sequence in K-theory for the short exact sequence

$$0 \longrightarrow I \longrightarrow A \longrightarrow B \longrightarrow 0,$$

the K_1^G terms are all zero by corollary 9.1.3. Therefore there is actually a short exact sequence

$$0 \longrightarrow K_0^G(I) \longrightarrow K_0^G(A) \longrightarrow K_0^G(B) \longrightarrow 0. \qquad (*)$$

Let $P \subset R(G)$ be any prime ideal not containing $I(G)$. Then the sequence (*) remains exact when localized at P. Furthermore, $K_0^G(A)_P = 0$, since α has locally discrete K-theory. Therefore $K_0^G(I)_P = 0$. Since I and P are arbitrary, as long as P does not contain $I(G)$, we conclude that α is K-free. Q.E.D.

9.1.5 Corollary. Let G be a finite group, and let (G,A,α) be an AF G-algebra. If $\alpha|_S$ is K-free for every cyclic subgroup S of G, then α is totally K-free.

Observe that, unlike in theorem 5.2.6, we don't need to assume that G is a p-group.

Proof of corollary 9.1.5. Let H be a subgroup of G. We have to prove that $\alpha|_H$ is K-free. By the previous proposition, it is sufficient to show that $\alpha|_H$ has locally discrete K-theory. (The restriction of an AF action to a subgroup is clearly AF.) By corollary 9.1.3, $K_*^H(A)$ is torsion-free. By hypothesis, $\alpha|_H$ has locally discrete K-theory for all cyclic subgroups S of H. Therefore, by proposition 5.2.2, the action $\alpha|_H$ has locally discrete K-theory. Q.E.D.

9.1.6 Proposition. Let G be a finite group, and let (G,A,α) be an AF G-algebra. Let K be the algebra of compact operators on a separable Hilbert space, and let

$\beta : G \longrightarrow Aut(K)$ be any action. Then the diagonal action $\beta\otimes\alpha$ on $K\otimes A$ is an AF action, and is totally K-free if and only if α is.

Proof. We will show that $\beta\otimes\alpha$ is an AF action. Then this proposition will follow from corollary 9.1.5 in exactly the same way that corollary 5.2.7, in which the hypothesis that (G,A,α) be AF is replaced by the hypothesis that G be a p-group, follows from theorem 5.2.6, which is the analog of corollary 9.1.5 for a similar change in the hypotheses.

It is obvious that the tensor product of two AF actions is AF. We need therefore only prove that (G,K,β) is AF. Let H be the Hilbert space on which K acts. If H is finite dimensional, there is nothing to prove. So assume H is infinite dimensional. Now each β_g is inner, and so β defines a projective unitary representation $u : G \longrightarrow PU(H) = U(H)/S^1$, where $U(H)$ is the unitary group of H and S^1 is the set of complex numbers of absolute value 1, regarded as multiples of the identity on H. Let G_0 be the inverse image of $u[G]$ in $U(H)$. Then G_0 is a compact group, and we can write $H = \bigoplus_{n=0}^{\infty} H_n$, where each H_n is a finite dimensional G_0-invariant subspace of H. It is obvious that $B_n = L(H_1 \oplus \cdots \oplus H_n)$ is a finite dimensional G-invariant subalgebra of $K(H)$, and we have $K(H) = \overline{\bigcup_{n=1}^{\infty} B_n}$. Therefore β is an AF action. Q.E.D.

Proposition 9.1.4 can be generalized to cover extensions of commutative unital algebras by AF algebras. The proof requires the following lemma, which may be of independent interest.

9.1.7 Lemma. Let G be a compact Lie group, and let (G,A,α) be a unital G-algebra such that there is an equivariant exact sequence

$$0 \longrightarrow I \longrightarrow A \longrightarrow C(X) \longrightarrow 0$$

for some compact G-space X. If α has locally discrete K-theory, then $\alpha_{(\cdot)}|_I$ has locally discrete K-theory and G acts freely on X.

Proof. Let $\mathbf{C}$ have the trivial action of G, and consider the equivariant commutative diagram

$$\begin{array}{ccccccccc} & & & & \mathbf{C} & \searrow & & & \\ & & & & \downarrow & & & & \\ 0 & \longrightarrow & I & \longrightarrow & A & \longrightarrow & C(X) & \longrightarrow & 0, \end{array}$$

in which the maps from $\mathbf{C}$ to A and to $C(X)$ send z to $z \cdot 1$. Let $P \subset R(G)$ be a prime ideal not containing $I(G)$, and apply the functor $K_*^G(\cdot)_P$ to obtain:

$$\begin{array}{ccccccccccc} & & & & & & R(G)_P & \searrow & & & \\ & & & & & & \downarrow & & & & \\ \cdots & \longrightarrow & K_G^1(X)_P & \longrightarrow & K_0^G(I)_P & \longrightarrow & K_0^G(A)_P & \longrightarrow & K_G^0(X)_P & \longrightarrow & \cdots \end{array} \qquad (*)$$

By assumption, $K_0^G(A)_P = 0$. Therefore the map from $R(G)_P$ to $K_G^0(X)_P$ is the zero map. Since this map is a unital ring homomorphism, we conclude that $K_G^0(X)_P = 0$. Since $K_G^1(X)$ is a module over $K_G^0(X)$, we conclude that $K_G^1(X)_P = 0$ also. Therefore $K_0^G(I)_P = 0$, and, since $K_1^G(A)_P = 0$, the part of the long exact sequence which doesn't appear in (*) now implies that $K_1^G(I)_P = 0$. Thus, the actions on I and $C(X)$ have locally discrete K-theory. Theorem 1.1.1 now implies that G acts freely on X. Q.E.D.

9.1.8 Proposition. Let G be a finite group. Let

$$0 \longrightarrow I \longrightarrow A \longrightarrow C(X) \longrightarrow 0$$

be an equivariant exact sequence, where (G,I,γ) is an AF G-algebra, (G,A,α) is a unital G-algebra, and X is a compact Hausdorff G-space. If α has locally discrete K-theory, then α is K-free.

Proof. By the previous lemma, γ has locally discrete K-theory and the action of G on X is free. By proposition 9.1.4, the action γ is K-free, and by proposition 4.2.7, the action on $C(X)$ is K-free. It now follows from proposition 4.2.2 that α is K-free. Q.E.D.

We now return to AF algebras. For the next result, we assume that the group is abelian, although this is probably not necessary.

9.1.9 Proposition. Let G be a finite abelian group, and let (G,A,α) be an AF G-algebra. If α has locally discrete K-theory, then $I(G)K_0^G(A) = 0$.

Thus, if for every $\eta \in K_*^G(A)$ there is n such that $I(G)^n\eta = 0$, then already $I(G)$ annihilates $K_*^G(A)$. (We actually need only consider $K_0^G(A)$, because $K_1^G(A)$ is zero anyway.) The proof proceeds by way of the following lemma.

9.1.10 Lemma. Let α be an AF automorphism of an AF algebra A. (An AF automorphism is one such that the action of **Z** by its powers is an AF action.) Suppose that for some $\eta \in K_0(A)$ and some $n > 0$, one has $(id - \alpha_*)^n(\eta) = 0$. Then $\alpha_*(\eta) = \eta$.

Proof. By assumption, $A = \varinjlim A_m$ where each A_m is finite dimensional and α-invariant. Therefore $K_0(A) = \varinjlim K_0(A_m)$, by theorem 2.8.3 (6). In particular, there is some k_0 such that η is the image of some $\lambda_0 \in K_0(A_{k_0})$, and there is $k \geq k_0$ such that, with λ being the image of λ_0 in $K_0(A_k)$, we have $(id - \alpha_*)^n(\lambda) = 0$. We may therefore reduce to the case in which A itself is finite dimensional.

Let $I_1, \dots, I_m$ be the minimal α-invariant ideals of A. Then $A = \bigoplus_{i=1}^m I_i$, whence $K_0(A) \simeq \bigoplus_{i=1}^m K_0(I_i)$. Under this isomorphism, η corresponds to an m-tuple $(\eta_1, \dots, \eta_m)$, with $\eta_i \in K_0(I_i)$. The equation $(id - \alpha_*)^n(\eta) = 0$ becomes $(id - \alpha_*)^n(\eta_i) = 0$ for $1 \leq i \leq m$. It is therefore sufficient to prove that $\alpha_*(\eta_i) = \eta_i$ for all i. That is, we may reduce to the case in which A has no nontrivial α-invariant ideals.

If A has no nontrivial α-invariant ideals, then the powers of α act transitively on its primitive ideal space. Therefore we can write A as a direct sum of simple algebras $A = \bigoplus_{j=1}^{k} A_j$, where $\alpha[A_j] = A_{j+1}$ for $j < k$, and $\alpha[A_k] = A_1$. The algebras A_j are all isomorphic. Since they are finite dimensional and simple, they are all isomorphic to a matrix algebra M_r. So $K_0(A_j) \simeq \mathbf{Z}$ for each j. We choose the isomorphism so that the class of a one-dimensional projection is identified with $1 \in \mathbf{Z}$. Then

$$K_0(A) \simeq \bigoplus_{j=1}^{k} K_0(A_j) \simeq \mathbf{Z}^k .$$

Since α sends a one-dimensional projection in A_j to a 1-dimensional projection in A_{j+1}, or in A_1 if $j=k$, it follows that α_* is a cyclic permutation of the factors of $\mathbf{Z}^k$.

We are given that $(id - \alpha_*)^n(\eta) = 0$, and we want to conclude that $(id - \alpha_*)(\eta) = 0$. That is, we want to show that

$$Ker\,((id - \alpha_*)^n) = Ker\,(id - \alpha_*) .$$

Now α_*, being a cyclic permutation of the factors of $\mathbf{Z}^k$, extends to a $\mathbf{C}$-linear map $u : \mathbf{C}^k \longrightarrow \mathbf{C}^k$, which is again a cyclic permutation of the factors, this time of $\mathbf{C}^k$. It is clearly sufficient to prove that

$$Ker\,((1-u)^n) = Ker\,(1-u) . \qquad (*)$$

Now u is unitary. Therefore $1-u$ is normal, and hence diagonalizable. The equation (*) follows immediately. Q.E.D.

Proof of proposition 9.1.9. We have an AF action α of a finite abelian group G on A. Since α has locally discrete K-theory, proposition 4.1.3 implies that for every $\eta \in K_*(C^*(G,A))$ and every $\tau \in \widehat{G}$, there is n such that $(id - (\widehat{\alpha}_\tau)_*)^n(\eta) = 0$. We know from proposition 9.1.2 that $\widehat{\alpha}_\tau$ is an AF automorphism. Therefore $(\widehat{\alpha}_\tau)_*(\eta) = \eta$ by the lemma just proved. Since the isomorphism $K_*^G(A) \simeq K_*(C^*(G,A))$ identifies $\tau\eta$ with $(\widehat{\alpha}_\tau)_*(\eta)$, by theorem 2.8.3 (7), we conclude that $(1-\tau)\eta = 0$ for all $\eta \in K^G(A)$ and $\tau \in \widehat{G}$. The elements $1-\tau$ of $R(G)$ generate $I(G)$ as τ ranges over $\widehat{G}$, so we conclude that $I(G)K_0^G(A) = 0$. Q.E.D.

We conclude this section by showing that the Tensor Product Theorem 6.6.3 can also be improved in the special case of AF actions.

9.1.11 Proposition. Let G be a finite group, and let (G,A,α) and (G,B,β) be AF G-algebras. If α is totally K-free, then so is $\alpha \otimes \beta$.

Proof. Since the hypotheses hold for the restriction to any subgroup of G, it is sufficient to show that $\alpha \otimes \beta$ is K-free. It is clear that the tensor product of AF actions is AF, so by proposition 9.1.4, it is actually sufficient to prove that $\alpha \otimes \beta$ has locally discrete K-theory.

Write $B = \varinjlim B_n$, where each B_n is finite dimensional and G-invariant. Fix n, and write B_n as the direct sum of its minimal G-invariant ideals $I_1, \ldots, I_k$. Since each I_j is finite dimensional, it is a direct sum of full matrix algebras, which are permuted by the action of G. If H is the stabilizer of one of them, say M_n, then we have $I_j \simeq C(G \times_H M_n)$ by proposition 2.9.6, relative to some action γ of H on M_n. By assumption, $\alpha|_H$ is K-free, whence $\alpha|_H \otimes \gamma$ is K-free by proposition 9.1.6. We have equivariant isomorphisms

$$A \otimes I_j \simeq A \otimes C(G \times_H M_n) \simeq C(G \times_H (A \otimes M_n))$$

by lemma 5.1.7, whence $\alpha \otimes (\beta|_{I_j})$ is K-free by theorem 4.3.4. Since this is true for all j, we see that $\alpha \otimes (\beta|_{B_n})$ is K-free, and in particular has locally discrete K-theory. Now $A \otimes B \simeq \varinjlim A \otimes B_n$, so $\alpha \otimes \beta$ has locally discrete K-theory by lemma 4.2.14. Q.E.D.

9.2. Locally Representable Actions

Throughout this section, G is a finite group. We will study the implications of K-freeness in the situations considered by Handelman and Rossmann in [44] and [45], where the actions are assumed respectively to be of product type and locally representable. One particular AF G-algebra will play a central role, namely the infinite tensor product $\bigotimes_1^\infty L(l^2(G))$, where the action is conjugation by the regular representation in each factor. For use in the examples in the next section, we state the results needed to analyze this action in somewhat greater generality. First, we need a lemma.

9.2.1 Lemma. Let $k > 0$ be an integer. Then the annihilator of $I(G)^k$ in $R(G)$ consists exactly of the (integral) multiples of the class λ of the regular representation of G.

Proof. The representation theory of compact groups enables us to identify $R(G)$ with the ring generated by the characters of G. Thus, $R(G)$ is regarded as a ring of complex-valued functions on G. Under this identification, $I(G)$ becomes the set of all functions in $R(G)$ which vanish at the identity e of G, and the regular representation is the function taking the value $n = |G|$, the order of G, at e, and the value zero elsewhere. So it is obvious that $m\lambda \cdot I(G)^k = 0$ for any $m \in \mathbf{Z}$.

Now let $\mu \in R(G)$, and suppose that $\mu \cdot I(G)^k = 0$. Let $g \in G - \{e\}$, let H be the cyclic subgroup of G generated by g, and let m be the order of H. Let $\eta \in R(H)$ be the character of the trivial one-dimensional representation of H, and set $\chi = \lambda - m\, Ind_H^G(\eta) \in R(G)$. Obviously $\chi \in I(G)$. The formula for the induced character ([99], theorem 12), together with the fact that $\eta(h) = 1$ for all $h \in H$, implies that $Ind_H^G(\eta)(g) > 0$. Since $\lambda(g) = 0$, we have $\chi(g) \neq 0$. Consequently $\chi^k(g) \neq 0$. Since $\chi^k \in I(G)^k$ and $\mu \cdot I(G)^k = 0$, it follows that $\mu(g) = 0$. Thus, μ vanishes on $G - \{e\}$, and so μ is some complex multiple of the regular representation. Now the standard formula (using inner products of characters) for the number of times the trivial one-dimensional representation of G occurs in the representation corresponding to μ yields $\frac{1}{n}\mu(e)$. Since this number must be an integer (positive or negative), it follows that μ is an integral multiple of λ. Q.E.D.

9.2.2 Proposition. Let G be a finite group, and let (G,V,u) be a finite dimensional representation space of G, where u is a unitary representation. Let α be the corresponding inner action of G on $L(V)$, that is, $\alpha_g(a) = u_g a u_g^*$. Then the infinite tensor product G-algebra

$$(G,B,\beta) = (G, \bigotimes_1^\infty L(B), \bigotimes_1^\infty \alpha)$$

is an AF G-algebra satisfying $K_0^G(B) \simeq R(G)[\eta^{-1}]$, where $\eta = [V]$. The action β is K-free if and only if u is a multiple of the regular representation.

Of course, $R(G)[\eta^{-1}]$ is the same as $S^{-1}R(G)$, where S is the multiplicative system consisting of all positive powers of η. The first part of this proposition is proved in [108], pages 99-100, but for completeness we give a proof here.

Proof of proposition 9.2.2. We have $B = \varinjlim B_n$, where B_n is the tensor product of n copies of $L(V)$, all carrying the action α. The map from B_n to $B_{n+1} = L(V) \otimes B_n$ is given by $\varphi_n(a) = 1 \otimes a$. Now each B_n is a full matrix algebra, and the action of G on B_n is inner. Lemma 2.7.5 therefore implies that $K_0^G(B_n)$ is canonically isomorphic to $R(G)$ for all n, and that $(\varphi_n)_*$ is multiplication by η for all n. Thus,

$$K_0^G(B) \simeq \varinjlim K_0^G(B_n) = \varinjlim_{\lambda_n} R(G),$$

where each $\lambda_n : R(G) \longrightarrow R(G)$ is multiplication by η. It is now easy to check that the last direct limit is just $R(G)[\eta^{-1}]$. (The required map from the n-th term in the direct limit to $R(G)[\eta^{-1}]$ is $\chi \longrightarrow \chi\eta^{-n}$.)

We now prove the part about K-freeness. If η is a multiple of the regular representation, then $\eta \cdot I(G) = 0$ by the previous lemma, whence

$$I(G) \cdot R(G)[\eta^{-1}] = I(G) \cdot \eta \cdot R(G)[\eta^{-1}] = 0.$$

Therefore the action β has discrete K-theory, and is thus K-free by proposition 9.1.4. For the converse, observe that $R(G)[\eta^{-1}]$ is in fact a unital $R(G)$-algebra. Consequently, if β has locally discrete K-theory then there is by lemma 1.2.2 an integer k such that $I(G)^k \cdot R(G)[\eta^{-1}] = 0$. Now $I(G)^k \cdot R(G)[\eta^{-1}] = S^{-1}I(G)^k$, where S is the multiplicative system consisting of all positive powers of η. Therefore, for every $\mu \in I(G)^k$ there is r such that $\eta^r \mu = 0$. Choosing r sufficiently large that $\eta^r \mu = 0$ for all μ in a finite set of generators for $I(G)^k$, we see that $\eta^r \cdot I(G)^k = 0$. So η^r is a multiple of the regular representation by the previous lemma. Regarding η

as a character, it follows that $\eta(g)=0$ for $g \in G-\{e\}$, whence η is a multiple of the regular representation.

Q.E.D.

In particular, the tensor product of infinitely many copies of conjugation by the regular representation is K-free. We now recall the definition of local representability.

9.2.3 Definition ([45]). Let (G,A,α) be a unital AF G-algebra. Then α is said to be locally representable if there is an increasing sequence (A_n) of finite dimensional G-invariant subalgebras of A whose union is dense in A and such that $\alpha_{(\cdot)}|_{A_n}$ is inner for all n. (That is, for each n there is a homomorphism $g \longrightarrow u_g$ from G to the unitary group of A_n such that $\alpha_g(a)=u_g a u_g^*$ for all $a \in A_n$.)

The equivalence of statements (1) and (5) of the following theorem, and the proof, were suggested by David Handelman.

9.2.4 Theorem. Let G be a finite group, and let $B=\bigotimes_1^\infty L(l^2(G))$, with the action $\beta: G \longrightarrow Aut(B)$ given by conjugation by the regular representation in each factor. Then for a locally representable action α of G on a unital AF algebra A, the following are equivalent:

(1) The action α has locally discrete K-theory.

(2) The action α is totally K-free.

(3) $I(G)K_0^G(A)=0$.

(4) The restriction map $K_0^G(A) \longrightarrow K_0(A)$ is an isomorphism of abelian groups.

(5) The G-algebras (G,A,α) and $(G,A\otimes B,\iota\otimes\beta)$ are equivariantly isomorphic, where ι is the trivial action of G on A.

(6) The G-algebras (G,A,α) and $(G,A\otimes B,\alpha\otimes\beta)$ are equivariantly isomorphic.

Before proving this theorem, we recall a few facts about order on K_0 groups and prove a lemma on tensor products. Recall that if A is a unital C^*-algebra, then an element $\eta \in K_0(A)$ is called positive if η is the class of some projection in some $M_n(A)$. (This is a much weaker notion than the concept of total positivity introduced in definition 7.3.1.) The set of positive elements is a subsemigroup of $K_0(A)$, but in general that is about all one can say. The situation for AF algebras is of course much better: $K_0(A)$ becomes a (partially) ordered group. (See [31], corollary 8.2 and section 3.) If now (G,A,α) is a unital G-algebra for a finite group G, then there are two possible notions of positivity, namely being represented by a projection in some $M_n(C^*(G,A))$ and being represented by a G-invariant projection in some $L(V)\otimes A$, where V is a finite dimensional representation space of G. It is easily seen that both notions coincide. Furthermore, if $\eta \in K_0^G(A)$ is positive and V is a finite dimensional representation space of G, then $[V]\eta$ is positive. Thus, in the case of an AF action, $K_0^G(A)$ is a partially ordered $R(G)$-module, for the obvious order on $R(G)$.

9.2.5 Lemma. Let G be a finite group, let (G,A,α) be a unital G-algebra, and let B be a unital AF algebra with the trivial action of G. Then $K_0^G(A\otimes B)\simeq K_0^G(A)\otimes K_0(B)$, where the $R(G)$-module structure on $K_0^G(A)\otimes K_0(B)$ is given by the action of $R(G)$ on the first factor. This isomorphism identifies the positive elements of $K_0^G(A\otimes B)$ with the subsemigroup of $K_0^G(A)\otimes K_0(B)$ generated by all $\eta\otimes\mu$ for which η and μ are both positive.

Proof. If B is finite dimensional, then B is a direct sum of matrix algebras, and $A\otimes B$ is a direct sum of matrix algebras over A. The result, including the statement about positive elements, is thus trivial. In the general case, we have $B=\varinjlim B_n$, where each B_n is finite dimensional. Since B is unital, we may assume that the maps $B_n \longrightarrow B_{n+1}$ are all unital. Then $A\otimes B\simeq\varinjlim A\otimes B_n$. Therefore

$$K_0^G(A\otimes B)\simeq\varinjlim K_0^G(A\otimes B_n)\simeq\varinjlim K_0^G(A)\otimes K_0(B_n)$$

$$\simeq K_0^G(A)\otimes\varinjlim K_0(B_n)\simeq K_0^G(A)\otimes K_0(B), \qquad (*)$$

where the first and last steps are proposition 2.5.4, the second step is the special case done already, and the third step follows from the fact that tensor products commute with direct limits.

It remains to verify the assertion concerning positive elements. The proof of the unital case of proposition 2.5.4 shows that the positive elements of $K_0(B)$ are exactly the images of the positive elements in the groups $K_0(B_n)$ under the maps $K_0(B_n)\longrightarrow K_0(B)$. Similarly, the positive elements of $K_0^G(A\otimes B)$ are the images of the positive elements in the groups $K_0^G(A\otimes B_n)$. The rest of the proof is straightforward algebra, and is omitted. Q.E.D.

Proof of theorem 9.2.4. We will prove the implications (5)$\Longrightarrow$(6)$\Longrightarrow$(2)$\Longrightarrow$(1) $\Longrightarrow$(4)$\Longrightarrow$(5) and (4)$\Longrightarrow$(3)$\Longrightarrow$(1).

(5)$\Longrightarrow$(6). This follows immediately from the existence of an equivariant isomorphism $(G,B,\beta)\simeq(G,B\otimes B,\beta\otimes\beta)$.

(6)$\Longrightarrow$(2). Let H be a subgroup of G. Since the restriction of the regular representation of G to H is a multiple of the regular representation of H, proposition 9.2.2 implies that β is totally K-free. Now proposition 9.1.11 implies that $\alpha\otimes\beta$ is totally K-free.

(2)$\Longrightarrow$(1). Trivial.

(1)$\Longrightarrow$(4). Write A as an equivariant direct limit $A=\varinjlim A_k$, where each A_k is finite dimensional and the action $\alpha^{(k)}$ of G on A_k is inner. Let $\varphi_k : K_0^G(A_k)\longrightarrow K_0^G(A_{k+1})$ be the map on equivariant K-theory induced by the map from A_k to A_{k+1}, let $\psi_k : K_0(A_k)\longrightarrow K_0(A_{k+1})$ be the corresponding map on ordinary K-theory, and let $r_k : K_0^G(A_k)\longrightarrow K_0(A_k)$ be the restriction map. We certainly have $r_{k+1}\circ\varphi_k=\psi_k\circ r_k$. Further, let $s(k)$ be the number of simple summands of A_k. Then clearly $K_0(A_k)\simeq\mathbf{Z}^{s(k)}$, and, since $\alpha^{(k)}$ is inner, it follows from theorem 2.8.3 (5) and remark 2.8.5 that $K_0^G(A_k) \simeq R(G)^{s(k)}$. Consequently we have a commutative diagram

$$\begin{array}{ccccccc} \cdots \longrightarrow & R(G)^{s(k)} & \xrightarrow{\varphi_k} & R(G)^{s(k+1)} & \longrightarrow \cdots \\ & r_k \downarrow & & r_{k+1} \downarrow & \\ \cdots \longrightarrow & \mathbf{Z}^{s(k)} & \xrightarrow{\psi_k} & \mathbf{Z}^{s(k+1)} & \longrightarrow \cdots \end{array} \qquad (*)$$

The direct limit of the top row is $K_0^G(A)$, the direct limit of the bottom row is $K_0(A)$, and the direct limit of the maps r_k is the restriction $r : K_0^G(A) \longrightarrow K_0(A)$.

Let $\varepsilon : R(G) \longrightarrow \mathbf{Z}$ be the augmentation map, assigning to each representation of G its dimension. Then it is easy to see that, with appropriate identifications of $K_0^G(A)$ with $R(G)^{s(k)}$ and $K_0(A)$ with $\mathbf{Z}^{s(k)}$, we have

$$r_k(\eta_1, \ldots, \eta_{s(k)}) = (\varepsilon(\eta_1), \ldots, \varepsilon(\eta_{s(k)})). \qquad (**)$$

Indeed, this is certainly true if $\alpha^{(k)}$ is trivial, and follows for arbitrary inner actions by the exterior invariance of equivariant K-theory. Therefore r_k is surjective for all k. Since direct limits are exact ([5], problem 2.19), it follows that r is surjective.

So far we have not used the hypothesis that α has locally discrete K-theory, but we do so now. Let d be the order of G, and let λ be the character of the regular representation. Then $d - \lambda \in I(G)$. Let $\eta \in K_0^G(A)$, and choose n such that $I(G)^n \eta = 0$. Then η is the image of an element $\mu \in K_0^G(A_k)$ for some k, and by choosing k sufficiently large we may assume that $I(G)^n \mu = 0$. In particular $(d - \lambda)^n \mu = 0$. Since $K_0^G(A_k) \simeq R(G)^{s(k)}$, we can write $\mu = (\mu_1, \ldots, \mu_{s(k)})$ for $\mu_j \in R(G)$, and we have $(d - \lambda)^n \mu_j = 0$ for all j. Regarding elements of $R(G)$ as functions on G, we see that $(d - \lambda)\mu_j = 0$ for all j, whence $(d - \lambda)\eta = 0$. We now have

$$d\eta = \lambda\eta = Ind_{\{e\}}^G(Res_{\{e\}}^G(\eta)) = Ind_{\{e\}}^G(r(\eta)),$$

where the middle equality follows from proposition 5.1.3 (2) and the fact that $Ind_{\{e\}}^G(1) = \lambda$. Consequently, if $r(\eta) = 0$ then $d\eta = 0$, whence $\eta = 0$ because $K_0^G(A)$ is torsion-free by corollary 9.1.3. Thus, r is injective.

(4)$\Longrightarrow$(5) and (4)$\Longrightarrow$(3). We use the notation from the previous part of the proof, in particular the diagram (*) with r_k given by (**) for all k, and as there we let d be the order of G. We further assume that $A_0 = \mathbf{C}$ and contains the identity of A. Fix k, and let $\varphi : R(G)^{s(k)} \longrightarrow K_0^G(A)$ and $\psi : \mathbf{Z}^{s(k)} \longrightarrow K_0(A)$ be the maps on K-theory coming from the map $A_k \longrightarrow A$. Then $r \cdot \varphi = \psi \cdot r_k$. Since r is an isomorphism, φ must vanish on $Ker(r_k)$. By standard properties of direct limits, and because $Ker(r_k)$ is finitely generated as an $R(G)$-module, the image of $Ker(r_k)$ in $R(G)^{s(l)} = K_0^G(A_l)$ is zero for sufficiently large l. By deleting some of the algebras A_m, we may therefore assume that $Ker(r_k) \subset Ker(\varphi_k)$ for all k. Since $Ker(r_k) = I(G)K_0^G(A_k)$, we conclude that the $R(G)$-module structure on $K_0^G(A)$ is given by $\rho\eta = \varepsilon(\rho)\eta$ for $\rho \in R(G)$ and $\eta \in K_0^G(A)$. This is (3).

Write $\varphi_k : R(G)^{s(k)} \longrightarrow R(G)^{s(k+1)}$ as an $s(k) \times s(k+1)$ matrix $(\eta_{ij}^{(k)})$ with entries in $R(G)$. The fact that φ_k is zero on the kernel of r_k implies that each $\eta_{ij}^{(k)}$ annihilates $I(G)$. Lemma 9.2.1 therefore implies that $\eta_{ij}^{(k)} = m_{ij}^{(k)} \lambda$ for some integer $m_{ij}^{(k)}$, where λ is the class of the regular representation. The commutativity of the diagram (*) now implies that ψ_k is given by the matrix $(\varepsilon(\eta_{ij}^{(k)})) = d\,(m_{ij}^{(k)})$, since $\varepsilon(\lambda) = d$. Since this is true for all k, it follows that multiplication by d in the direct limit $K_0(A)$ is surjective. Since $K_0(A)$ is torsion-free, multiplication by d is also injective, so it has an inverse, which we write as multiplication by d^{-1}.

We now compute $K_0^G(A \otimes B)$ relative to the action $\iota \otimes \beta$. We first need to compute $K_0^G(B)$. By proposition 9.2.2, we have $K_0^G(B) \simeq R(G)[\lambda^{-1}]$. Since $\lambda \cdot I(G) = 0$, this group is in fact $(R(G)/I(G))[\varepsilon(\lambda)^{-1}] = \mathbf{Z}[\frac{1}{d}]$, and its $R(G)$-module structure is given by $\rho m = \varepsilon(\rho) m$ for $\rho \in R(G)$ and $m \in \mathbf{Z}[\frac{1}{d}]$. The order is easily seen to be the one obtained by regarding $\mathbf{Z}[\frac{1}{d}]$ as a subset of $\mathbf{R}$. Now $K_0^G(A \otimes B) \simeq K_0(A) \otimes K_0^G(B)$ by the previous lemma, and it is easily seen that the map sending $\eta \otimes \frac{m}{d^s} \in K_0(A) \otimes K_0^G(B)$, where $m, s \in \mathbf{Z}$, to $m d^{-s} \eta \in K_0(A)$ defines an order isomorphism of $K_0(A) \otimes K_0^G(B)$ with $K_0(A)$. (Here the positive elements of $K_0(A) \otimes K_0^G(B)$ are those in the statement of the previous lemma. Multiplication by d^{-s} makes sense on $K_0(A)$ by the previous paragraph, and is order-preserving because $K_0(A)$ is unperforated as an ordered group — see section 3 of [31].) This map is in fact an isomorphism of $R(G)$-modules when $K_0(A)$ is given the $R(G)$-module structure $\rho\eta = \varepsilon(\rho)\eta$ for $\rho \in R(G)$ and $\eta \in K_0(A)$. It follows that $K_0^G(A \otimes B) \simeq K_0(A)$ as ordered $R(G)$-modules. Since the class of the identity in B is $1 \in \mathbf{Z}[\frac{1}{d}]$, this isomorphism also preserves the class of the identity.

We now return to the map $r : K_0^G(A) \longrightarrow K_0(A)$. It is known to be an isomorphism of $R(G)$-modules, and it obviously preserves the order and the class of the identity. Furthermore, if $\eta \in K_0(A)$ is positive, then so is $Ind_{\{e\}}^G(\eta)$. Since $(d - \lambda) r^{-1}(\eta) = 0$, we can use proposition 5.1.3 (2) to see that $Ind_{\{e\}}^G(\eta) = d\, r^{-1}(\eta)$. Since $K_0^G(A)$ is unperforated as an ordered group, being the K_0 group of the AF algebra $C^*(G, A)$, it follows that $r^{-1}(\eta)$ is positive. So we have shown that r is in fact an order isomorphism. Combining this with the results of the previous paragraph, we see that $K_0^G(A)$ and $K_0^G(A \otimes B)$ are isomorphic as partially ordered $R(G)$-modules via an isomorphism which preserves the class of the identity. (We recall that the action of G on A is α, but the action of G on $A \otimes B$ is $\iota \otimes \beta$.) By theorem III.1 of [45], there is an equivariant isomorphism $(G, A, \alpha) \simeq (G, A \otimes B, \iota \otimes \beta)$, which is (5).

(3)$\Longrightarrow$(1). Trivial. Q.E.D.

We now specialize further, to the case of actions on UHF algebras.

9.2.6 Theorem. Let G be a finite group, and let α be a locally representable action of G on a UHF algebra A. Then the following conditions are equivalent.

(1) The action α is K-free.

(2) $C^*(G,A,\alpha)$ is a UHF algebra.

(3) The action α can be written as an infinite tensor product of multiples of the regular representation.

Proof. (1)⟹(2). By the previous theorem, the restriction map $K_0^G(A) \longrightarrow K_0(A)$ is an isomorphism, and the last paragraph of the proof shows that it is in fact an order isomorphism. Consequently $K_0(C^*(G,A)) \simeq K_0(A)$ as ordered groups. It follows that $C^*(G,A)$ is also a UHF algebra. (See [34], 6.1.)

(2)⟹(3). If we knew that the action α is outer, then this would be exactly corollary III.3 of [45]. But outerness is not used in the proof of the relevant part of the conclusion of that corollary. (We note that the one thing it is used for, to show that $C^*(G,A)$ is strongly Morita equivalent to A^G, follows from corollary 7.1.5 anyway, since the crossed product is by assumption simple and the closed linear span of the elements in corollary 7.1.5 is always an ideal.)

(3)⟹(1). This is easily proved, either by a direct calculation using $K_0^G(\varinjlim A_n) \simeq \varinjlim K_0^G(A_n)$ and lemma 9.2.1, or by the previous theorem. Q.E.D.

9.2.7 Corollary. A K-free locally representable action of a finite group G on a UHF algebra A is of product type. A product type action α is K-free if and only if it is outer and satisfies the nine equivalent conditions of [44], theorem IV.3. In particular, α is K-free if and only if α is outer and A is a free A^G-module.

Proof. The first statement is part of the previous theorem. To verify the second statement, we note that condition (3) of the previous theorem is one of the equivalent conditions of [44], theorem IV.3, so we need only show that a K-free action α is outer. That is, we must show that α_g is not inner for any $g \in G - \{e\}$. So let $g \in G$ and assume that α_g is inner, that is, that there is a unitary $u \in A$ such that $\alpha_g(a) = uau^*$ for all $a \in A$. Let n be the order of G. Then u^n is in the center of A and is hence a multiple $\zeta \cdot 1$ of the identity of A, since A is a UHF algebra. Consequently, with H the subgroup of G generated by g and with $\zeta^{-\frac{1}{n}}$ some n-th root of ζ^{-1}, the action $\alpha|_H$ is given by conjugation by the representation $g^k \longrightarrow (\zeta^{-\frac{1}{n}}u)^k$ of H in the unitary group of A. So $\alpha|_H$ is inner, whence $K_0^H(A) \simeq R(H) \otimes K_0(A)$. However, $\alpha|_H$ is K-free by theorem 9.2.4, whence $K_0^H(A) \simeq K_0(A)$, again by theorem 9.2.4. Since $K_0(A)$ is of rank 1 as an abelian group, this is impossible unless H is the one element group, that is, $g = e$. Thus, α is outer.

The third statement is an immediate consequence of the second statement, because A being a free A^G-module is one of the equivalent conditions of [44], theorem IV.3. Q.E.D.

9.3. Examples

We now calculate some examples. Even though the results of the previous section have shown that locally representable actions behave very nicely in certain ways, we are able to find among them many of the counterexamples promised in earlier chapters. The following lemma will be useful.

9.3.1 Lemma. Let u be a unitary in a finite dimensional matrix algebra M_d, and suppose that n is a positive integer such that $u^n = 1$. Then u defines a product type action $k \longrightarrow \bigotimes_1^\infty \mathrm{ad}(u^k)$ of $G = \mathbf{Z}/n\mathbf{Z}$ on the UHF algebra $B = \bigotimes_1^\infty M_d$. (Here $(\mathrm{ad}\, v)(a) = vav^*$.) This action has locally discrete K-theory if and only if every n-th root of 1 appears as an eigenvalue of u with the same multiplicity, and in that case it is totally K-free.

Proof. This is immediate from proposition 9.2.2. Q.E.D.

9.3.2 Example. Using the notation of the previous lemma, let $G = \mathbf{Z}/2\mathbf{Z}$ and let $u = \begin{pmatrix} 1 & 0 \\ 0 & -1 \end{pmatrix}$. Then the resulting action β of G on $B = \bigotimes_1^\infty M_2$ is totally K-free. However, the induced action on *Prim* (B) is trivial. (*Prim* (B) consists of only one point, since B is simple.) In particular, the converse of theorem 4.3.8 is false, and theorem 8.2.5 is false for algebras which are not type I. Furthermore, theorems 7.2.8 and 7.4.3, together with proposition 7.3.3, imply that α is hereditarily saturated, and proposition 7.6.2 now implies that α is hereditarily K-saturated. This example therefore also shows that theorem 8.2.3 is false for algebras which are not type I: (G,B,β) satisfies *all* of the conditions (1), (2), and (3) of theorem 8.2.3, but the action of G on *Prim* (B) is not free.

9.3.3 Example. Again using the notation of lemma 9.3.1, let $G = \mathbf{Z}/4\mathbf{Z}$ and let $u = \begin{pmatrix} 1 & 0 \\ 0 & i \end{pmatrix}$. The resulting action β of G is not K-free. However, if H is the two element subgroup of G, then $\beta|_H$ is the action of the previous example, which is K-free. Thus, theorem 5.2.6 is false if prime order subgroups are considered instead of cyclic subgroups.

9.3.4 Example. Once again, we use the notation of lemma 9.3.1. Let $G = \mathbf{Z}/6\mathbf{Z}$, and let ζ be a primitive sixth root of 1, say $\zeta = \frac{1}{2} + \frac{i\sqrt{3}}{2}$. Let u be the diagonal matrix

$$u = \begin{bmatrix} 1 & & & & & \\ & \zeta & & & 0 & \\ & & \zeta & & & \\ & & & \zeta^2 & & \\ & 0 & & & \zeta^2 & \\ & & & & & -1 \end{bmatrix}.$$

Then the action β of G on $B = \bigotimes_1^\infty M_6$ is not K-free. However, the restriction of β to the subgroup of order 3 is generated as in lemma 9.3.1 by the unitary

$$u^2 = \begin{bmatrix} 1 & & & & & \\ & \zeta^2 & & & 0 & \\ & & \zeta^2 & & & \\ & & & \zeta^4 & & \\ & 0 & & & \zeta^4 & \\ & & & & & 1 \end{bmatrix},$$

and is hence K-free. Furthermore, the restriction of β to the subgroup of order 2 is generated by the unitary

$$u^3 = \begin{bmatrix} 1 & & & & & \\ & -1 & & & 0 & \\ & & -1 & & & \\ & & & 1 & & \\ & 0 & & & 1 & \\ & & & & & -1 \end{bmatrix},$$

and is thus also K-free. Therefore, even K-freeness of the restrictions of an action to all subgroups of prime power order does not imply that the action is K-free.

9.3.5 Example. Still using the notation of lemma 9.3.1, let $G = \mathbf{Z}/2\mathbf{Z}$ and let

$$u = \begin{bmatrix} 1 & 0 & 0 \\ 0 & 1 & 0 \\ 0 & 0 & -1 \end{bmatrix}.$$

Then the corresponding action β of G on $B = \bigotimes_1^\infty M_3$ is not K-free. However, we are going to prove below that β is hereditarily saturated. Thus theorem 8.3.5 is false for algebras which are not type I, and the converses of theorems 7.4.2 and 7.4.3 are false.

Since β is hereditarily saturated, theorem 7.2.8 and theorem 3.5 of [58] imply that $C^*(G,B,\beta)$ is simple. Let τ be the nontrivial element of $\hat{G}$. Then $\hat{\beta}_\tau$ acts trivially on the primitive ideal space of $C^*(G,B,\beta)$, but there is $\eta \in K_0(C^*(G,B,\beta))$ such that $(id - (\hat{\beta}_\tau)_*)^n(\eta) \neq 0$ for all n, since β is not K-free. Thus, lemma 8.3.2 is false for algebras which are not of type I.

The proof that β is hereditarily saturated is contained in the following proposition.

9.3.6 Proposition. Let β be a product type action of a finite abelian group G on a UHF algebra B, of the form described in proposition 9.2.2, for some finite dimensional representation space (G,V,u) of G. Assume that the Arveson spectrum $Sp(u)$ (see [79], 8.1.6; the definition is reproduced for compact G at the beginning of section 7.2) contains the identity of $\hat{G}$. Then β is hereditarily saturated if and only if $Sp(u)$ generates $\hat{G}$ as a group.

Proof. Since B is simple, it follows from theorem 7.2.8 that β is hereditarily saturated if and only if the Connes spectrum satisfies $\Gamma(\beta) = \hat{G}$. By corollary 2.3 of [13], we have $\Gamma(\beta) = \hat{G}$ if and only if whenever $(U_n)_{n\geq 1}$ and $(V_n)_{n\geq 1}$ are two sequences of subsets of $\hat{G}$, both satisfying the condition (*) below, then there exists n such that $U_n \cap V_n \neq \emptyset$. The condition these sequences must satisfy is:

$$\text{There is } n \text{ such that } U_n \neq \emptyset \text{, and for all } n, \bigcup_{\tau \in Sp(u)} \tau U_n \subset U_{n+1}. \tag{*}$$

(We do not need to assume explicitly that the sets U_n are open, as in [13], since $\hat{G}$ is finite. Also we write the group operation in $\hat{G}$ as multiplication rather than addition.)

Suppose $Sp(u)$ generates $\hat{G}$, and let $(U_n)_{n\geq 1}$ and $(V_n)_{n\geq 1}$ be two sequences of sets satisfying (*). By assumption there is k such that $U_k \neq \emptyset$. Since $\hat{G}$ is finite, there is an integer m such that every element of $\hat{G}$ is the sum of at most m elements of $Sp(u)$. Since $1 \in Sp(u)$, the second part of (*) implies that $U_j = \hat{G}$ for all $j \geq k+m$. By (*) for $(V_n)_{n\geq 1}$, there is also an integer l such that $V_l \neq \emptyset$. Choosing j greater than both l and $k+m$, we obtain $U_j \cap V_j \neq \emptyset$. This shows that β is hereditarily saturated.

Conversely, suppose $Sp(u)$ does not generate $\hat{G}$. Let H be the subgroup of $\hat{G}$ generated by $Sp(u)$, and let σ be an element of $\hat{G}$ which is not in H. Set $U_n = H$ and $V_n = H\sigma$ for all n. Then the sequences U_n and V_n satisfy (*), and $U_n \cap V_n = \emptyset$ for all n. So β is not hereditarily saturated. Q.E.D.

Of course, in example 9.3.5, we have $Sp(u) = \hat{G}$.

9.3.7 Remark. Proposition 9.3.6 actually covers the cases in which $1 \notin Sp(u)$ as well: replacing u by the representation $g \longrightarrow \tau(g)^{-1} u_g$ for some $\tau \in Sp(u)$ has the effect of replacing $Sp(u)$ by $\tau^{-1} Sp(u)$, and does not change β. Therefore, in the general case, β is hereditarily saturated if and only if $\tau^{-1} Sp(u)$ generates $\hat{G}$ for some (and therefore any) $\tau \in Sp(u)$.

9.3.8 Example. Let $G = S^1$, and for each positive integer m define a representation ρ_m of G on $\mathbf{C}^m$ by the formula

$$\rho_m(z) = \begin{bmatrix} 1 & & & & \\ & z & & 0 & \\ & & \cdot & & \\ & & & \cdot & \\ & 0 & & & \cdot \\ & & & & z^{m-1} \end{bmatrix}.$$

Let $B = \bigotimes_{k=1}^{\infty} M_{k!}$, and let $\beta_z = \bigotimes_{k=1}^{\infty} ad\, \rho_{k!}(z)$. Then β is a product type action of G on B. We can compute the equivariant K-theory: it is $\varinjlim_k R(G)$, where the map from the k-th term to the $(k+1)$-th term is multiplication by $[\rho_{k!}]$. Since $R(G)$ is an integral domain (it is a Laurent polynomial ring $\mathbf{Z}[t, t^{-1}]$ with t being the class of the identity representation), this direct limit can be identified with the $R(G)$-submodule of the field of fractions of $R(G)$ generated by the elements $1, [\rho_1]^{-1}, [\rho_1]^{-1}[\rho_2]^{-1}$, $[\rho_1]^{-1}[\rho_2]^{-1}[\rho_3]^{-1}, \ldots$. Clearly no power of the augmentation ideal can annihilate any nonzero element of the field of fractions of $R(G)$, so β does not have locally discrete K-theory.

Now let H be any closed subgroup of G. Then H is generated by a primitive n-th root of unity ζ for some n. The equivariant K-theory of $\beta|_H$ is given by $\varinjlim_k R(H)$, where now the map from the k-th term to the $(k+1)$-th term is multiplication by $[\rho_{k!}|_H]$. For $k \geq n$, the representation $\rho_{k!}|_H$ is a multiple of the regular representation H, and it follows that $I(H)$ annihilates the image of $K_0^H\left(\bigotimes_1^k M_{j!}\right)$ in $K_0^H(B)$. So $\beta|_H$ has locally discrete K-theory and is hence K-free. We have thus produced an action of S^1 which is not K-free but such that its restrictions to every nontrivial subgroup is K-free. In particular, an action of a compact Lie group need not be K-free even if its restrictions to all finite subgroups are K-free.

The examples we have considered so far have all involved product type actions, for which K-freeness implies total K-freeness. This is certainly not true for general actions on AF algebras — recall that in example 4.2.3, a K-free but not totally K-free action of $\mathbf{Z}/2\mathbf{Z} \times \mathbf{Z}/2\mathbf{Z}$ on the finite dimensional algebra M_2 was constructed. We are now going to produce an action of $\mathbf{Z}/4\mathbf{Z}$ which is K-free but such that its restriction to the two element subgroup is not K-free. This example should be considered as a companion to example 4.2.3. It shows that K-freeness does not imply total K-freeness even for cyclic groups. In particular, corollary 8.2.4 is false for algebras which are not of type I.

9.3.9 Example. Let $G = \mathbf{Z}/4\mathbf{Z}$. We construct an AF algebra as follows. Let $A_0 = \mathbf{C} \oplus \mathbf{C}$, and let $A_n = M_2 \otimes A_{n-1}$. We regard A_n as a direct sum of 2 copies of M_{2^n}. Before defining the embeddings and the action, we introduce the matrices defined recursively by $w_0 = 1 \in \mathbf{C}$ and

$$w_{n+1} = \begin{pmatrix} w_n & 0 \\ 0 & -w_n \end{pmatrix} \in M_{2^{n+1}}.$$

Thus, w_n is a unitary diagonal matrix for all n. Now define $j_n : A_n \longrightarrow A_{n+1}$ by

$$j_n(a,b) = \left(\begin{pmatrix} a & 0 \\ 0 & b \end{pmatrix}, \begin{pmatrix} b & 0 \\ 0 & w_n a w_n^* \end{pmatrix} \right)$$

and define $\alpha_n : A_n \longrightarrow A_n$ by $\alpha_n(a,b) = (w_n b w_n^*, a)$. Since $w_n^2 = 1$ for all n, it is easy to see that α_n^4 is the identity. Some calculations using the fact that $w_n^2 = 1$ also show that $j_n \cdot \alpha_n = \alpha_{n+1} \cdot j_n$ for all n. We now set $A = \varinjlim A_n$. (Notice that j_n is injective for all n.) Then there exists a unique automorphism α of A such that $\alpha|_{A_n} = \alpha_n$ for all n. Let $\beta : G \longrightarrow Aut(A)$ be the action of G which sends a generator g_0 of G to α. Then (G, A, β) is an AF G-algebra.

We now show that β has locally discrete K-theory. It is sufficient to show that $1 - \tau_0$ annihilates the image of $K_0^G(A_n)$ in $K_0^G(A_{n+1})$ for each n, where τ_0 is a generator of $\widehat{G}$. (This follows from the fact that $1 - \tau_0$ generates $I(G)$ and that $K_0^G(A) = \varinjlim K_0^G(A_n)$.) We may as well choose the generator τ_0 such that $\tau_0(g_0) = i$. Accordingly, let V be a finite dimensional representation space of G, and let $p \in L(V) \otimes A_n$ be a G-invariant projection. Since g_0 is a generator of G, the unitary representation of G on V can be written in the form $g_0^k \longrightarrow u^k$, where $u \in L(V)$ is a unitary such that $u^4 = 1$. We identify $L(V) \otimes A_n$ with the direct sum of two copies of $L(V) \otimes M_{2^n}$, and we make a similar identification for $L(V) \otimes A_{n+1}$. Write $\overline{w}_n$ for $1 \otimes w_n \in L(V) \otimes M_{2^n}$, and let

$$\overline{j}_n = id_{L(V)} \otimes j_n : L(V) \otimes A_n \longrightarrow L(V) \otimes A_{n+1}.$$

Let δ_0 be the automorphism $\delta_0(a) = uau^*$ of $L(V)$, and let $\overline{\alpha}_n = \delta_0 \otimes \alpha_n$ and $\overline{\alpha}_{n+1} = \delta_0 \otimes \alpha_{n+1}$. Thus $\overline{\alpha}_n$ and $\overline{\alpha}_{n+1}$ are the automorphisms given by the action of g_0 on $L(V) \otimes A_n$ and $L(V) \otimes A_{n+1}$. Further, let $\delta = \delta_0 \otimes id_{M_{2^n}}$.

Since $L(V) \otimes A_n \simeq (L(V) \otimes M_{2^n}) \oplus (L(V) \otimes M_{2^n})$, we can write $p = (q, r)$, where q and r are projections in $L(V) \otimes M_{2^n}$. The G-invariance of p implies that

$$(q,r) = \overline{\alpha}_n(q,r) = (\overline{w}_n \delta(r) \overline{w}_n^*, \delta(q)).$$

Thus, $r = \delta(q)$ and $\overline{w}_n \delta^2(q) \overline{w}_n^* = q$. Let $x = (u \otimes 1)q$, and let

$$v = \left(\begin{pmatrix} 0 & ix^* \\ ix & 0 \end{pmatrix}, \begin{pmatrix} 0 & \delta(x^*) \\ \delta(x) & 0 \end{pmatrix} \right),$$

which is an element of $L(V) \otimes A_{n+1}$. Now $x^* x = q$ and $x x^* = (u \otimes 1) q (u \otimes 1)^* = \delta(q)$. Therefore

$$v^*v = vv^* = \left[\begin{pmatrix} q & 0 \\ 0 & \delta(q) \end{pmatrix}, \begin{pmatrix} \delta(q) & 0 \\ 0 & \delta^2(q) \end{pmatrix}\right]$$

$$= \left[\begin{pmatrix} q & 0 \\ 0 & r \end{pmatrix}, \begin{pmatrix} r & 0 \\ 0 & \overline{w}_n q \overline{w}_n^* \end{pmatrix}\right] = \bar{j}_n(p).$$

Here, the fact that $\delta^2(q) = \overline{w}_n q \overline{w}_n^*$ follows from the equations $\overline{w}_n \delta^2(q) \overline{w}_n^* = q$ and $\overline{w}_n^* = \overline{w}_n$. We furthermore have

$$\bar{\alpha}_{n+1}(v) = \left[\begin{pmatrix} 0 & -\overline{w}_n \delta^2(x^*) w_n^* \\ -\overline{w}_n \delta^2(x) \overline{w}_n^* & 0 \end{pmatrix}, \begin{pmatrix} 0 & i\delta(x^*) \\ i\delta(x) & 0 \end{pmatrix}\right].$$

Now $\overline{w}_n \delta^2(x) \overline{w}_n^* = x$, since this holds for $u \otimes 1$ and q. Therefore also $\overline{w}_n \delta^2(x^*) \overline{w}_n^* = x^*$. Consequently $\bar{\alpha}_{n+1}(v) = iv$.

Now define an action γ of G on M_2 by

$$\gamma_g(a) = \begin{pmatrix} 1 & 0 \\ 0 & \tau_0(g) \end{pmatrix} a \begin{pmatrix} 1 & 0 \\ 0 & \tau_0(g) \end{pmatrix}^*$$

for $g \in G$ and $a \in M_2$. Further set

$$e_1 = \begin{pmatrix} 1 & 0 \\ 0 & 0 \end{pmatrix} \quad \text{and} \quad e_2 = \begin{pmatrix} 0 & 0 \\ 0 & 1 \end{pmatrix}.$$

Give $M_2 \otimes A_{n+1}$ the diagonal action $\gamma \otimes \beta$. Then by lemma 2.7.5, the canonical isomorphism $K_0^G(M_2 \otimes A_{n+1}) \simeq K_0^G(A_{n+1})$ identifies $[e_1 \otimes \bar{j}_n(p)]$ with $[\bar{j}_n(p)]$ and $[e_2 \otimes \bar{j}_n(p)]$ with $\tau_0[\bar{j}_n(p)]$. Set $y = \begin{pmatrix} 0 & v \\ 0 & 0 \end{pmatrix}$, which is an element of $M_2 \otimes L(V) \otimes A_{n+1}$. Then it is easily checked, using the formulas $vv^* = v^*v = \bar{j}_n(p)$ and $\bar{\alpha}_{n+1}(v) = iv$, that y is a G-invariant partial isometry such that $yy^* = e_1 \otimes \bar{j}_n(p)$ and $y^*y = e_2 \otimes \bar{j}_n(p)$. Consequently

$$[\bar{j}_n(p)] = [e_1 \otimes \bar{j}_n(p)] = [e_2 \otimes \bar{j}_n(p)] = \tau_0[\bar{j}_n(p)],$$

and thus $1 - \tau_0$ annihilates the image of $[p]$ in $K_0^G(A_{n+1})$. Since n and V were arbitrary, and $p \in L(V) \otimes A_n$ was an arbitrary G-invariant projection, we conclude that $1 - \tau_0$ annihilates the image in $K_0^G(A)$ of every $K_0^G(A_n)$. Thus, β has locally discrete K-theory, and, by proposition 9.1.4, it follows that β is K-free.

We now show that, for H the two element subgroup $\mathbf{Z}/2\mathbf{Z}$ of G, the restricted action $\beta|_H$ is not K-free. Let $p = (1,0) \in A_0$. Then p is an α^2-invariant projection in A. (Recall that $\alpha = \beta_{g_0}$.) We are going to show that $\alpha^2|_{pAp}$ is the identity. Let $i_n : A_0 \longrightarrow A_n$ be the inclusion map, thus $i_n = j_{n-1} \circ j_{n-2} \circ \cdots \circ j_0$. Write $i_n(p) = (q_n, r_n)$. Then it is easy to show by induction on n that $w_n q_n = q_n w_n = q_n$ and $w_n r_n = r_n w_n = -r_n$. If $(a, b) \in A_n$, then

$$\alpha^2(i_n(p)(a,b)i_n(p)) = (w_n q_n a q_n w_n^*, w_n r_n b r_n w_n^*) = i_n(p)(a,b)i_n(p).$$

So $\alpha^2|_{pA_np}$ is the identity for all n. Thus $\alpha^2|_{pAp}$ is the identity, as required. Since $K_0(pAp)$ is not a 2-torsion group (it is torsion-free — see corollary 9.1.3), lemma 4.1.5 implies that $\beta|_H|_{pAp}$ does not have locally discrete K-theory. Therefore proposition 4.2.13 implies that $\beta|_H$ is not K-free, as desired.

Notice that similar arguments show that $(1-p)A(1-p)$ is also invariant for α^2, and that α^2 is multiplication by -1 on $pA(1-p)$ and $(1-p)Ap$. Therefore α^2 is actually inner, and given by conjugation by the unitary $2p-1$.

9.3.10 Remark. This example is interesting for several other reasons as well. Since β is K-free but $\beta|_H$ is not, it follows from theorem 9.2.4 that β is not locally representable. (Note that to prove directly that an action is not locally representable, one must show that it does not have the correct form relative to *any* increasing sequence (A_n) of finite dimensional invariant subalgebras whose union is dense.) However, it is easily seen that the algebra A is the UHF algebra $\bigotimes_1^\infty M_2$. So we have produced an action of $\mathbf{Z}/4\mathbf{Z}$ on a UHF algebra which not only isn't of product type, but isn't even locally representable. Consequently theorem IV.1 of [45] fails without the assumption that the automorphism have prime order. (Note that all automorphisms of a UHF algebra are approximately inner, by theorem 2.3 of [9].)

We can also compute $K_0^G(A)$. It follows from proposition 2.9.6 that $A_n \simeq C(G \times_{\mathbf{Z}/2\mathbf{Z}} M_{2^n})$ for each n, and the action of $\mathbf{Z}/2\mathbf{Z}$ is inner by lemma 4.2.12. Therefore $K_0^G(A_n) \simeq R(\mathbf{Z}/2\mathbf{Z})$ by proposition 2.9.4, and the $R(G)$-module structure is given by restriction. The map from $K_0^G(A_n)$ to $K_0^G(A_{n+1})$ was shown in the example to annihilate $(1-\tau_0)\eta$ for every $\eta \in K_0^G(A_n)$. Since this map multiplies the dimensions of projections by 2 and is an order-preserving $R(G)$-module homomorphism, it must be multiplication by $\sigma(1+\tau_0)$ for some $\sigma \in \widehat{G}$. So we obtain in the limit

$$K_0^G(A) \simeq R(\mathbf{Z}/2\mathbf{Z})[(1+\tau_0|_{\mathbf{Z}/2\mathbf{Z}})^{-1}] \simeq \mathbf{Z}[\tfrac{1}{2}],$$

and the order is easily seen to be the usual one. It follows that $C^*(G,A,\beta)$ is UHF. So theorem 9.2.6 and corollary 9.2.7 are false without local representability, even for cyclic groups. (It is still possible, however, that K-freeness could be equivalent to the crossed product being UHF for arbitrary AF actions on UHF algebras.) This calculation also shows that corollary III.3 of [45] fails without local representability and outerness. (Note that local representability implies outerness under the other hypotheses of the corollary, but that the action β fails to be outer in the appropriate sense.)

Finally, we point out that $\widetilde{\Gamma}(\beta) = \widehat{G}$ by theorem 7.4.3, while $\Gamma(\beta|_{\mathbf{Z}/2\mathbf{Z}}) = \{1\}$ since $\beta|_{\mathbf{Z}/2\mathbf{Z}}$ is inner. Thus, this example exhibits the same behavior of both the ordinary and strong Connes spectrum under restriction to subgroups as is found in example 4.2.3.

9.4. The Main Theorem on K-Freeness of AF Actions

We start by formalizing a construction that has been used in several places already, namely in examples 7.4.6 and 9.3.9.

9.4.1 Definition. Let G be a compact abelian group, and let (G,A,α) be a G-algebra. Let G act on $H = l^2(\widehat{G})$ via the representation $u_g\xi(\tau) = \tau(g)\xi(\tau)$. (As is well known, this is just a disguised form of the regular representation.) Let $K(H)$ have the associated inner action of G, and let $K(H)\otimes A$ have the diagonal action. For $\tau \in \widehat{G}$, let e_τ be the projection on the one-dimensional subspace spanned by the vector ξ defined as follows:

$$\xi(\sigma) = \begin{cases} 1 & \sigma = \tau \\ 0 & \sigma \neq \tau. \end{cases}$$

Then e_τ is a G-invariant projection in $K(H)$. We define a homomorphism $\varphi_\tau^A : A \longrightarrow K(H)\otimes A$ by $\varphi_\tau^A(a) = e_\tau \otimes a$.

In the two examples mentioned above, we made use of the fact that under the identification $K_*^G(K(H)\otimes A) \simeq K_*^G(A)$, the map $(\varphi_\tau^A)_*$ becomes multiplication by τ. In example 7.4.6, we actually showed that φ_τ^A was G-homotopic to φ_1^A; this was needed to prove that a certain action had discrete K-theory. In fact, more can be said.

9.4.2 Lemma. Let G be a compact abelian group, and let (G,A,α) be a separable G-algebra. Suppose that for each $\tau \in \widehat{G}$, the maps φ_τ^A and φ_1^A of definition 9.4.1 are G-homotopic. Then α has discrete KK-theory; in fact, $I(G)KK_G^0(A,A) = 0$.

Proof. As in definition 3.2.13, the homomorphisms φ_1^A and φ_τ^A define elements $[\varphi_1^A]$ and $[\varphi_\tau^A]$ of $KK_G^0(A, K(H)\otimes A)$. It follows from theorem 3.5.10 and its proof that $[\varphi_1^A]$ is an invertible element of $KK_G^0(A, K(H)\otimes A)$. It further follows from remark 3.2.11 that $[\varphi_\tau^A] = \tau \cdot [\varphi_1^A]$, where τ is considered to be an element of $R(G)$. Since φ_τ^A and φ_1^A are G-homotopic, we have $[\varphi_\tau^A] = [\varphi_1^A]$. (See remark 3.2.11 again.) We therefore have, in $KK_G^0(A,A)$,

$$(1-\tau)\, 1_A = [\varphi_1^A][\varphi_1^A]^{-1} - [\varphi_\tau^A][\varphi_1^A]^{-1} = 0\,.$$

This is true for all $\tau \in \widehat{G}$, so by proposition 4.4.4, the action α has discrete KK-theory. In fact, since the elements $1-\tau$ generate $I(G)$, it is easily seen that $I(G)KK_G^0(A,A) = 0$. Q.E.D.

It is, of course, highly unlikely that there is any converse to this lemma.

To motivate the main theorem, we recall Blackadar's result ([10], theorem 3.1) that if α is an automorphism of an AF algebra such that α_* is the identity on $K_0^G(A)$, then α is homotopic to the identity. The most significant part of the main theorem asserts that if α is a totally K-free and AF action of a finite abelian group G on an AF algebra A, then in fact φ_τ^A is G-homotopic to φ_1^A for all $\tau \in \widehat{G}$. The proof of this implication requires the following two lemmas.

9.4.3 Lemma. Let G be a finite abelian group, and let (G,A,α) be an AF G-algebra. If α is totally K-free, then for every $\tau \in \widehat{G}$ there is a homotopy $t \longrightarrow \psi_t$ for $t \in [0,1]$ of endomorphisms of $C^*(G,A,\alpha)$ such that ψ_0 is the identity, $\psi_1 = \widehat{\alpha}_\tau$, and ψ_t commutes with $\widehat{\alpha}_\sigma$ for every $t \in [0,1]$ and every $\sigma \in \widehat{G}$.

Since $\widehat{G}$ is abelian, $\widehat{\alpha}_\tau$ is $\widehat{G}$-equivariant for every $\tau \in \widehat{G}$. Thus, this lemma really says that $\widehat{\alpha}_\tau$ is $\widehat{G}$-homotopic to the identity of $C^*(G,A,\alpha)$.

9.4.4 Lemma. Let G be a finite abelian group, and let (G,A,α) be any G-algebra. Let $\tau \in \widehat{G}$, and suppose that the automorphism $\widehat{\alpha}_\tau$ of $C^*(G,A,\alpha)$ is $\widehat{G}$-homotopic to the identity. Then φ_τ^A is G-homotopic to φ_1^A.

Lemma 9.4.3 will be proved in section 9.5, and lemma 9.4.4 will be proved in section 9.6.

9.4.5 Theorem. Let G be a finite abelian group, and let (G,A,α) be an AF G-algebra. Then the following are equivalent:

(1) The action α is totally K-free.

(2) $I(H)K_0^H(A) = 0$ for every subgroup H of G.

(3) $I(H)K_0^H(A) = 0$ for every cyclic subgroup H of G.

(4) The action α is KK-free.

(5) $I(G)KK_G^0(A,A) = 0$.

(6) For all $\tau \in \widehat{G}$, the automorphism $\widehat{\alpha}_\tau$ is $\widehat{G}$-homotopic to the identity map on $C^*(G,A,\alpha)$.

(7) For all $\tau \in \widehat{G}$, the map φ_τ^A of definition 9.4.1 is G-homotopic to φ_1^A.

Proof. We will prove (1) $\Longrightarrow$ (6) $\Longrightarrow$ (7) $\Longrightarrow$ (5) $\Longrightarrow$ (1), (1) $\Longleftrightarrow$ (4), and (1) $\Longrightarrow$ (2) $\Longrightarrow$ (3) $\Longrightarrow$ (1).

(1) $\Longrightarrow$ (6). Lemma 9.4.3.

(6) $\Longrightarrow$ (7). Lemma 9.4.4.

(7) $\Longrightarrow$ (5). Lemma 9.4.2.

(5) $\Longrightarrow$ (1). By lemma 5.3.1, the action $\alpha|_H$ has discrete KK-theory for every subgroup H of G. By proposition 4.4.6, $\alpha|_H$ therefore has discrete K-theory for every subgroup H of G. Proposition 9.1.4 now implies that α is totally K-free.

(1) $\Longrightarrow$ (4). Let I be a G-invariant ideal in A. By (1), $\alpha_{(\cdot)}|_I$ is also totally K-free. Since $\alpha_{(\cdot)}|_I$ is also an *AF* action, the implication (1) $\Longrightarrow$ (5), which we have already proved, shows that $I(G)KK_G^0(I,I) = 0$. Since this is true for all G-invariant ideals I of A, we conclude from proposition 4.4.4 that α is KK-free.

(4) $\Longrightarrow$ (1). By the KK-Subgroup Theorem 5.3.6, $\alpha|_H$ is KK-free for every subgroup H of G. By proposition 4.4.6, $\alpha|_H$ is therefore K-free for every subgroup H of

G, that is, α is totally K-free.

(1) $\Longrightarrow$ (2). This follows immediately from proposition 9.1.9.

(2) $\Longrightarrow$ (3). Trivial.

(3) $\Longrightarrow$ (1). By proposition 9.1.4, $\alpha|_H$ is K-free for every cyclic subgroup H of G. Therefore α is totally K-free by corollary 9.1.5. Q.E.D.

If G is a finite nonabelian group, then conditions (6) and (7) of the theorem no longer make sense. However, most of the rest of the theorem remains valid. The proof uses the results about subgroups in section 5.2 to derive the nonabelian result from the abelian result. The key step is the following lemma.

9.4.6 Lemma. Let G be a finite group, and let (G,A,α) be an AF G-algebra. Then $KK_G^0(A,A)$ is torsion-free as an abelian group.

This lemma will be proved in section 9.7, where a general method for the computation of equivariant KK-theory for AF-algebras will be developed.

9.4.7 Theorem. Let G be a finite group, and let (G,A,α) be an AF G-algebra. Then the following are equivalent:

(1) The action α is totally K-free.

(2) The restricted action $\alpha|_H$ has discrete K-theory for every subgroup H of G.

(3) $I(H)K_0^H(A)=0$ for every cyclic subgroup H of G.

(4) The action α is KK-free.

(5) The action α has discrete KK-theory.

Proof. We will prove (1)$\Longrightarrow$(4)$\Longrightarrow$(5)$\Longrightarrow$(2)$\Longrightarrow$(3)$\Longrightarrow$(1).

(1) $\Longrightarrow$ (4). Since α is totally K-free, it follows that $\alpha|_H$ is totally K-free for every cyclic subgroup H of G. The implication (1) $\Longrightarrow$ (4) of the previous theorem now shows that $\alpha|_H$ is KK-free for every cyclic subgroup H of G. If I is any G-invariant ideal in A, then $\alpha|_H|_I$ therefore has discrete KK-theory for every cyclic subgroup H of G. By lemma 9.4.6, $KK_G^0(I,I)$ is torsion-free as an abelian group, whence by proposition 5.2.10, we find that $\alpha_{(\cdot)}|_I$ has discrete KK-theory. Since I was arbitrary, it follows that α is KK-free.

(4) $\Longrightarrow$ (5). Trivial.

(5) $\Longrightarrow$ (2). Let H be a subgroup of G. By lemma 5.3.1, $\alpha|_H$ has discrete KK-theory. So by proposition 4.4.6, it follows that $\alpha|_H$ has discrete K-theory.

(2) $\Longrightarrow$ (3). If H is a cyclic subgroup of G, then $\alpha|_H$ has discrete K-theory by assumption. Since H is abelian, we obtain $I(H)K_0^H(A)=0$ from proposition 9.1.9.

(3) $\Longrightarrow$ (1). By proposition 9.1.4, $\alpha|_H$ is K-free for every cyclic subgroup H of G. Then α is totally K-free by corollary 9.1.5. Q.E.D.

9.4.8 Remark. If A and B are AF algebras then one can use the Universal Coefficient Theorem for KK-theory ([92], theorem 1.17) to show that there is an isomorphism

$$KK^0(A,B) \simeq Hom\,(K_0(A), K_0(B)), \tag{*}$$

and that furthermore the Kasparov product from $KK^0(A,B) \times KK^0(B,C)$ to $KK^0(A,C)$ corresponds to composition of homomorphisms of abelian groups in the opposite order. It is tempting to conjecture that such an isomorphism also exists in equivariant KK-theory, at least for AF actions. In fact, however, such a result cannot hold, at least for $G = \mathbf{Z}/4\mathbf{Z}$ or $G = \mathbf{Z}/2\mathbf{Z} \times \mathbf{Z}/2\mathbf{Z}$. Indeed, suppose that there were an $R(G)$-module isomorphism

$$KK_G^0(A,A) \simeq End_{R(G)}(K_0^G(A)). \tag{**}$$

(Here $End_R(M)$ is the set of all R-module endomorphisms of M.) If α is any K-free AF action of G on A, then $I(G)K_0^G(A) = 0$ by proposition 9.1.9. Therefore $I(G)End_{R(G)}(K_0^G(A)) = 0$. We would thus obtain $I(G)KK_G^0(A,A) = 0$, and theorem 9.4.5 would then imply that α is totally K-free. However, in example 4.2.3 there is a K-free action of $\mathbf{Z}/2\mathbf{Z} \times \mathbf{Z}/2\mathbf{Z}$ on a finite dimensional C^*-algebra which is not totally K-free, and in example 9.3.9 there is a K-free and AF action of $\mathbf{Z}/4\mathbf{Z}$ on an AF algebra which is not totally K-free. So (**) cannot hold for these G-algebras. However, it is shown in theorem 9.7.11 that the equivariant analog of (*) holds if the action of G on A is locally representable.

We have said nothing about actions of infinite groups on AF algebras. Indeed, we conjecture the following:

9.4.9 Conjecture. There does not exist a totally K-free AF action of a nontrivial connected compact Lie group on an AF algebra.

As evidence for the conjecture, we offer the following. First, it is certainly true for commutative AF algebras, since then total K-freeness implies freeness, and a commutative AF algebra has a totally disconnected maximal ideal space. Secondly, our attempt to construct such an action of the circle group on a UHF algebra resulted in an action whose restrictions to all proper subgroups are K-free, but which itself is not K-free. The resulting action appears in example 9.3.8. It is clear from the calculation there that no product type action of the circle on a UHF algebra can be K-free.

9.5. Proof of Lemma 9.4.3

The proof is inspired by the proof of the result of Blackadar mentioned in the previous section. Thus, we will need to construct certain unitaries in finite dimensional subalgebras which commute with everything in a given finite dimensional subalgebra and which satisfy a condition related to K-theory. The necessary condition turns out, in the unital case, to be that $\alpha_g(u) = \tau(g)u$ for $g \in G$, when τ is a given element of $\widehat{G}$. The construction of these unitaries takes up several preparatory lemmas. The first of them is:

9.5.1 Lemma. Let G be a finite abelian group, and let (G,A,α) be a finite dimensional G-algebra. Let $p \in A$ be a G-invariant projection, and let $\tau \in \widehat{G}$. If $[p] = \tau \cdot [p]$ in $K_0^G(A)$, then there is an element $u \in A$ such that $uu^* = u^*u = p$ and $\alpha_g(u) = \tau(g)u$ for all $g \in G$.

Proof. Since A is finite dimensional, it is a direct sum of its minimal G-invariant ideals. We may therefore consider only the case in which A is G-simple. Now A is the direct sum of its minimal ideals. Furthermore, if I is a minimal ideal, and H is the subgroup of $h \in G$ such that $\alpha_h[I] = I$, then $A = \bigoplus_{g \in R} \alpha_g[I]$ for any set R of coset representatives of H in G. By proposition 2.9.6, we have $A \simeq C(G \times_H I)$, where I carries the action $\beta = \alpha|_H|_I$ of H. Since p is G-invariant, p corresponds to a constant function $p(g) = p_0$, where $p_0 \in I$ is an H-invariant projection. Since there is, by proposition 2.9.4, an $R(G)$-module isomorphism $K_0^G(C(G \times_H I)) \simeq K_0^H(I)$ which sends $[p]$ to $[p_0]$, we find that $(\tau|_H) \cdot [p_0] = [p_0]$ in $K_0^H(I)$.

Give M_2 the H-action $\gamma_h(a) = u_h a u_h^*$, where $u_h = \begin{pmatrix} 1 & 0 \\ 0 & \tau(h) \end{pmatrix}$. Further let

$$e_1 = \begin{pmatrix} 1 & 0 \\ 0 & 0 \end{pmatrix} \quad \text{and} \quad e_2 = \begin{pmatrix} 0 & 0 \\ 0 & 1 \end{pmatrix}.$$

Then the maps $\psi_i : I \longrightarrow M_2 \otimes I$ defined by $\psi_i(a) = e_i \otimes a$ are H-equivariant when $M_2 \otimes I$ is given the diagonal action of H. By lemma 2.7.5, the isomorphism $K_0^H(M_2 \otimes I) \simeq K_0^H(I)$ identifies $[\psi_1(p_0)]$ with $[p_0]$ and $[\psi_2(p_0)]$ with $(\tau|_H) \cdot [p_0]$. Thus, $[e_1 \otimes p_0] = [e_2 \otimes p_0]$ in $M_2 \otimes I$.

Let q_i be the projection in $C^*(H, M_2 \otimes I)$ corresponding to $e_i \otimes p_0$ under the isomorphism of theorem 2.8.3 (7). Then $q_i C^*(H, M_2 \otimes I) \simeq (e_i \otimes p_0)(M_2 \otimes I)$ as right $C^*(H, M_2 \otimes I)$-modules. (Note that both $M_2 \otimes I$ and $C^*(H, M_2 \otimes I)$ are unital because they are finite dimensional.) Therefore $[q_1] = [q_2]$ in $K_0(C^*(H, M_2 \otimes I))$. Since $C^*(H, M_2 \otimes I)$ is finite dimensional, it follows that q_1 and q_2 are unitarily equivalent. So $q_1 C^*(H, M_2 \otimes I)$ and $q_2 C^*(H, M_2 \otimes I)$ are isomorphic as $C^*(H, M_2 \otimes I)$-modules. Therefore $(e_1 \otimes p_0)(M_2 \otimes I)$ and $(e_2 \otimes p_0)(M_2 \otimes I)$ are isomorphic as $C^*(H, M_2 \otimes I)$-modules, hence also as $(H, M_2 \otimes I)$-modules. Consequently there is an H-invariant partial isometry $v \in M_2 \otimes I$ such that $vv^* = e_1 \otimes p_0$ and $v^*v = e_2 \otimes p_0$. (See the proof of theorem 2.4.4.) We may replace v by $(e_1 \otimes p_0)v(e_2 \otimes p_0)$ and still have $vv^* = e_1 \otimes p_0$ and $v^*v = e_2 \otimes p_0$. (This calculation has been done before, in the proof of part (2) of lemma 2.6.3.) Therefore $v = \begin{pmatrix} 0 & u_0 \\ 0 & 0 \end{pmatrix}$ for some $u_0 \in I$ such that $u_0 u_0^* = u_0^* u_0 = p_0$. The fact that v is G-invariant implies that $\beta_h(u_0) = \tau(h) u_0$ for $h \in H$.

Define $u \in C(G \times_H I)$ by $u(g) = \tau(g)^{-1} u_0$. The equation $\beta_h(u_0) = \tau(h) u_0$ for $h \in H$ implies that u actually is in $C(G \times_H I)$. Clearly $uu^* = u^*u = p$. Finally, we have

$$\alpha_g(u)(g_0) = u(g^{-1} g_0) = \tau(g) u(g_0)$$

for $g, g_0 \in G$, as desired. Q.E.D.

The next lemma is the equivariant version of a well known factorization lemma. (See [30], part 1, chapter 2, proposition 5, for an analogous result for von Neumann algebras.)

9.5.2 Lemma. Let G be any topological group, and let A be a unital G-algebra. Suppose M_k is embedded in A as a G-invariant subalgebra containing the unit of A. Let $p \in M_k$ be a minimal projection. Then there is an action β of G on pAp such that A is equivariantly isomorphic to $M_k \otimes pAp$. For $g \in G$, the action β_g is given as follows: if $v \in M_k$ is any partial isometry with $vv^* = p$ and $v^*v = \alpha_g(p)$, then $\beta_g(a) = v\alpha_g(a)v^*$ for $a \in pAp$.

Proof. We first assume that the action of G on M_k is inner, that is, that $\alpha_g(x) = u_g x u_g^*$ for $x \in M_k$, where $g \longrightarrow u_g$ is a unitary representation of G in M_k. For $g \in G$ define $x_g \in M_k$ by $x_g = pu_g^*$. Then $x_g^* x_g = u_g p u_g^* = \alpha_g(p)$ and $x_g x_g^* = p$. One furthermore has the cocycle condition $x_g \alpha_g(x_h) = x_{gh}$ for $g, h \in G$. (To prove it, use the fact that $\alpha_g(x_h) = u_g x_h u_g^*$.)

Define β_g by $\beta_g(a) = x_g \alpha_g(a) x_g^*$ for $a \in pAp$. Since any partial isometry v in M_k from $\alpha_g(p)$ to p is a scalar multiple of x_g, we do have $\beta_g(a) = v\alpha_g(a)v^*$ as required. We first show that β_g is an automorphism of pAp. It is clear from the definition of x_g that $\beta_g(a) \in pAp$. Now let $a, b \in pAp$. Then

$$\beta_g(ab) = x_g \alpha_g(apb) x_g^* = x_g \alpha_g(a) x_g^* x_g \alpha_g(b) x_g^* = \beta_g(a)\beta_g(b).$$

where the middle step follows from the relation $x_g^* x_g = \alpha_g(p)$. An easier calculation shows that $\beta_g(a^*) = \beta_g(a)^*$. Next, we observe that for $g, h \in G$ and $a \in A$, we have

$$\beta_{gh}(a) = x_{gh} \alpha_{gh}(a) x_{gh}^* = x_g \alpha_g(x_h \alpha_h(a) x_h^*) x_g^* = \beta_g(\beta_h(a)),$$

where the middle step this time is the cocycle condition. Since $g \longrightarrow x_g$ is continuous, we also find that $g \longrightarrow \beta_g(a)$ is continuous for $a \in pAp$. Thus, we have shown that β is a continuous action of G on pAp.

Let $\{e_{ij} : i, j = 1, n\}$ be a set of matrix units for M_k with $e_{11} = p$. Define a homomorphism $\varphi : M_k \otimes pAp \longrightarrow A$ by $\varphi(e_{ij} \otimes a) = e_{i1} a e_{1j}$. It is a standard result that φ is in fact a homomorphism. (See for example [30], part 1, chapter 2, proposition 5, for the von Neumann algebra case.) Furthermore, if y and z are elements of M_k with $yp = y$ and $pz = z$, then we have $\varphi(yz \otimes a) = yaz$ for any $a \in pAp$. (To see this, write

$$y = \sum_1^n \lambda_i e_{i1} \quad \text{and} \quad z = \sum_1^n \mu_i e_{1i}$$

for appropriate $\lambda_i, \mu_i \in \mathbf{C}$.)

We next show that φ is equivariant. Let $g \in G$ and $a \in pAp$. Then

$$\varphi(\alpha_g(e_{ij}) \otimes \beta_g(a)) = \varphi((u_g e_{i1} e_{1j} u_g^*) \otimes (x_g \alpha_g(a) x_g^*)) = u_g e_{i1} x_g \alpha_g(a) x_g^* e_{1j} u_g^*$$

$$= u_g\, e_{i1} p u_g^* \alpha_g(a) u_g\, p\, e_{1j}\, u_g^* = \alpha_g(e_{i1}) \alpha_g(a) \alpha_g(e_{1j}) = \alpha_g(\varphi(e_{ij} \otimes a)),$$

as desired. Here, the second step follows from the end of the previous paragraph, since $u_g\, e_{i1} p = u_g\, e_{i1}$ and $p e_{1j}\, u_g^* = e_{1j}\, u_g^*$. This completes the proof in the special case in which the action of G on M_k is inner.

We now consider the general case. It is a standard result that there is a topological group T (obtained as an extension of G by the circle group S^1), a homomorphism $\pi: T \longrightarrow G$ which identifies G with a quotient of T, and a unitary representation $t \longrightarrow u_t$ of T in M_n such that $u_t\, x u_t^* = \alpha_{\pi(t)}(x)$ for $x \in M_n$ and $t \in T$. (The group T is given by

$$T = \{(g, u) \in G \times U : \alpha_g |_{M_n} = \mathrm{ad}(u)\},$$

where U is the group of unitaries in M_n and $\mathrm{ad}(u)$ is the automorphism of M_n given by $\mathrm{ad}(u)(a) = uau^*$. The homomorphism π is the restriction to T of the projection on the first factor.) Let $\bar{\alpha}$ be the action of T on A defined by $\bar{\alpha}_t = \alpha_{\pi(t)}$. The previous part of the proof then provides an action $\bar{\beta}$ of T on pAp such that there is an equivariant isomorphism of T-algebras $A \simeq M_k \otimes pAp$. If $t \in Ker(\pi)$, then u_t is a multiple of the identity, so that for $a \in pAp$, we have $\bar{\beta}_t(a) = pu_t^* \bar{\alpha}_t(a) u_t\, p = a$, since $\bar{\alpha}_t(a) = a$. Therefore $\bar{\beta}$ defines a continuous action β of $G \simeq T/Ker(\pi)$ on pAp, and the isomorphism $A \simeq M_k \otimes pAp$ is clearly G-equivariant. It is immediate that the formula for β in the statement of the lemma is correct in this case also. Q.E.D.

9.5.3 Corollary. Assume the hypotheses of the previous lemma. If (A_n) is an increasing sequence of finite dimensional α-invariant subalgebras of A which contain M_k and whose union is dense in A, then (pA_np) is an increasing sequence of finite dimensional β-invariant subalgebras of pAp whose union is dense in pAp.

Proof. That (pA_np) is an increasing sequence of finite dimensional subalgebras whose union is dense in pAp is immediate from the fact that $p \in A_n$. That pA_np is β-invariant follows from the fact that $M_k \subset A_n$ and the formula for β. Q.E.D.

9.5.4 Remark. It is presumably true that if α is an AF action, then so is β, at least if G is compact. To prove this, one would have to show that if α is an AF action, then the finite dimensional G-invariant subalgebras A_n can always be chosen to contain M_k. The proof would be even messier than the proof of the corresponding fact in the case in which G is trivial. (See [12] and [36].) Since we don't need this result, we don't attempt to prove it.

9.5.5 Lemma. Let G be a finite abelian group, and let (G, A, α) be an AF G-algebra, where α is totally K-free. Write $A = \overline{\bigcup_{n=0}^{\infty} A_n}$, where (A_n) is an increasing sequence of finite dimensional G-invariant subalgebras, and let p be the identity in A_0. Then for $\tau \in \widehat{G}$ there is an integer n and a unitary $u \in pA_np$ such that u commutes with all elements of A_0 and $\alpha_g(u) = \tau(g) u$ for all $g \in G$.

Proof. Replacing A by pAp, we may clearly assume that $p = 1$. (The induced action on pAp is totally K-free by proposition 4.2.13.)

We first consider the case in which A_0 is simple. Then $A_0 \simeq M_k$ for some k, and lemma 9.5.2 yields an action β of G on qAq, where q is a minimal projection in A_0, such that there is an equivariant isomorphism $\varphi : A_0 \otimes qAq \longrightarrow A$. By corollary 9.5.3, β is an AF action. By proposition 9.1.6, β is totally K-free. Proposition 9.1.9 now implies that $[q] = \tau[q]$ in $K_0^G(qAq)$. Since $qAq = \varinjlim qA_n q$, and the subalgebras $qA_n q$ are β-invariant by corollary 9.5.3, it follows that there is n such that $[q] = \tau[q]$ in $K_0^G(qA_n q)$. By lemma 9.5.1, there is $w \in qA_n q$ such that $w^*w = ww^* = q$ and $\beta_g(w) = \tau(g)w$ for $g \in G$. Let $u = \varphi(1 \otimes w)$. Since φ is equivariant, we clearly have $\alpha_g(u) = \tau(g)u$ for $g \in G$, and obviously u is a unitary in A_n. Finally, if $x \in A_0$, then $x = \varphi(x \otimes q)$, whence

$$uxu^* = \varphi((1 \otimes w)(x \otimes q)(1 \otimes w^*)) = \varphi(x \otimes 1) = x .$$

So u commutes with all elements of A_0, and the proof of the special case is complete.

Next we consider the case in which A_0 is G-simple but not necessarily simple. As in the proof of lemma 9.5.1, there is an ideal I of the form M_l and a subgroup H of G such that $A_0 \simeq C(G \times_H I)$, where I carries the totally K-free action $\beta = \alpha|_H|_I$ of H. Let p be the identity of I. By the special case we have already done, there is an integer n and an element $u_0 \in pA_n p$ such that u_0 commutes with the elements of I, $u_0 u_0^* = u_0^* u_0 = p$, and $\alpha_g(u_0) = \tau(g)u_0$ for $g \in H$. Let S be a system of coset representatives of H in G, and set $u = \sum_{g \in S} \overline{\tau(g)}\,\alpha_g(u_0)$. Then $u^*u = uu^* = 1$, and u clearly commutes with all elements of A_0. To see that $\alpha_g(u) = \tau(g)u$, we first prove that u does not depend on the choice of the system S of coset representatives. It is sufficient to show that the function $g \longrightarrow \overline{\tau(g)}\,\alpha_g(u_0)$ is constant on cosets of H in G, that is, if $g^{-1}h \in H$ then $\overline{\tau(g)}\,\alpha_g(u_0) = \overline{\tau(h)}\,\alpha_h(u_0)$. Since $\alpha_k(u_0) = \tau(k)u_0$ for $k \in H$, this is clear. It now follows that

$$\alpha_g(u) = \sum_{h \in S} \overline{\tau(h)}\,\alpha_{gh}(u_0) = \tau(g) \sum_{h \in gS} \overline{\tau(h)}\,\alpha_h(u_0) = \tau(g)u ,$$

as desired, because gS is also a system of coset representatives of H in G.

Finally, we consider the general case. Write A_0 as a direct sum $B_1 \oplus \cdots \oplus B_l$ of G-simple ideals, and let p_i be the identity in B_i. By the case just considered, there is for each i an integer n_i and a unitary $u_i \in p_i A_{n_i} p_i$ which commutes with all of the elements of B_i and satisfies $\alpha_g(u_i) = \tau(g)u_i$ for $g \in G$. Let n be the largest of the n_i, and set $u = u_1 + \cdots + u_l \in A_n$. Then u clearly satisfies the conclusion of the lemma.

Q.E.D.

We are now in a position to construct a $\widehat{G}$-homotopy from $\widehat{\alpha}_\tau$ to the identity of $C^*(G,A)$, roughly following Blackadar's method in section 2 of [9].

Proof of lemma 9.4.3. We are given a finite abelian group G, an AF algebra $A = \overline{\bigcup_{n=0}^{\infty} A_n}$, where (A_n) is an increasing sequence of finite dimensional subalgebras, and a totally K-free action $\alpha : G \longrightarrow Aut(A)$ such that $\alpha_g[A_n] = A_n$ for all n. We assume that $A_0 = \mathbf{C}$. (This does not imply that A_0 contains the identity of A — indeed, A need not have an identity.) We let e_n be the identity of A_n. The previous lemma implies that by deleting some of the A_n's, we may assume that for all n there is a unitary $u_{n+1} \in e_n A_{n+1} e_n$ such that $\alpha_g(u_{n+1}) = \tau(g) u_{n+1}$ for all $g \in G$ and u_{n+1} commutes with all elements of A_n.

For $n \geq 1$ let B_n be the set of G-invariant elements of $e_{n-1} A_{n+1} e_{n-1}$ which commute with all elements of A_{n-1}. Then B_n is a finite dimensional C^*-algebra, and $u_n^* u_{n+1} = u_n^* e_{n-1} u_{n+1} \in B_n$, since e_{n-1} commutes with u_{n+1} and

$$\alpha_g(u_n^* u_{n+1}) = \overline{\tau(g)}\, \tau(g)\, u_n^* u_{n+1} = u_n^* u_{n+1}.$$

We are now going to define a continuous path of unitaries $t \longrightarrow w_t \in A$ (or $w_t \in A^+$ if A doesn't have an identity) for $t \in [0, \infty)$. Let 1 be the identity of A (or A^+ if $1 \notin A$).

We start by letting $t \longrightarrow v_t$ for $t \in [0,1]$ be a continuous path of unitaries in $e_0 A_1 e_0$ such that $v_0 = e_0$ and $v_1 = u_1$. Then set $w_t = v_t + 1 - e_0$ for $t \in [0,1]$. For $t \in [n, n+1]$, let $t \longrightarrow v_t$ be a continuous path of unitaries in B_n such that $v_n = e_{n-1}$ and $v_{n+1} = u_n^* u_{n+1}$ (which is a unitary element in B_n). Further let $t \longrightarrow y_t$ be a continuous path of unitaries in $(e_n - e_{n-1}) A_{n+1} (e_n - e_{n-1})$ such that $y_n = e_n - e_{n-1}$ and $y_{n+1} = (e_n - e_{n-1}) u_{n+1}$. The paths $t \longrightarrow v_t$ and $t \longrightarrow y_t$ exist because the algebras B_n and $(e_n - e_{n-1}) A_{n+1} (e_n - e_{n-1})$ are finite dimensional, so that their unitary groups are connected. Now set $w_t = u_n v_t + y_t + (1 - e_n)$. This definition is consistent at the endpoints of $[n, n+1]$ because $w_n = u_n + (1 - e_{n-1})$ and

$$w_{n+1} = u_n^* u_n u_{n+1} + (e_n - e_{n-1}) u_{n+1} + (1 - e_n) = u_{n+1} + (1 - e_n).$$

It is obvious that w_t is a unitary element of A or A^+ (depending on whether A has a unit) for all $t \in [0, \infty)$, and that $t \longrightarrow w_t$ is continuous.

The important properties of $t \longrightarrow w_t$ are the following:

$$w_t^* a w_t = a \quad \text{for } a \in A_n \text{ and } t \geq n+1, \tag{*}$$

and

$$a\, \alpha_g(w_t) = \tau(g)\, a w_t \quad \text{for } g \in G,\ a \in A_n, \text{ and } t \geq n+1. \tag{**}$$

To prove (*), we observe that, for $t \in [n, n+1]$ and $a \in A_{n-1}$, we have $a u_n = u_n a$ by the definition of u_n; also $a v_t = v_t a$ because $v_t \in B_n$; furthermore, $a y_t = y_t a = 0$ because $a = a e_{n-1} = e_{n-1} a$ and $e_{n-1} y_t = y_t e_{n-1} = 0$; and finally $a(1 - e_n) = (1 - e_n) a = 0$. The definition of w_t now implies that $a w_t = w_t a$, so that $w_t^* a w_t = a$. Since $A_m \subset A_{n-1}$ for $m \leq n-1$, this equation holds for all $a \in A_m$ whenever $m \leq n-1$ and $t \in [n, n+1]$. The property (*) now follows. We next prove (**). If $t \in [n, n+1]$ and $a \in A_{n-1}$, then

$$a\, \alpha_g(w_t) = a e_{n-1}\, \alpha_g(w_t) = a\, \alpha_g(e_{n-1} w_t) = a\, \alpha_g(u_n v_t) = \tau(g)\, a u_n v_t = \tau(g) a w_t,$$

where the second to the last equality follows from the equations $\alpha_g(v_t) = v_t$ and $\alpha_g(u_n) = \tau(g)u_n$, while the last equality is obtained using the projection e_{n-1} in a way similar to its use in the first several steps. This proves (**).

Let $\overline{w}_t$ be the element of $C^*(G,A,\alpha)$ (or $C^*(G,A^+,\alpha)$ if A has no unit) defined by

$$\overline{w}_t(g) = \begin{cases} w_t & g = e \\ 0 & g \neq e. \end{cases}$$

Here we have $C^*(G,A,\alpha) = L^1(G,A,\alpha)$ as a *-algebra (since G is finite), and

$$xy(g) = \sum_{h \in G} x(h)\alpha_h(y(h^{-1}g)).$$

(Thus, we use counting measure on G, rather than a measure of total mass 1.) With these conventions, $\overline{w}_t$ is easily seen to be unitary. Furthermore, if $x \in C^*(G,A,\alpha)$, then

$$(\overline{w}_t^* x \overline{w}_t)(g) = w_t^* x(g)\alpha_g(w_t).$$

If $x \in C^*(G, A_n, \alpha)$ and $t \geq n+1$, then (*) and (**) imply that

$$(\overline{w}_t^* x \overline{w}_t)(g) = \tau(g)x(g) = \hat{\alpha}_\tau(x)(g).$$

Set $f_t(x) = \overline{w}_t^* x \overline{w}_t$ for $t \in [0,\infty)$. Then f_t is as automorphism of $C^*(G,A,\alpha)$ with f_0 equal to the identity. Furthermore, f_t depends continuously on t, and $\lim_{t \to \infty} f_t(x) = \hat{\alpha}_\tau(x)$ for x in the dense subset $\bigcup_{n=0}^{\infty} C^*(G, A_n, \alpha)$ of $C^*(G,A,\alpha)$. A standard argument involving the fact that $\| f_t \| = 1$ for all t now shows that $\lim_{t \to \infty} f_t(x) = \hat{\alpha}_\tau(x)$ for all $x \in C^*(G,A,\alpha)$. Consequently, if we set $f_\infty = \hat{\alpha}_\tau$, then $t \longrightarrow f_t$ for $t \in [0,\infty]$ is a homotopy of automorphisms connecting the identity to $\hat{\alpha}_\tau$.

To finish the proof, it remains only to show that $f_t \circ \hat{\alpha}_\sigma = \hat{\alpha}_\sigma \circ f_t$ for all $t \in [0,\infty]$ and $\sigma \in \hat{G}$. For $t = \infty$, this is trivial, so suppose $t < \infty$. Then the result follows immediately from the fact that $\hat{\alpha}_\sigma(\overline{w}_t) = \overline{w}_t$. Q.E.D.

9.6. Proof of Lemma 9.4.4

9.6.1 Notation. In this section, we use the notation of definition 9.4.1. However, we assume from the start that the group G is finite abelian. Thus, $H = l^2(\hat{G})$ with the representation $(u_g\xi)(\tau) = \tau(g)\xi(\tau)$, and $K(H) = L(H)$. Let $e_{\sigma,\tau}$ be the rank one partial isometry in $L(H)$ such that, with ξ_ρ being the function taking the value 1 at $\rho \in \hat{G}$ and 0 elsewhere, we have $e_{\sigma,\tau}\xi_\tau = \xi_\sigma$. In particular, $e_{\tau,\tau}$ is the projection which in definition 9.4.1 was called e_τ.

We further let n be the order of G. If (G,A,α) is a G-algebra, and $\tau \in \hat{G}$, then we write

$$A_\tau = \{a \in A : \alpha_g(a) = \tau(g)a \text{ for all } g \in G\}.$$

Then $A = \bigoplus_{\tau \in \hat{G}} A_\tau$ as a Banach space (but not as an algebra). Therefore, if $a \in A$, there is a unique decomposition of a relative to this direct sum, which we write as $a = \sum_{\tau \in \hat{G}} a_\tau$, where $a_\tau \in A_\tau$ for $\tau \in \hat{G}$.

Since the representation u of G on $l^2(\hat{G})$ is a disguised form of the regular representation, it follows that $C^*(\hat{G}, C^*(G,A,\alpha), \hat{\alpha}) \simeq L(H) \otimes A$ as G-algebras, where the double crossed product carries the second dual action $(\hat{\alpha})\hat{}$. (This is just Takai duality, [79], 7.9.3.) In order to prove lemma 9.4.4, we will need an explicit form of this isomorphism. First, we express the double crossed product in a convenient form.

9.6.2 Lemma. Let G be a finite abelian group, and let (G,A,α) be a G-algebra. Then the double crossed product $C^*(\hat{G}, C^*(G,A,\alpha), \hat{\alpha})$ is isomorphic as a *-algebra to the set of all functions $x : \hat{G} \times G \longrightarrow A$, where multiplication is given by

$$(xy)(\tau,g) = \sum_{\sigma \in \hat{G}} \sum_{h \in G} \sigma(h^{-1}g)\, x(\sigma,h)\, \alpha_h(y(\sigma^{-1}\tau, h^{-1}g)),$$

adjoint is given by

$$x^*(\tau,g) = \tau(g)\,\alpha_g(x(\tau^{-1},g^{-1})^*),$$

and the second dual action is given by

$$((\hat{\alpha})\hat{})_h(x)(\tau,g) = \tau(g)\, x(\tau,g)$$

for $h \in G$.

Proof. This is a straightforward calculation, starting from the definition of the crossed products obtained by using counting measure on G and $\hat{G}$. Note that we have

$$C^*(\hat{G}, C^*(G,A,\alpha), \hat{\alpha}) = L^1(\hat{G}, L^1(G,A,\alpha), \hat{\alpha})$$

as *-algebras, because both G and $\hat{G}$ are finite. Q.E.D.

Notice that this lemma says nothing about what the C^*-algebra norm on the double crossed product is.

9.6.3 Lemma. Let G be a finite abelian group, and let (G,A,α) be a G-algebra. Using the notation of 9.6.1, and the identification of the crossed product by the dual action in lemma 9.6.2, the function

$$\psi : L(H) \otimes A \longrightarrow C^*(\hat{G}, C^*(G,A,\alpha), \hat{\alpha})$$

defined by

$$\psi(e_{\sigma,\rho} \otimes a)(\tau,g) = \frac{1}{n}\rho(g)\tau(g)\, a_{\rho\sigma^{-1}\tau},$$

and extended to $L(H)\otimes A$ by linearity, is an equivariant isomorphism for the diagonal action of G on $L(H)\otimes A$ and the double dual action $(\hat{\alpha})\hat{}$.

Proof. Before starting the calculations required for the proof, we state several facts which will be needed, and introduce some notation. We set, for $g, h \in G$,

$$\delta_{g,h} = \begin{cases} 1 & g=h \\ 0 & g\neq h. \end{cases}$$

Similarly define $\delta_{\sigma,\tau}$ for $\sigma, \tau \in \hat{G}$. The following facts are now easy to check:

$$\sum_{g\in G} \sigma(g)\tau(g)^{-1} = n\delta_{\sigma,\tau} \quad \text{for } \sigma, \tau \in \hat{G}, \tag{1}$$

$$\sum_{\tau\in\hat{G}} \tau(g)\tau(h)^{-1} = n\delta_{g,h} \quad \text{for } g, h \in G, \tag{2}$$

$$(ab)_\lambda = \sum_{\sigma\tau=\lambda} a_\sigma b_\tau \quad \text{for } a, b \in A \text{ and } \lambda\in\hat{G}, \tag{3}$$

and

$$(a^*)_\lambda = (a_{\lambda^{-1}})^* \quad \text{for } a \in A \text{ and } \lambda\in\hat{G}. \tag{4}$$

We now prove that ψ is multiplicative. It is sufficient to prove that, if $\sigma,\rho,\mu,\nu\in\hat{G}$ and $a, b \in A$, then

$$\psi(e_{\sigma,\rho}\otimes a)\psi(e_{\mu,\nu}\otimes b) = \psi(e_{\sigma,\rho}e_{\mu,\nu}\otimes ab).$$

We have:

$$\begin{aligned}
&(\psi(e_{\sigma,\rho}\otimes a)\psi(e_{\mu,\nu}\otimes b))(\tau,g) \\
&= \sum_{\lambda\in\hat{G}}\sum_{h\in G} \lambda(h^{-1}g)\frac{1}{n}\rho(h)\lambda(h)a_{\rho\sigma^{-1}\lambda}\cdot\alpha_h\left[\frac{1}{n}\nu(h^{-1}g)(\lambda^{-1}\tau)(h^{-1}g)b_{\nu\mu^{-1}\lambda^{-1}\tau}\right] \\
&= \frac{1}{n^2}\nu(g)\tau(g)\left[\sum_{h\in G}\rho(h)\mu(h^{-1})\right]\left[\sum_{\lambda\in\hat{G}} a_{\rho\sigma^{-1}\lambda}b_{\nu\mu^{-1}\lambda^{-1}\tau}\right] \\
&= \frac{1}{n^2}\nu(g)\tau(g)n\delta_{\rho,\mu}(ab)_{\rho\sigma^{-1}\nu\mu^{-1}\tau} = \psi(\delta_{\rho,\mu}e_{\sigma,\nu}\otimes ab),
\end{aligned}$$

as desired, since $e_{\sigma,\rho}e_{\mu,\nu} = \delta_{\rho,\mu}e_{\sigma,\nu}$. (The second step in the calculation above is obtained by using the definition of a_τ for $\tau\in\hat{G}$, and rearranging the terms.) A similar but less complicated calculation shows that $\psi(x^*) = \psi(x)^*$ for $x\in L(H)\otimes A$.

Next, we prove that ψ intertwines the diagonal and the double dual actions. It is easily checked that $u_g e_{\sigma,\rho}u_g^* = \rho(g)^{-1}\sigma(g)e_{\sigma,\rho}$. Then we have

$$\begin{aligned}
\psi(u_g e_{\sigma,\rho}u_g^*\otimes\alpha_g(a))(\tau,g) &= \rho(g)^{-1}\sigma(g)\cdot\frac{1}{n}\rho(g)\tau(g)\alpha_g(a)_{\rho\sigma^{-1}\tau} \\
&= \frac{1}{n}\sigma(g)\tau(g)(\rho\sigma^{-1}\tau)(g)a_{\rho\sigma^{-1}\tau} \\
&= \tau(g)\cdot\frac{1}{n}\rho(g)\tau(g)a_{\rho\sigma^{-1}\tau} = ((\hat{\alpha})\hat{}\,)_g(\psi(e_{\sigma,\rho}\otimes a))(\tau,g),
\end{aligned}$$

as desired.

The next step is to prove that ψ is surjective. Let $y : \widehat{G} \times G \longrightarrow A$ be an element of $C^*(\widehat{G}, C^*(G,A,\alpha), \widehat{\alpha})$, and define $x \in L(H) \otimes A$ by

$$x = \sum_{\sigma,\rho \in \widehat{G}} e_{\sigma,\rho} \otimes x_{\sigma,\rho},$$

where

$$x_{\sigma,\rho} = \sum_{\lambda \in \widehat{G}} \sum_{h \in G} \overline{\rho(h)\lambda(h)}\, y(\lambda,h)_{\rho\lambda\sigma^{-1}}$$

for $\sigma, \rho \in \widehat{G}$. Then for $\mu \in \widehat{G}$, we have

$$(x_{\sigma,\rho})_\mu = \sum_{h \in G} \overline{\mu(h)\sigma(h)}\, y(\mu\sigma\rho^{-1}, h)_\mu .$$

Consequently

$$\begin{aligned}
\psi(x)(\tau,g) &= \sum_{\sigma,\rho \in \widehat{G}} \frac{1}{n} \rho(g)\tau(g)(x_{\sigma,\rho})_{\rho\sigma^{-1}\tau} \\
&= \sum_{\rho \in \widehat{G}} \sum_{h \in G} \frac{1}{n} \rho(g)\tau(g)\overline{\rho(h)\tau(h)} \left[\sum_{\sigma \in \widehat{G}} y(\tau,h)_{\sigma\rho^{-1}\tau} \right] \\
&= \sum_{\rho \in \widehat{G}} \sum_{h \in G} \frac{1}{n} (\rho\tau)(gh^{-1})\, y(\tau,h) \\
&= \frac{1}{n} \sum_{h \in G} \tau(h^{-1})\, y(\tau,gh) \left[\sum_{\rho \in \widehat{G}} \rho(h^{-1}) \right] = y(\tau,g).
\end{aligned}$$

Here the second last step is a change of variables, and the last step follows from the relation (2) at the beginning of the proof. Therefore ψ is surjective.

Finally, we show that ψ is injective. Let $x \in L(H) \otimes A$, and assume that $\psi(x) = 0$. Write

$$x = \sum_{\sigma,\rho \in \widehat{G}} e_{\sigma,\rho} \otimes x_{\sigma,\rho},$$

where $x_{\sigma,\rho} \in A$. Then we have, for $\tau \in \widehat{G}$ and $g \in G$,

$$0 = \psi(x)(\tau,g) = \sum_{\sigma,\rho \in \widehat{G}} \frac{1}{n} \rho(g)\tau(g)(x_{\sigma,\rho})_{\rho\sigma^{-1}\tau} = \sum_{\rho \in \widehat{G}} \left[\sum_{\sigma \in \widehat{G}} \frac{1}{n} \rho(g)\sigma(g)(x_{\sigma,\rho\sigma\tau^{-1}})_\rho \right].$$

For $\rho \in \widehat{G}$, let $a(\rho)$ be the expression in parentheses in the last sum above. Then we have $a(\rho) \in A_\rho$. Since $A = \bigoplus_{\rho \in \widehat{G}} A_\rho$, the relation $\sum_{\rho \in \widehat{G}} a(\rho) = 0$ shows that $a(\rho) = 0$ for all ρ. That is,

$$\sum_{\sigma \in \widehat{G}} \frac{1}{n} \rho(g)\sigma(g)(x_{\sigma,\rho\sigma\tau^{-1}})_\rho = 0.$$

This holds for every $g \in G$. So for $\lambda \in \widehat{G}$,

$$0 = \sum_{\sigma \in \widehat{G}} \sum_{g \in G} \frac{1}{n} \lambda(g)\rho(g)\sigma(g)(x_{\sigma,\rho\sigma\tau^{-1}})_\rho = (x_{\rho^{-1}\lambda^{-1},\lambda^{-1}\tau^{-1}})_\rho,$$

by (1). Changing variables, we find that $(x_{\lambda,\tau})_\rho = 0$ for all $\lambda, \tau, \rho \in \widehat{G}$. Therefore $x = 0$, and ψ has been shown to be injective.

The function ψ is continuous because homomorphisms are automatically continuous.

Q.E.D.

Proof of lemma 9.4.4. We are given a $\widehat{G}$-homotopy $t \longrightarrow f_t$ for $t \in [0,1]$ with $f_0 = id_{C^*(G,A)}$ and $f_1 = \widehat{\alpha}_\tau$, and we want to construct a G-homotopy $t \longrightarrow \varphi_t$ for $t \in [0,1]$ with $\varphi_0 = \varphi_1^A$ and $\varphi_1 = \varphi_\tau^A$, where φ_1^A and φ_τ^A are as in definition 9.4.1. In lemma 9.6.2, we wrote the elements of $C^*(\widehat{G}, C^*(G,A,\alpha), \widehat{\alpha})$ as functions from $\widehat{G} \times G$ into A. Here, we identify $C^*(\widehat{G}, C^*(G,A,\alpha), \widehat{\alpha})$ with the set of all functions $x : \widehat{G} \longrightarrow C^*(G,A,\alpha)$, with multiplication

$$(xy)(\tau) = \sum_{\sigma \in \widehat{G}} x(\sigma)\, \widehat{\alpha}_\sigma (y(\sigma^{-1}\tau))$$

and adjoint

$$x^*(\tau) = \widehat{\alpha}_\tau (x(\tau^{-1})^*).$$

We further regard the function ψ constructed in the previous lemma as a function from $L(H) \otimes A$ to functions from $\widehat{G}$ to $C^*(G,A,\alpha)$. We thus write $\psi(x)(\tau)(g)$ rather than $\psi(x)(\tau, g)$, for $x \in L(H) \otimes A$, $\tau \in \widehat{G}$, and $g \in G$.

For $t \in [0,1]$, define

$$\psi_t : L(H) \otimes A \longrightarrow C^*(\widehat{G}, C^*(G,A,\alpha), \widehat{\alpha})$$

by $\psi_t(x)(\sigma) = f_t(\psi(x)(\sigma))$. We claim that $t \longrightarrow \psi_t$ is a G-homotopy. We first show that ψ_t is a homomorphism for each t. We have, for $x, y \in L(H) \otimes A$,

$$\begin{aligned}(\psi_t(x)\psi_t(y))(\sigma) &= \sum_{\rho \in \widehat{G}} f_t(\psi(x)(\rho))\, \widehat{\alpha}_\rho (f_t(\psi(y)(\rho^{-1}\sigma))) \\ &= f_t \left[\sum_{\rho \in \widehat{G}} \psi(x)(\rho)\, \widehat{\alpha}_\rho (\psi(y)(\rho^{-1}\sigma)) \right] \\ &= f_t((\psi(x)\psi(y))(\sigma)) = f_t((\psi(xy)(\sigma)) = \psi_t(xy)(\sigma).\end{aligned}$$

Here the second step follows from the fact that f_t is a homomorphism from $C^*(G,A,\alpha)$ to $C^*(G,A,\alpha)$ which commutes with $\widehat{\alpha}_\rho$ for all ρ, and the second to the last step follows from the fact that ψ is a homomorphism. The equation $\psi_t(x^*) = \psi_t(x)^*$ follows similarly. That $t \longrightarrow \psi_t$ is a homotopy is now obvious.

We now complete the proof of the claim by showing that ψ_t is a G-homotopy. We write $\mathrm{ad}\, u_g(a) = u_g a u_g^*$. Then

$$\begin{aligned}\psi_t((\mathrm{ad}\, u_g \otimes \alpha_g)(x))(\sigma) &= f_t(\psi((\mathrm{ad}\, u_g \otimes \alpha_g)(x))(\sigma)) = f_t(((\widehat{\alpha})\widehat{\ })_g(\psi(x))(\sigma)) \\ &= \sigma(g) f_t(\psi(x)(\sigma)) = ((\widehat{\alpha})\widehat{\ })_g(\psi_t(x))(\sigma),\end{aligned}$$

as required. So $t \longrightarrow \psi_t$ is a G-homotopy.

Now define $\varphi_t : A \longrightarrow L(H)\otimes A$ by $\varphi_t = \psi^{-1}\cdot\psi_t\cdot\varphi_1^A$. Then $t \longrightarrow \varphi_t$ is a G-homotopy. We have $\varphi_0 = \psi^{-1}\cdot\psi_0\cdot\varphi_1^A = \varphi_1^A$, since $\psi_0 = \psi$. Also, for $a \in A$, $\sigma \in \widehat{G}$ and $g \in G$, we have

$$(\psi\cdot\varphi_1)(a)(\sigma)(g) = (\psi_1\cdot\varphi_1^A)(a)(\sigma)(g) = f_1(\psi(e_{1,1}\otimes a)(\sigma))(g)$$

$$= \tau(g)\frac{1}{n}\sigma(g)a_\sigma = \psi(e_{\tau,\tau}\otimes a)(\sigma)(g) = (\psi\cdot\varphi_\tau^A)(a)(\sigma)(g).$$

Since ψ is invertible, we have $\varphi_1 = \varphi_\tau^A$. Therefore $t \longrightarrow \varphi_t$ is a G-homotopy from φ_1^A to φ_τ^A, as desired.

Q.E.D.

9.7. Equivariant KK-Theory of AF Algebras and the Proof of Lemma 9.4.6

In this section we will describe a method of computing $KK_G^0(A,B)$, where G is compact and A and B are AF G-algebras, by reducing the computation to ordinary K-theory. This method will then be applied to prove lemma 9.4.6.

9.7.1 Lemma. Let G be a a second countable compact group, let H be a closed subgroup of finite index, let (H,A,α) be a separable H-algebra, and let (G,B,β) be a separable G-algebra. Then there is a natural isomorphism of $R(G)$-modules

$$KK_G^0(C(G\times_H A),B) \simeq KK_H^0(A,B).$$

We point out that there is an isomorphism

$$KK_G^0(B,C(G\times_H A)) \simeq KK_H^0(B,A),$$

without the hypothesis that H have finite index in G. However, we don't need that result, so we don't prove it.

Proof of lemma 9.7.1. We recall from definition 2.9.2 that

$$C(G\times_H A) = \{f \in C(G,A) : f(g) = \alpha_h(f(gh)) \text{ for all } h\in H \text{ and } g\in G\},$$

and that the G-action is defined by $\vartheta_g(f)(g_0) = f(g^{-1}g_0)$ for $g, g_0 \in G$. Define $\mu_A : A \longrightarrow C(G\times_H A)$ as follows:

$$\mu_A(a)(g) = \begin{cases} 0 & g\notin H \\ \alpha_g^{-1}(a) & g\in H. \end{cases}$$

Then μ_A is clearly a homomorphism. A computation shows that $\mu_A(a)$ is in fact in $C(G\times_H A)$ for $a\in A$. (Notice that $\mu_A(a)$ is a continuous function on G because H, having finite index, is open in G.) Another easy calculation shows that μ is H-equivariant. Now let r be the composite $r = \mu_A^*\cdot Res_H^G$, that is,

$$r: KK_G^0(C(G\times_H A), B) \xrightarrow{Res_H^G} KK_H^0(C(G\times_H A), B) \xrightarrow{\mu_A^*} KK_H^0(A, B).$$

Clearly r is a natural $R(G)$-module homomorphism. (Here, as usual, the $R(H)$-modules are regarded as $R(G)$-modules via the map $R(G) \longrightarrow R(H)$.) For convenience in later calculations, we will write the formula for r as $r(\lambda) = [\mu_A]\cdot Res_H^G(\lambda)$, where $[\varphi]$ as usual stands for the KK element defined by an equivariant homomorphism φ.

We will prove that r is an isomorphism by constructing an inverse. Let $V = l^2(G/H)$. We need the maps $d_B: B \longrightarrow L(V)\otimes B$ and $\pi_B: C(G\times_H B) \longrightarrow L(V)\otimes B$ constructed in definition 5.1.9, as well as the homomorphism

$$I_H^G: KK_H^0(A, B) \longrightarrow KK_G^0(C(G\times_H A), C(G\times_H B))$$

from proposition 5.1.8. We then define

$$i: KK_H^0(A, B) \longrightarrow KK_G^0(C(G\times_H A), B)$$

by $i(\lambda) = I_H^G(\lambda)[\pi_B][d_B]^{-1}$. We now have to show that both $r\circ i$ and $i\circ r$ are the identity map.

As preparation for the first of these, we prove that, for $\lambda \in KK_H^0(A, B)$, we have

$$[\mu_A]\cdot Res_H^G(I_H^G(\lambda)) = \lambda\cdot[\mu_B], \qquad (*)$$

where B is regarded as an H-algebra. Representing λ by a prequasihomomorphism

$$(\varphi, \psi): A \longrightarrow E \triangleright J \longrightarrow K\otimes B,$$

we see that $[\mu_A]\cdot Res_H^G(I_H^G(\lambda))$ is represented by the prequasihomomorphism

$$(\varphi\circ\mu_A, \psi\circ\mu_A): A \longrightarrow C(G\times_H E) \triangleright C(G\times_H J) \longrightarrow K\otimes C(G\times_H B), \qquad (**)$$

where we assume, as we may, that the action of H on K is the restriction of an inner action of G, as in the proof of lemma 5.1.8. Now we have a canonical direct sum decomposition $C(G\times_H E) = \mu_E[E]\oplus E_0$ as H-algebras, where E_0 is the set of functions in $C(G\times_H E)$ which vanish on H. Similarly define J_0 and B_0. Then the prequasihomomorphism $(**)$ defines the same quasihomomorphism as the prequasihomomorphism

$$(\varphi\circ\mu_A, \psi\circ\mu_A): A \longrightarrow \mu_E[E] \triangleright \mu_J[J] \longrightarrow K\otimes\mu_B[B] \longrightarrow K\otimes C(G\times_H B),$$

since the process of constructing the associated quasihomomorphism eliminates the summands E_0, J_0, and B_0. Since μ is a natural isomorphism onto its image, this last prequasihomomorphism represents the element $\lambda\cdot[\mu_B]$. Thus $(*)$ is proved. We now have

$$r\circ i(\lambda) = [\mu_A]\,Res_H^G(I_H^G(\lambda)[\pi_B][d_B]^{-1}) = \lambda[\mu_B]\,Res_H^G([\pi_B][d_B]^{-1}),$$

by $(*)$. Regarding π_B and d_B as maps of H-algebras via restriction, it is easily seen that $\pi_B\cdot\mu_B = d_B$. So $[\mu_B]\,Res_H^G([\pi_B][d_B]^{-1}) = 1_B$, and we have $r\circ i(\lambda) = \lambda$, as desired.

To show that $i \circ r$ is the identity, we first prove that, with $C = C(G\times_H A)$, the G-algebra maps d_C and $\pi_C \cdot I_H^G(\mu_A)$ are G-homotopic. Let ξ_{gH} be the standard basis vector $\xi_{gH} \in V = l^2(G/H)$ corresponding to $gH \in G/H$, let $\xi_0 = \xi_H$, and let

$$\xi_1 = \frac{1}{\sqrt{n}} \sum_{gH \in G/H} \xi_{gH},$$

where n is the index of H in G. Then ξ_0 and ξ_1 are both H-invariant unit vectors in V, and there is therefore a continuous path $t \longrightarrow \xi_t$ for $t \in [0,1]$ of H-invariant unit vectors with ξ_0 and ξ_1 as given. Let p_t be the orthogonal projection on $\mathbf{C}\cdot\xi_t$. Furthermore, for $g \in G$ let χ_{gH} be the characteristic function of gH as a subset of G, and let S be a set of left coset representatives for H in G. Finally, let $g \longrightarrow u_g$ be the usual representation of G on V. Define

$$\varphi_t : C(G\times_H A) \longrightarrow L(V)\otimes C(G\times_H A)$$

by

$$\varphi_t(f) = \sum_{g\in S} u_g p_t u_g^* \otimes \chi_{gH} f .$$

Notice that $\chi_{gH} \in M(C(G\times_H A))$, so that the image of φ_t really is in $L(V)\otimes C(G\times_H A)$. Furthermore, the χ_{gH} are mutually orthogonal central projections, from which it follows that φ_t is a homomorphism for every t. To see that φ_t is equivariant, we first observe that φ_t does not depend on the choice of the set of coset representatives S. Indeed, this follows from the fact that, for $h \in H$, we have $u_h p_t u_h^* = p_t$ and $(gh)H = gH$. Now let $g_0 \in G$. Write $\mathrm{ad}\, u_{g_0}$ for the automorphism $a \longrightarrow u_{g_0} a\, u_{g_0}^*$ of $L(V)$. Since $\vartheta_{g_0}(\chi_{gH}) = \chi_{g_0 gH}$, we obtain, for $f \in C(G\times_H A)$,

$$\begin{aligned} ((\mathrm{ad}\, u_{g_0})\otimes\vartheta_{g_0})(\varphi_t(f)) &= \sum_{g\in S} (u_{g_0 g}\, p_t\, u_{g_0 g}^*)\otimes(\chi_{g_0 gH}\vartheta_{g_0}(f)) \\ &= \sum_{g\in g_0 S} (u_g p_t u_g^*)\otimes(\chi_{gH}\vartheta_{g_0}(f)) = \varphi_t(\vartheta_{g_0}(f)), \end{aligned}$$

since $g_0 S$ is again a set of left coset representatives for H in G. So φ_t is G-equivariant.

It is clear from the definitions that $\varphi_1 = d_C$. We now show that $\varphi_0 = \pi_C \cdot I_H^G(\mu_A)$, and to do so we must compute $\pi_C \cdot I_H^G(\mu_A)$. We may clearly identify the algebra $C(G\times_H C(G\times_H A))$ with the set of continuous functions $f : G\times G \longrightarrow A$ satisfying the relations

$$f(g_1 h, h^{-1} g_2) = f(g_1, g_2) \quad \text{and} \quad \alpha_h(f(g_1, g_2 h)) = f(g_1, g_2)$$

for $g_1, g_2 \in G$ and $h \in H$. With this identification, the homomorphism $I_H^G(\mu_A)$ is given by the formula

$$I_H^G(\mu_A)(f)(g_1, g_2) = \mu_A(f(g_1))(g_2) = \begin{cases} 0 & g_2 \notin H \\ f(g_1 g_2) & g_2 \in H, \end{cases}$$

for $f \in C(G\times_H A)$ and $g_1, g_2 \in G$. (Here we have used the fact that $\alpha_{g_2}^{-1}(f(g_1)) = f(g_1 g_2)$ for $g_2 \in H$.) To compute π_C, we must identify $C(G\times_H C(G\times_H A))$ with $C(G/H)\otimes C(G\times_H A)$ via the isomorphism

$$\omega: C(G/H)\otimes C(G\times_H A) \longrightarrow C(G\times_H C(G\times_H A))$$

of lemma 5.1.7. Regarding $C(G/H)\otimes C(G\times_H A)$ as the set of continuous functions $f: G\times G \longrightarrow A$ satisfying

$$f(g_1 h, g_2) = f(g_1, g_2) \quad \text{and} \quad \alpha_h(f(g_1, g_2 h)) = f(g_1, g_2)$$

for $g_1, g_2 \in G$ and $h \in H$, a calculation shows that ω is given by the formula $\omega(f)(g_1, g_2) = f(g_1, g_1 g_2)$. It follows that

$$[\omega^{-1}\cdot I_H^G(\mu_A)](f)(g_1, g_2) = \begin{cases} 0 & g_1^{-1}g_2 \notin H \\ f(g_2) & g_1^{-1}g_2 \in H. \end{cases}$$

This formula can be rewritten as

$$[\omega^{-1}\cdot I_H^G(\mu_A)](f) = \sum_{g\in S} \chi_{gH}\otimes\chi_{gH} f$$

for $f \in C(G\times_H C(G\times_H A))$, where S is an arbitrary set of left coset representatives for H in G. It now follows from the definitions of π_C and φ_0 that

$$[\pi_C \cdot I_H^G(\mu_A)](f) = \sum_{g\in S} u_g p_0 u_g^* \otimes \chi_{gH} f = \varphi_0(f).$$

Thus, $t \longrightarrow \varphi_t$ is a G-homotopy from $\pi_C \cdot I_H^G(\mu_A)$ to d_C, as desired.

The existence of this homotopy shows that, as elements of $KK_G^0(C(G\times_H A), C(G\times_H A))$, we have

$$I_H^G([\mu_A])[\pi_{C(G\times_H A)}][d_{C(G\times_H A)}]^{-1} = 1_{C(G\times_H A)}.$$

Now let $\lambda \in KK_G^0(C(G\times_H A), B)$. Then

$$\begin{aligned} i\cdot r(\lambda) &= I_H^G([\mu_A]\, Res_H^G(\lambda))[\pi_B][d_B]^{-1} \\ &= I_H^G([\mu_A])\,\tau_{C(G/H)}(\lambda)[\pi_B][d_B]^{-1} = I_H^G([\mu_A])[\pi_{C(G\times_H A)}][d_{C(G\times_H A)}]^{-1}\lambda = \lambda, \end{aligned}$$

where the second step follows from proposition 5.1.8 (2) and (3), and the third step follows from lemma 5.1.11 (2b) and (2c). So $i\cdot r$ is the identity. Q.E.D.

9.7.2 Lemma. Let G be a second countable compact group, let V be a finite dimensional Hilbert space, and let G act continuously on $L(V)$. Then for any separable G-algebras A and B, there is a natural isomorphism of $R(G)$-modules

$$KK_G^0(L(V)\otimes A, B) \simeq KK_G^0(A, L(\tilde{V})\otimes B),$$

where $\tilde{V}$ is the complex conjugate space of V.

Proof. By lemma 4.4.9, there is a natural isomorphism

$$KK_G^0(L(V)\otimes A, B) \simeq KK_G^0(L(\tilde{V})\otimes L(V)\otimes A, L(\tilde{V})\otimes B).$$

As in the proof of that lemma, the action of G on $L(\widetilde{V})\otimes L(V)$ comes from a representation of G on $\widetilde{V}\otimes V$, so that

$$KK_G^0(L(\widetilde{V})\otimes L(V)\otimes A,\, L(\widetilde{V})\otimes B) \simeq KK_G^0(A,\, L(\widetilde{V})\otimes B),$$

by stability. Q.E.D.

9.7.3 Proposition. Let G be a second countable compact group, and let (G,A,α) be a finite dimensional G-simple G-algebra. Then there is a closed subgroup H of finite index in G, and a finite dimensional vector space V with an action of H on $L(V)$, such that, on the category of separable G-algebras, the functor $B \longrightarrow KK_G^*(A,B)$ is naturally isomorphic to the functor $B \longrightarrow K_*^H(L(V)\otimes B)$.

Proof. Since A is finite dimensional, A is the direct sum of finitely many minimal ideals. Let J be one of them, and let H be the stabilizer of J. Then H is a closed subgroup of finite index in G, since in fact it is easily seen that if S is a set of left coset representatives of H in G, then $A \simeq \bigoplus_{g\in S} \alpha_g\,[J]$. (We have used the G-simplicity of A here.) By proposition 2.9.6, we have $A \simeq C(G\times_H J)$. Since $J \simeq L(W)$ for some finite dimensional Hilbert space W, we obtain $A \simeq C(G\times_H L(W))$. Therefore

$$KK_G^0(A,B) \simeq KK_H^0(L(W),B) \simeq KK_H^0(\mathbf{C},\, L(\widetilde{W})\otimes B)$$

by lemmas 9.7.1 and 9.7.2. So, with $V = \widetilde{W}$, we have $KK_G^i(A,B) \simeq K_i^H(L(V)\otimes B)$ for $i = 0$, by theorem 3.6.12. Taking suspensions, we find that the same holds for $i = 1$. Q.E.D.

9.7.4 Corollary. Let G be a second countable compact group, and let A be a finite dimensional G-algebra. Then the equivariant homology theory $KK_G^*(A,\cdot)$ commutes with countable direct limits, that is,

$$KK_G^*(A, \varinjlim B_n) \simeq \varinjlim KK_G^*(A, B_n)$$

for any countable direct system (B_n) of separable G-algebras.

Proof. Writing A as the direct sum of finitely many G-simple ideals, and applying the previous proposition to each, we find that $KK_G^*(A,B)$ is naturally isomorphic to a direct sum of finitely many terms of the form $K_*^H(M_n\otimes B)$ for appropriate closed subgroups H of G and appropriate actions of H on M_n. Since tensoring with M_n and equivariant K-theory both commute with direct limits, so does $KK_G^*(A,\cdot)$. (We assume that the direct system (B_n) is countable in order to ensure that its direct limit is separable.) Q.E.D.

We point out that KK_G is of course always finitely additive (that is, commutes with finite direct sums) in its second variable, but that in general one does *not* have

$$KK_G^0(A, \bigoplus_{n=1}^{\infty} B_n) \simeq \bigoplus_{n=1}^{\infty} KK_G^0(A, B_n). \qquad (*)$$

For example, let G be trivial, take $A = C_0(\mathbf{Z}^+)$, where $\mathbf{Z}^+$ is the set of positive integers, and take $B_n = \mathbf{C}$ for all n. Then a calculation shows that (*) does not hold. (See [92], remark 7.12). However, KK_G *is* additive in its first variable, provided that the algebras appearing in the first variable are nuclear. For G trivial, this is shown in [90], theorem 4.8, in terms of *Ext*. It is easy to check that the arguments given there apply equally well to equivariant *Ext*, provided one is careful to work in the G-multiplier algebras and to average the strictly positive elements used in lemma 4.7 of [90] over the group so as to obtain G-invariant ones. Since $Ext_G(A,B) \simeq KK_G^1(A,B)$ by [55], section 7, theorem 1, we obtain:

9.7.5 Theorem. Equivariant KK-theory is additive in its first variable. That is, if G is a second countable compact group, and if A_n for $n \in \mathbf{Z}^+$ and B are separable G-algebras, with all A_n nuclear, then there is a natural isomorphism

$$KK_G^i\Big(\bigoplus_{n=1}^{\infty} A_n, B\Big) \simeq \prod_{n=1}^{\infty} KK_G^i(A_n, B).$$

9.7.6 Corollary. Let G be a second countable compact group, let (A_n) be a direct system of separable G-algebras indexed by $\mathbf{Z}^+$, and let B be a separable G-algebra. Then there is a natural short exact sequence

$$0 \longrightarrow \varprojlim{}^1 KK_G^{1-i}(A_n, B) \longrightarrow KK_G^i(\varinjlim A_n, B) \longrightarrow \varprojlim KK_G^i(A_n, B) \longrightarrow 0.$$

Proof. This is essentially theorem 7.1 of [96]. Note that the arguments given there for the non-equivariant case apply just as well to equivariant cohomology theories, provided that all of the maps are equivariant. Q.E.D.

We need one more lemma.

9.7.7 Lemma. Let G be a compact group, and let A be a finite dimensional G-algebra. Then $K_0^G(A)$ is torsion-free as an abelian group and $K_1^G(A) = 0$.

Proof. It was shown in lemma 2.2.4 that there is an increasing net (e_n) of central projections in $L^1(G)$ such that $\bigcup_n e_n L^1(g) e_n$ is dense on $L^1(G)$. Furthermore, each $e_n L^1(G) e_n$ is finite dimensional. Therefore $e_n L^1(G,A) e_n$ is also finite dimensional, of dimension $dim\,(e_n L^1(G) e_n) \cdot dim\,(A)$, and it is isomorphic to $e_n C^*(G,A) e_n$ as a *-algebra. Since (e_n) is an increasing approximate identity for $C^*(G,A)$ (see lemma 2.2.4 again), we have

$$C^*(G,A) = \overline{\bigcup_n e_n C^*(G,A) e_n}.$$

So $C^*(G,A)$ is an AF algebra, at least in the generalized sense that it is a direct limit of a net of finite dimensional subalgebras. Since K-theory commutes with arbitrary direct limits, and since the direct limit of torsion-free groups is torsion-free, we conclude that $K_0(C^*(G,A))$ is torsion-free and $K_1(C^*(G,A)) = 0$ from the corresponding facts about finite dimensional C^*-algebras. Since $K_*^G(A) \simeq K_*(C^*(G,A))$ by theorem 2.8.3 (7), we are done. Q.E.D.

We can now describe the procedure for calculating $KK_G^*(A,B)$ for AF G-algebras A and B. We start by writing $A = \varinjlim A_n$ and $B = \varinjlim B_n$ for finite dimensional G-invariant subalgebras A_n and B_n of A and B respectively. Then each group $KK_G^0(A_n, B_k)$ can be computed in terms of equivariant K-theory by writing A_n as a direct sum of G-simple algebras and applying proposition 9.7.3. By lemma 9.7.7, we have $KK_G^1(A_n, B_k) = 0$ for all n and k. We then have

$$KK_G^0(A_n, B) \simeq \varinjlim_k KK_G^0(A_n, B_k) \quad \text{and} \quad KK_G^1(A_n, B) = 0$$

by corollary 9.7.4. Finally, corollary 9.7.6 shows that

$$KK_G^0(A, B) \simeq \varprojlim KK_G^0(A_n, B) \quad \text{and} \quad KK_G^1(A, B) \simeq \varprojlim{}^1 KK_G^0(A_n, B).$$

We apply this procedure to prove the following theorem, of which lemma 9.4.6 is a special case.

9.7.8 Theorem. Let G be a second countable compact group, and let A and B be AF G-algebras. Then $KK_G^0(A,B)$ is torsion-free as an abelian group.

Proof. Using the notation of the discussion preceding the theorem, we have

$$KK_G^0(A, B) \simeq \varprojlim_n \left[\varinjlim_k KK_G^0(A_n, B_k) \right].$$

Writing A_n as a direct sum of G-simple algebras, and using proposition 9.7.3 and lemma 9.7.7, we find that each $KK_G^0(A_n, B_k)$ is torsion-free as an abelian group. It is well known that the direct limit of torsion-free groups is torsion-free. We therefore only need to show that the inverse limit of torsion-free groups is torsion-free. Now the inverse limit of an inverse system can be identified with a certain subgroup of the product of the groups appearing in the system. (See, for example, [5], page 103.) Since the product of torsion-free groups is torsion-free, we are done. Q.E.D.

9.7.9 Remark. The procedure given above also enables one to prove that total K-freeness implies KK-freeness for AF actions of finite abelian groups without constructing homotopies as in conditions (6) and (7) of theorem 9.4.5. Indeed, let α be a totally K-free action of G on A, and let $A = \varinjlim A_n$ where the A_n are finite dimensional G-invariant subalgebras of A. If A_n is G-simple, then by proposition 9.7.3, we have $KK_G^0(A_n, A) \simeq K_0^H(L(V) \otimes A)$ for an appropriate subgroup H of G and an appropriate finite dimensional representation space V of H. Since the action of G on A is totally K-free, the action of H on A is also. Therefore, by proposition 9.1.6, so is the action of H on $L(V) \otimes A$. By proposition 9.1.9, we conclude that

$I(H)\,K_0^H(L(V)\otimes A)=0$. Since the restriction of $I(G)$ is contained in (in fact, equal to) $I(H)$, we obtain $I(G)\,KK_G^0(A_n,A)=0$. This result now clearly holds even if A_n is not G-simple, just by writing A_n as the direct sum of its minimal G-invariant ideals. We now have

$$KK_G^0(A,A)=\varprojlim KK_G^0(A_n,A).$$

Since $I(G)$ annihilates each $KK_G^0(A_n,A)$, it also annihilates the inverse limit. Thus, $I(G)\,KK_G^0(A,A)=0$. But the same argument applies to any subgroup H of G and any H-invariant ideal I of A. So the action of G on A is KK-free.

We point out that it is not sufficient to consider localizations in the argument above, because localization need not commute with inverse limits. This argument also yields no information about the existence of homotopies as in theorem 9.4.5.

The procedure for calculating $KK_G^0(A,B)$ takes a particularly nice form if the action of G on A is locally representable, and this result can be used to slightly improve theorem 9.4.7 in this case. We need an algebraic lemma.

9.7.10 Lemma. Let R be a unital ring with identity, let N be an R-module, and let $(M_\lambda)_{\lambda\in D}$ be a directed system of R-modules. Then there is a natural isomorphism

$$Hom_R(\varinjlim M_\lambda, N)\simeq \varprojlim Hom_R(M_\lambda, N).$$

Proof. Let $M=\varinjlim M_\lambda$, let $\varphi_\lambda: M_\lambda \longrightarrow M$ be the canonical map, and for $\lambda<\mu$ let $\varphi_{\lambda\mu}: M_\lambda \longrightarrow M_\mu$ be the maps of the directed system. The construction of the inverse limit obtains $\varprojlim Hom_R(M_\lambda, N)$ as

$$\{\psi\in \prod_{\lambda\in D} Hom_R(M_\lambda,N): \psi_\mu\circ\varphi_{\lambda\mu}=\psi_\lambda \text{ for } \lambda,\mu\in D \text{ with } \lambda<\mu.\}$$

By the universal property of the direct limit, the elements of this set are in one-to-one correspondence with homomorphisms from $\varinjlim M_\lambda$ to N. It is now easily seen that this isomorphism is natural and is an R-module homomorphism. Q.E.D.

9.7.11 Theorem. Let G be a second countable compact group, let (G,A,α) be a locally representable G-algebra, and let (G,B,β) be an AF G-algebra. Then there is an isomorphism

$$F: KK_G^0(A,B)\longrightarrow Hom_{R(G)}(K_0^G(A), K_0^G(B)).$$

If we identify $K_0^G(A)$ with $KK_G^0(\mathbf{C},A)$, via theorem 3.6.12, then F can be defined, using the Kasparov product, by $F(\sigma)(\eta)=\eta\otimes_A\sigma$ for $\sigma\in KK_G^0(A,B)$ and $\eta\in KK_G^0(\mathbf{C},A)$.

Proof. It is obvious that F is well defined and is an $R(G)$-module homomorphism. We first prove bijectivity for the case in which $A=M_n$ and the action of G is conjugation by some unitary representation of G. In theorem 3.5.10, we proved the existence of elements $\lambda\in KK_G^0(\mathbf{C},A)$ and $\mu\in KK_G^0(A,\mathbf{C})$ such that $\lambda\otimes_A\mu=1_{\mathbf{C}}$ and $\mu\otimes_{\mathbf{C}}\lambda=1_A$. For

$\eta \in KK_G^0(\mathbf{C}, A)$, we then have $\eta = \eta \otimes_A \mu \otimes_{\mathbf{C}} \lambda = r\lambda$, where $r = \eta \otimes_A \mu \in KK_G^0(\mathbf{C}, \mathbf{C})$, which we identify with $R(G)$. Now define

$$Z: Hom_{R(G)}(K_0^G(A), K_0^G(B)) \longrightarrow KK_G^0(A,B)$$

by $Z(f) = \mu \otimes_{\mathbf{C}} f(\lambda)$. Then we have $Z \circ F(\sigma) = \mu \otimes_{\mathbf{C}} \lambda \otimes_A \sigma = \sigma$ for $\sigma \in KK_G^0(A,B)$, by the associativity of the Kasparov product. To show that $F \circ Z$ is the identity, let $\eta \in KK_G^0(\mathbf{C}, A)$, and write $\eta = r\lambda$ for $r \in R(G)$. Then

$$F \circ Z(f)(\eta) = \eta \otimes_A \mu \otimes_{\mathbf{C}} f(\lambda) = r(\lambda \otimes_A \mu \otimes_{\mathbf{C}} f(\lambda)) = rf(\lambda) = f(\eta),$$

since f is an $R(G)$-module homomorphism. So we have shown that F is an isomorphism.

Now let A be an arbitrary AF algebra with a locally representable action of G. Then A can be written as an equivariant direct limit $A = \varinjlim A_n$, where each A_n is a finite direct sum of algebras of the form considered above. It follows immediately that F is an isomorphism for each pair (A_n, B), that is, that

$$KK_G^0(A_n, B) \simeq Hom_{R(G)}(K_0^G(A_n), K_0^G(B)).$$

By corollary 9.7.6 and lemma 9.7.7, we have

$$KK_G^0(A,B) \simeq \varprojlim KK_G^0(A_n, B),$$

while the fact that equivariant K-theory commutes with direct limits and the previous lemma imply that

$$Hom_{R(G)}(K_0^G(A), K_0^G(B)) \simeq \varprojlim Hom_{R(G)}(K_0^G(A_n), K_0^G(B)).$$

So we obtain

$$KK_G^0(A,B) \simeq Hom_{R(G)}(K_0^G(A), K_0^G(B)). \qquad (*)$$

The naturality of the two previous isomorphisms and of the Kasparov product implies that the isomorphism in (*) is given by F. Q.E.D.

The proof actually shows, using the general case of corollary 9.7.6, that if α is locally representable and B is any separable G-algebra, then F is surjective.

9.7.12 Corollary. Let α be a locally representable action of a finite group G on an AF algebra A. If α has locally discrete K-theory then $I(G)\,KK_G^0(A,A) = 0$.

Of course, this really doesn't tell us much more than the combination of the implications (1)$\Longrightarrow$(2) of theorem 9.2.4 and (1)$\Longrightarrow$(5) of theorem 9.4.7, from which we can conclude that $I(G)^n\,KK_G^0(A,A) = 0$ for some n. However, it still seems to have some interest.

Proof of corollary 9.7.12. By theorem 9.2.4, we have $I(G)\,K_0^G(A) = 0$, whence $I(G) \cdot Hom_{R(G)}(K_0^G(A), K_0^G(A)) = 0$. So $I(G) \cdot KK_G^0(A,A) = 0$ by the previous theorem.

Q.E.D.

References

[1] C. A. Akemann, G. K. Pedersen, and J. Tomiyama, Multipliers of C^*-algebras, *J. Functional Anal.* **13** (1973), 277-301.

[2] M. F. Atiyah, *K-Theory*, Benjamin, New York, 1967.

[3] M. F. Atiyah, Bott periodicity and the index of elliptic operators, *Quart. J. Math. Oxford* Ser. 2, **19** (1968), 113-140.

[4] M. F. Atiyah, Global theory of elliptic operators, in: *Proc. Intern. Conf. on Funct. Analysis*, 21-30, Univ. of Tokyo Press, 1970.

[5] M. F. Atiyah and I. G. Macdonald, *Introduction to Commutative Algebra*, Addison-Wesley, Reading, Mass., 1969.

[6] M. F. Atiyah and G. B. Segal, Equivariant K-theory and completion, *J. Differential Geometry* **3** (1969), 1-18.

[7] M. F. Atiyah and I. M. Singer, The index of elliptic operators I, *Ann. of Math.* **87** (1968), 484-530.

[8] P. Baum and A. Connes, Geometric K-theory for Lie groups and foliations, preprint.

[9] B. Blackadar, A simple C^*-algebra with no nontrivial projections, *Proc. Amer. Math. Soc.* **78** (1980), 504-508.

[10] B. Blackadar, A simple unital projectionless C^*-algebra, *J. Operator Theory* **5** (1981), 63-71.

[11] B. Blackadar, Shape theory for C^*-algebras, *Math. Scand.* **56** (1985), 249-275.

[12] O. Bratteli, Inductive limits of finite dimensional C^*-algebras, *Trans. Amer. Math. Soc.* **171** (1972), 195-234.

[13] O. Bratteli, Crossed products of UHF algebras by product type actions, *Duke Math. J.* **46** (1979), 1-23.

[14] G. E. Bredon, *Introduction to Compact Transformation Groups*, Academic Press, New York, London, 1972.

[15] L. G. Brown, Stable isomorphism of hereditary subalgebras of C^*- algebras, *Pacific J. Math.* **71** (1977), 335-348.

[16] L. G. Brown, R. G. Douglas, and P. A. Fillmore, Extensions of C^*-algebras and K-homology, *Ann. of Math.* **105** (1977), 265-324.

[17] L. G. Brown, P. Green, and M. A. Rieffel, Stable isomorphism and strong Morita equivalence of C^*-algebras, *Pacific J. Math.* **71** (1977), 349-363.

[18] R. C. Busby, Double centralizers and extensions of C^*-algebras, *Trans. Amer. Math. Soc.* **132** (1968), 79-99.

[19] M.-D. Choi and E. G. Effros, Nuclear C^*-algebras and injectivity: the general case, *Indiana Univ. Math. J.* **26** (1977), 443-446.

[20] A. Connes, An analogue of the Thom isomorphism for crossed products of a C^*-algebra by an action of **R**, *Advances in Math.* **39** (1981), 31-55.

[21] J. Cuntz, Simple C^*-algebras generated by isometries, *Comm. Math. Phys.* **57** (1977), 173-185.

[22] J. Cuntz, A class of C^*-algebras and topological Markov chains II: reducible chains and the Ext-functor for C^*-algebras, *Inventiones Math.* **63** (1981), 25-40.

[23] J. Cuntz, K-theory for certain C^*-algebras, *Ann. of Math.* **113** (1981), 181-197.

[24] J. Cuntz, K-theoretic amenability for discrete groups, *J. Reine Ang. Math.* **344** (1983), 180-195.

[25] J. Cuntz, Generalized homomorphisms between C^*-algebras and KK-theory, in: *Dynamics and Processes* (Proceedings, Bielefeld 1981), 31-45, Lecture Notes in Math. no. 1031, Springer-Verlag, New York, Heidelberg, Berlin, 1983.

[26] J. Cuntz, K-theory and C^*-algebras, in: *Algebraic K-Theory, Number Theory, Geometry and Analysis* (Proceedings, Bielefeld 1982), 55-79, Lecture Notes in Math. no. 1046, Springer-Verlag, New York, Heidelberg, Berlin, 1984.

[27] J. Cuntz and G. Skandalis, Mapping cones and exact sequences in KK-theory, *J. Operator Theory* **15** (1986), 163-180.

[28] J. Dixmier, Points séparés dans le spectre d'une C^*-algèbre, *Acta. Sci. Math. Szeged* **22** (1961), 115-128.

[29] J. Dixmier, *C^*-Algebras*, North-Holland, Amsterdam, New York, Oxford, 1977

[30] J. Dixmier, *Von Neumann Algebras*, North-Holland, Amsterdam, New York, Oxford, 1981.

[31] E. G. Effros, *Dimensions and C^*-Algebras*, CBMS Regional Conf. Ser. in Math., no. 46, Amer. Math. Soc., 1981.

[32] E. G. Effros and J. Rosenberg, C^*-algebras with approximately inner flip, *Pacific J. Math.* **77** (1978), 417-443.

[33] S. Eilenberg and N. Steenrod, *Foundations of Algebraic Topology*, Princeton University Press, 1952.

[34] G. A. Elliott, On the classification of inductive limits of sequences of semi-simple finite-dimensional algebras, *J. Algebra* **38** (1976), 29-44.

[35] J. M. G. Fell, *Induced Representations and Banach *-Algebraic Bundles*, Lecture Notes in Math. no. 582, Springer-Verlag, New York, Heidelberg, Berlin, 1977.

[36] J. Glimm, On a certain class of operator algebras, *Trans. Amer. Math. Soc.* **95** (1960), 318-340.

[37] P. Green, The local structure of twisted covariance algebras, *Acta Math.* **140** (1978), 191-250.

[38] P. Green, The structure of imprimitivity algebras, *J. Functional Anal.* **36** (1980), 88-104.

[39] P. Green, Equivariant K-theory and crossed product C^*-algebras, *Proc. Symposia Pure Math.* vol. 38, part 1, 337-338, Amer. Math. Soc., 1982.

[40] P. Griffiths and J. Harris, *Principles of Algebraic Geometry*, Wiley-Interscience, New York, 1978.

[41] A. Guichardet, Tensor products of C^*-algebras I, Aarhus University Lecture Notes 12 (1969).

[42] P. R. Halmos, *A Hilbert Space Problem Book*, Van Nostrand, New York, 1967.

[43] D. Handelman, Homomorphisms of C^*-algebras to finite AW^*-algebras, *Michigan Math. J.* **28** (1981), 229-240.

[44] D. Handelman and W. Rossmann, Product type actions of finite and compact groups, *Indiana Univ. Math. J.* **33** (1984), 479-509.

[45] D. Handelman and W. Rossmann, Action of compact groups on AF C^*-algebras, *Illinois J. Math.* **29** (1985), 51-95.

[46] J. Hilgert, Foundations of K-theory for C^*-algebras, preprint München.

[47] P. J. Hilton and U. Stammbach, *A Course in Homological Algebra*, Graduate Texts in Math. no. 4, Springer-Verlag, New York, Heidelberg, Berlin, 1971.

[48] L. H. Hodgkin, The equivariant Künneth theorem in K-theory, in: *Topics in K-Theory*, 1-102, Lecture Notes in Math. no. 496, Springer-Verlag, New York, Heidelberg, Berlin, 1975.

[49] K. H. Hofmann and K. Keimel, Sheaf theoretic concepts in analysis: bundles and sheaves of Banach spaces, Banach C(X)-modules, in: *Applications of Sheaves*, 415-441, Lecture Notes in Math. no. 753, Springer-Verlag, New York, Heidelberg, Berlin, 1979.

[50] W. Iberkleid and T. Petrie, *Smooth S^1 Manifolds*, Lecture Notes in Math. no. 557, Springer-Verlag, New York, Heidelberg, Berlin, 1976.

[51] P. Julg, K-theorie equivariante et produits croisés, *C. R. Acad. Sci. Paris* Ser. I, **292** (1981), 629-632.

[52] P. Julg, Induction holomorphe pour le produit croisé d'une C^*-algebre par un groupe de Lie compact, *C. R. Acad. Sci. Paris* Ser. I, **294** (1982), 193-196.

[53] I. Kaplansky, *Rings of Operators*, Benjamin, New York, 1968.

[54] G. G. Kasparov, Hilbert C^*-modules: theorems of Stinespring and Voiculescu, *J. Operator Theory* **4** (1980), 133-150.

[55] G. G. Kasparov, The operator K-functor and extensions of C^*-algebras, *Izv. Akad. Nauk. SSSR*, Ser. Mat. **44** (1980), 571-636 (in Russian), English translation in *Math. USSR Izvestija* **16** (1981), 513-572.

[56] G. G. Kasparov, K-theory, group C^*-algebras, and higher signatures (conspectus), part 1, preprint Chernogolovka.

[57] J. L. Kelley, *General Topology*, Van Nostrand, New York, 1955.

[58] A. Kishimoto, Simple crossed products of C^*-algebras by locally compact abelian groups, *Yokohama Math. J.* **28** (1980), 69-85.

[59] A. Kishimoto, Outer automorphisms and reduced crossed products of simple C^*-algebras, *Comm. Math. Phys.* **81** (1981), 429-435.

[60] A. Kishimoto, Freely acting automorphisms of C^*-algebras, *Yokohama Math. J.* **30** (1982), 39-47.

[61] E. C. Lance, Automorphisms of separable postliminal C^*-algebras, *Pacific J. Math.* **23** (1967), 547-555.

[62] E. C. Lance, On nuclear C^*-algebras, *J. Functional Anal.* **12** (1973), 157-176.

[63] E. C. Lance, K-theory for certain group C^*-algebras, *Acta Math.* **151** (1983), 209-230.

[64] S. Lang, *Algebra*, Addison-Wesley, Reading, Mass., 1970.

[65] S. Lang, *Algebraic Number Theory*, Addison-Wesley, Reading, Mass., 1970.

[66] H. Matsumura, *Commutative Algebra*, Benjamin, New York, 1970.

[67] J. A. Mingo and W. J. Phillips, Equivariant triviality theorems for Hilbert C^*-modules, *Proc. Amer. Math. Soc.* **91** (1984), 225-230.

[68] A. S. Miščenko and A. T. Fomenko, The index of elliptic operators over C^*-algebras, *Izv. Akad. Nauk. SSSR*, Cer. Mat. **43** (1979), 831-859 (in Russian), English translation in *Math. USSR Izvestija* **15** (1980), 87-112.

[69] C. C. Moore, Group extensions and cohomology for locally compact groups IV, *Trans. Amer. Math. Soc.* **221** (1976), 35-58.

[70] D. Olesen, A classification of ideals in crossed products, *Math. Scand.* **45** (1979), 157-167.

[71] D. Olesen and G. K. Pedersen, Applications of the Connes spectrum to C^*-dynamical systems, *J. Functional Anal.* **30** (1978), 179-197.

[72] D. Olesen and G. K. Pedersen, Applications of the Connes spectrum to C^*-dynamical systems II, *J. Functional Anal.* **36** (1980), 18-32.

[73] D. Olesen and G. K. Pedersen, Applications of the Connes spectrum to C^*-dynamical systems III, *J. Functional Anal.* **45** (1982), 357-390.

[74] D. Olesen, G. K. Pedersen, and E. Størmer, Compact abelian groups of automorphisms of simple C^*-algebras, *Inventiones Math.* **39** (1977), 55-64.

[75] R. Palais, *Seminar on the Atiyah-Singer Index Theorem*, Ann. of Math. Study 57, Princeton, 1965.

[76] W. L. Paschke, Inner product modules over B^*-algebras, *Trans. Amer. Math. Soc.* **182** (1973), 443-468.

[77] W. L. Paschke, K-theory for actions of the circle group on C^*-algebras, *J. Operator Theory* **6** (1981), 125-133.

[78] W. L. Paschke, $\mathbf{Z}_2$-equivariant K-theory, in: *Operator Algebras and their Connections with Topology and Ergodic Theory*, 362-373, Lecture Notes in Math.

no. 1132, Springer-Verlag, New York, Heidelberg, Berlin, 1985.

[79] G. K. Pedersen, *C^*-Algebras and their Automorphism Groups*, Academic Press, London, New York, San Francisco, 1979.

[80] G. K. Pedersen, Remarks on the Connes spectrum for C^*-dynamical systems, *J. Operator Theory* **3** (1980), 143-148.

[81] N. C. Phillips, K-theoretic freeness of finite group actions on C^*-algebras, in: *Group Actions on Rings*, Contemporary Mathematics vol. 43, 227-243, Amer. Math. Soc., 1985.

[82] M. Pimsner and D. Voiculescu, Exact sequences of K-groups and Ext-groups of certain crossed product C^*-algebras, *J. Operator Theory* **4** (1980), 93-118.

[83] M. Pimsner and D. Voiculescu, K-groups of reduced crossed products by free groups, *J. Operator Theory* **8** (1982), 131-156.

[84] M. A. Rieffel, Induced representations of C^*-algebras, *Advances in Math.* **13** (1974), 176-257.

[85] M. A. Rieffel, Unitary representations of group extensions: an algebraic approach to the theory of Mackey and Blattner, *Advances in Math. Supplementary Studies* **4** (1979), 43-82.

[86] M. A. Rieffel, C^*-algebras associated with irrational rotations, *Pacific J. Math.* **93** (1981), 415-429.

[87] M. A. Rieffel, Morita equivalence for operator algebras, *Proc. Symposia in Pure Math.* vol. 38, part 1, 285-298, Amer. Math. Soc., 1982.

[88] M. A. Rieffel, Applications of strong Morita equivalence to transformation group C^*-algebras, *Proc. Symposia in Pure Math.* vol. 38, part 1, 299-310, Amer. Math. Soc., 1982.

[89] J. Rosenberg, Appendix to O. Bratteli's paper on "crossed products of UHF algebras," *Duke Math. J.* **46** (1979), 25-26.

[90] J. Rosenberg, Homological invariants of extensions of C^*-algebras, *Proc. Symposia in Pure Math.* vol. 38, part 1, 35-76, Amer. Math. Soc., 1982.

[91] J. Rosenberg and C. Schochet, The classification of extensions of C^*-algebras, *Bull. Amer. Math. Soc. (New Series)* **4** (1981), 105-110.

[92] J. Rosenberg and C. Schochet, The Künneth and the universal coefficient theorem for Kasparov's generalized K-functor, preprint MSRI.

[93] J. Rosenberg and C. Schochet, The Künneth theorem and the universal coefficient theorem for equivariant K-theory and KK-theory, preprint MSRI.

[94] R. L. Rubinsztein, Restriction of equivariant K-theory to cyclic subgroups, *Bull. Acad. Pol. Sci.* **29** (1981), 299-304.

[95] C. Schochet, Topological methods for C^*-algebras II: geometric resolutions and the Künneth formula, *Pacific J. Math.* **98** (1982), 443-458.

[96] C. Schochet, Topological methods for C^*-algebras III: axiomatic homology, *Pacific J. Math.* **114** (1984), 399-445.

[97] G. Segal, The representation ring of a compact Lie group, *Publ. Math. Inst. Hautes Etudes Sci.* **34** (1968), 113-128.

[98] G. Segal, Equivariant K-theory, *Publ. Math. Inst. Hautes Etudes Sci.* **34** (1968), 129-151.

[99] J.-P. Serre, *Linear Representations of Finite Groups*, Graduate Texts in Math. no. 42, Springer-Verlag, New York, Heidelberg, Berlin, 1977.

[100] G. Skandalis, Some remarks on Kasparov theory, *J. Functional Anal.* **56** (1984), 337-347.

[101] R. W. Swan, Vector bundles and projective modules, *Trans. Amer. Math. Soc.* **105** (1962), 264-277.

[102] M. Takesaki, On the cross-norm of the direct product of C^*-algebras, *Tôhoku Math. J.* **16** (1964), 111-122.

[103] M. Takesaki, Covariant representations of C^*-algebras and their locally compact automorphism groups, *Acta Math.* **119** (1967), 273-303.

[104] M. Takesaki, Theory of Operator Algebras I, Springer-Verlag, New York, Heidelberg, Berlin, 1979.

[105] J. L. Taylor, Banach algebras and topology, in: *Algebras in Analysis*, 118-186, Academic Press, London, New York, San Francisco, 1975.

[106] J. Tomiyama, Applications of Fubini type theorems to tensor products of C^*-algebras, *Tôhoku Math. J.* **19** (1967), 213-226.

[107] J. W. Vick, *Homology Theory*, Academic Press, London, New York, San Francisco, 1973.

[108] A. J. Wasserman, Automorphic actions of compact groups on operator algebras, Thesis, University of Pennsylvania, 1981.

Author / Reference Index

The notation $x(n)$ (where $n > 1$) means that the reference occurs n times on page x.

Index of Notation

For each notation, we give a capsule description of its meaning (if possible) and a page on which it is defined or explained, or a reference number in square brackets (if there is one). The symbol "—" indicates a repetition of the previous entry (with a different meaning). Notation is arranged alphabetically, with Roman letters first, then Greek, and miscellaneous symbols at the end.

Subject Index

Boldface numbers indicate where a concept is explained, defined, or introduced, or where a theorem is stated or proved. Other numbers indicate where the concept or theorem is used. If no other numbers are listed, then it was not felt to be feasible or useful to indicate all the places the concept is used.

LECTURE NOTES IN MATHEMATICS
Edited by A. Dold and B. Eckmann

Some general remarks on the publication of monographs and seminars

In what follows all references to monographs, are applicable also to multiauthorship volumes such as seminar notes.

1. Lecture Notes aim to report new developments - quickly, informally, and at a high level. Monograph manuscripts should be reasonably self-contained and rounded off. Thus they may, and often will, present not only results of the author but also related work by other people. Furthermore, the manuscripts should provide sufficient motivation, examples and applications. This clearly distinguishes Lecture Notes manuscripts from journal articles which normally are very concise. Articles intended for a journal but too long to be accepted by most journals, usually do not have this "lecture notes" character. For similar reasons it is unusual for Ph.D. theses to be accepted for the Lecture Notes series.

   Experience has shown that English language manuscripts achieve a much wider distribution.

2. Manuscripts or plans for Lecture Notes volumes should be submitted either to one of the series editors or to Springer-Verlag, Heidelberg. These proposals are then refereed. A final decision concerning publication can only be made on the basis of the complete manuscripts, but a preliminary decision can usually be based on partial information: a fairly detailed outline describing the planned contents of each chapter, and an indication of the estimated length, a bibliography, and one or two sample chapters - or a first draft of the manuscript. The editors will try to make the preliminary decision as definite as they can on the basis of the available information.

3. Lecture Notes are printed by photo-offset from typed copy delivered in camera-ready form by the authors. Springer-Verlag provides technical instructions for the preparation of manuscripts, and will also, on request, supply special staionery on which the prescribed typing area is outlined. Careful preparation of the manuscripts will help keep production time short and ensure satisfactory appearance of the finished book. Running titles are not required; if however they are considered necessary, they should be uniform in appearance. We generally advise authors not to start having their final manuscripts specially tpyed beforehand. For professionally typed manuscripts, prepared on the special stationery according to our instructions, Springer-Verlag will, if necessary, contribute towards the typing costs at a fixed rate.

   The actual production of a Lecture Notes volume takes 6-8 weeks.

.../...

4. Final manuscripts should contain at least 100 pages of mathematical text and should include

   - a table of contents
   - an informative introduction, perhaps with some historical remarks. It should be accessible to a reader not particularly familiar with the topic treated.
   - subject index; this is almost always genuinely helpful for the reader.

5. Authors receive a total of 50 free copies of their volume, but no royalties. They are entitled to purchase further copies of their book for their personal use at a discount of 33 1/3 %, other Springer mathematics books at a discount of 20 % directly from Springer-Verlag.

   Commitment to publish is made by letter of intent rather than by signing a formal contract. Springer-Verlag secures the copyright for each volume.

Vol. 1117: D.J. Aldous, J.A. Ibragimov, J. Jacod, Ecole d'Été de Probabilités de Saint-Flour XIII – 1983. Édité par P.L. Hennequin. IX, 409 pages. 1985.

Vol. 1118: Grossissements de filtrations: exemples et applications. Seminaire, 1982/83. Edité par Th. Jeulin et M. Yor. V, 315 pages. 1985.

Vol. 1119: Recent Mathematical Methods in Dynamic Programming. Proceedings, 1984. Edited by I. Capuzzo Dolcetta, W.H. Fleming and T. Zolezzi. VI, 202 pages. 1985.

Vol. 1120: K. Jarosz, Perturbations of Banach Algebras. V, 118 pages. 1985.

Vol. 1121: Singularities and Constructive Methods for Their Treatment. Proceedings, 1983. Edited by P. Grisvard, W. Wendland and J.R. Whiteman. IX, 346 pages. 1985.

Vol. 1122: Number Theory. Proceedings, 1984. Edited by K. Alladi. VII, 217 pages. 1985.

Vol. 1123: Séminaire de Probabilités XIX 1983/84. Proceedings. Edité par J. Azéma et M. Yor. IV, 504 pages. 1985.

Vol. 1124: Algebraic Geometry, Sitges (Barcelona) 1983. Proceedings. Edited by E. Casas-Alvero, G.E. Welters and S. Xambó-Descamps. XI, 416 pages. 1985.

Vol. 1125: Dynamical Systems and Bifurcations. Proceedings, 1984. Edited by B.L.J. Braaksma, H.W. Broer and F. Takens. V, 129 pages. 1985.

Vol. 1126: Algebraic and Geometric Topology. Proceedings, 1983. Edited by A. Ranicki, N. Levitt and F. Quinn. V, 423 pages. 1985.

Vol. 1127: Numerical Methods in Fluid Dynamics. Seminar. Edited by F. Brezzi, VII, 333 pages. 1985.

Vol. 1128: J. Elschner, Singular Ordinary Differential Operators and Pseudodifferential Equations. 200 pages. 1985.

Vol. 1129: Numerical Analysis, Lancaster 1984. Proceedings. Edited by P.R. Turner. XIV, 179 pages. 1985.

Vol. 1130: Methods in Mathematical Logic. Proceedings, 1983. Edited by C.A. Di Prisco. VII, 407 pages. 1985.

Vol. 1131: K. Sundaresan, S. Swaminathan, Geometry and Nonlinear Analysis in Banach Spaces. III, 116 pages. 1985.

Vol. 1132: Operator Algebras and their Connections with Topology and Ergodic Theory. Proceedings, 1983. Edited by H. Araki, C.C. Moore, Ş. Strătilă and C. Voiculescu. VI, 594 pages. 1985.

Vol. 1133: K.C. Kiwiel, Methods of Descent for Nondifferentiable Optimization. VI, 362 pages. 1985.

Vol. 1134: G.P. Galdi, S. Rionero, Weighted Energy Methods in Fluid Dynamics and Elasticity. VII, 126 pages. 1985.

Vol. 1135: Number Theory, New York 1983–84. Seminar. Edited by D.V. Chudnovsky, G.V. Chudnovsky, H. Cohn and M.B. Nathanson. V, 283 pages. 1985.

Vol. 1136: Quantum Probability and Applications II. Proceedings, 1984. Edited by L. Accardi and W. von Waldenfels. VI, 534 pages. 1985.

Vol. 1137: Xiao G., Surfaces fibrées en courbes de genre deux. IX, 103 pages. 1985.

Vol. 1138: A. Ocneanu, Actions of Discrete Amenable Groups on von Neumann Algebras. V, 115 pages. 1985.

Vol. 1139: Differential Geometric Methods in Mathematical Physics. Proceedings, 1983. Edited by H. D. Doebner and J. D. Hennig. VI, 337 pages. 1985.

Vol. 1140: S. Donkin, Rational Representations of Algebraic Groups. VII, 254 pages. 1985.

Vol. 1141: Recursion Theory Week. Proceedings, 1984. Edited by H.-D. Ebbinghaus, G.H. Müller and G.E. Sacks. IX, 418 pages. 1985.

Vol. 1142: Orders and their Applications. Proceedings, 1984. Edited by I. Reiner and K. W. Roggenkamp. X, 306 pages. 1985.

Vol. 1143: A. Krieg, Modular Forms on Half-Spaces of Quaternions. XIII, 203 pages. 1985.

Vol. 1144: Knot Theory and Manifolds. Proceedings, 1983. Edited by D. Rolfsen. V, 163 pages. 1985.

Vol. 1145: G. Winkler, Choquet Order and Simplices. VI, 143 pages. 1985.

Vol. 1146: Séminaire d'Algèbre Paul Dubreil et Marie-Paule Malliavin. Proceedings, 1983–1984. Edité par M.-P. Malliavin. IV, 420 pages. 1985.

Vol. 1147: M. Wschebor, Surfaces Aléatoires. VII, 111 pages. 1985.

Vol. 1148: Mark A. Kon, Probability Distributions in Quantum Statistical Mechanics. V, 121 pages. 1985.

Vol. 1149: Universal Algebra and Lattice Theory. Proceedings, 1984. Edited by S. D. Comer. VI, 282 pages. 1985.

Vol. 1150: B. Kawohl, Rearrangements and Convexity of Level Sets in PDE. V, 136 pages. 1985.

Vol 1151: Ordinary and Partial Differential Equations. Proceedings, 1984. Edited by B.D. Sleeman and R.J. Jarvis. XIV, 357 pages. 1985.

Vol. 1152: H. Widom, Asymptotic Expansions for Pseudodifferential Operators on Bounded Domains. V, 150 pages. 1985.

Vol. 1153: Probability in Banach Spaces V. Proceedings, 1984. Edited by A. Beck, R. Dudley, M. Hahn, J. Kuelbs and M. Marcus. VI, 457 pages. 1985.

Vol. 1154: D.S. Naidu, A.K. Rao, Singular Pertubation Analysis of Discrete Control Systems. IX, 195 pages. 1985.

Vol. 1155: Stability Problems for Stochastic Models. Proceedings, 1984. Edited by V.V. Kalashnikov and V.M. Zolotarev. VI, 447 pages. 1985.

Vol. 1156: Global Differential Geometry and Global Analysis 1984. Proceedings, 1984. Edited by D. Ferus, R.B. Gardner, S. Helgason and U. Simon. V, 339 pages. 1985.

Vol. 1157: H. Levine, Classifying Immersions into $\mathbb{R}^4$ over Stable Maps of 3-Manifolds into $\mathbb{R}^2$. V, 163 pages. 1985.

Vol. 1158: Stochastic Processes – Mathematics and Physics. Proceedings, 1984. Edited by S. Albeverio, Ph. Blanchard and L. Streit. VI, 230 pages. 1986.

Vol. 1159: Schrödinger Operators, Como 1984. Seminar. Edited by S. Graffi. VIII, 272 pages. 1986.

Vol. 1160: J.-C. van der Meer, The Hamiltonian Hopf Bifurcation. VI, 115 pages. 1985.

Vol. 1161: Harmonic Mappings and Minimal Immersions, Montecatini 1984. Seminar. Edited by E. Giusti. VII, 285 pages. 1985.

Vol. 1162: S.J.L. van Eijndhoven, J. de Graaf, Trajectory Spaces, Generalized Functions and Unbounded Operators. IV, 272 pages. 1985.

Vol. 1163: Iteration Theory and its Functional Equations. Proceedings, 1984. Edited by R. Liedl, L. Reich and Gy. Targonski. VIII, 231 pages. 1985.

Vol. 1164: M. Meschiari, J.H. Rawnsley, S. Salamon, Geometry Seminar "Luigi Bianchi" II – 1984. Edited by E. Vesentini. VI, 224 pages. 1985.

Vol. 1165: Seminar on Deformations. Proceedings, 1982/84. Edited by J. Ławrynowicz. IX, 331 pages. 1985.

Vol. 1166: Banach Spaces. Proceedings, 1984. Edited by N. Kalton and E. Saab. VI, 199 pages. 1985.

Vol. 1167: Geometry and Topology. Proceedings, 1983–84. Edited by J. Alexander and J. Harer. VI, 292 pages. 1985.

Vol. 1168: S.S. Agaian, Hadamard Matrices and their Applications. III, 227 pages. 1985.

Vol. 1169: W.A. Light, E.W. Cheney, Approximation Theory in Tensor Product Spaces. VII, 157 pages. 1985.

Vol. 1170: B.S. Thomson, Real Functions. VII, 229 pages. 1985.

Vol. 1171: Polynômes Orthogonaux et Applications. Proceedings, 1984. Edité par C. Brezinski, A. Draux, A.P. Magnus, P. Maroni et A. Ronveaux. XXXVII, 584 pages. 1985.

Vol. 1172: Algebraic Topology, Göttingen 1984. Proceedings. Edited by L. Smith. VI, 209 pages. 1985.